Total Quality Management Handbook

Other McGraw-Hill Quality Books of Interest

Total Quality Management Handbook

Jack Hradesky

McGraw-Hill, Inc.

New York San Francisco Washington, D.C. Auckland Bogotá
Caracas Lisbon London Madrid Mexico City Milan
Montreal New Delhi San Juan Singapore
Sydney Tokyo Toronto

Library of Congress Cataloging-in-Publication Data

Hradesky, John L.
    Total quality management handbook / John L. Hradesky.
    p.   cm.
    Includes index.
    ISBN 0-07-030511-0 (acid-free paper)
    1. Quality control. 2. Total quality management. I. Title.
TS156.H73 1995
658.5'62—dc20   94-26973
                                                           CIP

Sci
TS
156
H73
1995

1 2 3 4 5 6 7 8 9 0   DOC/DOC   9 0 9 8 7 6 5 4

ISBN 0-07-030511-0

*The sponsoring editor for this book was Harold Crawford, and the production
supervisor was Suzanne W. B. Rapcavage. It was set in Palatino by North Market
Street Graphics.*

Printed and bound by R. R. Donnelley & Sons Company.

McGraw-Hill books are available at special quantity discounts to use as
premiums and sales promotions, or for use in corporate training programs.
For more information, please write to the Director of Special Sales,
McGraw-Hill, Inc., 11 West 19th Street, New York, NY 10011. Or contact
your local bookstore.

I dedicate this book to my wife Trish, for her tolerance of my time-consuming obsession with the subject, my clients, and this textbook. Also, an acknowledgment to my father, Louis A. Hradesky, P.E., the late Albert Holzman, Ph.D., the late W. Edwards Deming, Ph.D., and my many supervisors at GM, Xerox, Kodak, American Hospital Supply, and Johnson & Johnson who had their influences through the years.

Contents

5. Recognition and Rewards   177

6. Leadership/Team Building   193

Preface

This book is the result of a genuine need to make United States manufacturers and service companies competitive in the twenty-first century. Survival of these companies is achieved through three objectives: (1) retaining customers; (2) improved profits; and (3) increased business. The book deals with quantum leaps of improvement as well as continuous improvement of companies' business performances. It addresses companies that are successful, but have stalled on a plateau; companies that are successful, but whose outlook includes intense competition or substitute products; companies that have failed in previous TQM attempts; and companies that are in need of a compelling turnaround of their business performances for survival.

Besides the companies just described, this book addresses firms that have implemented TQM. According to the *Wall Street Journal*, many companies have adopted TQM and up to 80 percent of them have failed. These are companies who did not achieve their goals; they have adopted the jargon of TQM, SPC, QFD, etc., but have done very little implementation, or did poorly in their execution of implementation. These companies have quality veneer! To talk with them is like talking to the king who wore no clothes. These companies are heading off a cliff; some know it, but others are going in the same direction and are unaware of it.

Since TQM, SPC, QFD, etc., is not rocket science, many people believe that they can either develop their own curriculum or attend a seminar to become equipped to successfully implement TQM, SPC, etc. This may be true for 2 to 3 percent of the people; however, the other 97 to 98 percent of the people will require broad training, with sessions extending over a lengthy period of time.

Employees must receive expectations from their management to actively participate and achieve significant results. A critical mass of TQM champions must be developed. Effective role modeling by senior management, extending to all levels in the organization, is crucial. Recognition of individual and group results and behaviors must be a common practice in every organization.

Among the reasons for the high TQM failure rate are:

- Lack of commitment by the president or CEO
- Lack of commitment of the senior staff (which is due to lack of commitment by the senior executive)
- Lack of a defined corporate strategic direction due to not conducting strategic business planning
- Lack of a tactical plan to achieve any specific critical business success factors, i.e., creation of a road map to achieve success
- Lack of a training curriculum tailored to the company's needs and the identification of specific business results to be achieved
- Lack of a proven process to implement
- Lack of review of progress with appropriate support and/or corrective action
- Lack of recognition of performances, or the opposite, consequences for the nonperformers

Unfortunately, the high rate of failure of TQM programs has created many casualties. Many seasoned companies have gone by the wayside, careers have ended, and layoffs have been prevalent. Those who survived are either skeptical or total nonbelievers.

These companies' inadequate thrust to become competitive, coupled with the indifference of our government to control the shifting of manufacturing jobs abroad, has created a "shark feeding frenzy" in the elimination of jobs in the United States. The only bright spot is that some states, such as California and Texas, reimburse companies to improve their productivity, quality, and competitiveness.

Hence, the purpose of this book is to help companies become competitive and profitable, and to enjoy growth. This book will provide a proven process, a road map of what it takes to prevent becoming trapped in the pitfalls, and the "how to's" of surviving and thriving. This process has a 99 percent success experience.

Intended Audience

This book is addressed principally to owners, executives, managers, and employees of service and manufacturing companies. In addition, investors, bankers, venture capitalists, students, and scholars of management who are keenly interested in the success and failure of service and manufacturing companies may find this text helpful. This text focuses on a path to success. It discusses the need to tailor the training to the company's specific objectives and to integrate the training with implementation to achieve those objectives. It identifies 10 tracks of curriculum needed to make these companies competitive, profitable, and to create sales growth.

The scope of this text is not limited to the standard TQM objectives (e.g., customer satisfaction, continuous improvement, process improvement, and empowerment of the organization). It also addresses development of personnel, determination of strategic business direction, deployment of critical business success factors, their measurements, projects required for success, and the development of new products and markets. This book is also appropriate for those companies that think of themselves as "professionally managed" but have lost some of their momentum and much of their entrepreneurial spirit. It can show them what must be done to reverse the situation.

Background

Many of today's buzz words or popular techniques were utilized at American companies years ago, as everyday activities without fanfare. For example, continuous improvement of quality and reduced labor hours, just-in-time, model mix, and cells were utilized at Harrison Radiator at GMC, in the 1960s. These technical concepts were considered Industrial Engineering 101 in the 1960s. Critical Business Success Factors, concurrent engineering and continuous improvement, and quality policies and practices (a tougher version of ISO 9000) were practiced at Xerox in the 1970s. Teams, measurement of performance appraisals based on results, and rewards and recognition were practiced at the Edwards Laboratory Division of American Hospital Supply, resulting in a 24 percent productivity improvement—the highest productivity improvement of 29 divisions in the late 1970s. Likewise, these techniques and approaches were used at Kodak and Johnson & Johnson. Thus, this text's techniques were practiced before being espoused by recent gurus.

In addition, NSG's 13 years of consulting and training experience include companies whose products ranged from food products to high-technology aerospace and electronics; firms with small, $6 million sales to billions of dollars of sales; and from entrepreneurs to professionally managed companies such as Solectron, Malcolm Baldrige National Quality Award Winner, 1991; Safariland, Ltd., Toyota's "Most Improved Supplier" Award Winner, 1993; and Arral Industries, who converted from Aerospace products to commercial products.

Consulting and training for such a diversity of clients has enabled NSG to develop a program ranging from "What worked?" to "What's not working, and why?" This experience is the mainstream of this text: not theory, not education, not talent, but the experience of what works and achieves results with sustained cultural changes.

This book focuses on three significant objectives for survival, namely, retaining customers, improving profits, and increasing business. They are addressed through three major thrusts: (1) *customer satisfaction,* both internal and external, (2) *cultural change,* and last but not least, (3) *Critical Business Success Factors.* All three thrusts are supported by an integrated training and implementation undertaking.

Jack Hradesky

Acknowledgments

Although a book may bear the name of a particular individual as author, I doubt whether many manuscripts are ever completed without the assistance and significant contribution of a great number of people. This is certainly the case with this text, and I am indebted to several individuals for their support.

First, this book is a product of my having had the privilege of working in many challenging positions utilizing the concepts, principles, and techniques espoused in the textbook at General Motors, Xerox, Eastman Kodak, American Hospital Supply, and Johnson & Johnson.

Second, the application of these strategies, principles, and techniques was the product of training and consulting at many different organizations since 1981. These companies ranged from start-ups to members of the *Fortune* 500. The experience acquired enabled this development, and proved the process within to be effective.

I thank Richard Bush, Scott Hay, Hank Rogers, Dan Pitkin, and Steve Wernick for their contribution to several of the chapters, especially Steve Wernick, who drafted many of the chapters, and Hank Rogers, who proofed the chapters.

I would also like to acknowledge the consultants and trainers who have trained and implemented these concepts and techniques and refined the training: Larry Bieller, Richard Bush, Dan Dunahay, Scott Hay, Bert Lee, Tom Levitt, Terry Mercer, Dan Pitkin, Eric Rees, Hank Rogers, Annette Sumrall, and Nick Testa.

Jack Hradesky

1
Introduction

Crisis

There is clearly a crisis in the world marketplace that will dictate change for corporations in the United States and elsewhere. Technology is shrinking the distances between supplier and potential customer at an exponential pace. It is apparent to many that a new emphasis on customer satisfaction will be a prerequisite to competition in the global economy.

The crisis is not new; rather, it is in fact a slowly evolving revolution of the customer. Customers now demand and expect that their requirements for quality in product and service be met. American automotive manufacturers can testify that suppliers exist who are ready, able, and willing to meet customers needs. To stay competitive and profitable, all businesses will have to pay attention to this new attitude.

The European Community represents a much larger potential market for goods and services than either the United States or Japan. After 1992, suppliers who want to compete in this arena will have to conform to the ISO 9000 standards for quality management systems, another indication that product quality is becoming a given. Businesses cannot compete without quality designed and built into products and services. Customers will not accept less. "Buy American" programs will not provide the answer. This is a market-driven situation, not one that can be regulated by government intervention.

The solution to the crisis is Total Quality Management (TQM), more aptly named in this time Total Survival Management (TSM). It is an undertaking which, in one form or another, has been adopted by successful world-class businesses and will be mandatory for those competing without a technological or geographical monopoly. To stay in business, business must listen to the voice of the customer. Companies do not have to do it—survival is not mandatory.

Purpose

The purpose of this book is to provide the reader with a set of requirements to successfully implement companywide Total Quality Management. Manufacturing, service, and government organizations can use the approach presented in this text. The business objectives are to retain customers, improve profits, and generate new business. The focus is on achieving Critical Business Success Factors, attaining internal customer satisfaction, and creating a cultural change which will provide a competitive edge. This approach has a proven track record of success regardless of the nature of the companies' processes or products. It can be implemented in companies with recent awareness to quality demands or in companies which have had one or more false starts or in companies that have successfully implemented TQM and hit a plateau.

Definition of Quality

The first thing to be decided before implementing changes in business is the definition of quality. If the customer wants quality, what does that mean? How do you define quality? In today's competitive environment, the answer is not simple. A straightforward paradigm of the last 10 to 15 years defined *quality* as "conformance to requirements." The only problem with this definition is that the requirements are subjective. Only the customers can really determine the requirements by their expectations and acceptance of the product or service provided. Every parameter of the service or product you provide needs to be related in some way to the customer's needs and expectations. Whether it is a landing strut for a jet fighter or dinner at a restaurant, every variable in the material, process, and workmanship needs to be viewed with regard to the final output and how well that output will satisfy the final customer. Quality may be defined as the customer's expectations and requirements; it is determined by the customer and the marketplace and includes all products and service attributes. Quality addresses defects, errors, and complaints, but goes well beyond such traditional values: quality includes anything the customer expects and requires, and is ever changing.

What Is TQM?

Total Quality Management (TQM) is a philosophy, a set of tools, and a process whose output yields customer satisfaction and continuous improvement. This philosophy and process differs from traditional philosophies and processes in that everyone in the company can and must practice it. It espouses "win-win" attitude, differentiates cost versus price, and provides added value.

TQM combines cultural-changing tactics and structured technical techniques whose focus is on satisfying the needs of internal customers and, hence, external customers. TQM requires that the executives are involved and committed, not just interested, and that the focus is on implementation. Results of TQM include error-free processes which deliver products and services fit for use, on time, with competitive pricing and good value. Above all, TQM ensures satisfaction of all customer requirements to retain those customers; improving internal processes to increase profits; and generating new business from new products, services, and markets.

When properly carried out, TQM becomes integrated into all aspects of the corporate identity. TQM's scope covers all functions within a company from sales and marketing through design, production, and service. The formula for success is one part "effective training," two parts "effective implementation," and three parts "executive involvement." The training is analogous to a football team's practice; the implementation is the real "game action."

Broadening the concept of quality is the aim of TQM, so that quality moves from a product appraisal function to a corporate imperative for excellence and the refusal to be satisfied with the status quo.

There are infinite approaches to undertaking TQM. However, this text deals with one which has a consistent and successful track record of implementation. This is not an approach that worked once or twice with superinordinate-effective executives, but one that is effective, providing the president/CEO is truly committed.

Cultural change tactics include:

Creating Commitment
- Provides focus and sustained, committed effort to achieve a desired outcome. It emphasizes the difference between being passionately committed and having an "interest of convenience" commitment.

Creating Cultural Change
- Implementation of values that produce the behaviors which will achieve the mission and visions; hence, success.
- Creation of questionnaires that will measure the effects of the culture change; i.e., behaviors of the organization
- Cultural-fostering group monitors the effectiveness of the cultural change effort and recommends what adjustments to make in regard to the implementation and when to make them.

Effective Communication

- Serves as a basis for interchange, both for information and in relationships.
- Establishes the foundation for verbal and nonverbal interaction.

Empowering the Workforce

- Enables everyone to participate in making the vision a reality.

Taking Responsibility

- Prepares the workforce to take ownership of the processes; the flip side of empowerment.

Technical techniques include:

Statistical Process Control

- The implementation of an SPC system, which includes process control charts, events logs, corrective/preventive-action matrices, and process control procedures for the purpose of controlling and reducing variation around a customer requirement, thus improving customer satisfaction.

Effective Problem-Solving Techniques

- Utilizes appropriate techniques and resources to solve various problems, prevent future problems, and create opportunities for improvement.

Design of Experiments

- Provides for finding causes of significant and complex problems (i.e., to identify factors causing variation and to establish set points for design functions and processes).

Quality Function Deployment

- Determines customer requirements and translates them into design requirements, process requirements, and manufacturing inputs.
- Identifies customer satisfiers, dissatisfiers, and delighters.

World-Class Manufacturing Techniques

- Provides for shorter cycle times, reduced work-in-progress, and reduced setup times. Emphasis is on cellular manufacturing, just-in-time, reduction of inventory, measurements, etc.

Quality/Productivity Improvement Process

- A road map that implements quality productivity improvements for nonmanufacturing as well as manufacturing processes, combining all of the various techniques into an effective approach and producing significant business performance improvement and financial results.

ISO 9000 Plus
- Implements an international standard quality management system which provides for consistent quality with a slant toward having effective procedures and policies in lieu of documentation only.

World-Class Quality Planning
- Provides a formula for planning and producing quality in lieu of inspecting it in the product or service.

External and Internal Customer Satisfaction

In any business there are both internal as well as external customers. Companies with a commitment to excellence need a commitment to satisfying their customer needs at every level, i.e., internal as well as external. They should develop a corporate (or organizational) culture of continuous improvement and a customer-driven attitude.

Once the need for customer satisfaction is recognized and the quality of the product or service are defined for your particular process (i.e., requirements and expectations), it is necessary to choose a medium for change. How, exactly, can a company go about establishing a process for interpreting and meeting customer requirements? The most successful method requires a basic foundation of total commitment from top management and an integrated approach to training and implementation.

Integrated Training and Implementation Approach

The scope of the text includes 10 tracks, each of which creates a desired outcome. A track is a composite of training knowledge (what to do and why to do it), skills (how to achieve the techniques), and behaviors (the desire to achieve the techniques) tailored to each company's needs. Implementation of task projects produce the desired results, which are defined as *objectives*.

Tracks unite the training and implementation to other tracks in order to provide a comprehensive, yet focused, approach to achieve the desired business objective. This integrated training and implementation approach excels over the alternative approaches because the integrated training and implementation produces significant business performance improvement or transformation of culture. In this case, the whole is greater than the components.

Any TQM undertaking will start with an assessment and introduction of the TQM undertaking to company employees. Gaps between today's state and the desired state will be identified. Plans will be developed to close these gaps. The desired state will be expressed in Vision statements that describe how the organization desires to be characterized in 3 to 5 years. Vision, Mission, and Values statements are not just the starting point for the undertaking; they provide the direction which must be sustained and reinforced continually as the company evolves into world-class competition.

The Tracks

There are 10 tracks in the program.

1. The initial track is a *Foundation Track* whose purpose is to develop the groundwork for launching a TQM undertaking. The Foundation Track develops the vision, mission, and values statements, "determines the gap," and creates a two-year tactical plan to close the gap. During the Foundation Track, Critical Business Success Factors are identified, the departments are aligned, their internal customers are identified, measured, and addressed by launching project teams. They are performance-monitored and Internal Customer Satisfaction Agreements are initiated.

Once the vision, mission, and values statements are established, communicated throughout the organization, and understood by all employees, management can launch a two-year tactical plan. The tactical plan is developed with the intent of developing a structure for continuous improvement and a process management system that will measure, review, establish accountability, and provide feedback and appropriate consequences.

At this time the CEO, president, and senior staff participate in making declarative promises to the workforce in person, in writing, and/or on video tape. The objective is to share the Vision, Mission, and Values by cascading the communication from level to level in a manner appropriate for the organization.

2. The second track is the *Implementation Track* which dovetails with the Foundation Track under the auspices of creating Internal Customer Satisfaction Agreements between departments. Internal Customer Satisfaction Agreements include identification of all functions: departmental mission statements, identification of internal customers and their requirements, measurement of customer satisfaction, and agreement on meeting customer expectations. These agreements are created to align all the departments together. Projects are prioritized and are undertaken to support the Critical Business Success Factors. These projects will provide significant

impact to the Critical Business Success Factors when implemented. ISO 9000's implementation, SPC implementation, and world-class manufacturing concepts would fall in the scope of this track. All activities in the Implementation Track are reviewed by resource committees in their process management function.

3. The next track is the *Cultural Track*. Its purpose is to create commitment to TQM and transform the culture of the company. This is accomplished by developing implementation plans and survey questionnaires for each of the values that were chosen by the executives and key managers during strategic business planning. The desired new attitudes, beliefs, and behaviors are then monitored by a cultural-fostering group. For example, the training and application of empowerment is emphasized in the Cultural Track as well as being monitored throughout the entire undertaking.

The Cultural Track addresses the values which determine group behaviors and support the performance objectives required to achieve the Critical Business Success Factors and internal customer satisfaction agreements. The Cultural Track also addresses the forces which impede change, the benefits of change, the risk of failing to change, and the rewards for change.

Management will take responsibility for responding to Critical Business Success Measurements and will empower the workforce to improve their processes and achieve the objectives. Everyone in the new corporate culture must assume ownership of their process and the quality of their deliverables. The workforce cannot assume that someone else will catch the problems and that anything going to their internal customer will be good enough. Once the performance and behaviors have been recognized and rewarded, the values will be sustained in the culture.

4. At this point, a fourth track, *Recognition and Rewards System*, is initiated to respond to the departments, teams, and individual successes. It addresses and reinforces the business results and desired behaviors. Suggestion programs and a "performance appraisal system" are elements of the Recognition and Rewards Track. They are installed early in the undertaking. This system is designed to develop goals for the Critical Business Success Factors and reward the organizations and individuals for their achievement of job performances, improvement of their processes, implementation of behaviors, and the effectiveness of their role modeling.

5. A *Leadership/Team-Building Track* is included as a separate track because of the importance of identifying leadership effectiveness and utilizing these traits to support the implementation program. Doing the right thing at the right time is emphasized. Each individual has his or her critical success factors for a given position. In addition to making the best use of available talent, it will be necessary to train all personnel in effective

team behaviors. Total survival management involves the total organization. Emphasis is also made on team-building and managing-conflict within the teams since teams are the basic vehicle of the TQM effort.

In regard to involvement, no one is to be excused from participating, and the final responsibility cannot be delegated. Employees, including managers at all levels, must understand and share the objectives in order to achieve total success. The Leadership/Team-Building Track is heavily linked to the Implementation Track in all activities and at all levels—from the executive board to the resource committee to mentors and team leaders.

6. The *Management Skills Track* emphasizes efficiency in management/operations. Training encompasses selecting topics such as the talent that is needed, identifying areas for developing and coaching, and measuring success through supervision skills. Once the leadership has set the direction, the management's role is to achieve it. This track deals with the skill to achieve the desired results.

7. Once the Leadership/Team-Building Track and Management Skills Track are set, the *Core Techniques Track* is initiated. This track deals with establishing Statistical Process Control (SPC) at all levels of the company on measurements which are meaningful. In addition, the organization is brought to a high level of competency in problem solving. The philosophy is that problem solving will be done by the lowest competent levels of the organization. This track is likewise enmeshed with the Implementation Track and provides the core skills to improve processes.

SPC is considered to be the core of any improvement program because it provides management with meaningful data to assess the current situation and to measure the effectiveness of the improvement activities. Training in all of the core SPC techniques is critical. Most companies will start with training management and continue until training is viewed as an integral part of the job and a prerequisite to advancement within the organization.

The Core Techniques Track describes a specific process for implementing SPC in administration, suppliers, software, production, and engineering development processes. Actual experience with firms involved in the quality and productivity improvement process has demonstrated certain common requirements and common pitfalls that are of value to those wishing to install change and improvement. The Implementation Track has been active since the inception of the program. As part of the Implementation Track, the Productivity/Quality-Improvement Process is adopted, which ties all the techniques together in a road map for improvement. The productivity/quality-improvement process is initiated on all processes that have unfavorable performances to provide opportunities for improvement.

8. Following the Core Techniques Track is the *Advanced Techniques Track* where more involved tools of analysis and problem solving are applied. Included are courses on "Design of Experiments," "World-Class Manufacturing Concepts," "World-Class Quality Planning," "Design for Manufacturability," "Quality Function Deployment," and "ISO 9000 Plus." These techniques become useful as the TQM Implementation program gains sophistication. Advanced statistical skills provide engineers with the means of modeling a process and determining the effect of various inputs. The most important factors can then be controlled for optimum productivity while the insignificant factors can be essentially ignored so that process cost is minimized. Quality Function Deployment (QFD) brings together all of the disciplines in the organization and, through careful analysis of customer needs and product characteristics, ensures that every aspect of design, manufacture, and delivery is aligned with the customers requirements.

9. The ninth track is *Customer Focus* which addresses hearing the customer needs with a focus on those needs in order to achieve maximum satisfaction and creation of new business. Among the areas addressed in this track are retention of existing customers through effective customer satisfaction; new business is achieved through marketing planning, sales, new product development, and new product introduction.

10. The last track is *Train the Trainer Track*. This provides the company with internal trainers for the business to become self-sufficient in continuing the education of their workforce.

The overall focus of this handbook is to offer an avenue for survival in the highly competitive world marketplace of today. Performance improvement has finally been recognized as a necessity and not just another "management gimmick." Product life cycles are becoming shorter while reputations for poor quality are hard to erase. To realize success, companies have to involve everyone in the effort to meet their internal and external customer's requirements and expectations and to continuously look for ways to improve. Successful TQM implementation is not a quick fix, and requires a continuous integrated training and implementation approach. Companies should not have to rediscover the wheel—but should pursue a proven process. As Dr. Deming says, "You don't have to do it; survival is not mandatory."

2
Foundation Track

Purpose

The purpose of the Foundation Track is to prepare the company to implement TQM.

Objective

The objective of this track is to achieve the following:

- Identify strengths, weaknesses (areas for improvement), opportunities, and risks and exposures in each function and the company as a whole. An assessment is conducted; the findings and recommendations are then evaluated and provided in a report.

- Determine a strategic business direction by consensus of the executives and key managers. This direction is expressed in Vision, Mission, and Values statements.

- Develop a two-year tactical plan which clarifies priorities, milestones, responsibilities, and expectations. The assessment is one source of input for the plan.

- Generate awareness by everyone in the company of the necessity of performance improvement emphasis and a definition of what TQM means. Employees learn that each person has a stake in the program, and what they can do to make a difference.

- Create Internal Customer Satisfaction Agreements for each department with departmental mission statements supporting the corporate mission statement, hence aligning all departments together.

- Identify Critical Business Success Factors (CBSF); identify and initiate project teams to support the CBSF. The attainment of goals for CBSFs yields success for the company. The required training needs must be identified and prioritized in order to have success on the project teams and the Critical Business Success Factors.

- Initiate the Implementation Track, which addresses the Internal Customer Satisfaction Agreements, and develop project teams to address the Critical Business Success Factors and to address Internal Customer Satisfaction as a continuous process.

Definitions

The words used to describe the initial formulation of a TQM process are all very similar. For the purposes of this text the following definitions are offered:

Vision. The *state* of *being* that the company desires to achieve in 3 to 5 years. Visions characterize what the company will be in 3 to 5 years.

Mission. The Mission statement describes: what products or services the company provides; who the customers are; what the major thrusts are in the future markets, and the appropriate strategy for their products and services; what is distinctive about the company; how the company desires to be identified; and the company's purpose in the market.

Values. These are the beliefs, attitudes, and behaviors of the organization that are observable and are required to achieve their mission and vision. These values will shape the future culture.

Critical Business Success Factors. These are factors which determine the company's probability of success in customer satisfaction, profit, growth, and competitiveness. If the success factor targets are not achieved, the success of the company will be impacted unfavorably.

Objectives. These are the immediate goals to be achieved. All teams, departments, groups, or task forces should have a charter which includes expectations (clearly stated and measurably objective), guidelines, authority, and resources. They should include a statement of their objectives at every meeting and in every progress report. The attainment of objectives should directly impact the Critical Business Success Factors. (See Figure 2-1.)

Assessment

The initial actions of the assessment are as follows:

- Identify the company's strengths, weaknesses, and areas needing improvement—functional areas as well as positioning of the company.

EXECUTIVE FOUNDATION TRACK

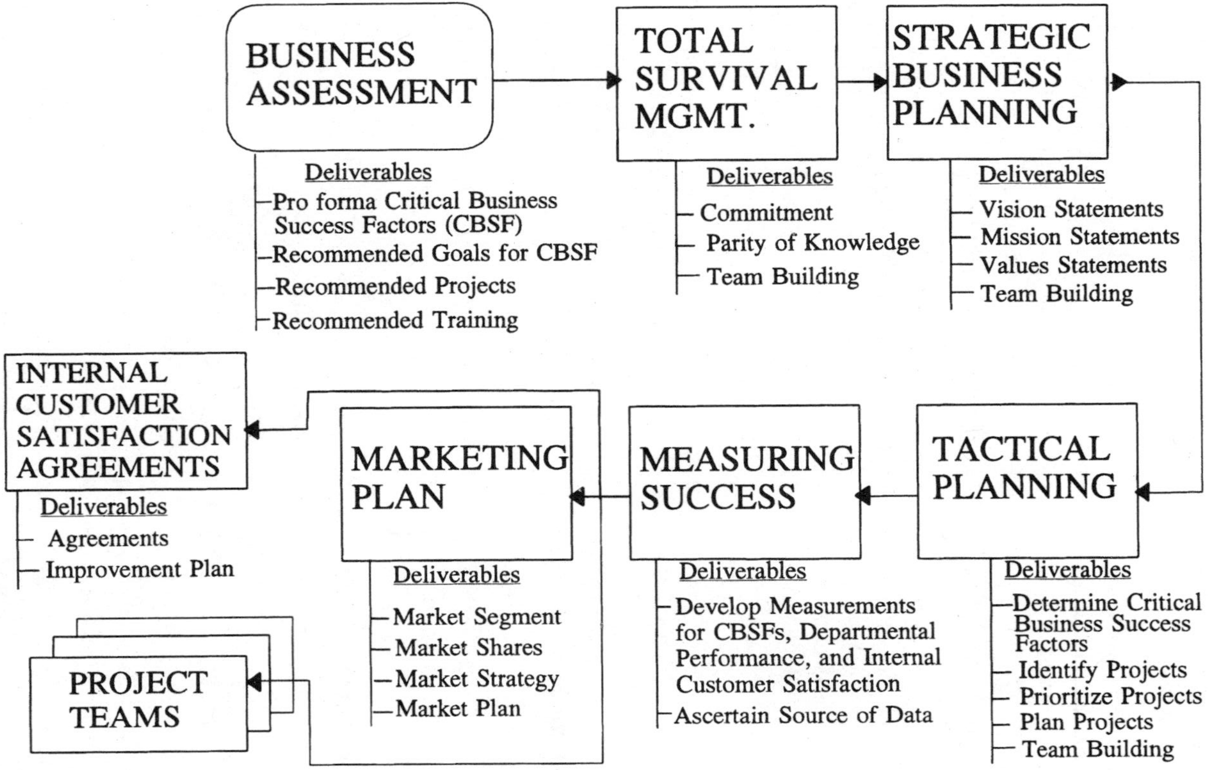

BUSINESS ASSESSMENT

Deliverables
- Pro forma Critical Business Success Factors (CBSF)
- Recommended Goals for CBSF
- Recommended Projects
- Recommended Training

TOTAL SURVIVAL MGMT.

Deliverables
- Commitment
- Parity of Knowledge
- Team Building

STRATEGIC BUSINESS PLANNING

Deliverables
- Vision Statements
- Mission Statements
- Values Statements
- Team Building

TACTICAL PLANNING

Deliverables
- Determine Critical Business Success Factors
- Identify Projects
- Prioritize Projects
- Plan Projects
- Team Building

MEASURING SUCCESS

Deliverables
- Develop Measurements for CBSFs, Departmental Performance, and Internal Customer Satisfaction
- Ascertain Source of Data

MARKETING PLAN

Deliverables
- Market Segment
- Market Shares
- Market Strategy
- Market Plan

INTERNAL CUSTOMER SATISFACTION AGREEMENTS

Deliverables
- Agreements
- Improvement Plan

PROJECT TEAMS

Figure 2-1.

- Identify opportunities for customer satisfaction, growth, profit, etc.
- Identify threats (i.e., risks/exposures, possible downsides for the business).

It is recommended that the assessment be conducted by an independent outside consulting firm which will view the business from a dispassionate viewpoint and be capable of benchmarking the various significant aspects of the company with other companies. The qualifications and experience of the consulting firm are essential in its selection.

Contents of the assessment are: strategic business direction and what it takes to close the gap, organizational dynamics, customer satisfaction, quality policy and practices, financial and business systems, operational/MRP issues, marketing and sales, and new product introduction. The assessment is conducted through one-on-one interviews with executives and key managers, discussions with focus groups at all levels of personnel, and distribution of select questionnaires to determine the company's attitudes, beliefs, and behaviors. Results of the assessment are written in the form of the consultants' findings and recommendations. This information is reviewed with the executives and utilized in the tactical planning.

Business Assessments

Purpose
- Understand the current state of the company and its environment
- Understand the desired future state of the company
- Understand the business
- Determine the Critical Business Success Factors (CBSF)
- Identify and evaluate critical business processes
- Identify requirements to close the gap between current and desired state
- Present analysis of findings and recommended action plan
- Determine training and consulting requirements

Scope of Business Assessments
- Strategic business direction evaluation
- Customer satisfaction and value perception
- Market performance and competitive position
- Current and recent financial performance

Business Functions and Current State
- Organizational structure, culture, and dynamics

- Marketing and sales, product definition
- Product development and design engineering
- Operations/manufacturing/systems (MRP)
- Information systems and document management
- Financial and business systems
- Quality and reliability: policies, practices, and performance
- Software development

Analysis of Findings
- Cause and nature of current problems
- Projected outlook for the company and the industry
- SWOT analysis—strengths, weaknesses, opportunities, threats
- Gap analysis—current state versus desired future state
- Pro forma Critical Business Success Factors

Recommendations
- Report
  Executive summary
  Critical Business Success Factors
  Issues, impact, recommendations, benefit, action plan
- Presentation

Strategic Business Planning

The purpose of strategic planning is to establish the direction and course of the company. It is accomplished by a planning process that requires participation of the executives and key managers using consensus management. The deliverables for strategic business planning are the vision, mission, and values statements. All executives and key managers as a consensus team must participate in the development of the statements. Hence, they have ownership of the company's vision, mission, and values.

It is crucial not only to establish these corporate statements, but to periodically review the company's direction relative to the vision, mission, and values. Feedback on any variations to plan for correction is essential. Realization of the mission must be the passion of top management and threaded through every communication line. It must also be clearly understood and supported by all employees. If you walk out to the production line and ask someone about the company's vision, mission, and values, they should be able to respond and add what they can do in their department to achieve it. Remember, the vision, mission, and values provide a theme for all members of the company to rally around. It is essential that people have a theme they can focus on, to give direction to their energies. This can be accomplished by top management and departmental supervision continuously discussing the points and "walking the talk."

Not only is it important that the vision, mission, and values statements are appropriately prepared, but equally important is the process that the executives and managers use to develop the statements and the commitment that is created by themselves. If the commitment is lacking, the total effort is useless. Top management must imprint vision, mission, and values statements in their minds and hearts as well as documenting them. (See Figure 2-2.)

An example of vision, mission, and values statements follows.

Sierracin/Sylmar

Vision Statement

To be recognized as the industry leader worldwide and preferred supplier to our customers.

Mission Statement

We will design and manufacture high-performance transparent and related products which provide total customer satisfaction.

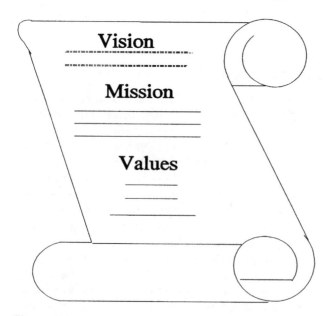

Figure 2-2.

Together, we will achieve this by encouraging creativity, innovation, and commitment to continuous improvement.

Through profitability, we will reinvest resources to provide an environment conducive to personal growth, and assure our continued leadership position.

Values Statement

We recognize and value our customers and employees by emphasizing:

- Customer-driven attitude
- Trust, integrity, and honesty
- Empowering the workforce
- Achievement of plans and objectives
- Teamwork
- Pride in product and company
- Proactivity
- Employee recognition and development
- Safe and positive work environment

By working all together, we will take the company from its present state to being the leader in the industry worldwide. We will be number one. We will all be proud that we deliver the best quality product in the world, the best on-time delivery record and the most affordable price. The customers will be running to Sylmar to give us their orders. We will assure ourselves a good future with full employment in a place we would like to come to every day to exercise the empowerment given to us.

Tactical Plan

The purpose of the tactical plan is to provide a road map with objectives that will ultimately achieve the desired goals. A two-year tactical plan consists of many tactical plans which delineate completion milestones, milestone responsibility, and expectations for projects and tasks needed to support the vision and mission. Tactical planning is a necessary function which schedules the integration of training and implementation of teams to achieve the desired business results. Effective tactical plans designate action items by priority, training seminars, committees, project team functions, responsible individuals, and the outcome expectations for each issue. These expectations serve as a milestone to measure progress.

Once the tactical plan has been completed, the company has a direction, a focus, and a road map to achieve the end goal. With the plan in place, Total Quality Management awareness can be introduced to the workforce. The introductory session should be kicked off by the president stating its purpose and how each person holds a stake in the TQM outcome. Objectives of the session are to create quality awareness, initiate an understanding of the impact of the internal customers, express the TQM philosophy, and present an overview of the skills and techniques required to succeed.

How TQM supports Vision, Mission, and Values Statements

When you set goals and objectives or establish vision, mission, and values statements, you have only scratched the surface. An airplane pilot will put considerable effort into developing a detailed flight plan designed to assure that pilot and aircraft arrive safely at the desired location (objective). The pilot may not be successful in reaching the objective if the flight plan is incomplete or inaccurate, and, by the same token, the pilot surely will not reach the objective if he or she does not know how to fly the airplane.

Vision, mission, and values statements should be developed carefully. They should take into consideration the business and organizational parameters that will influence the company's character. TQM provides the analytical tools that turn data into a lever for improvement and a guideline for action. The vision, mission, and values serve the organization as the flight plan does for the pilot. Likewise, the company must understand how to achieve the goals as a pilot must understand how to fly.

TQM Components

Tactical planning is the function which schedules the integration of training and implementation to achieve the following desired business results:

- Design Critical Business Success Factors, taken from the *tactical planning* session.

- Implement cultural change, which is the implementation of values which are chosen in the *strategic planning* session.

- Align the organization with Internal Customer Satisfaction Agreements to define: departmental missions, identification of customers, their requirements, measurements of customer satisfaction, agreements, and corrective action as applicable.

- Instigate right-sizing which is matching of human resources to functional needs.

- Choose projects whose objectives directly impact the performance of Critical Business Success Factors.

- Develop high-return-on-investment projects.

- Design projects required to *close the gap* between the present state and the *vision*. These projects are identified in the assessment and in the *tactical planning* session.

- Design all training to complement the implementation and achievement of projects.

- Implement technical techniques which consist of topics such as *SPC, ISO 9000, Design of Experiments, Cell Manufacturing,* and *Quality Function Deployment.*

- Establishment of quality management system which includes policies, procedures, and work instructions. (See Figure 2-3.)

Internal Customer Satisfaction Agreements

The purpose of tactical planning is to provide a road map with objectives that will ultimately achieve the desired objectives.

Charters align all departments in the company to support the company's Mission statement and to achieve customer satisfaction agreements between departments. Customer satisfaction agreements will be based on deliverables to internal customers, which meet or exceed internal customer requirements and expectations, measurements of satisfaction, and corrective actions to make the performance satisfactory if the performance is not satisfactory or improvement plans to create improvement even though the internal customer is satisfied.

Definitions

Internal Customer Satisfaction Agreement. The whole thing. A "job description" for the department. Includes *department Mission statement,* plus *activities, deliverables, customer requirements,* and *measurements.*

Departmental Mission. Purpose, focus of the department, that for which it will be held accountable in support of the company's Mission, Vision, and Values.

Functions. Distinct activities a department performs. This includes administrative, service (e.g., legal, financial, human resources), etc.

Task. Elements of an activity. Normally an activity will be made up of several tasks.

Figure 2-3. Two-year tactical plan model. (*Note:* There are more than twelve weeks to the process.)

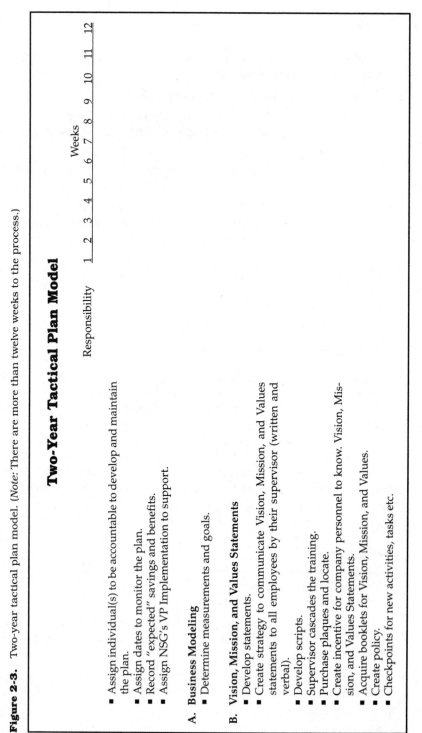

Two-Year Tactical Plan Model

	Responsibility	Weeks
		1 2 3 4 5 6 7 8 9 10 11 12

- Assign individual(s) to be accountable to develop and maintain the plan.
- Assign dates to monitor the plan.
- Record "expected" savings and benefits.
- Assign NSG's VP Implementation to support.

A. Business Modeling
- Determine measurements and goals.

B. Vision, Mission, and Values Statements
- Develop statements.
- Create strategy to communicate Vision, Mission, and Values statements to all employees by their supervisor (written and verbal).
- Develop scripts.
- Supervisor cascades the training.
- Purchase plaques and locate.
- Create incentive for company personnel to know. Vision, Mission, and Values Statements.
- Acquire booklets for Vision, Mission, and Values.
- Create policy.
- Checkpoints for new activities, tasks etc.

Figure 2-3. Two-year tactical plan model. (*Continued*)

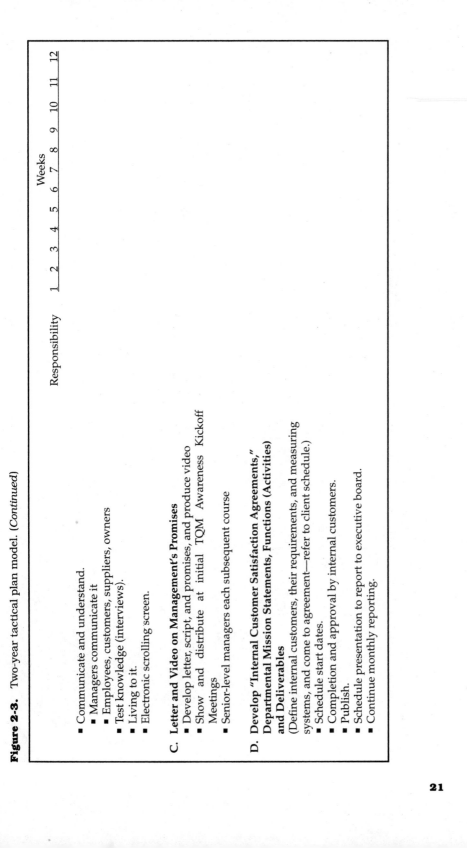

Responsibility

Weeks
1 2 3 4 5 6 7 8 9 10 11 12

- Communicate and understand.
  - Managers communicate it
  - Employees, customers, suppliers, owners
  - Test knowledge (interviews).
  - Living to it.
  - Electronic scrolling screen.

C. **Letter and Video on Management's Promises**
  - Develop letter, script, and promises, and produce video
  - Show and distribute at initial TQM Awareness Kickoff Meetings
  - Senior-level managers each subsequent course

D. **Develop "Internal Customer Satisfaction Agreements," Departmental Mission Statements, Functions (Activities) and Deliverables**
     (Define internal customers, their requirements, and measuring systems, and come to agreement—refer to client schedule.)
  - Schedule start dates.
  - Completion and approval by internal customers.
  - Publish.
  - Schedule presentation to report to executive board.
  - Continue monthly reporting.

Figure 2-3. Two-year tactical plan model. (*Continued*)

	Responsibility	Weeks											
		1	2	3	4	5	6	7	8	9	10	11	12

E. Critical Business Success Factors Project Teams

(Determined from tactical planning second session; "quick success" teams—insert projects.)

- Select Critical Business Success Factors based on customer retention, improving profits, and gaining new business.
- Examples:
  - % on-time proposals
  - % RFQ win ratio
  - Productivity
  - % on-time shipments
  - Quality targets
  - Determine measurements
  - Sales and marketing
  - Quality
  - % on-time material receipt
  - Customer satisfaction index
  - Before-tax profits
  - Develop new products
- Examples:
  - Develop new product introduction policy and procedure.
  - Prepare budget performance analysis report and action plan.
  - Develop customer satisfaction index.
  - Create activity-based cost accounting system.
  - Reengineer the estimating process.

Figure 2-3. Two-year tactical plan model. (*Continued*)

	Responsibility	Weeks											
		1	2	3	4	5	6	7	8	9	10	11	12

Guidelines

Estimated project span

First wave — 2- to 4-week project

Second and subsequent waves — 8-week project (subdivide project goals in 8-week maximum duration) Prioritize projects: quick successes, urgent, important, and as convenient.

F. Resource Committee—Scope of Teams and Dollar Savings/Team
- Select chairperson (usually the Critical Business Success Factor Champion).
- Select members if applicable.
- Adopt a charter (modify NSG's standard).
- To identify projects, issues which impede customer satisfaction, productivity, and quality improvements, opportunities, weaknesses.
- To identify team leaders.
- To identify team members.
- To identify mentor.
- Issue team meeting reports.

G. Executive Board/Committee
(for larger companies.)
- Select chairperson.
- Select members.
- Adopt a charter.

23

Figure 2-3. Two-year tactical plan model. (*Continued*)

	Responsibility	Weeks
		1 2 3 4 5 6 7 8 9 10 11 12

H. Recognition and Reward System
- Assign team leaders and mentor.
- Team members.
- Assign NSG's VP Implementation for support.
- Implement team meetings.
- Develop charter and produce budget.
- Provide recommendations for recognition and rewards.
- Suggestion plan.

I. Training Advisory Committee
- Assign chairperson.
- Assign members.
- Assign NSG's VP Implementation.
- Develop policy.
- Senior manager kicks off each course.

J. Performance Appraisal System
(This should begin once departmental charters are complete—dates are derived from the training schedule.)
- Define position accountability—start.
- Assign team, team leader, and mentor.
- Assign facilitator and conduct precourse assessment.
- Start meetings—to develop policy.
- Complete policy.
- Develop appraisal system, including format.
- Complete.
- Implement.

Figure 2-3. Two-year tactical plan model. (*Continued*)

Responsibility

Weeks

1 2 3 4 5 6 7 8 9 10 11 12

K. Alignment of Workforce to Requirements (Right Sizing)
(Determine cost drivers and determinate indicators of business activity to workforce.)
- Select organizations to be analyzed.
- Conduct activity based budgeting.
- Develop position requirements.
- Conduct training on observing and interpreting behavior for interviewing and coaching.
- Match existing personnel to position requirements.

L. Creating Culture Change
(The dates are derived from the training schedule.)
- Develop *implementation plan* at workshop for values.
- Develop survey questionnaire checklist to verify implementation of values.
- Create policy for monitoring the cultural implementation.
- Assign team leader and mentor to the cultural-fostering team.
- Select personnel for the cultural-fostering team.
- Assign NSG's VP Implementation
- Implement meetings for the cultural-fostering team.
- Execute *implementation plan* (itemize all action items).
- Cultural-fostering team to monitor implementation.

M. TQM Teams (Plans to Achieve the Vision)
(Derived from the assessment and driving forces to "close the gap" —third or fourth wave of teams.)
- Examples:
  - Develop customer satisfaction policy.
  - Develop customer satisfaction questionnaire.
  - New project development plan.

Figure 2-3. Two-year tactical plan model. (*Continued*)

	Responsibility	Weeks
		1 2 3 4 5 6 7 8 9 10 11 12

- Departmental hiring.
- Develop marketing plan.
- Develop new production development.
- Develop estimating process and policy.
- Develop purchasing policy.
- Develop department budget and report.
- Cost of quality measurement.
- Enhance supplier project.
- Reduction of scrap and rework.
- Roles and responsibilities of the team leader.

N. Implementation of Specific WCM Techniques (SPC, DOE, DFM, JIT, Cells, ISO 9000, SetUp Time Reduction) by Department, Machine, Process

(Derived from the training schedule.)

- Assign teams (list each major task and its respective itemized tasks).

O. Intradepartmental Competition for Performance Improvement

P. Operator Certification Program

Q. Benchmarking Status

R. Quality

- Quality statement.
- Quality policies.
- Quality procedures.
- Audit.

Primary activity. One that contributes directly to the mission of the department; it is associated with value-added processes to a product or service.

Support activity. One that supports a primary activity (supervision, administration, training, etc.).

Value-added activity. A primary activity that directly adds value to the final product.

Non-value-added activity. A primary activity that is associated with rework, lack of quality, etc.

Deliverable. Primary product of an activity (activities can have secondary or by-products).

Customer (internal or external). Whoever receives the deliverable. Internal customer is the one performing the next value-added step.

Customer requirements. Customer expectations for the deliverable.

Measurements. Metrics to ascertain adherence to customer requirements (scorekeeping).

Supplier. Whoever provides input, knowledge, or service to the person(s) performing the activity.

Activity measure. Measure of number of occurrences of an activity.

Internal/External Customers

Identification of internal and external customers is an important activity in the early phases of TQM. People are usually able to identify external customers—these are the people who buy our services or product! Frequently, there is little thought given to identification of internal customers.

An internal customer is the next individual or functional group in the process to receive the output or deliverables, and to act on them. Proper application of TQM will create the realization throughout the company that all employees are both customers and suppliers. Suppliers are, of course, groups or individuals who provide material, data, information, or services to others.

Every process or function, from the mail room to the production floor, from the sales force to the facilities engineering group has an identifiable output required by an individual or group. Accepting this concept is how you will identify your internal customers. This concept may initially be unclear and require facilitation from a third party who is removed from day-to-day operations and procedures. Training people to recognize the characteristics of their process and to identify their internal customers will enhance their confidence and support as process design improvements are implemented.

In evaluating internal customer/supplier relationships the following questions need to be answered.

Who is the customer?

- The next person (individual or functional group) in the work process.
- The recipient of the material, data, information, and/or service you produce.

Who is the internal supplier?

- The person or group responsible for preparing and developing an output (material, data, information, or service) that you will receive and act upon.

Who are the internal customers?

Who are the external customers?

What are the true customer requirements?

- These requirements are pulled from the written customer specifications, drawings, and standards and must be discussed in a two-way conversation between the customer and supplier. The goal is to meet the customer's needs in the most efficient and effective manner.

How do your customers measure their satisfaction?

Is there agreement on the state of customer satisfaction?

What is the corrective action plan to address unsatisfactory performance?

The most effective way to determine how the customer feels about output from the supplier is to ask. Customer interviews and reviews of their requirements provide the most direct answers. To initiate this review process for a department or function, do the following:

- List the process end products.
- Identify the user of each end product.
- Evaluate the process or department performance with indicators that the customer has identified.

Sit down face-to-face with the identified customer and determine the following:

- What are the customer's requirements?
- How is the performance or customer satisfaction measured? Typically, measurements are accuracy/completeness, on time delivery, elapsed time, cost, etc.

- Where is there agreement between supplier and customer, and where are there gaps?

A design function serves as a clear model of the internal customer/supplier relationship. There are typically discrete steps in the design process that take specifications as an input and provide documents to the next step as output. Each successive step is the customer for the previous step's output. For example, the top design level (systems) is the deliverable to the internal customer, subsystem design. The measure of customer satisfaction may be as simple as the number of error-free designs opportunities and whether the design document arrived on time to support master design schedules. If subsystem design finds that drawings are not properly annotated or that excessive dimensions are not specified with tolerances, his or her supplier is not meeting expectations. The corrective action in this case could be just a matter of communicating the errors to the system designer of the previous step, so that the designer can perform the verification of acceptable product before the drawing is sent forward to its next step.

Examples of Internal Customers and Deliverables

Department	Output or deliverable	Customer
Receiving	Material counted accurately with appropriate quality documentation	Stockroom
Sales	Order entry information	Scheduling
Personnel	Suitable employment candidates	Dept. managers
Engineering	Manufacturable design done with concurrent engineering	Manufacturing
Manufacturing	Products that meet specification; on-time delivery	External customers
Purchasing	Material on time and meeting manufacturing requirements	Manufacturing
Accounting	Accurate expense reports	Dept. managers

Vision, mission, and values statements should have been written for the whole corporation, and each department must develop its mission to align with the corporate mission.

If a function or process does not seem to have a real customer, then that function should be reevaluated. The function may be unnecessary, or it may simply be part of a larger process. Flow charts and diagrams help define the boundaries of a given process and define the contribution of all subfunctions to the final end product seen by the customer. Functions or processes which do not add value to the output should not be done.

Departments or people without an internal customer should find one whom they can support. If they cannot find an internal customer, they should get their résumés out before they are discovered.

Each department needs to develop its mission statements, which include performance objectives, and a list of functions with internal customers, requirements, and measurements. The internal customers approve their internal supplier mission statements. The reporting of Internal Customer Satisfaction Agreement measurements should be on the forms shown in Figures 2-4 through 2-6.

Pitfalls to a Successful TQM Implementation

Pitfall Number One. The executives believe that commitment is interest. They do not appreciate what commitment is and how their role modeling will influence the outcome of the TQM implementation. They believe commitment is interest, in that they delegate TQM responsibilities or do that when it is convenient.

Impact. Many TQM implementations do not have any leadership, and thus do not achieve their expectations and potential.

Solution. Ensure that the executives are trained and understand what commitment is and that they buy in to the responsibilities of their roles. Commitment is giving no excuse and accepting no excuses in pursuing the goals and objectives.

Pitfall Number Two. Doing a self-assessment by internal personnel.

Impact. The assessment is viewed internally through the paradigm of internal personnel and may not possess the necessary objectivity. The personnel may also be lacking the knowledge and experience necessary to conduct a proper assessment. When external personnel conduct the assessment, if they are experienced, they in effect benchmark the assessed company to other companies.

Solution. It is best to have an outside consulting firm who possesses the necessary knowledge and experience conduct the assessment. This will allow objectivity and an in-depth analysis of the company.

Pitfall Number Three. Failure to educate the executives on the concepts of Total Quality Management (TQM), therefore failing to create a parity of knowledge among them, gain their commitment, and build their team.

ILLUSTRATION

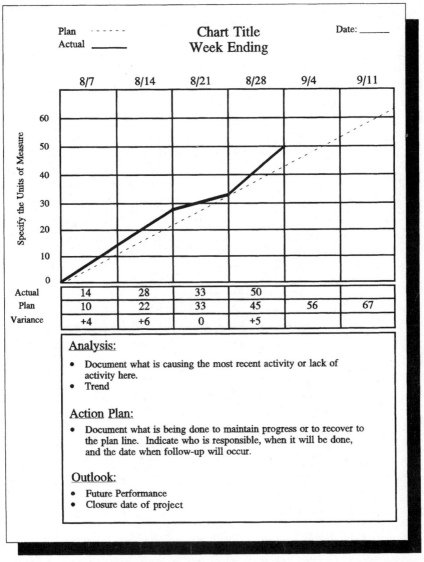

Plan   - - - - -

Actual ————

Chart Title
Week Ending

Date:_____

	8/7	8/14	8/21	8/28	9/4	9/11
Actual	14	28	33	50		
Plan	10	22	33	45	56	67
Variance	+4	+6	0	+5		

Specify the Units of Measure

Analysis:
- Document what is causing the most recent activity or lack of activity here.
- Trend

Action Plan:
- Document what is being done to maintain progress or to recover to the plan line. Indicate who is responsible, when it will be done, and the date when follow-up will occur.

Outlook:
- Future Performance
- Closure date of project

Figure 2-4. This chart format is recommended for near-term quantifiable objectives. It can be used to indicate 2- to 4-week objectives with several prior weeks of history for reference.

EXAMPLE

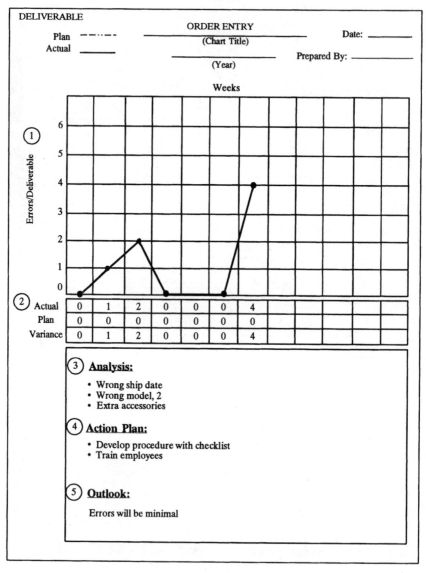

Figure 2-5.

EXAMPLE

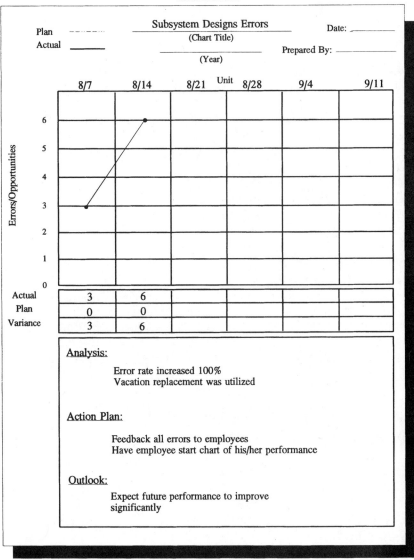

Figure 2-6.

Impact.   If the executive staff is not educated in TQM techniques, then the program start-up will flounder. Also, sustained success with TQM will be difficult to achieve without the executive staff's understanding and support and their commitment.

Pitfall Number Four.   Not conducting strategic business planning to set the future direction of the company prior to starting the TQM undertaking.

Impact.   Without the participation of the total executive staff, an effective vision, mission, and values statement cannot be produced. The executive staff will also fail to buy into the TQM undertaking. Without their participation, the company will not know what they want to be; what products or market they will focus their thrust on; and what values they want to use to guide their company's culture to retain existing customers, improve profits, or create new business. Further opportunities to build a stronger team are lost.

Solutions.   Conduct a strategic planning session with an experienced consultant, the executive staff, and key managers. The consultant will help the executives reach consensus on the issues which will guide the executives in developing vision, mission, and values. A by-product of the session will be the creation of bonding between the members of the staff.

Pitfall Number Five.   Failure of the senior staff to produce a two-year plan (hence, Critical Business Success Factors) for the company.

Impact.   Without a comprehensive plan establishing all the milestones of TQM goals, the company's champions could not be assigned to attain the necessary results to make TQM a success.

Solution.   Conduct a tactical planning session with the executive staff and an experienced consultant. This meeting will establish a two-year plan, and create Critical Business Success Factors and teams with prioritized goals. Again, the participation by the members will reinforce their buy-in and commitment.

Pitfall Number Six.   A lack of internal customer satisfaction agreements between various departments.

Impact.   Departments will continue to function at the level of cooperation they had before TQM was adopted. Consequently, the departments do not become aligned to meet the needs of the external customers.

Solution. Establish departmental teams to identify their internal customers, what their requirements are, and how to measure satisfaction and come to agreement with the internal customers. In the event the internal customers are not satisfied, a recovery plan should be developed. The status of the department's internal customer satisfaction agreement should be reviewed monthly.

3
Implementation Track

Purpose

The purpose of the Implementation Track is to utilize the knowledge, skills, and behaviors gained in training to produce measurable business results. In order to understand the role of the Implementation Track, let's use an analogy: gasoline in the automobile's gas tank is potential energy. The engine represents the process that transforms the potential energy into kinetic energy. However, without sparks from the spark plugs, the process cannot be initiated.

In this light, training is potential energy; the organization, improvement process, and competent individuals represent the process to transform potential energy into kinetic energy; and the executives provide the sparks which energize the process. The kinetic energy in this case is implementation. It is implementation which produces significant business results.

Objective

The objective of the Implementation Track is to produce results which will retain existing customers, improve profits by increasing productivity, and create new business. The approach the Implementation Track addresses is to focus on achieving the Critical Business Success Factors, creating a management cultural change, and installing Internal Customer Satisfaction Agreements and measurements.

Introduction

The Implementation Track addresses the actual implementation of Internal Customer Satisfaction Agreements, project teams to support Critical Business Success Factors, cultural change tactics, statistical process control, quality-function deployment, cellular manufacturing concepts, and any of the other techniques.

Where and How to Start

One of the first questions posed by managers exploring the possibility of TQM is "How do we start?" Either they have been swayed by the undeniable logic of the process, quality and profit improvements attainable through TQM, their customer's demands, or the desire to survive when current performances are not adequate. Now they want a clearly defined process for installing Total Quality Management in their organizations.

Many quality experts believe that in five years there will be only two types of companies: those who believe in business performance improvement and those who are not in business. Quality means error free, delivery on time, price as a value, and meeting whatever requirements and expectations the customer may have.

Requisites for TQM are a "tactical plan" to provide project expectations which relate to the Critical Business Success Factors, an appropriate organizational structure, a well-defined process (i.e., set of guidelines and responsibilities), and a schedule of implementation and measurement points. The cross-functional team approach has proven successful in providing these requisites, and has a good track record of success. It is important to build strong teams with horizontal and vertical communication channels in order to establish the requisites as efficiently as possible.

If your TQM team is made up of only a single department, the exercise will be viewed as a task with only marginal relevance to the rest of the company. If your team is made up of only the senior executives, the process of dissemination will not be effective in all areas, because each executive will choose his or her own method of implementation and extent of delegation. However, if senior executives are not part of the process, then TQM will fail. This text provides the roles of senior executives' involvement and how they impact the organization.

Commitment

In order to be successful, it is paramount that the CEO or president of the company is 100 percent committed to and supportive of TQM (or whatever name he or she wishes to call the process). The CEOs and presidents place their integrity on the line, and their actions must support TQM objectives. The number one cause of failed TQM programs is inadequate commitment and support of the CEO/president. Likewise, success of the TQM implementation can be related to the CEO/president's commitment.

Commitment to TQM means the following:

1. Major objectives/issues of the company are under the TQM umbrella. This is a consequence of the development of the "Critical Business Success Factors" (CBSF).

2. The executives are involved. They are active on the "executive board"; participate as champions of CBSFs and values; and chair resource committees to review and monitor the teams or mentor the teams.

3. Time for TQM is utilized by the executives and members of the organization. Once members are scheduled to attend meetings, only a customer crisis would have the authority to have members pulled out of the meetings.

4. Resources are made available in keeping with the size of projects and their potential benefits.

Now that the spark has been provided, the engine and drive train are energized to perform their process of providing kinetic energy. Likewise, the Implementation Track's process (the business performance improvement process) is energized with competent, committed personnel and objectives to achieve business results.

Organizational Structure

Once the direction of the company has been established in "strategic business planning," and projects have been identified and prioritized for the "Critical Business Success Factors" during *tactical planning*, the next step is to establish an organization to implement the plan. Like other endeavors, it is easy to acquire knowledge ("what to do" and "why to do it") and relatively easy to learn the skills through training ("how to do it"). However, to show measurable improvement through effective implementation of skills, techniques, and behaviors is the real challenge. As in football, an excellent game plan and mediocre execution results in disaster. What is required is a good plan with excellent execution. In TQM, we call the execution *implementation*.

To meet the challenge of implementation, certain elements must be present. The first element is organizational structure; the second is a medium in which to operate: an improvement process (i.e., an organized approach); the third element is techniques; and the fourth is competent personnel.

Competent personnel are created by training for knowledge, skills, and behaviors. These qualities are implemented through training and selection from management. Management also provides them role modeling, coaching, and proper assignments which match their behaviors and skills to the required behavior and skills of their position.

An effective organizational structure comprises an executive board, Critical Business Success Factors' champions; one or more resource committees, each with multiple project teams; a training committee; recognition and reward committee; culture foster committee; and values champions. Four levels of job roles arise in a TQM implementation: (1) executive board; (2) resource committee chairperson and members, plus mentors; (3) team leaders; and (4) team members. The TQM organizational structure does not replace the existing company structure, but overlaps the existing structure. (Power, influence, and priority are transferred to all vital and important tasks operated under the TQM umbrella.) (Refer to Figure 3-1.)

Executive Board Tasks

An executive board is the highest authority within the structure. Its purpose is to plan, review, and be accountable for the success of the overall TQM implementation. The executive board is responsible for adherence to the improvement process as well as the results.

Principal members are the CEO, president or VP/general manager, and the resource committee chairpersons (the latter is optional). Depending on the size of the company, the resource committee chairperson is usually the champion of a Critical Business Success Factor (CBSF), the highest-level functional manager, and has teams operating with regard to CBSFs. For example, a VP of design would be chairperson if there were team projects dealing with new product development, improving designs, or reducing new production introduction time.

The executive board should meet monthly to review and critique the following:

- Progress compared to the tactical plan

- Progress of each Critical Business Success Factor's measurement by its champion, a senior manager

- Progress of the cultural change by input from the cultural-fostering team

- Status of the implementation plan for each value by its champion, typically a senior manager

TQM Organization

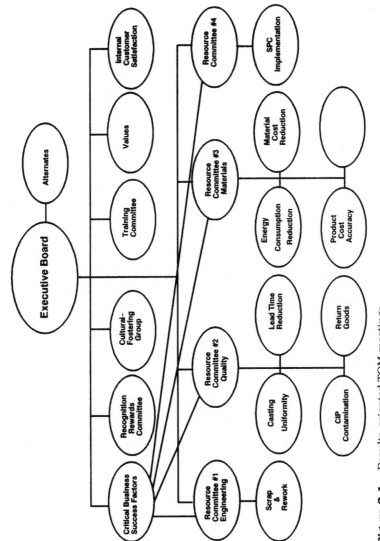

Figure 3-1. Results-oriented TQM meetings.

41

- Celebration, communication, recognition, and rewards generated by the recognition and reward committee
- Training customization input by the training committee
- Status of Internal Customer Satisfaction by department managers

Senior management's role in the Implementation Track is to provide the environment and resources necessary for cultural change and achievement of desired results. Management should never be satisfied with current performance status, but should always be asking for the next level of improvement and the plan to get there. The CEO or president or general manager must understand and support the process and take time to demonstrate support through actions and words. Having the president or general manager review the process-control charts on the line and ask well informed questions has a positive impact that is unequaled by any training program or quality improvement campaign. Having the president or general manager, on the other hand, allow shipment of a known inferior product to meet quarterly revenue goals sends a signal to every employee and can sink the most elaborate TQM plan. In this case, the president becomes "the stealth TQM destroyer."

Management at all levels is responsible to empower all employees and to encourage change, cooperation, and continuous improvement. Each manager, regardless of departmental function, must participate in the system of coaching the teams and rewarding success. To create this atmosphere, every manager must be thoroughly versed in the tools of TQM, both cultural and technical. Furthermore, managers must demonstrate commitment by applying those tools throughout the organization. All employees have a responsibility for quality improvement, but the primary responsibility resides with management. If results are not being achieved by the workforce, the executives must look to themselves for the answers.

Resource Committee

Resource committees report to the executive board. Depending upon the size of the company and the nature of the project team's objectives, one or more resource committees should be established. Resource committees are typically chaired by the champion for a CBSF. When there are four or more teams, it is advisable to have more than one resource committee. There are two reasons for this. First, it involves more senior or middle-level managers in the process; thereby the workload can be shared. Second, a small span of control for the resource committee will result in better focus on specific objectives and better communication between all the participants. Third, the limited number of teams limits the involvement and participation of the team members.

The purpose of the resource committee is to be responsible for identification and selection of new projects, and to oversee implementation of TQM through project teams. It is typically accomplished by having teams addressing projects. Resource committees provide the heartbeat of the TQM undertaking.

The resource committee assigns a team leader and mentor, if available, to the teams. A team leader is typically the manager of the function in which the teams are operating, and whose performance will benefit from the project's completion. Team leaders may be at lower levels, but will have challenging objectives in keeping with their position and background. Team leaders and mentors will be discussed later in this chapter.

Resource committees are empowered with the authority to select projects, change priorities, reallocate resources, abort projects, and increase resources without an encumbering approval process. Members define and manage the resource committee charter and provide advice and consent to team charters. In the event a special situation arises which exceeds the authority of the resource committee, they will address the executive board for approval.

The resource committee should review the teams' progress twice a month. In the early stages of a TQM program, weekly committee meetings will help create momentum for the process. The objective of the reviews is for the team leader to present the team's plans and progress through a standard progress report. Reviews are limited to between 10 and 15 minutes per team. The frequency of resource source committee meetings will establish the "heartbeat" for the TQM undertaking.

The responsibilities of the resource committee are:

- To provide the teams with written team charters, prior to the teams' launch, which state the following:

  Expectations, outcome, or results of the project

  Guidelines on how to achieve the above

  Limits of authority

  Adequate resources

  The requisite skills that the personnel are provided or can acquire in order to assure a competent team

- To negotiate with the team in the event the team doesn't agree with the team charter

- To ensure the team's objectives are achieved, while staying linked with the Critical Business Success Factors

- To provide reinforcement of clear expectations when appropriate

- To provide an agenda for each meeting

- To review teams' progress so that continuous and incremental progress is achieved at each review session

- To question team leaders in order to identify problems, constraints, and risks so that for the resource committee can assist them in solving the issues

- To focus the teams' outlook toward performance, problem resolution, completion dates, and results

- To provide resources to break constraints

- To provide written reports for each meeting

- To recognize, celebrate, and review the teams' success

Note: Experience has demonstrated that failure to hold regular resource committee meetings usually results in stagnation of the teams' progress. Also, an effective performance appraisal system, which recognizes personnel for their contributions on teams, must be initiated to supplement the motivation and rewards for team members. This sustains the interest, energy, and focus on individuals and teams. Without such a stimulus, TQM programs slowly fade away.

The executive board should have a policy that meetings be held on a regular schedule (i.e., same day, same time). The only acceptable cancellation should be an external customer problem. The meeting should then be rescheduled for the day before or after the original meeting date.

Implementing with Teams

Projects will be organized to support the Critical Business Success Factor's goals and recovery plans or improvement plans for the Internal Customer Satisfaction Agreements. Their projects' completion will close the gap between the existing state and their vision. Teams can be cross-functional or intrafunctional.

The team concept is not new. It has been used by businesses that have to introduce a new program or product. The systematic productivity/ quality-improvement process or business performance improvement process presented in this book is not complex, and one individual *could* perform every step; however, individual contribution does not lead to rapid culture change. For this reason, and to emphasize that productivity and quality improvement is everybody's business, businesses have preferred the team approach, and it has proven successful. Teams of this nature have been used for 20 years. As early as the 1970s, Xerox utilized cross-functional teams.

A *team* is two or more individuals who must coordinate their activities in pursuit of a common goal. This definition can help determine when the use of a team is appropriate. A team is most appropriate in the following situations:

- Programs/projects call for wide exposure in order to promote awareness throughout the organization.
- A problem spans departmental boundaries.
- Sequencing or coordination of effort between departments is necessary to accomplish the proposed solution or goal.
- Knowledge is needed from different specialists, or joint decisions are required. No one person has total control, influence, or authority over the problem.
- No single department has full ownership of the project.
- Departmental barriers exist due to the company's culture.

The use of teams is especially effective because of the synergism created by the mixing of the members' various skills. *Synergism* is the simultaneous action of various elements which have a greater total effect than the sum of their individual contributions. For example, consider two people walking on train track rails. The first one walks 10 yards before falling off the rail, the second 15 yards before falling. However, when both walk together, one on each rail, while holding hands—teamwork—they are able to walk for miles.

Synergism in a team provides the following benefits:

- The effectiveness of decision making is enhanced.
- More innovative ideas are generated through group interaction.
- The diverse talents and experiences of the members are best utilized.
- The group makes more accurate assessments of situations.
- Decisions are supported by all parties.
- Efforts of different specialists are effectively coordinated.
- Team members and managers look at problems more objectively and reexamine their own biases.

The team concept is a powerful approach because of its ability to produce results as well as its developmental contributions. In an effective team, the potential of weaker, less experienced team members will be developed through their interactions with more experienced and knowledgeable members. Before beginning team activity, all members must be trained in the systematic, 12-step productivity/quality-improvement process so that they will understand and effectively carry out their responsibilities. This process serves as a road map for the teams and will be discussed later.

Team Members

Collectively, an effective project team has all the necessary knowledge about the specific objective in order to achieve the desired outcome. The team leader is first selected to match the project's complexity. Leaders are usually first-level managers, but frequently can be the highest-level managers, depending upon the project objectives. Team members are then selected to bring a cross-disciplined approach to achieve the objectives.

In one case, the president of Arral Industries, Christi Brown, assumed the responsibility of team leader for developing an estimating process. This project was the number one priority project in the company, as their success rate for winning competitively bid jobs was low. The president's leadership of the team enabled their success rate to triple in six months.

Team members are assigned from functions considered necessary to contribute to solving the problem. Members should be of two types: those who can solve the problems and/or those who can benefit from solving the problems. It is not desirable to have excess people on the team who do not contribute; they only slow down the team. Usually four to six people is an ideal size.

In order to avoid team members from feeling burdened with extra work associated with project teams, it is important to select projects that are compelling in need and/or are part of their workload. The resource committee should establish how much time each member should spend supporting the team. Experience indicates an average of 10 percent of each member's time will allow for one hour of weekly meeting time and three hours of performing assignments. However, when teams' objectives are of significant impact, these guidelines can be altered.

Project teams are responsible for executing the productivity/quality-improvement process for production processes, administrative processes, supplier processes, design and development processes, and software processes.

The team must:

- Attend weekly meetings to report progress on assignments, receive assignments, and participate in problem solving.
- Identify obstacles to their progress.
- Develop recovery plans to keep the project on target with time and performance expectations.

- Communicate obstacles to the resource committee when support is required by describing the actual constraint, actions previously taken, and recommendations.
- Maintain documentation for each step that has been approved.
- Identify and present future potential projects to the resource committee.

In summary, the team is the ultimate unit in the organizational structure, having control over the project. When the executive board and resource committees are fulfilling their responsibilities, the project team is 100 percent accountable for the success or failure of the assigned project.

Teams will meet weekly and report progress biweekly to the resource committee. Members should understand that the productivity/quality-improvement process program requires a commitment of time and energy from everyone. The productivity/quality-improvement process program will become a significant part of their job activity during the course of the project. Using all available TQM and problem-solving tools, the team will, with practice, integrate these techniques into their normal function.

From a quality improvement perspective, one of the most significant elements of a successful program is the team meeting. Training in effective team behaviors may seem unnecessary to team members, but it can mean the difference between success and failure. Productivity/quality-improvement process meetings must take place regularly, on time, and must not be interrupted by any other priority in the company. Remember, the team has been assembled to improve processes and performances that improve quality, delivery, competitiveness, and profitability. There are no other activities that are more important than team meetings. Even the president should not have the authority to reschedule or preempt these meetings. No activity, including shipments, closing the books, or inventory should be considered more important than productivity/quality-improvement process meetings. These are current fires; productivity/quality-improvement process meetings prevent future fires.

Each team will be expected to establish control of their assigned process and to establish mechanisms to sustain improvements once the team has completed their project and moved on to other areas. Unsustained improvements to processes are worse than totally unimproved processes, because they are a constant reminder of a TQM "failure."

Team Leader

The team leader inspires and manages the project team; it is the most important position in the process. Team leaders inspire by doing the right thing, and manage by doing things right. Success of a project is usually directly proportional to the caliber of the team leader. A strong team leader and a strong team has a high chance of meeting expectations. A weak team leader and strong team will be mediocre, at best. However, a strong team leader and a weak team will improve, and the whole will be significantly more effective than the individuals.

A competent team leader, like a competent manager, must have a balance of leadership, technical skills, and managerial skills. The leader may come from any department within the organization. It is necessary that the leader have the following characteristics:

- Results-oriented, a sense of urgency
- Proactive
- People skills—capable of creating a climate for teamwork
- Effective role model
- Skilled at coaching
- Leadership skills and a clear vision of the objective
- Skilled in time management and follow-up
- Ability to conceptualize solutions and expected outcome
- Decisive in decisions
- High energy and enthusiasm

A team leader must be biased toward action—in other words, be a risk taker whose main approach is moving the team forward through planning and implementation. The team leader:

- Chooses and collaborates in selecting team members
- Schedules team meetings and keeps attendance
- Prepares meeting agendas with expected outcomes and time schedules
- Assigns tasks/goals to team members, with specific expectations
- Develops a milestone plan and follows it
- Prepares weekly progress reports to a prescribed format
- Identifies constraints, risks, and evaluations for the resource committee
- Oversees problems, performance, and completion dates
- Develops recovery plans to stay on target
- Follows up on assignments and ensures they are completed on time
- Complies to the productivity/quality-improvement process

- Evaluates team members' performance and provides input to performance appraisals
- Conducts presentations to the resource committee every other week

The team leader has a most demanding job. They must be many things to many people. If the team leader does not demonstrate the requisite characteristics, the resource committee has two choices: develop the team leader, or assign that individual to a more appropriate position. If the development will require time, it is desirable to replace the team leader in order not to deviate from the team's objective. This is one of the many side benefits this process offers—it is a vehicle for developing future leaders and managers.

One of the best ways to ensure that team leaders are functioning properly is to evaluate them. Following is a multipage critique sheet to be used for just that purpose. Team leaders should be critiqued regularly to ensure that they are staying focused on the teams goals.

Mentors

Mentors play the unique role of coach, teacher, and active participant. They provide a direct channel for action, encouragement, and support. They provide a "devil's advocate" position, and give advice and direction when necessary, but must also allow the team to learn to resolve problems on their own. The mentor is a catalyst for team success and is the prototype manager of the future.

Mentors ensure team commitment. They help to set the team mission and objectives and, through the justification process, help determine the actual values that drive the team performance. They must establish and maintain an atmosphere of focus, and dedication to improvement, to ensure that team meetings are held regularly. The mentor must foster the attitude that team meetings and assignments have appropriate priority with other activities. There is nothing more important to company survival than the continuous improvement process.

The mentor provides the team with encouragement and enthusiasm. He or she gives insight into the importance of the project to the company and draws parallels between team and business objectives. The mentor generates a sense of urgency and ensures that the team members never lose sight of their goal.

The mentoring process is to assist the team leader with options, strategies, and feedback for team leadership. A mentor does not have any direct authority, but does have significant opportunity to influence. The mentor can provide his or her assistance to teams, groups, or individuals. Typically, the mentor is selected from middle or senior management. They are usually those managers who are not CBSF champions, values champion, or resource committee chairpersons.

Mentors offer guidance, empower the team, remove roadblocks, and ensure that the process is being followed. Their first task is to assist the resource committee's chairperson in developing the teams and charters. Remember, charters include expectations, guidelines on how to achieve these expectations, the process that will be followed, boundaries of authority, and adequacy of resources. Mentors are usually found in organizations that are multilevel. Flat organizations do not have mentors, and the resource committee chairperson assumes the role of the mentor. Mentors are the means to involving all of the managers in a multilevel organization with meaningful achievement of business objectives and supporting the process of cultural change.

As with team leaders, the critique sheet in Figure 3-2 will help you evaluate mentors. As with others in the TQM process, mentors need to stay focused on team goals. This critique will help them do just that. This critique should be given on a regular basis.

Team Charter (Administration)

The purpose of a team charter is to clarify the goals of the teams project. It also establishes boundaries of authority for the members of the team. First, the resource committee will prepare the charter and give it to a team leader. Then, the team leader meets with the team, and ensures that all members commit to the charter. Following is an example of a team charter. This will give you an idea of how to set up your own charter.

Team Title: Customer Returns

1. *Objective*

   - To satisfy customers relative to returns for any products. The results of the team would favorably impact any customer satisfaction survey for the Critical Business Success Factor.

2. *Expectations* (outcomes of deliverables)

   - Resource committee prepares written charter and gives it to the team leader.
   - Team leader meets with team and commits to the charter.
   - Develop a system to meet customer expectations.
   - Develop a policy and procedure to support the system.
   - Develop a measurement to monitor the system.
   - Project will be closed in four weeks.

Figure 3-2.

Mentor's Critique Sheet

This form enables you to record your judgments regarding the mentor. A checklist of critical elements has been included for each criterion, as well as space for written comments. After using the checklist and reviewing your notes, you should rate the mentor on each factor.

CRITERIA			RATING		
	Well Below Average 1	Below Average 2	Average 3	Above Average 4	Well Above Average 5
1. Commitment	COMMENTS:				

___ Helps determine team's number one purpose
___ Helps determine importance of the purpose
___ Helps determine the actual values that drive the team's performance
___ Helps determine the preferred values that drive the team's performance

	Well Below Average 1	Below Average 2	Average 3	Above Average 4	Well Above Average 5
2. Encouragement and Enthusiasm Determines:	COMMENTS:				

___ What is the need?
___ What is at stake?
___ Who are the stake-holders?
___ Chosen team/ individuals
___ Generates a sense of urgency
___ Continuously recog-nizes real progress

(Continued)

Figure 3-2. (*Continued*)

CRITERIA	RATING				
	Well Below Average 1	Below Average 2	Average 3	Above Average 4	Well Above Average 5
3. High Expectations	COMMENTS:				

___ Influences selection
of challenging, but
attainable, objectives
___ Assists in concep-
tualization of the
project
___ Critiques performance
measurement indicators

	Well Below Average 1	Below Average 2	Average 3	Above Average 4	Well Above Average 5
4. Empowers the Team	COMMENTS:				

___ Gets people to believe
they can impact the
system
___ Creates a vision
___ Teams know resources
and support will be
available

	Well Below Average 1	Below Average 2	Average 3	Above Average 4	Well Above Average 5
5. Provides Sounding Board	COMMENTS:				

___ Listens to approaches
Explores alternatives
___ Probes for obstacles
and recovery plan
___ Reinforces good
judgment, indepen-
dent action, etc.

	Well Below Average 1	Below Average 2	Average 3	Above Average 4	Well Above Average 5
6. Critiques and Gives Feedback to Team Leader	COMMENTS:				

___ Validates team's
objectives and
performance

Figure 3-2. *(Continued)*

___ Concurs on mile-
   stone plan
___ Monitors team's
   direction
___ Requires team
   leader to forecast
___ Makes assessments
___ Coaches/develops
___ Reviews progress
   reports
___ Identifies exceptions
   to plan and recovery
___ Identifies opportun-
   ities for plans
___ Identifies opportun-
   ities for risk and
   exposure

7. Coaches/Develops	Well Below Average 1	Below Average 2	Average 3	Above Average 4	Well Above Average 5
	COMMENTS:				

___ Team to have
   command of destiny
___ Commitment
___ No limits to problem
   solving
___ Independent actions
___ Creates energy and
   enthusiasm
___ Knowledge of PQI
   process

8. Removes Roadblocks	Well Below Average 1	Below Average 2	Average 3	Above Average 4	Well Above Average 5
	COMMENTS:				

___ Probes team for
   roadblocks
___ Encourages teams
   to be innovative
___ Presses team to take
   action
___ Removes roadblocks
   by taking action and
   advising team leader

(Continued)

Figure 3-2. *(Continued)*

CRITERIA	RATING				
9. Recognition	Well Below Average 1	Below Average 2	Average 3	Above Average 4	Well Above Average 5
	COMMENTS:				
___ Acknowledges accomplishments					
___ Asks for and considers opinions					
___ Identifies and reinforces company values					
___ Acts as catalyst for getting company recognition for achievements					
___ Celebrates victories!					
10. Openness	Well Below Average 1	Below Average 2	Average 3	Above Average 4	Well Above Average 5
	COMMENTS:				
___ Works like he or she talks					
___ Is responsive to support					
___ Is on time					
___ Demonstrates commitment					

- Closure criteria is the completion of a proposed system, policy and procedure approved for returned products, and implementation of measurements which are approved by the resource committee.

3. *Guidelines* to clarify the project and establish boundaries of authority (how to achieve the expectations)

- Find out customer's expectations.
- Collect data for reasons of the returns.
- Develop a milestone and check status weekly.
- Prepare weekly progress reports on the meetings.
- Review the expectations and commit as is or modify to what you can achieve.

- Conduct weekly team meetings.
- Issue assignments to team members who represent all departments involved.
- Establish measurements that measure total time from receipt of shipment from customer to date of shipment to customer; provide visibility to each incremental time through each department.

4. *Boundaries of authority*

- None in developing the system, policies, and procedures.
- Resource committee will approve the system, and policies and procedures, prior to implementation.

5. *Resources*

- Team members shall be representatives of all departments involved.
- Each member will be assigned by management.

6. *Skills required*

- The necessary skills are provided or the learning of the skills must be made available. The mentor's next task will be to collaborate with the resource committee to select the proper caliber of team leader commensurate with the project requirements. Mentors can either participate with the team leader in selecting team members, or have the team leader recommend their team members and the mentor will review and concur.

A mentor should be present when the resource committee chairperson gives the project objectives to the team leader. The mentor should be in weekly contact with the team leader to discuss status problems, constraints, outlook, and what the mentor can contribute. In fulfilling the mentor's role of monitoring the process, the mentor should periodically audit the team meeting, especially whenever the team appears to lose momentum.

The processes that the mentor are involved with cover:

- Teamwork behaviors
- Problem-solving techniques
- Behavior reflecting values of the company
- The productivity/quality-improvement process

Key points the mentor should audit in the team's process are:

- A clear and quantifiable objective that states the existing level of performance and the desired level of performance
- Measurements which will reflect progress and realistic successes

- Compliance to a milestone plan
- Compliance to the progress reporting format
- Identification of problems being addressed, action taken, and actions planned

Other responsibilities of the mentor include:

- Reviewing the problem-solving techniques for applicability and adherence to the process procedure
- Identifying and removing road blocks by personally taking action to resolve them or escalating them to the resource committee
- Observing when team members are absent from meetings and not fulfilling their commitment
- Reviewing the progress report prior to the team leader presenting it to the resource committee meeting; be the devil's advocate
- Giving feedback to the team leader about any pertinent comments from team members
- Coaching the team leader as necessary
- Coaching the team members if the team leader requests it
- Being a cheerleader at any opportunity
- Ensuring timely celebration and recognition of the team upon closure of their project

Facilitator/Adviser

Role of the Facilitator. The facilitator is indispensable to the systematic productivity/quality-improvement process. The facilitator bridges the gap between theory and practice by nurturing the development, leadership, and effectiveness of the team leader and members. A facilitator expedites the productivity/quality-improvement process to obtain the best possible results. The facilitator must:

- Be committed to the success of the project.
- Be a recognized expert in teams; e.g., SPC, QFD, or whenever the technology is applicable, and be experienced in introducing it to organizations.
- Be results-oriented, with a sense of urgency.
- Be intimately knowledgeable with the 12-step productivity/quality-improvement process.

- Be a skilled motivator and coach.

- Be a person separate from the team or organization, and critically assess the members' behaviors and compliance to the process.

- Be a role model.

The facilitator assists the team leader in planning the project and organizing the meetings to ensure that the outcome is clearly identified. Furthermore, the facilitator sees to it that everyone participates and maintains the integrity of the systematic process. A facilitator must educate, demonstrate, coach, and audit results of the TQM implementation. Effective facilitators are truly a rare breed.

The facilitator is:

- A guide to circumvent the minefields and pitfalls on special applications not covered in textbooks

- A catalyst, to assist in developing a plan and to provide follow-up to all management levels, thus maintaining continuity

- An objective evaluator and auditor of team progress, identifying any roadblocks to success and opportunities to improve performance

The need for facilitation diminishes as the team leaders become seasoned. Specifically, the facilitator performs these responsibilities:

1. Provides training, immediate demonstration of the application of training, coaching of team members through a process, and audits after the team has performed a process independently of the facilitator.

2. Modifies forms to meet project needs.

3. Ensures that the team leader

   Plans each meeting with an agenda, expected outcomes, and time schedules.

   Documents each step.

   Prepares a weekly progress report with an updated milestone plan.

   Requests resource committee approval to modify systematic approach steps. (*Note:* The process can be customized to any situation, but only with approval of the resource committee.)

   Submits to the resource committee adequate team closure documentation.

4. Recommends to the team leader and adviser solutions to disagreements about the project's direction, scope, and progress.

5. Verbally critiques team leader's performance biweekly, works with the team leader to overcome weaknesses, and gives written critiques to the chairperson of the resource committee regarding the performance of the team leader and team.

6. When facilitating SPC implementation, conducts weekly audits of process control charts, corrective/preventive-action matrix, and events logs with supervisors and team leaders in order to make suggestions for improvement.

7. Each week reviews the team leader's progress reports and critiques them for accuracy, concurrence, and compliance to the format.

8. Each week reviews the milestone plan for accuracy and the need for a recovery plan.

9. Advises team leader of risks that the leader is unaware of and consults with them about those risks. If the facilitator does not concur with the proposed course of action, he or she must inform the adviser verbally and in biweekly progress reports.

When using a facilitator, either a person from within or without the organization, the company must clarify the working arrangement by establishing mutually agreed-upon guidelines. These guidelines should be consistent with the roles and responsibilities of the team leader and the facilitator. An effective facilitator is invaluable because that individual will see to it that everything is done right the first time. In the words of Bob Kliert, the CEO of Printronix in 1983, "Doing it right the first time is the only way to increase profits."

The facilitator for each team is typically a consultant or individual from within the company who is familiar with the TQM process. He or she will provide guidance in the application of new skills that the organization has learned and will advise the team of pitfalls during the productivity/quality-improvement process. A facilitator will also provide the training and reinforcement of prior training, as necessary.

Once the teams are experienced through a successful implementation, the consultant phases out, and the in-house facilitator phases in. A facilitator is used less as the team leader and team mentor become seasoned veterans of the productivity/quality-improvement process.

The facilitator will serve as the catalyst to develop and maintain inertia by reviewing team activities, assisting the team leader to establish weekly expectations, and making sure that all the problems have been addressed. In addition to weekly meetings, the facilitator will audit the procedures used to control the process and will evaluate the effectiveness of these procedures. While the team is working on continuous improvement of the process, the facilitator should be working on continuous improvement of the team. The facilitator must recognize when the team leader needs help and must fill in any gaps that threaten team continuity.

Role of the Adviser.   The adviser, like the facilitator, is typically a manager of the consultants and trainers or a respected statesperson from within the company. If the adviser is from within the company, then he or she must be capable of safely and credibly critiquing upper management. The adviser interacts with the executive boards and resource committees. The adviser must:

- Provide guidance to the resource committees and president.
- Guide both the executive board and resource committees through their start-ups.
- Call for strategy sessions when required to address any issues.
- Critique performance of both executive boards and resource committees.
- Observe behaviors of the executive board and resource committees, provide feedback and coaching, as applicable.
- Keep companies focused on CBSF and values.

Departmental Mission Statements and Internal Customer Satisfaction Agreements

The initial task of the Implementation Track is to develop *departmental mission statements and Internal Customer Satisfaction Agreements*. Departmental mission statements and Internal Customer Satisfaction Agreements serve the following purposes:

- To align all major functions and departments in support of the company's vision, mission, and values.
- To assist implementation of TQM principles throughout the organization.
- To measure internal customer expectation and attain internal customer satisfaction. If internal customer satisfaction is achieved, external customer satisfaction is achieved.

The internal customer approves the mission statements and the Internal Customer Satisfaction Agreements (ICSA). Occasionally, right sizing is a product of the ICS agreement.

Customer Satisfaction Agreements will be assessed on deliverables which meet or exceed customer requirements and expectations according to predetermined measurements.

Traditional definitions of quality include meeting specifications, conformance to requirements, absence of variation, zero defects, etc. Service is often considered an adjunct to quality and customer satisfaction. It is a worthy, but not always practical, objective. Total Quality Management embraces service and customer satisfaction as important determinants of quality. In fact, in a customer-focused environment, where competition and alternative choices abound, service and customer satisfaction often become key competitive differentiators.

The uneasy realization is that quality is ultimately determined by the customer. Whatever fails to meet or exceed customer expectations, whether expressed or implied, whether realistic or unrealistic, is—by definition—not acceptable quality. The challenge is to discover and understand the full range of customer expectations, to create measurements, and then to develop processes to do what is necessary to consistently meet or exceed these expectations. Success is measured by performance measurements and customer satisfaction.

An example of an organizational commitment to customer satisfaction can be seen at Safariland, Ltd., Inc., based in Ontario, California. Safariland is a producer of a wide array of sewn products ranging from automotive car masks to duty gear (the belts and holsters worn by police officers).

Under the leadership of CEO Neil Perkins and company president Scott O'Brien all departments at Safariland have undergone an internal customer analysis. Departments have had to identify the following factors: deliverables (e.g., completed paperwork and cut material), their internal customers' accounting (the sewing lines), and how well they measure their internal customers' satisfaction.

By focusing on quantitatively measuring their customers' needs, and by regularly measuring how they meet those needs, Safariland has experienced measurable improvements. These improvements include:

- A reduction in purchase order (PO) errors per month from 14 percent (POs with errors ÷ total POs) to 5 percent

- A decrease in cycle time (special orders) from over a week to less than a day

- A reduction in customer orders' data entry errors from 7 percent (orders with errors ÷ total orders) to under 1 percent

Safariland's focus on customer satisfaction includes regular meetings between internal suppliers and internal customers. Internal suppliers have been trained to ask, "How are we doing?" and "What more can we do for you?" The results speak for themselves. Safariland won Toyota's "Most Improved Supplier Award" for 1993.

The Need for Measurement

A ruler can be used to measure the length of an object. If that length were the length of a piece of metal being cut to a precise dimension, one could determine how close to that target dimension the cut had taken us. When we measure something, the comparison can be expressed in either quantitative or qualitative terms. The difference between the measurement and the target or goal is the variation. Knowing the variation between the target we wish to achieve and the item being measured is the first step in being able to determine what additional action will be required to achieve our goal.

For a clear understanding of the result desired, target, or goal, or the variation involved, we strive to express measurements in quantitative terms by using specific numerical values whenever possible; e.g., 2.5-inch diameter, 6.000 inches in length. Some things are more difficult to measure numerically, so we describe the variation in qualitative terms, such as *better, worse, bigger,* or *smaller.* The problem with describing things in qualitative terms is that people may disagree on exactly what terms like "better," "worse," "bigger," or "smaller" really mean. To avoid this problem, it is important that we express our business goals and objectives in quantitative terms.

When we use quantitative terms, we describe things with a clarity that allows others to understand our message. This can be a two-edged sword at times. When we express ourselves in quantitative terms, many times we make a commitment or set expectations with such clarity that any result less than the quantity expressed is considered a disappointment. This is a particular problem in cultures where commitment is avoided because the fear of failure is high. In the business world, however, little credibility is given to statements which are not backed up with quantitative data.

A corporate culture based on the principles of Total Quality Management will reduce the fear factor and promote an environment of individual commitment and accountability that encourages the use of specific quantitative goals and objectives. Although the qualitative statements made by our leaders can be inspiring when they express their vision or the organization's mission or values, when describing specific goals and objectives we must use quantitative terms if we expect them to be clearly understood by ourselves and others. In today's complex, time-critical work environment there is little chance for the success of any goal or objective that is not expressed in crystal-clear quantitative terms.

Quantifying the Right Things. Simply measuring things is not enough. It is crucial that you measure the right things. There is an old saying that says, "Getting things done is efficiency, getting the right things done is effective." The key is to make sure that you are effective. One way to view the difference is to refer to Figure 3-3. In this figure, we have a model which presents four options.

Option one, in the upper left-hand corner, presents the idea of doing the right things wrong. This is when we concentrate on the Critical Business Success Factors, but we do it in a very inefficient manner. Option two, in the upper right-hand corner, depicts the optimum, doing the right things right—concentrating on the important things, and doing them well.

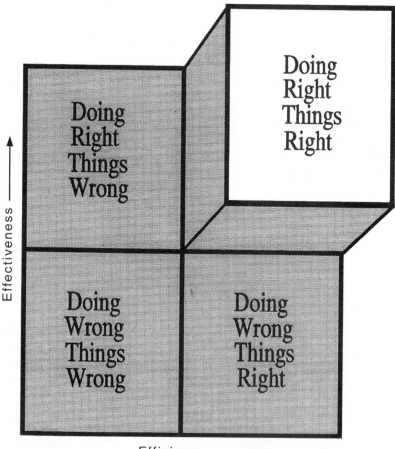

Figure 3-3.

Option three, in the lower left-hand corner, shows the worst of all worlds. You are concentrating on the wrong things, and you are doing it in a very inefficient manner. Finally, we have one of the most dreaded conditions, doing wrong things right. An example might be a factory which was very efficient at making product but has no orders.

Development of Measurable Success Factors.   World-class organizations create strategies to turn business goals into corporate or department success factors based on teamwork and consensus. This flow-down process maintains a high degree of individual buy-in to the goals and success factors which are critical to the success of the company. An overview of this process is shown in Figure 3-4.

The organization's focus must be on its vision, mission, and values. From these are developed the elements of the business processes that are critical to the overall operation. These Critical Business Success Factors (CBSFs) identify the areas in which the department's individuals and teams should focus their activities. Progress toward the objectives and goals that support these CBSFs is measured to keep activity on track.

Departments develop their own success factors relating to their operations and monitor progress to their objectives. Teams can operate within departments or on the broader CBSF level to complete projects that fit within the CBSFs or department success factors. In every case the measurement alone is not enough. The results must be fed back into the planning process to allow interpretation of results, course corrections, and replanning of the process.

The progress and performance measures monitor the group and process performance and are aimed at the improvement of the process. Individuals are also measured in the form of individual performance appraisals.

The measurement process is the foundation process throughout the organization to control the processes. A plan is established, activities are undertaken to support the plan, the measures are made of the activities, and corrective action is taken to replan and begin the cycle again. This is graphically represented in the Shewhart cycle.

Developing Key Performance Measures.   Department managers identify success factors for their areas of responsibility which maintain linkage and support the hierarchy above. This measurement must provide a measurement of their mission. These measurements should measure progress toward achieving department success factors. This part of the process requires the manager to do some hard thinking to develop quantitative measures for each department success factor.

These two stages provide a critical conversion of quantitative business objectives into quantitative requirements that more clearly communicate

Success Factor Creation & Flow

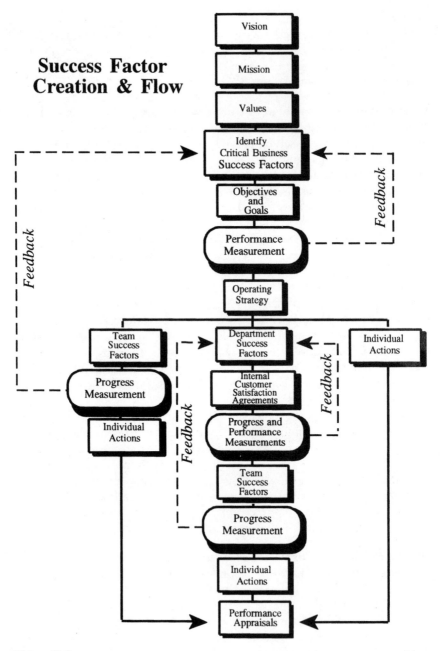

Figure 3-4.

the *what* and *when* necessary to provide evidence of the achievement of the success factor(s). This step is of vital importance in providing direction for the teams or individuals responsible for accomplishing action(s) to satisfy each success factor.

New Approaches to Performance Measurement. An effective approach to continuous improvement requires looking beyond the conventional management accounting practices and adopting more applicable methods of performance measurement. These new measures must be linked to each business success factor of the enterprise. The new performance measures of world-class manufacturing organizations place higher emphasis on cost reduction than on cost control. Seven characteristics of these new measures are:

- Directly related to the company's competitive strategy
- Primarily use nonfinancial measures
- Vary between location as necessary
- Change over time as needed
- Simple—easy to use
- Provide fast feedback to operators and managers
- Foster improvement rather than just monitor

These measures are typically aimed at understanding process costs, direct production costs, and actual costs. This approach has given rise to new accounting methods such as Activity-Based Costing (ABC), Throughput Accounting, and Life-Cycle Costing. But effective performance measurements can be obtained without the need for a new accounting system. Measurements which focus on capturing data in a direct and timely way from graphs, charts, signals, and bulletin boards in the work areas can be very effective if directly linked to the organization's business success factors. Measures that can be used effectively are:

- Statistical process control charts (X-bar and R charts)
- Timely feedback on customer satisfaction/period
- Inventory accuracy levels
- Actual hours/specific work tasks/volumes
- Average setup times
- Reject/scrap rates
- Production rates and adherence to schedule

- Absenteeism/turnover
- Elapsed times

Developing Performance Measures. Creating performance measures to track business success factors is the responsibility of management. Effective performance tracking provides the manager with an instrument panel to understand where the organization is, where it is going, and when it will get there.

By monitoring performance measurement trends, managers more clearly understand the performance of individuals or teams responsible for implementing or improving business success factors. This visibility is critical to fostering an environment of empowered employees by allowing management to clearly understand individual or team progress and to help determine when or if management intervention is necessary. In order to create an effective management instrument panel, performance measures should be:

- Specific
- Timely
- Achievable
- Realistic

After a functional manager identifies those corporate or department success factors for which he or she is responsible, then that manager selects performance measures which will clearly specify the desired result(s) to the teams or individual contributor responsible for implementation. Some typical measures are provided below to assist you in developing effective measures of performance. These lists are not all-inclusive, but rather are intended to act as thought starters.

General Measures for Most Departments

- Percent of time equipment down or downtime percentage
- Adherence to schedule:

$$\frac{\text{Items on time}}{\text{Total items}}$$

- Requests per hour, day, week, month
- Queue time for internal/external customer
- Times report/plan is revised after release
- Percent of budget expended

- Square feet of floor space required per employee
- Data processing cost per employee
- Pages of procedures eliminated/total pages
- Hours or dollars saved/total hours
- Value of cost reduction achieved
- Percent cost saved/total cost
- Cost of service or product
- Temporary employee hours required
- Scrap reduced:

$$\frac{\text{Scrap \$ or units}}{\text{Total \$ or units}}$$

- Interruptions per hour, day, week, month
- Forecast accuracy:

$$\frac{\text{Actual sales}}{\text{Forecast sales}} \quad \text{or} \quad \frac{\text{Actual demand}}{\text{Forecast demand}}$$

- Invoicing errors/invoice
- Data terminal response time
- Computer downtime
- Service/product lead time
- On-time performance:

$$\frac{\text{Line items on time}}{\text{Total line items}}$$

- Cycle time reduction:

$$\frac{\text{Units}}{\text{Hours}}$$

- Head count/month
- Equipment repair costs/hour of operation
- Process steps eliminated/total steps
- Planned improvements/$ of sales
- Overtime as a percent of total labor
- Computer downtime
- Tab run pages required

- File cabinets in use
- Telephone expense
- Telephone lines
- Rework eliminated:

$$\frac{\text{Rework during period}}{\text{Total product during period}}$$

- Tab run errors/page
- Training hours/total hours
- Errors per page
- Cost of quality
- Percent charge in cost of quality/month

Measures That Assess Organizational Culture

- Awards presented per month
- Grievances per month
- Attendance or employee turnover:

$$\frac{\text{Percent}}{\text{Month}}$$

- Safety problems observed or accident per month
- Number of employees on improvement teams average per month
- Improvement suggestions per employee per month
- Time required to resolve workers' compensation claims per month
- Employee complaints per month
- Budget allocated for training
- Percent employees on layoff/total employees
- Employee suits/total number of employees
- Rumors per week, month
- Internal promotions/external
- Benchmarking, employee surveys

Measures for Purchasing

- Orders placed per week or month
- Average time required to place order per week or month

- Expedite charges per order as a percent of order
- Number of partners in excellence
- P.O. changes per P.O.
- Number of suppliers
- Number of unique supplier packages
- Number of items returned to suppliers/total items received
- $$\frac{\text{Number of suppliers MRB actions}}{\text{Number of supplier time items received}}$$
- Percent of suppliers certified
- Percent of line items dock to stock
- Number of suppliers per purchasing agent
- Average supplier lead time
- Lead time performance:

$$\frac{\text{Actual time to deliver}}{\text{Promised lead time}}$$

- Percent of deliveries made on time
- Percent of deliveries made early or late
- Average time to place order
- Percent of lots rejected

Measuring for Material Control
- Inventory value
- Inventory turns
- Inventory value WIP
- Inventory value raw materials
- Inventory value finished goods
- Number of stocking locations for the same part number that is stored
- Square feet warehouse space/line item stored
- Inventory accuracy percent from cycle count
- Percent of kits with component shortages
- Number of line items/planner average
- Percent throughput time/product

- Orders issued on time
- Percent orders completed on time
- Age of shop order/average age
- Time since last transaction for each part number
- Percent of inventory with no transaction within the last year
- Number of days on hand "A" items
- Number of days on hand "B" items
- Number of days on hand "C" items

Measures for Marketing. These can be applied to internal and/or external customers. Warranty measurements should be assigned to a specific date the product was shipped to improve correlation.

- Customer Complaints/Line Items Shipped
- Customer Compliments/Line Items Shipped
- Customer Phone Calls Not Returned Each Day
- Customer Calls Recorded on Answering Machines per Day
- Market Share
- Customer Survey Results
- Customer Interview Results
- Questionnaire Results
- Customer Returns
- Customer Satisfaction Index
- Warranty cost as a Percent of Sales
- Number of Warranty Units as a Percentage of Units shipped over a specific period
- New Product Revenue as a percent of Total Revenue
- Percent of repeat orders
- Average Dollars per order
- Percent change in $ per order
- Advertising as a percent of sales

Measures for Sales
- New customers per month
- Orders canceled per month

- Options exercised or not exercised per month
- Manufacturing schedule changes per month
- Quote to book ratio
- Number of errors per order processed
- New orders per month
- Performance-to-sales forecast:

$$\frac{\text{Actual}}{\text{Forecast}}$$

- Sales calls made
- RFQ/RFPs won
- Book-to-bill ratio

Measures for Engineering

- BOM items reduced
- Mean time between failure (MTBF)
- Errors per square foot of drawing
- New product introduction time
- Break-even time/break-even after release
- Design changes for regulatory compliance
- Engineering change response time
- Time from concept to release
- Percent stand., common, unique parts used
- Engineering changes per drawing
- Indenture levels on BOM
- Errors/bill of material (single level)/month
- Software corrections per month
- Product recalls per month
- Number of new products launched each year
- Number of engineering changes per product per month

Measures for Manufacturing

- Quantity shipped (Qty. and mix)
- Lot size reduction

- Tools in storage/in use
- Skills per employee (cross training)
- Cost of energy per unit of product
- Amount of hazardous waste generated
- Percent of production capacity in use
- Number of different employees required
- Percent yield per assembly
- Percent yield per component
- Percent yield per operation
- Percent DOA or out-of-box failure
- Dollar volume shipped per employee
- Average setup time
- Delivery reliability
- Repair cost per hour of operation
- Hours in use/total hours possible
- Water used per unit of product or day
- Feet of product flow per day
- Parts per million defective
- Number of process steps per product
- Number of days material in WIP
- Hours of downtime per week
- Percent of rejects by product
- Time booked to shipped
- Time first received to shipped
- Time booked to payment received

Measures for Teams
- Progress of primary goal(s)
- Progress of key objectives
- Adherence to milestone schedule
- Action items created/closed
- Team member attendance
- Meetings start/stop on schedule
- Team member participation

- Timely distribution of action items
- Team process evaluation

Measures for Individuals
- Action items completed
- Training assignments completed
- Meetings started/attended/finished on time
- Study hours per week
- Personal goals or objectives planned/accomplished
- Improvement projects started/completed
- Times you "walked the talk"
- Visits to the line
- Work days planned vs. not planned
- Procrastination hit-list improvement

Measures for Business Performance Improvement or TQM Undertaking
- Number of TQM projects linked to strategic goals
- Management attendance at executive boards and resource committees
- Hours of training per employee
- Number of formal quality service agreements established with customers
- Percentage of projects sourced from customers, management, teams, suggestions
- Percentage of employees on teams
- Number of internal customer-supplier agreements
- Number of quality goals mutually agreed to by managers/employees
- Number of TQM projects completed
- Number of quality-related standard operating procedures
- Percentage of quality solutions applying to multiple departments/ functions

Delegating Performance.　Developing success factor measurements that are specific, timely, achievable, and realistic permit clear communication of the expectations of team or individual performance. Experienced managers recognize this as a vital step in achieving high performance from teams or individuals. Clear delegation of success factors and the expected outcome to teams and individuals occur at the stages shown below. By establishing the measures for success factors in advance, the team or individual can more clearly understand the level of effort required to achieve the desired outcome and can better communicate the resources required for success more accurately.

Reporting Onward and Upward. Performance monitoring for success factors starts at the individual contributor level and progresses upward to the applicable team and/or department manager. Periodic reports are given to senior management and the organizational leader. At predetermined times (usually quarterly or annually), senior managers meet with the organizational leader to assess the outcomes and compare them to expectations. At this stage senior managers and the leader measure the effectiveness of the operating strategy to satisfy expectations. The cycle of continuous improvement begins again when the leader and senior managers set new expectations for the organization.

Examples of Key Measures

$$\text{Hours per unit} = \frac{\text{total hours used}}{\text{total units produced}}$$

$$\% \text{ rework} = \frac{\text{quantity reworked}}{\text{total quantity produced}}$$

$$\% \text{ scrap} = \frac{\text{quantity scrapped}}{\text{total quantity produced}} \times 100$$

$$\% \text{ absenteeism} = \frac{\text{days absent}}{\text{total days planned}} \times 100$$

$$\text{Square feet per employee} = \frac{\text{total square feet}}{\text{total employees}}$$

$$\% \text{ on time (volume)} = \frac{\text{number on time}}{\text{total number promised}} \times 100$$

$$\% \text{ on time (single item)} = \frac{\text{days before or after promise date}}{\text{total days promised}} \times 100$$

$$\% \text{ of planned cost} = \frac{\text{planned cost of work}}{\text{actual cost of work performed}} \times 100$$

$$\% \text{ budget expended} = \frac{\text{Planned cost of work}}{\text{actual cost of work performed}} \times 100$$

$$\% \text{ of planned lead time} = \frac{\text{planned lead time}}{\text{actual lead time}} \times 100$$

Percent on Time Mix and Volume

Product	Plan	Actual	Qty. volume	Qty. mix (+10%)
A	200	327	+127	Miss
B	20	18	−2	OK
C	50	80	+30	Miss
D	500	440	−60	Miss
	770	865	+95	1/4

$$\frac{865}{770} = 112\%$$

$$\frac{1}{4} = 25\%$$

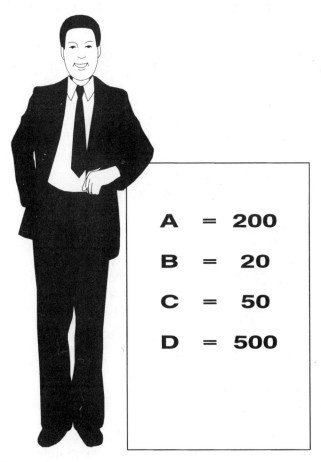

A = 200

B = 20

C = 50

D = 500

Figure 3-5.

What Should Be Measured? Keep in mind a very simple principle: *that which gets measured gets done.* It is a very simple but powerful principle. Therefore measure the important things—things that count. Examples include the Critical Business Success Factors, the things that just are not getting better, problems, and the things in which your internal and external customers are interested. What gets measured and reviewed gets done faster. Therefore a system for review is essential. The higher the level of the reviewing manager, the faster the results are achieved.

Charting Techniques. In order to make your charting effective, keep in mind several points. First, it is not necessary to account for all of the performance gap. Many times individuals or organizations have fiddled around trying to measure every last bit of rework or scrap. Accounting for a minimum of 90 percent of the issue is sufficient. More than anything, we are interested in getting a data point from which we can see a trend. Second, timeliness of measurement is very important. If you are reporting on last month's problems, then your measures are not sensitive enough to tell you what to do tomorrow. Finally, underlying measurement, there must be an element of accountability. When the chart goes the right way, who do you congratulate? When the chart goes the wrong way, who do you stimulate? Figures 3-6 and 3-7 show some techniques on how to chart effectively.

Success Factor Improvement Plan. Figures 3-8 and 3-9 provide examples of a very effective charting method. The key to using this method is the fact that it contains five crucial elements. First, it contains a graphical depiction of the situation. Second, the situation is also described via numeric data in actual, plan, and variance. Third, the chart contains an analysis of what influenced the results for the period and what the trend (if any) is. Fourth, the chart identifies an action plan, that is, what actions will be taken, who is responsible, when is the target date, and what is the follow-up date. Finally, the chart provides an outlook of the next period's performances.

Milestone Plan. Implementing measurement would be useless without implementing projects to cause the measurements to move in the right direction. In order to get projects off of the ground, milestone plans are crucial. Milestone plans force us to nail down what we are going to do and when we are going to do it. The milestone plan in Figure 3-10 is an example of a tool to use to force the numbers to move in the right direction. Organize a sequence of tasks that will be followed in achieving a specific target. As progress is made, the lines between Start (S) and completion (C) are changed to asterisks to indicate percent completion of that segment.

* Good things go up as they get better

* Negative things go down

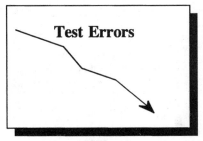

* A flow chart helps explain where the measurement is taken

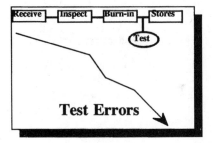

Figure 3-6.

Charting Techniques

• **Bar chart shows discrete events**

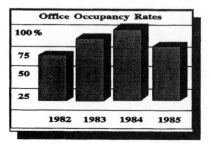

• **Combination charts can be used to show related events**
(e.g., reduction in rework and increase in output)

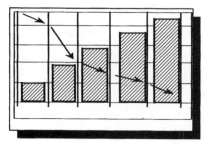

• **Anatomy of a chart**

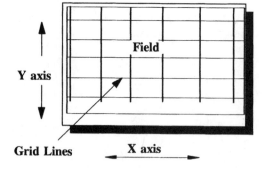

Figure 3-7.

Performance Report

EXAMPLE

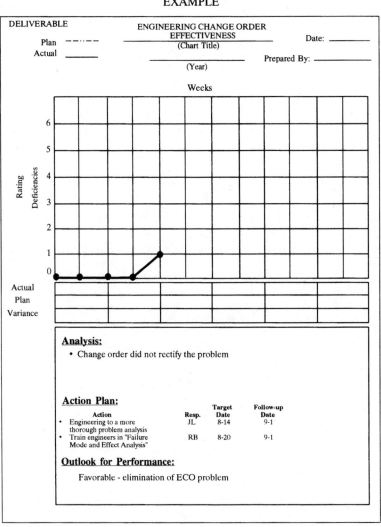

Figure 3-8.

Performance Report

EXAMPLE

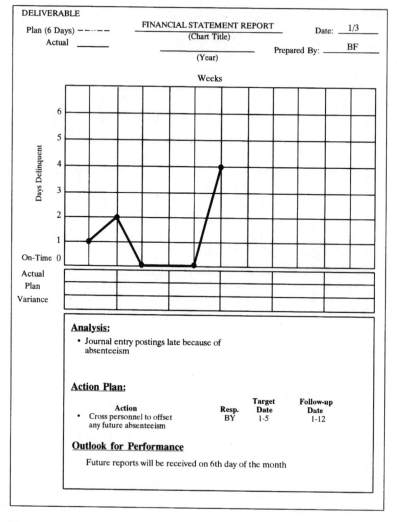

DELIVERABLE

Plan (6 Days) ―――・――

Actual _____

FINANCIAL STATEMENT REPORT
_____
(Chart Title)

_____
(Year)

Date: ___1/3___

Prepared By: ___BF___

Weeks

Days Delinquent

6
5
4
3
2
1

On-Time 0

Actual

Plan

Variance

Analysis:

• Journal entry postings late because of absenteeism

Action Plan:

Action	Resp.	Target Date	Follow-up Date
• Cross personnel to offset any future absenteeism	BY	1-5	1-12

Outlook for Performance

Future reports will be received on 6th day of the month

Figure 3-9.

Milestone Plan

Project Activities	Implementing the 12 Step Process																
	August					September				October				November			
	3	10	17	24	31	7	14	21	28	5	12	19	26	2	9	16	23	30
1. Define Problem	S••••••••C																
2. Select Goal		S••••••C															
3. Identify Measurements				S•••••C													
4. Problem Analysis - Root Causes - Solution				S•••••••••C•••RC													
5. Perform Inspection Capability Study							S•••——C										
6. Identify Process Capability (Variation)									S——C								
7. Develop Corrective and Preventative Action Measures										S——C							
8. Implement Process Control Procedures												S——C					
9. Process Control Implementation Review and Checklists													S—C				
10. Develop Problem Prevention Measures														S——C			
11. Measure Effectiveness of Corrective Action															S———C		

- Use to plot activities or events against time.
- Helps plan and communicate sequence of events and time required.
- Legend:

S	= Original planned start date	RS	= Replanned start date
C	= Original complete date	RC	= Replanned complete date
--	= Planned activity duration	**	= Percent complete
(S)	= Actual start date	(C)	= Actual completion date

Figure 3-10.

Productivity/Quality-Improvement (PQI) Process

One of the common pitfalls in implementing TQM with teams is that they do it without a road map. Even with clear expectations and objectives, teams without an improvement process run the risk of missing their objectives in their expected time frames. Teams without an improvement process frequently meet their objectives in a random manner. Some teams never finish, others do it with a great variability in time, and only a few finish on time. These teams are dependent on the personal strength and experience of the team leader.

A solution to offset this pitfall is to use the proven productivity/quality-improvement processes. These processes are a series of steps used to identify, measure, and solve problems; they may be customized to any problem. The team leader may select from five possible productivity/quality-improvement processes: production, administration, developing the processes, new product development, and software development. In this text, the Productivity/Quality-Improvement Process—Production will be reviewed.

Productivity/Quality-Improvement Process—Production

Once the specific prerequisite training for implementation is complete and the organizational structure is in place, the Implementation Track can begin. No time should be wasted in launching the implementation of the new concepts and techniques, as continuity and commitment are crucial to the success of the program.

It is also crucial to follow a well-defined approach to implementation. Experienced facilitators agree that new project teams have a tendency to stay with what they know and may dismiss the recommended procedures and methods as just another program that will go away in six weeks. People naturally are wary of unfamiliar ground, especially when it comes to making high-visibility, risky (in their eyes) changes to established processes in the organization.

People also tend to believe that no one knows their business as well as they do, and to some degree this is true. Facilitators in the productivity/quality-improvement process are sure to hear, "That's fine for the production department, but I don't think SPC (TQM, productivity/quality-improvement process, etc.) will work here in an administrative function." When team members begin to understand the anatomy of the process, and that most functions can be characterized with inputs, outputs, and customer requirements, they will agree that TQM works in any area: administration, materials, design, software, administration, and sales, as well as production.

The productivity/quality-improvement process recommended in this text is based on years of experience of numerous successful programs. It has been continually refined to meet the ever-changing nature of global competition in all industries and markets. Users may find it useful to tailor the program to their specific business environment, but there is no value in reinventing the process or looking for shortcuts. Failure to adopt and follow a well-defined program is one,of the reasons that there are so many false starts in Total Quality Management improvement programs.

Even the process of productivity and quality improvement requires a well-defined, integrated set of activities that result in a permanent change in the way a company operates. The productivity/quality-improvement process program should be flexible enough to adapt to variations in the nature of the business or the requirements of the customer, but it should not omit any part of the TQM tool kit that may be beneficial.

Specific steps used for "productivity/quality-improvement process—production" implementation will vary also from consultant to consultant. The 12 steps described in this text are as follows:

1. Project identification
2. Planning and reporting
3. Performance measurements
4. Problem analysis and solution
5. Inspection/test-capability study
6. Process-capability study
7. Corrective/preventive-action matrix
8. Process control procedure
9. Process control implementation
10. Problem prevention
11. Performance accountability
12. Measurement of effectiveness

Experience has shown that none of these steps should be omitted without the careful review and approval of the resource committee. Productivity/quality-improvement process teams are involved in a project to effect improvement within a specified period of time, but they are also involved in a learning process that should not be cut short.

Step 1: Project Identification. The first step in the productivity/quality-improvement process is to select a project, clearly define goals, and justify the commitment of resources to the project. Remember that you will be assigning from four to six of your most effective people to the productivity/quality-improvement process team, and they will be spending a minimum of four hours per week on the project. It is a sizable investment in time that demands a commensurate return.

Team members must understand that management has made a commitment to devote time to the project. Project teams are not just one more thing to take care of in their already busy schedule. The productivity/quality-improvement process project will positively affect company productivity, profitability, and customer satisfaction. Therefore, it should be given top priority in resource allocation.

Opportunities for improvement include reduction of rework or scrap, reduction of the time necessary for production or processing, reduction of paperwork, and reduction of errors. The obvious place to start, then, is where these improvement opportunities will be maximized. Choose an area with high volume, high product value, or highest "apparent" defect rate. If this is an administrative project, choose the area that has the greatest impact on internal or external customers.

Once the project has been selected, collect the necessary quality and financial data to verify the selection. Three to six months of data should provide the team with a reasonable estimate of the current conditions. The team, along with the team mentor and facilitator, can estimate the percent improvement that is possible and then calculate the expected annual savings that will result from successful productivity/quality-improvement process implementation. Avoid tendencies to overwork or underwork these estimates. This should be regarded as a defensible estimate that will allow the resource committee to evaluate a selection of projects and assign appropriate emphasis on those with the highest return.

Project justification may, in some instances, be expressed in terms other than ROI (Return on Investments). Customers may require implementation of statistical process control for their product as a condition of doing business. The company may also choose a specific project for strategic reasons that are not currently supported by financial or quality data. One of the common pitfalls is not to put quantifiable objectives on the projects.

Step 2: Planning and Reporting. For many team members, the productivity/quality-improvement process team will be their first exposure to the planning and monitoring function that is such a critical part of project management. They will need to have a good idea of all the steps required to complete the project, and based on current conditions, should be able to estimate time of completion. As a guideline, the average project should be complete within three months.

The team should develop and maintain a milestone chart that graphically depicts the steps they will take and their progress toward completion goals. In addition, they should submit a weekly progress report to the resource committee.

A recommended milestone chart is shown in Figure 3-11. The format of the chart is up to the individuals involved in the process, but maintaining a standard format from team to team will help the resource committee in analyzing the team's situations quickly.

A progress report should be published and distributed within 24 hours following each team meeting. Its purpose is to advise the resource com-

MILESTONE PLAN

PROJECT:	PREPARED BY:		Current Date:
OBJECTIVE:	DATE REVISED:		
DATE:	APPROVED BY:		
	APPROVAL DATE:		

LEGEND

| S - Planned Start
C - Planned Completion | RS - Rescheduled Start
RC - Rescheduled Completion
N/A - Nonapplicable | (S) - Actual Start
(C) - Actual Completion |

WEEK #

MILESTONE	1	2	3	4	5	6	7	8	9	10	11	12	13	14	15	16
1. Project Identification - Definition - Justification - Flow Charting - Decision Point																
2. Planning and Reporting - Milestone - Progress Report - Quarterly Presentation																
3. Performance Measurements - Measurements																
4. Problem Analysis and Solution																
5. Inspection/Test Capability Study - Plan and Gather Samples - Conduct Study																
6. Process Capability Study - Plan and Gather Samples - Conduct Study																
7. Corrective/Preventive Action Matrix																
8. Process Control Procedure																
9. Implement Process Control																
10. Problem Prevention																
11. Quality Targets/Defect Accountability - Quality Targets - Defect Accountability																
12. Measurement of Effectiveness - Process - Product - System - Financial																

MILESTONE	1	2	3	4	5	6	7	8	9	10	11	12	13	14	15	16

WEEK #

Figure 3-11.

85

mittee of items requiring immediate attention and to serve as a reminder of action items and completion dates assigned to team members. The report should be distributed to the team members, resource committee members, and managers of the team members. The suggested format for a team progress report is shown in Figure 3-12.

Step 3: Performance Measurements.   In addition to the control charts that may be used to monitor the process improvement and state of control, a set of measurements should also be established that monitor the team's performance against the established objectives. The steps for developing performance measurements are:

1. Determine the measurement criteria—answer the questions *when* and *how much*. What is the measure of success?

2. Establish a measurement system—assign responsibility for ongoing data collection and reporting.

3. Collect the data.

4. Report the data in a form that will be easily and quickly understood by management. Avoid complicated graphs with too many parameters. Keep the graphs or charts simple and relevant. However, each graph should have an "analysis" section and "action taken" section.

5. Correlate the data to financial data so management can become excited. As Dr. Juran says, "Top management speaks in the language of money." The workforce speaks in things; therefore, the teams will have to be interpreters.

Performance measurement data should become part of the team's weekly report to the resource committee. Performance measurement should continue even after the team project has been completed. The established measurement system will become a management tool for assessment of continued improvement.

Step 4: Problem Analysis and Solution.   Once the project and objectives are well defined and approved by the resource committee, the team can begin the process of problem analysis and solutions. There are a number of problem-solving tools available that should be applied according to the degree of complexity involved. These tools range from simple brainstorming to design of experiments that test the response from different combinations of input parameters. Problem analysis may be a very short exercise in the case of implementing SPC in a well-established process, or it may span the entire time of the project in cases where significant investigation and data collection is required.

Team Progress Report
Format

DATE: __/__/__
TO: Team Members, Resource Committee Members, and Managers
FROM: Team Leader
SUBJECT: Team Progress Report

Team Members

   I. **Objective**
- Statement
- Quantified target

  II. **Status**
- Steps completed
- Steps in-process
- Performance to objective
- Major problems identified

 III. **Accomplishments**
- In-process steps completed
- Assignments completed

 IV. **Exceptions to Milestone Plan**
- Statement of problem
- Recovery plan

  V. **Risk and Exposure**
- Identified constraints
- Causes of constraints
- Recommended action

 VI. **Outlook**
- Completion date of current step
- Overall project completion data
- Projected value of savings and performance

VII. **Detailed Assignments**
- Activity
- Responsible team member
- Target date
- Status

VIII. **Milestone Plan**

 IX. **Performance Measurement Graph/Chart**

Figure 3-12.

Problem-solving methods that have been found most effective in the productivity/quality-improvement process, not excluding other techniques, include:

- Events logs
- Diagnostic process audits
- Cause-and-effect diagrams
- Failure mode and effects analysis
- Design of experiments

These methods will be covered in depth in "Core Techniques," Chapter 8.

Step 5: Inspection Capability Study.   An inspection-capability study is a specific form of designed experiment intended to quantify the capability of the measurement system in place for the process. It can deal with either variable or attribute inspection systems. If the variation in the inspection measurement system is large enough to show up in monitoring the process output, then action should be taken to correct the measurement system before final implementation of process control. For variable data, the instruments may be out of calibration or may not have sufficient resolution to provide the necessary control. For attribute data, inspection methods may be inadequate, specifications unclear, or inspectors may be improperly trained. The methodology for an inspection/test-capability study is covered in depth in Chapter 8.

Step 6: Process Capability Study.   When the inspection-capability study is complete and acceptable, the productivity/quality-improvement process team can proceed with a process-capability study. This important statistical technique is used to compare the process control limits to the specifications. A process may be in statistical control but may contain so much natural variation or may be centered in such a way that there is a high probability that the output will be beyond specified limits. A process must be capable and centered within the specifications to the desired indices.

A number of indices are available to describe process capability. The most common are Cr and Cpk. Please refer to the Chapter 8 for the hands-on details of how to perform a process-capability study.

Step 7: Corrective and Preventive Action Matrix.   The true test of an effective SPC/TQM program is how the organization responds to the data collected and displayed on control charts. Defining corrective and preventive action to be taken assigns the responsibility for process control to those who actually run the process. This is management's true opportunity to empower the operator.

The first corrective action that many managers will suggest when an operator or inspector identifies an out-of-control condition is to "call for help." This may be the correct response in certain situations, but for truly effective process control, everyone in the company must understand and feel comfortable with their ability to take action when the data indicates a need.

The Corrective and Preventive Action Matrix (CPAM) spells out exactly what to do when a given set of conditions and causes are identified. Corrective action is taken to remove defects from the product and may include sorting or rework. Preventive action is taken within the process to ensure that the defects do not reoccur.

The corrective and preventive action matrix contains the following information:

1. *Defect or condition.* Known or potential defects that could occur at the process control point.

2. *Probable cause.* This serves to refine the definition of the defect or condition. One defect may have a number of causes, and they should be listed in their order of severity or likelihood of occurrence.

3. *Defect type.* Classification of the defect or condition into workmanship, product design, process, or components. This helps the operator understand the cause and suggested action, and sets accountability for defects.

4. *Station responsible.* Usually where action will be taken.

5. *Corrective action.* What the operator should do, in order of priority, to remove defective product from the process.

6. *Preventive action.* What the operator should do to remove the cause of defects and bring the process back into statistical control. These may include machine adjustment, machine repair, or operator instruction.

The productivity/quality-improvement process team should establish responsibility to continually review and update the CPAM. New problems will be added when they are identified, and old problems may be removed when their frequency of occurrence declines to minimal levels. The corrective and preventive action matrix is a powerful SPC tool that empowers operators to solve problems with a minimum of assistance from the engineering staff. Over time, it also provides consistency in problem solving. (See Figure 3-13.)

CORRECTIVE/PREVENTIVE ACTION MATRIX

INSPECTION STATION:____	PREPARED BY: _____	PAGE OF
PRODUCT:_____	DATE EFFECTIVE: _____	REVISED BY: _____
INSPECTION INSTRUCTION:____	APPROVED BY:_____	DATE EFFECTIVE: ___
DATE:_____ REVISION: __	APPROVED BY: _____	

DEFECT/ CONDITION	PROBABLE CAUSE	DEFECT TYPE W, D, P, C	STATION RESP.	CORRECTIVE ACTION (1) DEFECT	PREVENTIVE ACTION		
					OPER	INS/TST	OTHER

Figure 3-13.

Step 8: Process Control Procedure.   Once the SPC methodology has been established for the team's project, a *process control procedure* should be written. This procedure will serve as a training aid for new operators, a standard of performance for future audits, and a reference for all personnel involved in the process.

When developing the process control procedure, the team must ensure that it includes the requirements for:

- The history of the process and background on critical defects
- Selecting the type of chart to be used in new control stations
- Determining if the process is in or out of control
- Use and application of the CPAM
- Sustaining the SPC program

The procedure will specify details of the people responsible for taking action, their duties, and the methods they will use. (See Figure 3-14 for a more thorough breakdown of the process control procedure.)

Step 9: Process Control Implementation.   When the procedure for SPC has been fully defined and documented, formal implementation should be conducted. This ensures that all responsible personnel are

Figure 3-14. Process control procedures.

Date: Revision _____   Control No. _____
Subject: Process control procedure
1.0. Purpose. To establish the system for introducing and maintaining statistical process control (SPC), and to communicate instructions and requirements for identifying out-of-control situations.

    1.1. To provide rationale for the selection of the control chart, subgroups, and subgroup frequency.

    1.2. To determine if the measured production process is in a state of SPC.

    1.3. To identify when to search for the cause of defects and when to take corrective and preventive action.

    1.4. To maintain the process within statistical control.

2.0. Scope. This procedure applies to the manufacture of (specify product) which is tracked by and controlled with SPC methods.

3.0. Organizational units affected.

    3.1. Production.

    3.2. Inspection (quality control).

    3.3. Manufacturing engineering.

    3.4. Quality engineering.

    3.5. Quality assurance.

    3.6. Test engineering.

4.0. Definitions.

    4.1. Process control chart. A graph which plots the statistic $\bar{X}$, R, u, p, or c, calculated from subgroups (samples) collected periodically over time. The process control chart is characterized by a center line (process average) and upper and lower control limits. A process is judged in control if all points fall within the control limits, which are located three standard deviations away from the process average. A process is judged to be out of control if any point falls outside the control limits or there is an indication of a trend or shift.

    4.2. State of statistical control. An expression which describes a process which exhibits natural variation and reflects a constant cause system. Also known as a process which is in control.

    4.3. Corrective action. Actions taken to assure that all suspect material is identified, verified, and corrected; this is, purge, sort, rework, reinspect, use as is.

    4.4. Preventive action. Actions taken to address the cause of out-of-control conditions and to bring the process back to a state of statistical control.

(Continued)

Figure 3-14. (*Continued*)

4.5. Quality control action notice (QCAN). A document issued to initiate corrective action and/or preventive action on a process that is out of control. It may be issued to shut down or correct the process. (See Section 10.0.)

4.6. Corrective and preventive (CP) action matrix. A document which lists the defects expected to be seen at an inspection and test station and the respective CP actions to be taken if the process goes out of control.

4.7. Process capability study. A technique used to determine the capability of the natural variation of the process and its limits for ongoing process control.

4.8. Purge. Actions taken to identify suspect units, examine the units for specific defects and their causes, and segregate units with defects from regular production.

5.0. History.

5.1. A statement of current and the past three months' quality performance should be provided. This should be quantified and be expressed in percent defective, defects per unit or preferably defects per 1000 units, or defects per million opportunities. The selection should be the most applicable for the quality indicator.

5.2. A list of all defects produced by the process should be provided.

5.3. The defects should be referenced to a quality document which contains criteria for acceptance.

6.0. Rationale.

6.1. Selection of the type of control chart. A statement is required which identifies the type of control chart selected and the reasons for and advantages of its selection.

6.2. Selection of rationale subgroups. A statement is required which identifies how the subgroups were selected (i.e., progressively, randomly) and the reasons for their selection. The size of the subgroup and reasons that the size was selected should be stated.

6.3. Frequency of subgroups. A statement of what the original frequency should be and the strategy for reducing the frequency are required. The reasons should support the original selections and strategy.

6.4. Revisions to rationale. A change of control chart type and revisions to subgroup rationale, subgroup size, or frequency of subgroups should be recorded and explained.

Figure 3-14. (*Continued*)

7.0. Requirements and responsibilities.
   7.1. Initial requirements for establishing current and future SPC.
      7.1.1. Initiate events log as described in the problem analysis procedure. Responsibility: Manufacturing engineer.
      7.1.2. Select appropriate control chart ($\bar{X}$, R, p, c, u, np) and subgroup size. Responsibility: Quality engineer.
      7.1.3. Establish ongoing center line and control limits as described in Section 8.0. Responsibility: Quality engineer.
      7.1.4. Prepare a CP action matrix. Responsibility: Manufacturing engineer.
      7.1.5. Set up and post initial process control charts. Responsibility: Area supervisor.
      7.1.6. Provide a place to which control charts can be attached. Responsibility: Manufacturing engineer.
   7.2. Ongoing requirements (data collection and use).
      7.2.1. Update and post control charts at appropriate stations. Responsibility: Inspector, tester, and/or operator.
      7.2.2. Obtain and inspect samples. Record results on data collection forms and plot data on control charts as required. Responsibility: Inspector, tester, and/or operator.
      7.2.3. Check for out-of-control conditions. Notify inspection and test supervisor.
         7.2.3.1. Out-of-control conditions can be detected through one or more of the tests shown in Fig. 8.4 with normal control limits. Responsibility: Inspector, tester, operator.
         7.2.3.2. A point outside the control limits with modified control limits indicates an out-of-control condition. When a pattern of out-of-control points exists, revert to normal control limits.
         7.2.3.3. Continuing defects are an indication that previous preventive action was not effective; see Section 9.3. Responsibility: Inspection supervisor, assembly supervisor, test supervisor.
         7.2.3.4. Special situation. When the process is in control, improvement is desired, and none is being observed, and when two identical defects are observed in the same subgroup, three identical defects are observed in the same day, or four identical defects are observedin the same week, action on attribute control charts should be initi-

(*Continued*)

Figure 3-14. (*Continued*)

ated. Further improvement of the process can be undertaken even with other frequencies of defects. Common sense should prevail.

7.2.4. Initiate actions for out-of-control conditions as described in Section 9.0. Responsibility: Inspection supervisor, test supervisor, and/or production supervisor.

7.2.5. Record actions taken for out-of-control points in the legend of the control chart. Responsibility: Supervisor, inspector, tester, and/or operator.

7.3. Ongoing review and analysis of process control charts.

7.3.1. Review data on control charts and initial.

7.3.1.1. Supervisors are to review and initial charts a minimum of two times a day. Responsibility: Inspection supervisor, test supervisor, assembly supervisor.

7.3.1.2. Engineers are to review and initial charts a minimum of once a day. Responsibility: Manufacturing engineer, quality engineer.

7.3.2. Submit completed data collection forms and control charts to quality engineering for review and analysis. Responsibility: Inspection supervisor, test supervisor, assembly supervisor.

7.3.3. Review completed charts and analyze as described in Section 8.2. Responsibility: Quality engineer.

8.0. Establishing control limits.

8.1. Initial control limits.

8.1.1. Preferred method.

8.1.1.1. Data for the initial limits are obtained by conducting a process capability study. Responsibility: Manufacturing engineer, test engineer, quality engineer.

8.1.1.2. The process must be in control and capable. The center line and control limits found in the process capability study are then adopted. These values or modified limits, if appropriate, are the initial values used for the ongoing control chart. Responsibility: Manufacturing engineer, quality engineer, and test engineer.

8.1.2. Alternate method. Data for initial control limits may also be established by obtaining samples from the ongoing process for a minimum of 25 subgroups. If the process is

Figure 3-14. (*Continued*)

in control, the center-line value and control limits may be used for the initial ongoing control chart. If the process is out of control, correct the process and redo 25 subgroups of data. An in-control process must be achieved before establishing limits for an ongoing chart. Responsibility: Quality engineer or manufacturing engineer.

8.2 Ongoing control limit revisions. The completed charts are reviewed a minimum of once a month, preferably once a week, to determine if the center-line and control limit values have changed. Responsibility: Quality engineer.

8.2.1. Attribute control chart center-line values will only be modified downward. The center-line values (p, c) will not be increased to reflect a deteriorating process. Appropriate CP action should be taken to return the process to control when the chart indicates the process has gone out of control or shifted. Responsibility: Manufacturing engineer, test engineer, quality engineer, and/or production supervisor.

8.2.2. The range for variable control charts should be reviewed closely. If R has increased, the process needs correction, If R had decreased, the limits should be changed. Consider use of modified limits if the estimate of process spread 6 ($\bar{R}/d_2$) is 60% or less of total tolerance. Responsibility: Quality engineer.

8.2.3. Distribute a matrix of current, calculated, and new limits to plant management and supervision. Responsibility: Quality engineer.

9.0. Corrective/preventive action.

9.1. CP action matrix exists: defects are listed on the matrix.

9.1.1. Supervisor or designee takes action according to the CP action matrix. Responsibility: Test supervisor, inspection supervisor (if needed), assembly supervisor.

9.1.2. Supervisor or designee verifies that the problem has been corrected by auditing subsequent units.

9.1.2.1. If a problem has been corrected, proceed with normal process.

9.1.2.2. If a problem had not been corrected, repeat steps 9.1.1 and 9.1.2 until all actions on the CP action matrix have been taken.

9.1.2.3. If the problem persists after taking all actions on the CP action matrix, proceed with step 9.3.1.

(*Continued*)

Figure 3-14. (*Continued*)

>   Responsibility: Inspection supervisor, test supervisor.
>
> 9.2. CP action matrix does not exist or defect is not listed on any existing matrix.
>
>   9.2.1. Have the quality engineer verify the defect. If he or she is not available, notify the appropriate manager immediately. Responsibility: Inspection supervisor, test supervisor, and/or assembly supervisor.
>
>   9.2.2. Verify the defect and determine if the rejection is valid. Responsibility: Quality engineer.
>
>   9.2.2.1. If the rejection is not valid, notify the appropriate supervisor and proceed with the normal process (remove data from the chart). Responsibility: Quality engineer.
>
>   9.2.2.2. If the rejection is valid, have the appropriate supervisor determine cause. Responsibility: Quality engineer.
>
>   9.2.3. Determine if cause of defect is known. Responsibility: Inspection supervisor, test supervisor, assembly supervisor.
>
>   9.2.3.1. If cause is due to workmanship, take appropriate action to correct the problem and proceed with step 9.2.6. Responsibility: Assembly supervisor, inspection supervisor, test supervisor.
>
>   9.2.3.2. If the cause is known and it is not due to workmanship, notify the responsible individual (proceed with step 9.2.5). Responsibility: Assembly supervisor.
>
> | Cause | Responsibility |
> |---|---|
> | Process | Manufacturing engineer |
> | Component | Quality engineer |
> | Design | Design engineer |
>
>   9.2.3.3. If the cause is unknown, notify the manufacturing engineer immediately. If he or she is unavailable, notify manufacturing engineering management. Responsibility: Assembly supervisor.

Figure 3-14. (*Continued*)

9.2.4. Determine criticality of the defect and recommend continuation or shutdown of the process to production management. Responsibility: Quality engineer, manufacturing engineer.
  9.2.4.1. When required, shut down the operation. Responsibility: Production manager.
  9.2.4.2. Issue a QCAN to the responsible individual (see Section 10.0). Responsibility: Quality engineer.
9.2.5. Identify cause of defect, determine CP action, and implement CP action. Responsibility: Manufacturing engineer.
9.2.6. Workmanship defects. (This section may be included on a separate policy statement and not on the copy on the production floor.) The following steps should be taken.
  9.2.6.1. Make operator or inspector aware of defect. Note his or her name or employee number.
  9.2.6.2. If defects continue, conduct a process audit and correct the process, material, or operator, as applicable.
  9.2.6.3. If defects continue and process audit has been conducted and is determined to be acceptable, then give the operator a verbal warning.
  9.2.6.4 If defects continue and operator has had a verbal warning, then give the operator a written warning.
  9.2.6.5. If defects continue and operator has had a written warning, then suspend the operator.
  9.2.6.6. If defects continue and operator has been suspended, then fire the operator.
9.2.7. Verify effectiveness of preventive action.
9.2.8. Initiate or update CP action matrix. Responsibility: Manufacturing engineer.
9.3 CP action matrix exists: actions listed partially solve the problem; CP action matrix does not solve the problem; two points out of control within last eight subgroups; same defect on three consecutive subgroups which are in control.
  9.3.1. Have the quality engineer and manufacturing engineer verify the defect. If they are unavailable, notify the appropriate manager immediately. Responsibility: Production supervisor.

(*Continued*)

Figure 3-14. (*Continued*)

9.3.2. Verify the defect and determine if rejection is valid. Responsibility: Quality engineer, manufacturing engineer.

    9.3.2.1. If the rejection is not valid, notify the assembly supervisor and proceed with the normal process. Make the reject acceptable on the control chart by crossing out the date and initialing and writing in the correct data. Responsibility: Quality engineer, manufacturing engineer.

9.3.3. If defect is verified, determine its criticality and recommend continuation or shutdown of the process to production management. Responsibility: Quality engineer, manufacturing engineer.

    9.3.3.1. When required, shut down the operation. Responsibility: Production manager.

    9.3.3.2. Prepare a QCAN and review problem with Production supervisor. Responsibility: Quality engineer.

9.3.4. Determine if defect cause is known. Responsibility: Quality engineer, manufacturing engineer.

    9.3.4.1. If defect cause is known, issue QCAN to the responsible individual and verify effectiveness of preventive action. Responsibility: Quality engineer.

    9.3.4.2. If defect cause is unknown, issue QCAN to manufacturing engineering for identification of cause. Responsibility: Quality engineer.

9.3.5. Identify cause and determine CP action (return to step 9.2.7). Responsibility: Manufacturing engineer.

10.0. Quality control action notice (QCAN).

  10.1. Prepare and issue a QCAN to request CP action.

    10.1.1. When a defect with an unknown cause exists.

    10.1.2. When a problem persists after taking CP action (see Section 9.3 for criteria).

  10.2. Shut down the process when the criticality of the defect will have a major impact, as described below. Responsibility: Production manager.

    10.2.1. Safety. If defect will jeopardize the safety of those working on, or those who will be working on, the product.

Figure 3-14. (*Continued*)

> 10.2.2. Rework cost. If defect will be costly to repair in terms of both disassembly/assembly time and scrap.
>
> 10.2.3. Defect interaction. If defect is suspected of leading to a chain reaction of failures.
>
> 10.3. Distribute copies of QCAN.
>
> 10.3.1. Production manager.
>
> 10.3.2. Engineering manager.
>
> 10.3.3. Quality manager (QCAN file).
>
> 10.3.4. Test engineering
>
> 10.4. Maintain shutdown status until CP action is taken.
>
> 10.5. Take CP action.
>
> 10.6. Purge line of suspected nonconforming material. Sort off line according to established procedure.
>
> 10.7. Perform a first-article inspection. Inspect the equivalent of one subgroup. If in-control,
>
> 10.8. Resume operation.
>
> 10.9. Follow up to assure that CP action is effective.
>
> 11.0. Audit for effectiveness.
>
> 11.1. Review complete QCANs for compliance and effectiveness of CP action.
>
> 11.2. Review process control charts according to process audit procedure (step 12 of 12-step program).

aware of their responsibilities. The team will prepare a checklist to verify that all required SPC tools are in place for the appropriate control station and that all personnel have received adequate training.

A preparation and coordination meeting will be held to review the implementation process and to make sure that everyone is ready to begin operation in accordance with SPC methods and requirements. The final step in implementation is for the team to facilitate and monitor the process as SPC is initiated.

Step 10: Problem Prevention. Have you ever introduced a new product or process and thought you were fully prepared, just to discover an unforeseen problem that delays deliveries? Problem prevention is especially important with a new product or new technology, but many times the pressure for output does not allow sufficient opportunity for engineers to test new designs.

One approach is to build a problem prevention system into the design/introduction process. If potential problems and customer requirements are considered from the inception of the development cycle, there should be no impact to the schedule. In fact, if management charts the time and number of changes required for introduction or modification, average overall time to release will be shorter when problem prevention is used.

A problem prevention plan must:

1. Identify potential critical problems and determine if they are controllable.

2. Identify likely causes of the critical problems.

3. Identify preventive actions to address the causes.

4. Implement the preventive actions.

5. Prepare contingent actions in the event that preventive actions are not entirely successful. Contingent actions operate regardless of the problem.

6. Identify "triggering mechanisms" for contingent actions.

Step 11: Performance Accountability. Performance accountability is a data collection system that assigns responsibility for defects or performance problems. Some corporate cultures discourage accountability programs, either from fear of criticism or from lack of management's ability to be tactfully assertive.

Other firms will have already developed a performance accountability system because it seems like the natural thing to do. If there is inspection, it makes sense to collect the inspection results in such a manner that they can be analyzed according to the type of defect, where it happened, and who was responsible. If there is no such system in place, the team should develop a method for data collection, complete with forms, and then institute a means for the data to be reported during and beyond the duration of the project.

Performance accountability may or may not be integrated into the SPC program.

Step 12: Measurement of Effectiveness. The final step in the productivity/quality-improvement process is to look back and assess how effective the program has been. The best way to complete this task is by conducting an audit designed to evaluate the team's performance against the original objectives.

Types of audits that may be performed are:

1. *Product audits* where the product conformance to specification is used to judge the effectiveness of the improvement program. This type of audit is simply a reinspection performed by the productivity/quality-improvement process team at the appropriate control point.
2. *Process/system* audits that assess conformance to established procedures and methods.
3. *Financial audits* to verify that estimated savings and improvements have been fully realized.

This is an excellent calibrator for the team's estimating skills.

Pitfalls to Total Quality Management

Pitfall Number One. Total lack of leadership from the executive level (i.e., the president and/or his or her staff). This lack of leadership or commitment is caused by not understanding TQM, i.e., what the benefits of TQM are, and how to implement TQM. These executives believe that their management styles of the past will work for the future, but this is a false assumption. Modern business is facing challenges that it has never had to deal with before. To meet these challenges business must change, and continue to change, to meet the challenges of an ever changing business environment.

Impact. Without executive support TQM cannot be implemented. It is not surprising that several companies who decided not to use TQM have gone out of business. Without the competitive edge that TQM offers (i.e., customer retention, improved profits, and creation of new business) in an ever-increasing competitive market, a company may find it impossible to survive.

Solution. Have an expert in TQM present a seminar to the executive staff with the express purpose of influencing them to accept the TQM idea. In this situation it would be best to have the instructor be someone who has worked at the executive level. This individual will then be able to address the executives' concerns from his or her own actual experience.

If you are successful in influencing these executives to accept the TQM concept, then immediately involve them in tactical planning of their destiny and future successes. If the instructor is unable to sell the executive staff on a full-fledged TQM program, there is one other option. Convince them to use TQM on a pilot-program basis, and hope the results will convince them of the value of TQM.

The only problem with this idea is that the pilot success will provide few beneficial results for the company. Each member of the executive staff must experience the benefits of TQM personally, or they will not be convinced. This option has a low probability of success. However, with enough tenacity in providing information to the executives, eventually they will adopt TQM—"when the student is ready, the instructor will appear."

Pitfall Number Two. Not formalizing an organizational structure and accountability.

Impact. Implementation is either very slow, or does not happen at all. People do not know their roles or they are not held accountable for the results.

Solution. Take the time to develop an organizational structure, with accountabilities, policy, and operating procedures. Encourage team leaders to identify any situation where they need support from the resource committee.

Pitfall Number Three. The executive board does not meet monthly to review the progress of the TQM implementation or, if the board does meet, they do not review all of the elements involved.

Impact. Business results will usually fall below expectations. People will perform as expected only in areas that are inspected. Areas that are not inspected, or are underinspected, will experience haphazard follow-through. The culture will not change unless it is being monitored; that is, through values, recognition and reward committee, training committees, and cultural-fostering committee.

Solutions. The executives must monitor the areas for which they are responsible, and not delegate this task to others. The executives should hold an executive board meeting once a month.

Pitfall Number Four. Resource committee does not meet every other week, or on some other consistent basis.

Impact. The TQM process slows begins to slow down, and eventually stops completely, due to a lack of follow-through.

Solution. Develop a policy where nothing will permit a resource committee to miss a meeting. If the meeting must be rescheduled due to customer needs, then it must be rescheduled for the day before or after the original meeting date. In the event the chairperson cannot attend, a designee must substitute for the chairperson.

Pitfall Number Five. A resource committee takes on more teams than it can handle.

Impact. The teams do not receive the time and attention they need to perform properly.

Solution. Limit the number of teams per resource committee to four.

Pitfall Number Six. Resource committee does not give the teams a written charter.

Impact. Teams sometimes get confused about expectations and consequently lose focus on the goals and how to achieve them.

Solution. The resource committee must provide the teams with written charters. The charters can be modified to suit the needs of either the resource committee or the teams.

Pitfall Number Seven. The resource committee does not require the teams to present their progress using a standard format and milestone plan.

Impact. The teams' progress becomes fuzzy, and is difficult to track. This causes slips in milestones and lack of focus on the objectives.

Solution. Furnish the teams with a standard report format, including milestone plan, and require the teams to utilize the reports.

Pitfall Number Eight. The resource committee consistently allows the teams to miss their target dates.

Impact. A team project may never be completed, or lost opportunities occur when the milestones are achieved behind schedule. The teams' morale will also drop rapidly, and they will lose interest in the project.

Solution. Require the team leaders to report any delinquent action items. When a delinquency is reported, follow these steps:

- For the first occurrence of a missed target date, allow the person a one-week grace period.

- If the person misses the target date again, then assign the action item to his or her immediate supervisor. (This procedure must be approved by the executive board prior to its implementation and enforcement.)
- Hold the immediate supervisor accountable.

Pitfall Number Nine. Resource committee does not require true measurements as evidence of a project's success.

Impact. For all the effort expended, no one really knows if any progress is being made.

Solution. Measurements must be meaningful and linked to the CBSFs. Measurements must be taken before a project is begins in order to establish a baseline. Measurements must then be taken regularly to monitor progress.

Pitfall Number Ten. Resource committee does not require the teams to follow up PQI process step by step. Teams are also allowed to deviate from the process without permission from the resource committee.

Impact. Progress can become more difficult to measure; the team may lose sight of their goal after not fully achieving it and become significantly late in coming to closure.

Solution. The resource committee must be sure the teams are closely following the PQI process. If the team must deviate, then the resource committee should be notified immediately. Teams must receive permission from the resource committee to deviate from the process. This is acceptable only if the problem has unique conditions where the PQI process does not apply.

Pitfall Number Eleven. Resource committee does not critique or provide feedback or consequences to the team leaders on performance and behaviors.

Impact. Team leaders do not feel rewarded for personal growth. Opportunities for personal performance improvement are missed, which results in little or no improved performance.

Solution. The resource committee charter should include a provision stating that team leaders will be critiqued and provided with feedback and appropriate consequences for performance and behaviors by their mentors and/or resource committee chairperson.

Productivity/Quality-Improvement
Process—Suppliers

In developing world-class quality organizations, firms have to worry about satisfying internal and external customers from a supplier's point of view. For most businesses, there is an additional need to become a good customer. A "good customer" makes its needs known to its suppliers in clear, concise, and specific terms. A good customer works with those suppliers as a partner to ensure on-time delivery of conforming product and materials.

Purchasing and quality departments can achieve significant gains for the company by leading a quality/productivity-improvement process for suppliers. Supplier quality improvement can be spread across all providers of goods and services, or can be focused on specific key suppliers or critical commodities. The most costly and most complex components or materials used by a manufacturer may deserve special attention due to long lead times or inventory constraints. "Nuts and bolts" kinds of commodities that are readily available anywhere and that have no direct impact on the product function may be ignored until the most significant items are under control. The decision of how to proceed will depend on the resources available for quality improvement efforts and the number of suppliers involved.

Supplier quality management takes on true significance to world-class organizations as they begin to explore worldwide sources. New markets in Europe, China, and the developing nations in Eastern Europe and South America, and shifting economies in Asia and the Pacific Rim present new opportunities and call for new sourcing strategies. Quality improvement will not be enough in this arena; quality needs to be in the product at the beginning.

Basic manufacturing companies begin supplier quality management by setting up an incoming inspection function to screen out lots of material with excessive defects. Most use the lot sampling plan developed by the military during World War II, and most use it incorrectly.

As the inspection function evolves, there comes the realization that the data from this inspection could be useful in understanding problems and avoiding production delays and inefficiencies that result from material shortages and poor quality. Of course, some companies never get beyond the incoming inspection phase, but they probably won't be reading this book.

Eventually a new level of awareness blossoms and quality engineering support and a Material Review Board (MRB) are introduced. Quality engineering is thrown into problem solving (fire fighting) as the problems mount and often become material expediters with a slightly different point of view. Material review board is a group brought together essentially to rationalize why it is OK for the supplier to send in defective materials. These additions to overhead have little impact on product quality until a structured approach to supplier measurement and improvement is implemented. Without doing any supplier corrective action, acceptance rates for an incoming inspection will settle at around 70 percent. As production volumes rise, the inspection/MRB process can become a significant bottleneck.

Errors leading to incoming inspection rejects may stem from either internal or external causes. Internal causes may include inadequate, missing, or incorrect specifications. There can be miscommunication between purchasing, engineering, and quality departments that leads to confusion at even the most well-intentioned vendor. An engineer might call the supplier and ask for a change that he or she then fails to document. A buyer may place an order for something that is close to, but does not exactly meet, the original specification. In a rare instance, inspection may use the wrong method or equipment for inspection or may fail to obtain the most recent drawing. Quality engineering and purchasing efforts directed at internal problems alone will generally bring the acceptance rate up to 80 or 90 percent.

External causes relate to a supplier's process and quality management system. Quality engineering can drive supplier corrective action and help purchasing identify suppliers that should be thrown out.

A structured and well-planned supplier quality management program will address both internal and external factors. An overall objective of the program should be to get out of the inspection business through shared data and visibility to the supplier's process control and product quality. Suppliers can be certified to ship directly to the production floor with no incoming inspection once confidence has been established in the supplier's product.

The supplier PQI program described here is composed of 14 steps, similar to the in-house PQI programs. A team is formed consisting of active participants from purchasing, quality, and other departments that can contribute to improvement—not only in the incoming yield, but also in the efficiency of the inspection process.

The 14 steps to supplier productivity/quality improvement are:

1. Project identification
2. Planning and reporting
3. Performance measurements
4. Internal review
5. Joint review
6. Joint action plan

7. Problem analysis and solution

8. Inspection capability

9. Process capability

10. Corrective/preventive-action matrix

11. Process control procedure

12. Process control implementation

13. Defect accountability

14. Measurement of effectiveness

Step 1. Project Identification. The basic purpose of project identification is to determine and establish the need for a project team. Once the need is established, it serves a second purpose of defining the magnitude and scope of the project. This is critical for effectively planning and allocating resources. Better decisions can be made about who should be on the team and how much time the team can be allowed to accomplish each task. In the majority of cases the time required to complete the project is directly proportional to the magnitude and scope. Therefore, how well a project is identified sets the stage for the future success or failure of the team.

Emphasis in project identification should be placed on three elements:

1. Where the best opportunity for improvement or concern exists

2. What part(s) of the operation should be improved to take advantage of the opportunity or eliminate the concern

3. What the impact of the improvement will have on the organization, customer satisfaction, added sales, cost reductions, etc.

Project identification is composed of three separate and distinct sections

- Project definition
- Project justification
- Process flow chart

In project identification, the team drafts a document for presentation to management or to the resource committee that includes the following information:

1. *Statement of problem*
   a. *Description.* The problem description includes part numbers, names, model renumbers operations, process source or customer plant, and responsible team.

b. Impact. Provides a general statement of the effect of the problem such as cost to service, number of repairs per 100 assemblies, percent defective, or number of late orders. Attempts to assign a dollar value to the problem if sufficient information is available.

c. Source. A statement defining where the problem is occurring: the shift, 2 of 5 machines, the plant location, geographic location, or location in the field.

2. *Summary.* Identifies area for significant improvement from the opportunities checklist. Summarize the checklist and fill in the blanks with potential cost reduction areas.

3. *Opportunities checklist.* The opportunities checklist is also provided to assist in checking all possibilities. The possibilities are then summarized in the summary section.

a. Productivity. Identify any opportunities related to productivity improvement that are concerned with labor, manufacturing methods, and inspection techniques (100 percent or sampling).

b. Quality. Identify any opportunities related to quality that are concerned with scrap or rework due to defective material. It is advisable to identify the most critical operations and characteristics so that the greatest impact can be realized in the shortest time.

Product characteristics can be classified as critical, major, or minor. A critical characteristic is defined as any attribute of the product which will result in defects or failure of the product or that will be perceived by the customer as a defect or failure. Working on control of the critical characteristics first can yield some savings in quality appraisal costs. Extremely close tolerances, for example, require more care and more expensive inspection equipment. Manufacturers need to recognize the difference between a precision bearing for the space shuttle and a plastic cover on the back of a television set. Quality is important for both, but the different applications call for different manufacturing and inspection techniques.

Another example of savings through characteristic classification would be soldering of electronic components. There are hundreds of defects that can be observed in a solder joint, and they are all important in setting up a soldering process. Not all of these defects, however, will actually result in the failure of the device, and so can be treated differently when observed in the final product.

c. Schedule impact. Identify any opportunities related to schedule, such as missed scheduled, shortages, or backlog.

d. Final customer. Identify any opportunities related to final customer, such as installation quality, service response, and reliability.

Opportunities Checklist

	Applicable	
	Yes	No

1. *Productivity*

	Yes	No
Rework labor—source	_____	_____
Rework labor—customer* plant	_____	_____
Manufacturing methods	_____	_____
Inspection techniques (labor)	_____	_____
Scrap (labor)—source plant	_____	_____
Scrap (labor)—customer plant	_____	_____
Amount of inspection (labor)	_____	_____
Reinspection labor—source plant	_____	_____
Reinspection labor—customer plant	_____	_____

2. *Quality*

	Yes	No
Critical operations	_____	_____
Scrap (material)—source plant	_____	_____
Scrap (material)—customer plant	_____	_____
Classification of characteristics	_____	_____

3. *Schedule impact*

	Yes	No
Variances	_____	_____
Shortages	_____	_____
Backlog	_____	_____
Cycle time through inspection	_____	_____

4. *Customer impact*

	Yes	No
Customer installation quality—DOAs	_____	_____
Product reliability—MTBF	_____	_____
Customer perception of defects and failure modes	_____	_____
Customer service response	_____	_____
Service calls	_____	_____
Cost to repair—maintainability	_____	_____
Shipping and handling	_____	_____
Spares inventory	_____	_____

Project Justification. The intent of project justification is to determine the expected level of improvement. Project justification includes the computation of annual savings for each project. Performance history, esti-

---

* Customer is the purchaser—not necessarily the final customer.

mated improvements, and customer complaints should be taken into consideration.

The team should document details of performance for at least the last 3 months, including reject rates, scrap rate, productivity, rework, schedule compliance, and customer complaints. Once again, this ties back to the evolution of a department that is responsible for collecting and assembling the necessary data. The role generally falls to quality assurance, but could be assigned to any appropriate group.

A worksheet format is suggested here to compute costs and ROI. There is nothing complex about these calculations, but justification is often completely ignored by engineering "fire fighters." The estimated cost improvements from each area are combined to arrive at a potential annual savings and, in keeping with the accepted PQI process, the project justification should be approved by the resource committee.

Step 2: Planning and Reporting. The elements of the planning and reporting step establish the time-phased objectives for the program and are generally of vital interest to senior management. The true return on investment for a project is directly proportional to the time in which it can be accomplished. Estimates of the time required to implement supplier quality improvement programs are slightly more difficult than internal programs because of the different organizations involved and the roadblocks to communications that can arise between companies.

Milestone Plan. The milestone plan indicates the planned start and completion dates for each of the 14 steps. A suggested schedule for each step is shown here. (See Figure 3-10.)

The initial plan is developed by the PQI team and is submitted to the resource committee for approval. The plan allows for revisions to start and completion dates, as well as current status. The milestone plan with current status should be presented as a part of the team's progress reports.

Some steps take much longer than others, and in some cases will require agreement from the supplier. Travel times and the supplier's production schedule should be taken into consideration when establishing milestones. If the supplier is available and willing to participate in the PQI program, the benefits will be even greater.

Progress Reports. There are two types of progress reports: (1) team progress reports and (2) resource committee progress reports.

The team progress report primarily assists the team leader in evaluating the progress of his or her team and in keeping that individual focused on the objective. Of secondary importance is its value in communicating with management. The team progress report is prepared weekly, within 24 hours of completion of the team's weekly meeting. The 24-hour rule is significant in that it provides a written reminder of action items with sufficient time for completion, and it prompts the team members to prepare for the next meeting.

The team report is distributed to resource committee members, team members, and team leaders.

The resource committee progress report is prepared by the resource committee chairperson and issued within 24 hours of the meeting to senior management, team leaders, and team members. This reconfirms any assignments in writing to steering committee and team members.

Team Progress Report Contents

1. *Objectives.*   Objective of team expressed as measurable improvements.

2. *Status.*   Indicates which steps have been completed and which have not.

3. *Accomplishments.*   Statement of assignments completed and closure of any of the 14 steps. This is not a list of team member activities. An updated chart of the selected performance measurements should be attached. Any savings actually achieved during the report period should be stated. For example if recalculated control charts show a statistical improvement in yield, a dollar value should be assigned and reported.

4. *Exception to plan.*   Indicates assignments of planned steps that are delinquent or had an inadequate response. An action plan, including action, responsible person, and target completion date is required.

5. *Risks and exposures.*   Identifies any constraints to progress which require management action or support. Indicates causes of constraints along with a recommended action plan to the resource committee.

6. *Outlook.*   Forecasts the dates of completion for steps in progress and for the overall project. Also forecasts the expected values of performance measurements that will be achieved. If any unfavorable items are "outlooked," the team should provide a recovery plan (controllable items) or a contingency plan (uncontrollable items). It is much better to provide management with up-to-date status and expectations than to reach the due date without completing the assigned task.

7. *Detailed assignments.*   For each detailed assignment the report should state the activity, responsible team member, and the targeted completion date. Completed items are retained in the progress report for only one week after completion.

Resource Committee Progress Report Contents

1. *Status.*   Identify teams reviewed. State progress of each team to milestone plan, performance to objective, and outlook for project completion.

2. *Accomplishments.* Indicate resolutions to risks and exposures and exceptions to plan.

3. *Assignments.* Describe any assignments for resource committee member related to resource committee activities.

4. *Planned activities.* State date, time, and location of next meeting along with agenda of teams to be reviewed.

Presentations. The teams should be encouraged to present status and accomplishments to the resource committee and to senior management. Presentations can be in a format that is suitable for the organization and that effectively communicates the progress and improvements that have been realized. Some opportunity for give-and-take between team members and management will be helpful in setting directions consistent with the established vision, mission, and values statements.

The team should also consider presentations to individual or groups of suppliers. The team may formally close step 2 when the following criteria are met:

1. The milestone plan for their project is completed and approved by the resource committee.

2. Progress reporting must have begun.

3. Management presentations must be scheduled.

Step 3: Performance Measurements. The objective of this step is to develop and implement a method to measure performance in quality, productivity, or schedule relative to the target objective.

Performance measurements will enable the team leader and the resource committee to assess the team's progress toward meeting the stated objectives and to measure the team's overall effectiveness upon completion of the project.

In regard to suppliers, the classic measurements of success are percentage of on-time delivery and percent lot acceptance at incoming inspection. In developing a working partnership with suppliers there should soon be no need to verify material quality through incoming inspection. Suppliers can be certified to deliver material directly to stock or production, and the performance measurements applied at that time will be the supplier's own process yield and time to resolve corrective action items.

In effect the team is charged with motivating and assisting the suppliers in their own productivity and quality improvement programs. For critical commodities, suppliers' representatives may be included on the internal teams.

Step 3 may be closed when the following criteria are met:

1. The performance measurement criteria or parameters relative to the target objective must be selected (i.e., percentage of lots rejected, percentage of delinquencies within the delivery window's requirements).

2. The measurement system of method (including data collection) must be established and implemented, with assignment of responsibilities for the system.

3. Reports of performance are being issued monthly.

Step 4: Internal Review. Step 4 established a method to review customer internal information about the product or assembly to be addressed by the customer's personnel only. This is the point where the improvement process for suppliers deviates from the internal improvement process. If nothing else, teams are formed to conduct the internal and joint reviews. The teams are extremely effective at improving communications between the customer and suppliers.

Prior to the review of a product or assembly with a supplier, it is mandatory that customer internal information regarding performance history, as well as specifications and requirements for that product is current, are available and agreed upon among all responsible internal disciplines. The best way to ensure that your suppliers meet your expectations is to define those expectations in detail and share this information with the supplier. Inadequate or misunderstood specifications should never be the cause for lack of material quality.

Methodology. A meeting is convened by quality engineering with representatives from purchasing, incoming inspection, manufacturing engineering, design engineering, production, cost accounting, and other appropriate departments. The agenda of the meeting will be a review of the specific part(s) or assemblies, using the internal/joint-review checklist as the format. The review will cover elements in the following areas:

Quality history

Delivery history

Part/assembly drawings and/or artwork

Product specification

Cost

Process

Inspection

Packaging requirements

Manufacturing schedule

Procurement administration

Any issues that are unresolved will be documented and assigned to a responsible member with the appropriate action required and a target date for completion. Follow-up meetings, if required, will be scheduled to review open-action items until all items are completed. Use the checklist and cover each item that applies. Allow adequate time at review meetings to explore all areas of concern.

The requirements of this step are satisfied with a meeting of the appropriate individuals and completion of the internal/joint-review checklist. Resolution must be obtained on all issues and outstanding action items in the checklist. (See Figure 3-15.)

Step 5: Joint Review.   Step 5 establishes a common understanding and agreement between the customer and supplier regarding the following items:

Drawings and specifications

Procurement administration

Processes

Capacity

Manufacturing feasibility and capability

Acceptance criteria

Past performance

Delivery expectations

The joint review is best held at the supplier's facility and should include discussion of the results of the internal review, comparison of answers to the internal/joint-review checklist, and assessment of the supplier's capability to meet the specified requirements.

Many times the customer's engineering or quality department will believe that the specifications are clear and understandable, but the supplier will have an entirely different view. From lack of either positive or negative feedback the supplier may produce nonconforming material unknowingly.

Step 6: Joint Action Plan.   This step provides a mechanism for obtaining resolution to unresolved issues of the joint customer/supplier review. During the review meeting issues are likely to arise that will require further investigation, data collection, approval by others, or some other action before resolution or agreement can be established. This step ensures that all the unresolved issues are addressed and action is taken to arrive at an agreed-upon result.

Figure 3-15.

SUPPLIER QUALITY IMPROVEMENT
INTERNAL/JOINT-REVIEW CHECKLIST

		RESP	REQD	RESP	REQD
1.0	**Quality history (past 6 months for this P/N or a similar part)**	___	___	___	___
1.1	How many lots were received with this part number?	___	___	___	___
1.2	What % of lots were rejected with this part number?	___	___	___	___
1.3	What types of defects were identified on DMRs in	___	___	___	___
	Receiving inspection?	___	___	___	___
	Work-in-process?	___	___	___	___
1.4	What was the disposition of parts by defect as indicated on DMR's?	___	___	___	___
1.5	Were previously committed corrective/preventive actions effective?	___	___	___	___
2.0	**Delivery history (past 6 months for this P/N or a similar part)**	___	___	___	___
2.1	What % of lots were on time for this part number (plus or minus 3 days)?	___	___	___	___
2.2	How many days early or late were the lots for this part number?	___	___	___	___
2.3	Were previously committed recovery plans effective in improving delivery?	___	___	___	___
3.0	**Part/assembly drawings and artwork**	___	___	___	___
3.1	Are all drawings available, including component detail?	___	___	___	___
3.2	Are all drawings complete (dimensions, tolerances, material finishes, etc.), signed, darted, and released through document control?	___	___	___	___
3.3	Does the drawing include the latest revision?	___	___	___	___
3.4	Is the bill of material complete and accurate?	___	___	___	___
3.5	Are the next-higher assembly drawing numbers listed and available?	___	___	___	___

(Continued)

Figure 3-15. (*Continued*)

3.6	Are dimension/tolerances excessively tight for normal processing, and will special processing be required?				
3.7	Are dimension/tolerance changes required to support tooling and process?				
3.8	Has an assembly tolerance stack-up or statistical tolerancing been conducted?				
3.9	Are there drawing notes, dimensions, tolerances, representing areas where the supplier does not currently have manufacturing experience?				
3.10	Does the artwork meet all parameters as called out in the manufacturing specification and fabrication drawing?				
3.11	Does the supplier have any questions regarding the print interpretation?				
3.12	Does the supplier have recommendations to improve the clarity of the drawings?				
4.0	**Product specifications**				
4.1	Are all specifications available (product, functional assemblies, electronic components)?				
4.2	Are specifications complete, signed, dated, and released?				
4.3	Are specification changes required to support tooling and processes?				
4.4	Are there specifications that exceed the suppliers current technology?				
4.5	Are there specifications that appear unrealistic and costly to meet?				
4.6	Are the specifications and artwork compatible?				
4.7	Are there any questions or recommendations regarding the specifications?				
4.8	Does the supplier have any difficulty in meeting the specifications?				
4.8	Has engineering completed design verification testing?				

Figure 3-15. (*Continued*)

4.9	Has regulatory testing been completed and have regulatory approvals been obtained?	___	___	___	___
4.10	Has reliability and environmental testing been completed?	___	___	___	___
4.11	Are there recommendations for design changes that will improve product reliability?	___	___	___	___
5.0	**Cost**	___	___	___	___
5.1	Does the design meet target costs for manufacturing, capital, and maintenance?	___	___	___	___
5.2	Are there recommendations for design changes that will reduce cost?	___	___	___	___
6.0	**Process**	___	___	___	___
6.1	Is the design capable of being manufacturing with the suppliers current process?	___	___	___	___
6.2	Are there any special tooling requirements to build or assemble this design?	___	___	___	___
6.3	Is there any special equipment required?	___	___	___	___
6.4	Is the process capable of making the part as designed?	___	___	___	___
6.5	Are the manufacturing methods documented?	___	___	___	___
6.6	Does the supplier need assistance in developing the process?	___	___	___	___
6.7	Are the material-handling procedures adequate to prevent damage to the part?	___	___	___	___
7.0	**Inspection**	___	___	___	___
7.1	Is the drawing clear and understandable?	___	___	___	___
7.2	Are datum selection complete and logical?	___	___	___	___
7.3	Are material, finish, and form adequately called out?	___	___	___	___
7.4	Are current inspection methods and gages capable of measuring the part as designed?	___	___	___	___
7.5	Are the standard measuring tools calibrated?	___	___	___	___
7.6	Are the calibration procedures documented?	___	___	___	___

(*Continued*)

Figure 3-15. (*Continued*)

7.7	Is special or unique equipment required at the supplier and in-house correlated?	___	___	___	___
7.8	Are inspection instructions available for each part number? Are they comprehensive?	___	___	___	___
7.9	Are critical/major/minor dimensions called out on the drawing or in or in the inspection instructions?	___	___	___	___
7.10	Has an inspection/test capability-study been performed? Are the results acceptable?	___	___	___	___
7.11	What are the current sampling plans?	___	___	___	___
7.12	Does the supplier need assistance in developing an inspection plan?	___	___	___	___
7.13	Does the supplier have a quality management system?	___	___	___	___
7.14	Is an ATP required? Is it available and released?	___	___	___	___
7.15	Is source inspection required—plant location? Source inspector trained?	___	___	___	___
8.0	**Packaging requirements**	___	___	___	___
8.1	Is shipping protection against moisture/humidity required?	___	___	___	___
8.2	Is there a special finish on the product that requires special handling?	___	___	___	___
8.3	Is thermal protection required?	___	___	___	___
8.4	Are there special aging and shelf-life limitations?	___	___	___	___
8.5	Is there a delicate nature to the product which requires special packaging to allow for handling in-house?	___	___	___	___
8.6	Will the product require special packaging and shipping due to size, weight, mounting, or cooling?	___	___	___	___
8.7	Has testing of the packaging been completed (drop test, road test, other environmental)?	___	___	___	___
9.0	**Manufacturing schedule**	___	___	___	___
9.1	What is the projected annual volume?	___	___	___	___
9.2	What is the projected daily volume?	___	___	___	___

Figure 3-15. (*Continued*)

9.3	When is the product scheduled to start production?	___	___	___	___
9.4	Is the supplier's process capable of manufacturing the required volume and meeting the delivery schedule?	___	___	___	___
9.5	This part number and quantity will represent what percent of the supplier's capacity?	___	___	___	___
10.0	**Procurement administration**	___	___	___	___
10.1	Verbal instructions are unacceptable and changes must be backed up by documentation, including approved marked drawings. Is this procedure being followed internally and with the supplier?	___	___	___	___
10.2	Are change orders incorporated only after receipt of documentation and in an acceptable time by the supplier?	___	___	___	___
10.3	Do drawings and purchase orders reflect the same part number and revision?	___	___	___	___
10.4	Are the requirements for the first article specified in the purchase documents?	___	___	___	___
10.5	Have the quality assurance policies and requirements been included in the purchase documents?	___	___	___	___
10.6	Has the first article been received, inspected, and approved?	___	___	___	___
10.7	Is the supplier on the Approved Manufacturers List?	___	___	___	___
10.8	Does the supplier's part number match the approved in-house part number?	___	___	___	___
10.9	Are tooling charges made at the supplier only after receipt of purchase orders?	___	___	___	___
10.10	Are suppliers aware of and given the opportunity to submit deviation requests prior to shipment of the product?	___			

Specific action required, the responsible individuals, and the time for completion should be documented. A process for follow-up should be established to ensure that commitments are met. This step may be closed when a joint-action plan is completed and approved by the resource committee. (See Figure 3-16.)

Step 7: Problem Analysis and Solution. In problem analysis and solution, the team and the selected suppliers are concerned with identifying and testing potential causes of the performance variance identified in step 1. Appropriate corrective/preventive action will be taken to reach and sustain the desired levels of performance. Even before the project was launched, some analysis should have been conducted to find the best place to start. If Pareto charts show that most of your defects are in printed circuit boards or if most of your purchased material comes from one source or if most of your dollars are spent on one commodity, then you have already done some problem analysis in project identification.

This is the first step directed at actual improvement of the process. It is geared toward making immediate improvements without having to wait for the implementation of control charts. The team collectively identifies the most likely causes and assumes responsibility for verifying and implementing the corrective and preventive action.

This step is iterative and stays in effect as new problems are identified throughout the 14 steps of the process. In some cases team leaders or department managers will employ this step to attack a pet peeve or perceived chronic problem. That is a good approach as long as solution of the pet peeve contributes to reaching the team's stated objective. If problem-solving activities have only a marginal relation to the project or begin to preclude completion of other team activities, the resource committee should redirect the team's efforts back to the intended course.

SUPPLIER QUALITY IMPROVEMENT
JOINT-ACTION PLAN

ASSY/PART NAME _____ OPERATION/PROCESS _____

ASSY/PART NO. _____ SUPPLIER _____

MODEL NO. _____ CUSTOMER PLANT _____

RESPONSIBLE TEAM _____

ACTION REQUIRED	RESPONSIBILITY	TARGET DATE	STATUS

Figure 3-16.

As an example, it may be common knowledge that the design of a certain part leads to excessive rework in production, but no one has time to fix the problem or release new drawings. Getting appropriate members of the team together to walk the change through the drawing control system may provide a quick and relatively simple improvement for both the supplier and the customer.

The techniques in this step have been selected for their practicality and effectiveness; they are the how-to's of analyzing and solving problems. Since each situation is unique in magnitude and complexity, so too are the techniques employed. They are:

Events log

Diagnostic process audit

Cause-and-effect diagram

Design of experiments

These are the same tools used for internal improvement programs, and they are described in detail in Chapter 8. For supplier quality improvement, these tools may either be applied in the supplier's process or in the customer's incoming inspection and test areas. In either case the supplier should be fully aware of expectations, measurement techniques, and results of each analysis.

By far the most common problem-analysis technique used with suppliers is the diagnostic process audit. A process audit is intended to investigate the adequacy of a process and the use of the process by operators and inspectors. The process audit verifies the proper implementation of procedures and work instructions and provides assurance that the process documentation, tools, and material support have been maintained current for optimum producibility.

With the growing acceptance of *ISO 9000 Quality Management System Standards*, the audit is becoming a major component of customer/supplier relations. Audits are most effective when performed by trained auditors, but team members can generate a checklist based on the existing process documentation and simply verify that instructions are being followed at each station.

The basic elements of a process audit include the following:

Current and correct documentation (work instructions)

Tools, gages, and measuring equipment in good condition, used as required, and calibrated

Material control and handling to prevent damage or degradation

Station layout to optimize productivity and quality

Adequate personnel training (do the operators know and follow the documented procedures?)

Control of nonconforming material (is defective material identified and kept out of the production flow?)

Corrective/preventive action (is data collected and acted on to correct the process when necessary?)

Proper application of Statistical Process Control

Step 8: Inspection Capability. The objective of inspection capability is to evaluate and quantify a measurement of inspection system for variable or attribute data, both at the supplier's site and at the customer's receiving inspection.

In this step a statistical study is conducted to determine the inherent error of inspection methods or equipment and to calculate whether or not the error is within acceptable limits. For variable data, study results are expressed as a percent of tolerance consumed by inspection capability. Obviously, if 25 percent or more of your part tolerance is taken up by error in the measuring system, it will be very difficult to tell good parts from bad parts. In many cases the study uncovers tolerances that have been set too tight for the application.

For attribute data, inspection-capability studies provide information on effectiveness and bias of inspection methods and personnel. Some companies use inspection-capability studies for attributes as a training aid for new operators or inspectors.

Suppliers will have varying degrees of sophistication in measuring and test techniques. The PQI team must design and conduct studies that are appropriate to the supplier's process. Sharing the results of studies from the customer's incoming inspection department with the supplier will also be of considerable value in aligning perceptions of quality requirements.

Step 9: Process Capability. If there was a way that process-capability studies could be made part of the supplier selection process, the customer could do away with incoming inspection very quickly. The process capability step establishes a systematic procedure for studying a process by means of control charts. Through this study the team can compare the true process variation against the established specifications and determine the likelihood of producing conforming material.

A quality product can be made only when the production processes are able to consistently meet specific targets, such as design tolerances or specified AQLs (Acceptable Quality Levels). The purpose of this step is to determine the total process variability and stability over time under normal operating conditions. This study indicates when unnatural variation is present so the cause can be identified and removed.

A process capability study can:

- Improve productivity and quality
- Determine the design tolerances that can be met and held with the current process
- Determine if new equipment is capable of meeting the requirements
- Compare the capabilities of alternative equipment/machines

The difficulty in conducting process-capability studies for suppliers is that the study will take some time and may be expensive. Unless you are a really good customer, the supplier may not be willing to cooperate. The customer may have to pay for the study, including the purchase of all the material produced unless it is part of an open purchase order.

The procedure for process capability must be followed closely, and some training of supplier personnel will be necessary. If results are not acceptable, the study will have to be repeated. All of this may disrupt the supplier's production schedule if unacceptable parts are not produced.

The corrective action may require new equipment or other changes that could represent a considerable investment. You may even find out that the supplier's process is not at all suitable to your needs and that another supplier would be better. For some commodities (nuts and bolts, common electronic components, etc.), it may not be practical to conduct studies, and then it is up to engineering to select the grade of component required to achieve the desired results at the customer plant.

If process-capability results are not acceptable, corrective action must be taken and the study redone. This cycle must be repeated until acceptable results are achieved or the decision is made to:

Reestablish the tolerances or specifications

Redesign the component to conform to the process capability

Purchase new equipment with the required capability

The methodology for process-capability studies is described in Chapter 6.

Step 10: Corrective/Preventive Action. The objective of step 10 is to eliminate guesswork and delay in taking action when a process goes "out of control."

The Corrective/Preventive-Action Matrix (CPAM) is the means for "closing the loop" in the application of SPC. It identifies the specific actions which must be taken to bring the process back in control, and the action that must be taken to address defective (or marginal) product. It also provides documented instructions that can be used to train new operators and engineers. The matrix is a listing of all known defects and conditions which can exist in the process being measured and their respective corrective and preventive actions. It is a prioritized troubleshooting guide with a specific and clear road map to follow. Delays and guessing at the action to be taken directly relates to additional cost.

Corrective and preventive action may be defined as follows:

Corrective action. Addresses the defective, nonconforming, or questionable product. Actions taken to identify, sort, and correct any defective product.

Preventive action. Addresses the process or cause of the defects or out-of-control condition.

These actions are taken to remove the cause of the out-of-control condition and therefore prevent the nonconformity from appearing on subsequent subgroups.

The CPAM should be developed by the people responsible for the design and application of the process. Thus, with supplier quality improvement, the location and control of the CPAM will be determined by the manner in which the process is measured and controlled. For fabricated parts, control will typically reside with supplier, but for standard components the control may be best applied at source or incoming inspection.

The steps to develop the CPAM are as follows:

1. Identify and list defects and other out-of-control conditions. This can be done with the help of defect summary reports or Pareto analysis.

2. Determine the probable cause(s) of each item listed. Since each item may have several causes, list in order of likelihood (the most likely first).

3. Determine the defect type. Classify into either workmanship, design, process, or component.

4. Identify the responsible station or stations where the condition originates.

5. Establish corrective action for the product; list in order of effectiveness, with the most likely to correct the problem first.

6. Establish preventive action to remove the cause of the problem. As with corrective action; list in order of effectiveness and minimum cost.

Many times the same action will be specified for different defects/conditions or probable causes. For this reason coding of actions is recommended. The codes and definitions should be listed at the bottom of the matrix.

The CPAM is dynamic. It requires periodic update to document new-found defects, conditions, causes, and actions.

It is designed to be used with either variable or attribute control charts. A CPAM is a means of empowering employees to control their own process with no need for direct management interventions.

The requirements of step 10 are satisfied with the completion and introduction of a corrective/preventive-action matrix at the appropriate control station.

Step 11: Process Control Procedure. When appropriate, the suppliers involved in the improvement project should be encouraged to document their process control procedure. Methods for selecting and maintaining control charts should be documented as a medium for training and a means of maintaining consistency in the case of personnel turnover.

Step 11 may be closed when the supplier has documented a process control procedure and reviewed it with all appropriate individuals to ensure their understanding and acceptance of the responsibilities.

Step 12: Process Control Implementation. This step provides the criteria and instructions for implementing the process control procedure developed in the preceding step. This is the stage where all of the team's prior activities are brought together, and any uncompleted activities are assigned to be completed. This step will begin to internalize process control for the supplier.

The team should generate a checklist of "prerequisites for SPC implementation" that will include:

Satisfactory completion of all 14 steps

Inspection-capability study completed and acceptable

Process-capability study completed and acceptable

Types of control charts selected for appropriate stations

Training completed for all personnel

A preparation meeting should be held to review the schedule and process for SPC implementation. With supplier programs it is critical to coordinate implementation so that all necessary support personnel are available on both sides.

At the designated time, begin SPC implementation and monitor activities closely in order to head off any misunderstandings or incorrect use of the SPC procedure. It is often beneficial, from a psychological standpoint, to phase in SPC with new programs, new products, or at new facilities. The change to SPC then becomes part of an overall change in work environment and is more readily accepted.

Step 13: Defect Accountability. Defect accountability is a reporting system that is established to facilitate problem solving in the production line. Accountability for a defect/cause combination should be assigned to a workstation, machine, operator, or segment of the process so that quality and manufacturing engineers know where to expend their efforts for the greatest return in improvements.

All that is required is some form of data collection that will tie defect/cause combinations back to the accountable source. In the case of supplier quality improvement, defect accountability is commonly applied at incoming inspection where records are maintained of lots inspected and lots rejected by supplier and part number. Incoming inspection may also use supplier rating reports or control charts to track the defects back to the responsible vendors.

Defect accountability should also be established by those suppliers targeted for the improvement program. The customer should ask for specific yield and defect data from stations throughout the supplier's process and should insist that appropriate corrective/preventive action be taken in a timely manner.

This step of the supplier quality improvement program is considered closed when the defect accountability system has been defined and installed.

Step 14: Measurement of Effectiveness. This is not only the last step of the PQI process, but is also the step that may cycle back to step 1 if further opportunities are identified. The primary purpose of this step is an audit to ensure that the desired results were achieved. It will assess the success of the team's efforts against the objectives of productivity and quality that were defined in step 1. It will also assess overall implementation of the techniques used throughout the first 11 steps.

The team should select an appropriate audit technique and develop a checklist that objectively probes the areas of performance measurement. Following the audit, a report should be generated to the resource committee, and the step can be closed when a minimum of 90 percent of the audit elements are in the acceptable range.

When all of the 14 steps have been closed, the team should make a final presentation to the resource committee with emphasis on the current values of the selected performance measurements. The resource committee can, at that time, decide if further action is required or if a new project should be initiated.

When implementing other techniques such as project teams, statistical process control, any of the world-class manufacturing concepts or quality-function development, a plan is developed which includes a project charter, milestone plans, and responsibilities.

4

Cultural Track

Purpose

The purpose of this chapter is to discuss methods and techniques to manage the cultural change. Ultimately, the aim is to create the desired culture which will be the force that generates the actions to achieve the vision and goals of the Critical Business Success Factors (CBSFs) and, ultimately, the vision. Changing the culture within an organization is one of the tracks required to move an organization from a hierarchical, traditional organization to an empowered, TQM organization.

Objective

The objective of the cultural track within the TQM implementation process is to create commitment to the company, the improvement process, and accomplishment of the CBSFs. Creating a cultural change means: to identify the values, both those to retain and those to adopt; to define the values; to convert the values to behaviors; to develop the implementation plans for these values; execution of the value implementation plans; measurement of the implementation of the values; and follow-up to ensure achievement and sustainability of the values. Creating a cultural change means promoting open and effective communications through the empowerment process. The process of empowering employees will encourage employees to take responsibility and ownership of problems and their solutions.

Introduction

A new era in the business community is starting. Instead of an elite group of executives running a company, all company employees need to be involved. Everyone uses their knowledge and experience to focus on a company's goals, Critical Business Success Factors (CBSFs), and to solve problems. Companies that do not use the talents of all the employees, combined with new techniques, will begin to falter and eventually fail in today's highly competitive marketplace. No company can survive without tapping into all of its available resources. To achieve these goals a new culture must be established in the company. This new culture will promote and support TQM, and it will allow a company to survive new challenges.

Success with TQM will depend on the following:

- Achieving customer satisfaction, both internal and external
- Creating and sustaining cultural change
- Achieving the goals of the Critical Business Success Factors
- Continuously improving performance measurements

This chapter will focus on four major topics required to effectively manage cultural change:

- Creating commitment
- Creating a cultural change
- Empowering the workforce and taking responsibility or ownership
- Managing committed workers in an empowered organization

Why Change?

An important question which needs to be addressed is, "Why change?" Many companies have been successful for years by using a rigid hierarchical structure. What is different today that makes it necessary to move from the old but previously successful methods to new methods? Let's use an allegory of two dinosaurs to explain why change is necessary.

Many millions of years ago, one dinosaur recognized that the competition for foliage was becoming keen and that dinosaurs were slowly dying off. This dinosaur realized he had to change in order to survive. First, he shrank his body and dropped his bulky tail. He became lighter and faster. Next, he thinned down his legs and made them more flexible. These changes gave him the speed to run away from predators and to reach food in places too small for the other bulky dinosaurs.

One day the new and improved dinosaur fell asleep under a tree. He had not realized that his smaller size had both good and bad qualities. A while later, when an old bulky dinosaur came up to the tree to eat, she didn't even notice her small cousin sleeping by the trunk. She also didn't notice when she stepped on her little cousin as she walked up to the tree. This brought a quick end to the new and improved dinosaur.

Later, the old dinosaur who stepped on her cousin also realized the intense competition for foliage. She contemplated the changes her cousin had made and decided to overcome the competition by changing too.

She reduced her size and weight. Her hind legs stretched into long slender supports, tipped with a set of strong claws. Her head grew smaller and her mouth stretched into a long beak, lined with sharp teeth. With these new teeth she stopped eating plants, and made meat her new diet. She flattened her arms into wings, and grew feathers over her entire body.

When she next encountered an enormous dinosaur in her newly altered state, she escaped by flying away. She became one of the few dinosaurs to inhabit the sky and treetops where competition was greatly reduced. She had successfully changed, and put herself well ahead of her competition. This was a complete metamorphosis.

To survive in today's competitive environment, companies in the United States must evaluate two points about change. First, they must evaluate what led to past successes and failures. (Will the attempt to slim down and become nimble make an organization vulnerable to old threats?) Second, they must take what they learn from these evaluations and see how they apply to today's marketplace. (What can organizations learn from the success and failures of other firms, even competitors?)

Most business leaders and writers agree that traditional ways of doing business are not likely to be successful in today's marketplace. Companies need to be flexible and able to adapt to quickly changing market conditions. In order to develop the survival traits of flexibility and quick adaptation, companies must create a culture which is committed to the organization's articulated goals. Besides creating commitment to the organization's goals, companies must delegate decision-making authority to the lowest feasible level within the organization (empowerment). The next section addresses one of the keys to creating a culture with survivability—commitment to the organization's goals.

Creating Commitment

The first requirement for cultural change is all-out organizational commitment to the company, with the goal being economic survival of the company. As part of the commitment to the company, there must be an organizational commitment to TQM, with the goal being achievement of the CBSFs, Internal Customer Satisfaction Agreements, and the cultural change. Commitments are the actions in continuous pursuit of the goals. Commitment differs from interest, as interest depicts involvement when convenient, whereas commitment depicts a tenacity to pursue the goals in spite of obstacles and knows no excuses.

Commitment to the company's survival means commitment to the TQM process; commitment to the TQM process means commitment to achievement of the CBSFs; commitment to achieving the CBSFs means commitment to achieving each team's project objectives; commitment to achieving project objectives means that individuals commit to achieve their goals and to constantly do their job better. The rolldown of commitment, from the survival of the organization down to maintaining the restrooms better, ultimately boils down to a devotion to excellence. Perhaps it can be best expressed in the words of Vince Lombardi:

> The quality of a person's life is in direct proportion to their commitment to excellence, regardless of their chosen field of endeavor.

Simply stated, commitment is what transforms a promise into reality. It is words that speak boldly of your intentions—words which are then backed up with action. It is the making of time when there is none, and the daily triumph of integrity over skepticism. Commitment is the stuff character is made of, and it has the power to change the face of things to meet your needs. In the words of an anonymous author, commitment means "coming through—time after time, year after year after year." Commitment to leaders also means giving no excuses and accepting no excuses—just making it happen.

Barriers to Achieving Commitments

The instant a commitment is made, obstacles will surface. When people make a commitment, they must understand the price of failure and the rewards of success. They also need sufficient motivation and support if they are to succeed. Without the various consequences for failure and success, no one becomes committed.

Usually, when team members commit to an objective, they develop a plan to approach the achievement of the objective, but are not confident as to what the solution to the problem will be. As soon as the action is initiated, the problems and breakdowns begin. Team members must work through these problems, and must confront the fact that they may not know how to complete the action.

Keep in mind that in the attainment of an objective, mistakes should be viewed as a normal part of the learning process. In other words, people must have the freedom to make mistakes. Unfortunately, in most corporate cultures, mistakes aren't viewed this way. Mistakes are valuable when there is something to learn. Repetition of the same mistakes can also be valuable, as it indicates the individual doesn't care. Managers and individual contributors are expected to know exactly how to reach objectives without stumbling. The real point is confronting and overcoming obstacles while learning how to achieve objectives. The idea that action, even if it is initially misdirected, is better than inaction is best summed up by quoting Conrad Hilton:

> Success . . . seems to be connected with action. Successful men keep moving. They make mistakes, but they don't quit.

If the benefits of the successes outnumber and outweigh the mistakes, the individual or team performance is an asset.

Typical obstacles to attaining "committed to" objectives include:

1. *Conflicting goals.* If the company is in a cash-poor position, it is the wrong time to select projects that require extensive capital expenditures.

2. *Procrastination.* "We have until the next meeting to look at those control charts." There is no sense of urgency, or people tend to stay in their comfort zones.

3. *Lack of focus.* No priority has been set for any activities or projects. Instead a myriad of "emergencies" consume employees' time, preventing them from concentrating on improvement projects.

4. *Inflexibility and resistance to change.* "I don't understand this program; besides, it's just another management attempt to get blood out of a turnip." "Why should I change?"

5. *Excuses.* "Don't blame me." "It's not my job." "They won't let us do that."

6. *Excessive pride.* "The idea didn't come from engineering; it can't be much good."

7. *Self-defeatism.* "I don't have the authority to change this." "What can one person do?" "This will never work." "I'm really not very good at solving problems."

8. *Lack of enthusiasm.* This is usually caused by lack of involvement from top management and is easily fixed when the top players "walk like they talk."

9. *Not enough time to work on the project.* There is a perceived lack of time to devote to the goals. Many people claim they are too busy to attend training courses or participate in improving their process. This results from a subculture dedicated to sustaining the existing culture. They see no reason to change, because they are satisfied with the way things

are and would only be exposed to the unknown if things should change. The management is responsible for the organization's behavior, as they have allowed this behavior to persist.

Many companies whose organizations claim they are too busy are actually activity-oriented, rather than results-oriented. Only management can transform this behavior.

10. *Commitment misinterpreted as interest.* Employees believe that they will do it when it is convenient. "We'll improve when we have time."

Where Does Commitment Start?

Commitment is needed from the entire corporate structure. This concept is not easily embraced by most top executives. A natural tendency is to delegate the responsibility for commitment. If a CEO or president buys into TQM, he or she must commit to the entire chart, represented by Figure 4-1. This chart depicts the organization. The traditional organization is turned upside down. In lieu of the executives' traditional role of giving directions, their new role is one of giving support to the managers, work force, and ultimately the customers; likewise, the manager's role is to support the work force and customers and the work force to support the customers.

Organizations and companies cannot guarantee employment—only customers can. Therefore all the focus is on the customers.

The support of the various groups can only be accomplished by empowerment and process management.

The single most frequent cause of the failure to implement TQM is the lack of commitment by the CEO and other senior staff. Since our behavior

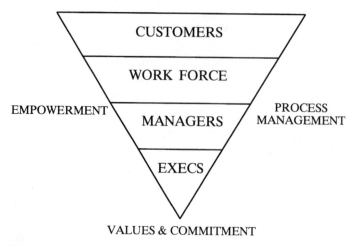

Figure 4-1.

is a clear indication of our commitment level, a leader lacking commitment to the success of the organization behaves in ways that indicate a lack of desire for long-term improvement. They have a domino effect on the organization. They are just going through the motions—they are not really involved. Characteristics of an uncommitted leader include the following:

- Personally solves daily problems and makes instant decisions
- Does not meet with immediate subordinates, either individually or as a group
- Is aloof, rational, critical, and "cold" to people and ideas for Total Quality Improvement
- Does not provide resources for TQM functions
- Pulls people from team meetings
- Ignores ongoing systematic problems
- Does not spend any time on TQM
- Does not reward or recognize TQM accomplishments or behaviors

Measuring Commitment to TQM

Commitment to TQM can and must be measured. If an organization commits to economic survival, then implementing TQM comes with the commitment. After all, companies measure other indicators of economic survival such as sales and net profit.

Meaningful measures of TQM need to address the following issues:

- Accomplishment of TQM issues vital to the company; must be quantifiable and are best expressed as Critical Business Success Factors
- Involvement (measured by action) of all levels of personnel
- Percent of time expended on TQM issues

The checksheet in Figure 4-2 provides examples of quantifiable measures of TQM. Notice that although it is impossible to measure the level of employee commitment, it is possible to measure the results of that commitment in terms of tangible results.

Required Management Commitments to the TQM Process

Measuring TQM-related activities will help spur the implementation and institutionalization of TQM. In addition, management must provide the

Use of Quantifiable Measures to Track TQM			
Does your organization track and report . . .	No	Yes	Not sure
Achievement of Critical Business Success Factors and Goals?			
Number of Total Quality Management (TQM) projects completed?			
Results for TQM projects quantified, i.e., dollars saved, cycle time reduction?			
Management attendance at resource committees?			
Number of quality-related standard operating procedures?			
Percentage of employees on TQM teams?			
Number of quality goals mutually established by managers/employees?			
Number of internal customer-supplier agreements?			
Number of TQM projects initiated at suggestion of employees?			
Percentage of quality solutions applying to multiple departments/functions?			
Hours of TQM training per employee?			
Measurements of values?			
Number of projects rewards?			

Figure 4-2.

leadership to transform TQM from simply record keeping (i.e., projects started versus projects completed) to an organizational commitment to excellence. Management's activities can be broken down into four arenas: permitting, supporting, managing, and leading by example.

1. *Permitting.* Each manager must permit people in the organization to attend TQM training. (Management should also attend TQM training.) These actions show the people that you support the TQM undertaking. People must be given the opportunity to be proactive and involved in TQM.

2. *Supporting.* All managers must give expectation to their personnel, including guidelines on how to accomplish the expectation and clear

boundaries of authority. Senior managers should introduce the TQM training sessions for the people in their organizations. Managers should establish a cultural-fostering committee to monitor the progress of the cultural as articulated by the vision, mission, and values statements. By monitoring the success of these efforts through quantifiable measures, TQM will be more easily attained. Senior managers must set up a resource committee(s) to monitor the teams' progress and provide resources to the teams. Finally, a training committee should be established to customize the training and hold managers accountable for the training of their personnel.

3. *Managing.* The manager should either chair a resource committee or be a mentor, thereby ensuring that projects are reviewed, barriers to success removed, and adequate resources provided. People must be held accountable for supporting quality goals, and their projects must be measurable. Next, the manager needs to establish a formal recognition and reward system for successful TQM efforts and observation of the values. (The recognition and reward system installs successful teams of individuals in a wall of fame and celebrates success. A formal recognition and reward system publicizes and spreads the TQM message to all departments within the organization.) Managers must provide regular feedback to all teams to promote TQM goals.

4. *Leading by example.* Executives and key managers must develop vision, mission, and values statements. These statements must be clearly communicated to all personnel. Vision, mission, and values statements become the rallying cry for the company. It is necessary to monitor the organization to see that the intent of these statements is being implemented. The manager must commit to becoming a champion of CBSFs and values, and ensure that TQM is used in all significant issues and decision-making processes to achieve organizational goals. By effective role modeling, the organization can be influenced to conform to the new values and models of behavior.

By providing leadership in the arenas of permitting, supporting, managing, and leading, TQM implementation and success will be assured. One way to evaluate how an organization's management is doing in the arenas of permitting, supporting, managing, and leading is to ask the following questions:

1. Are the company's major issues handled under the banner of TQM?
2. Are TQM projects assigned the highest priority? Are they related to CBSFs?

3. Is time made available to address TQM projects?

4. Do managers and senior management review TQM teams' progress? Regularly?

5. Are individuals held accountable for results?

6. Is recognition and reward linked to TQM performance and values?

Managing Commitments

Another word for commitment is *promise*. Promises are a vital part of the TQM process. Any promises made must be kept because they are a tangible demonstration of one's commitment to TQM. You, as a leader, will be constantly evaluated by how you fulfill your commitments. Figure 4-3 displays one possible consequence of not fulfilling commitments.

Impact of Commitments

■ Any organization is essentially a network of agreements.

■ When agreements are kept, the impact is positive and widespread.

■ When agreements are violated, the organization is weakened—violated commitments are like termites in the woodwork.

Figure 4-3.

When you are asked to promise to do something, you have four options.

1. *Accept.* This means you agree with all the conditions of the request and are willing to fulfill those requirements.

2. *Decline.* This will let the person requesting the promise know that you cannot meet all the requirements. It is better to state this up front than to break a promise at a later date.

3. *Counteroffer.* If you cannot meet the conditions of a promise, but still want to offer assistance, then this is your best course of action. By offering to meet the goal, but with altered conditions, the goal can still be attained. This is an excellent option as long as the altered conditions are agreeable to everyone.

4. *Promise to Promise.* If you can fulfill a promise, but not at this time, then offer to help in the future. This shows your commitment to the goal, and the fact that you are reliable, because you don't make false promises.

Few things are more disastrous for a TQM program than a rash of broken promises. Broken promises (read "broken commitments") are the antithesis of what TQM represents. The following displays the effect of kept versus broken commitments.

Effect of Commitments		
	Outcome	
Characteristic	Kept	Broken
External customer satisfaction	Improves	Declines
Company reputation	Asset	Liability
Internal customer satisfaction	Improves	Declines
Organizational integrity	Strengthened	Weakened
Personal integrity	Stronger	Weaker
Predictable outcomes	More	Less
Communications	Get better	Get worse
Morale	Higher	Lower
Productivity	Increases	Decreases
Teamwork	Improves	Declines
Trust	Higher	Lower

Commitment and Leadership

People, especially leaders, who consistently meet their commitments, have learned to anticipate problems and their potential impact. (They meet their commitments because of their ability to anticipate obstacles and take preventative action.) Individuals who find it difficult to meet commitments usually have not yet developed the skills of anticipation and proactivity. Lacking these important skills, they find themselves constantly reacting to problems in their environment. (See Figure 4-4.)

Leaders who fulfill their commitments believe it is in their own self-interest to do so. Fulfilling their commitments provides a personal sense of satisfaction. By fulfilling their commitments, they make a positive contribution to the success of the company as well as to their own career. When a leader is committed to the future survival of this company he or she consistently demonstrates a passion for improvement. Characteristics of committed leaders include:

- A constructive dissatisfaction with the status quo
- Articulation of the TQM philosophy
- Development of an inspiring vision of the future
- Translation of the vision into reality by concentrating on CBSFs (i.e., Internal Customer Satisfaction Agreements and values)

Earlier in this chapter we presented a series of characteristics of an uncommitted leader. Contrast those characteristics with the characteristics presented in the previous paragraph.

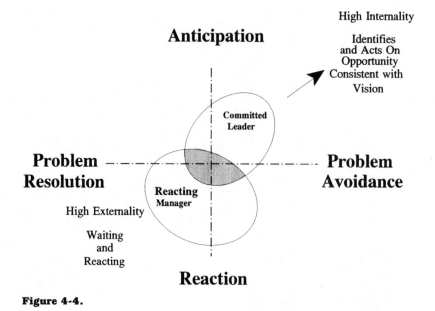

Figure 4-4.

Tangible Indicators
of Commitment

One indication of commitment is the realization of improvement of performances. From an observer's point of view, the best indication of commitment comes when individuals start repeating the fundamental concept that goes with the program: a personal commitment to continuous improvement. Examples of individual's commitments include: people consistently showing up on time for meetings; team members fulfilling their promises; and an intolerance of poor performance.

In addition to observing and judging the commitment of subordinates, leaders must be constantly aware that their actions provide a looking glass to their level of commitment. After all, we can judge a person's commitment only by their actions. The higher the level of commitment, the more obvious are the leader's actions. When a leader is committed to a positive vision of the future, he or she is involved in making it a reality every day on the job. Indications of commitment by leaders include:

- Remaining deeply involved at the very heart of things, spurring the actions necessary to carry out the vision via attending executive board and resource committee meetings

- Motivating employees to embrace the vision, becoming a cheerleader and celebrating success

- Constantly articulating the vision so that it permeates all organizational levels and functions, taking the organization to levels of performance that it has never reached before via meetings and one-on-one contact

Improving the Chances
for Achieving Commitment

To improve commitment, try planning for small wins. Establish "quick success" teams by setting one-month goals instead of six- or twelve-month goals. Celebrate all measurable improvement at every opportunity and publicize the program and its progress. Develop an atmosphere which encourages collaboration between individuals, teams, and departments. Remember to consider the human needs of the employees and continually evaluate the general feeling about the program.

Commitments and Cultural
Change

Quite often management will recognize the need to initiate improvement programs. They can appoint trainers to provide instruction and suggest ways in which behaviors could be aligned with values. Such attempts will fail, however, if there aren't management and people within the organization committed to sustaining and expanding the proper behavior. This is because, once the trainers leave, the incentive to practice the new behavior disappears and the employees will go back to the old, more familiar behavior. The next section of this chapter deals with achieving cultural change within an organization. As you read the following sections, keep in mind that obtaining commitment to the process of continual improvement is the key to implementing and sustaining cultural change.

Creating Cultural Change

Every company has a culture. It is made up of the ideas, customs, and skills possessed by its employees. This culture is passed on to new employees as they are brought into the company, and the culture remains basically unchanged for the life of the company. An organization's culture is determined by a variety of elements such as:

- *Business environment.* This covers the relation between the company and its marketplace, which in turn depends on what it sells, its competitors, threats of substitutes, technologies, and government relations.

- *Values.* These are the basic concepts and beliefs of an organization; as such they form the heart of the corporate culture. Values define "success" in concrete terms for employees—"if you do this, you too will be a success"—and establish standards of achievement within the organization.

  Strong culture companies openly speak of their values, and do not tolerate deviance from those values and company standards.
  Weak culture companies have values which are subculture issues, not spoken about or defended openly, and rarely seen in print.

- *Heroes.* These people personify the company's cultural values and as such provide tangible role models for employees. Some heroes are born, some are made by a few memorable moments, and some are the result of the day-to-day occurrences of leadership, respect, and trust.

- *Rites and rituals.* These are the systematic programmed routines of daily life in the company. In their mundane manifestations, which we call *rituals,* they show employees the kind of behavior that is expected. They are visible and potent examples of what the company stands for, what it takes to "belong," and how to be successful.

- *The cultural network.* This is the primary, yet informal, means of communication within an organization. The cultural network is the "carrier" of the corporate values and heroic mythology. Working the network effectively is the only way to get things done, or to understand what's "really going on" in the organization.

A strong culture is a system of informal rules that spells out how people are to behave most of the time. By knowing what is expected of them, employees will waste little time in deciding how to act in a given situation.

The Process of Cultural Change

The process of changing an organization's culture involves a series of steps. Before one can begin the process of cultural change, it is important to recognize that there are certain preconditions which must exist if cultural change is to take place. The six essential preconditions are:

1. There must be a compelling reason to change—such as economic survival. Executive management must forecast the future.

2. There must be a well-defined, implemented plan which indicates what to change, i.e., specific actions, responsibilities, target dates, and the benefits to be realized by the change. (The plan, along with the rationale behind the plan, must be communicated to all employees.) Top management must be committed to the vision or desired values in both words and actions.

3. There must be regular measurement of the desired values being promoted. Whenever reality does not meet the plan, a recovery plan must be developed.

4. Employees must be given expectations and trained in the new values, behaviors, and skills desired.

5. The organization must establish recognition, rewards, incentives, and promotions to encourage behavior compatible with the values and beliefs being promoted. Behaviors incompatible with the values must not be tolerated and be dealt with appropriately.

As mentioned earlier in this section, implementing cultural change is a process which can take place if certain preconditions exist. Assuming that the preconditions exist, the process of cultural change is depicted in Figure 4-5.

Defining the Present Culture

The process of changing a culture starts by identifying gaps between the existing culture state and the desired culture state. Management must determine how the employees perceive the existing corporate culture,

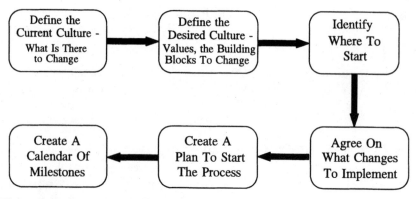

Figure 4-5. Creating cultural change.

since perceptions shape people's behavior. (For example, employees will see individuals rewarded for working toward the company's goals; they will then make assumptions about the reasons for the reward and behave in a manner perceived as accepted and desired by the organization.) The values of the existing environment must be identified.

Defining the Desired Culture

When you are the customer, and think of expected quality, certain companies come to mind for their reputation of quality, service, or accomplishments that other companies try to emulate.

Which one or two companies would you like your organization to be like? These do not have to be companies that are in your industry. Choose the company or companies that represent the qualities that you would most like to have in your organization. (If you choose your own company as one of your choices, select another company as well.)

You cannot hit the targeted goal without a plan for getting there. That plan is designed to close the gap between the current culture and the desired culture.

Once the organization has identified what needs to be changed, it is important to clearly define and articulate the values to the organization. The building blocks of culture are the values which prevail within an organization.

People seldom think about values, but values are critical in helping to manage change. Values, and people's belief in them, create attitudes that influence behavior. Value statements reflect management's intentions; however, they do not create a change in behavior on their own. This can be achieved only by marrying "managerial behaviors" to specific value statements, and this does not happen by accident. Managers and employees require constant coaching and positive reinforcement to ensure that they understand and practice the established values. In turn, these values will produce the desired behaviors.

Defining the Desired Culture: Changes Required for a TQM Organization. Using the typical organization as a starting point, the cultural changes most important to TQM implementation are:

From	To
Bottom-line emphasis	"Quality first" equal in importance to delivery and cost
Short-term objectives	Long-term view as well as short-term objective
"Limits" satisfaction	Continuous improvement by achieving stepped goals
Product delivery	Customer satisfaction, on-time delivery and 100 percent fill rate of orders
Focus on product	Focus on process and its input variables; if the process is correct, the product will be correct
Quality-delegated responsibility	Management-led improvement process
Inspection orientation	Prevention orientation
Sequential engineering	Concurrent engineering; all departments functioning parallel instead of in a series
Minimum-price suppliers	Customers partnering with suppliers, focusing on lowest costs, not lowest price
Compartmentalized activities	Cooperative team efforts
Management by edict	Stakeholders, i.e., employee participation, focus on common goals

From this brief list, it becomes more obvious that management commitment is the key to changing the company's culture.

Emphasis on quarterly bottom-line (profit or revenue) results is one of the first "old dogs" that must be modified; however, a plan to achieve the desired profits over the long term is mandatory. This cultural element is not only common to individual companies, but also to entire industries in the United States. Quality and long-term profitability have been historically associated with industry leaders in all areas. This isn't to imply that profit should be ignored in favor of quality. There isn't any reason to be in business if the company isn't making a profit. But, when long-term continuous improvement is integrated with customer satisfaction, product development, production, and marketing, then quarterly profits will exceed expectations. Conversely, shipping defective or suspect product just to "make revenue" for the quarter will negatively impact profit when the product is returned from the customer. TQM is a tool for optimizing profitability, not just something that the quality department uses to compile statistics.

Emphasis must be refocused on shipping acceptable (defect-free) product and sustaining a competitive edge by minimizing the cycle time from order entry through customer receipt of service or product.

Service or product delivery is often viewed to be more important than any quality or customer issues. However, quality and delivery must work together to deliver exactly what the customer expects on time, every time, which means that the two go hand in hand. One senior executive explained the key to survival for his company was elimination of mistakes through the product life cycle.

Focus on product or corrective action says that there is only time to fix defects after the fact, or that the company is willing to sort through piles of nonconforming product to find the acceptable parts for shipment. The culture has to change so that employees are constantly looking for ways, through motivation, suggestion plans, ownership, and recognition, to improve the process and prevent the processing of defective product or the perception of bad service. The same argument holds for inspection-oriented processes that rely on product appraisal to find defects. Unfortunately, no inspection is 100 percent effective, and the customer will ultimately find the problems that an inspection department misses. The real key to success is a prevention-oriented system that removes the possibility for recurring defects.

Sequential engineering, the traditional mode for most manufacturing companies, needs to be replaced by concurrent engineering. This would enable customer requirements to be integrated into the product/process design early in the development cycle. Quality partner suppliers should be nurtured instead of relying on minimum cost suppliers. Departmental barriers should be broken down in favor of cooperative team efforts.

Management by edict works in the short term but will often result in "malicious obedience," whereby employees carry out management directives with full knowledge that there is a better, cheaper solution. As knowledge of the TQM process grows within the organization, it will be possible to identify more specific changes in culture that would enhance the program. Cultural change must be continuous, just as process and quality improvement is continuous. Management should always seek out ways to improve the company, and by this example all employees will soon adopt this same attitude.

One company developed a cultural characteristic of shipping product from the plant to the distribution warehouse at the end of the quarter, regardless of quality. The plant management was routinely rewarded and praised for making scheduled production quantities. Defective product found at the warehouse was then reworked by a large force of temporary employees. Management seemed oblivious to the unnecessary cost incurred in this operation. In fact, the manager in charge of rework was promoted to vice president. The cultural change made was to install accountability as a value. Hence, rework costs were reduced by charging all rework costs back to the plant that was the source of the nonconforming product. This changed the attitudes of the plant's personnel, and turned out to be a powerful quality improvement tool.

Closing the Culture Gap

The cultural gap is closed using three approaches. One is to change the culture by implementing the desired values. The second approach is to align the organization by developing Internal Customer Satisfaction Agreements and measuring the success of internal customer satisfaction. The third approach is to launch project teams designed to influence the performance of the CBSFs.

The favorable outcomes of each of the approaches must be recognized and rewarded to sustain the change. The recognition and reward committee plays a key role, in that it ensures that the favorable outcomes are recognized by management. A discussion of the three approaches follows.

The identification of the existing and desired values by the executives occurs by consensus management during strategic planning sessions. The executives select the existing values to be retained and the desired values to be acquired. They also define the company's definition of the values. During a creating-cultural-change session, an implementation plan for each value is developed with a champion identified. This champion is responsible for developing measurements for the value, creating a questionnaire, and executing the implementation plan for the values.

Each person's role, both executive and workforce, is to embrace the values through actions. This "walking the talk" acts like a turbocharger and causes the values to be contagious.

The cultural-fostering committee then recognizes the individuals who are performing the desired behaviors.

While the aforementioned thrust is taking place, two other forces are under way. One is the Internal Customer Satisfaction Agreements activity between departments whose purpose is to align the departments in becoming customer-driven, both internal and external.

The other force is the launching of teams to support the favorable performance of the Critical Business Success Factors.

The activities of all three thrusts are to be monitored by the executive board and resource committees, as appropriate. The results are to be recognized, rewarded, and celebrated. The recognition and reward committee role is to ensure management recognition to sustain the culture change.

To create an implementation plan for each value in definition, actions, responsibility, target dates, and follow-up dates:

- Identify where to start.
- Agree on what changes to implement.
- Create a plan.
- Start the process.
- Create a calendar of milestones.

After the values have been established as part of "defining the desired culture," the organization must begin the task of closing the gap between the existing culture and the desired culture. The steps involved in closing the gap are the final four steps depicted in Figure 4-5. When planning a change, it is best to identify the areas of values that are currently satisfactory and build on them. The gap exists when there are values which need to be addressed. Figure 4-6 depicts what has to happen to close the culture gap.

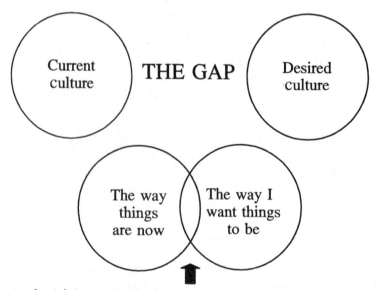

The overlap is bringing into the future what worked in the past. At the same time, it involves implementing new values.

Figure 4-6.

**Resistance to Cultural Change
and Desired Behaviors**

Although it is true that creating a cultural change involves a process, don't be fooled into thinking that it is an easy process. Most people have a natural fear of change, as they wish to stay in their "comfort zone." They will not want to be the first to adopt new standards of behavior, and change will require repeated education and positive reinforcement of desired behavior. Management must align employee perceptions of acceptable behaviors with the vision, mission, and values statements to achieve cultural change. Management's role in establishing perceptions is paramount in creating a cultural change.

As the chart in Figure 4-7 indicates, the difficulty of effecting a change is proportional to the amount of time needed to implement the change. Different types of changes require different amounts of time to affect. Knowledge is the easiest change to attain and can be acquired in a short time. Attitudes will be more difficult to change and will require more time. Individual behaviors will be even more difficult and require more time. It has

Change in an Organization

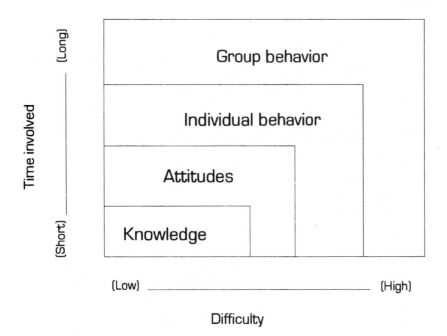

Figure 4-7.

been said that it takes at least 21 days to internalize a positive behavior. Group behavior is the most difficult to change and takes the longest of the change types indicated. The key elements to effective change are role modeling by the leader, an urgent need to change, and consequences if the change doesn't occur.

The types of changes sought in TQM implementation are sometimes lumped together by general buzz words, but these individual changes should be understood and integrated into all daily activities. Individuals and teams must become "proactive," anticipating problems and solving them before they occur. Most people in the typical organization are just the opposite, reacting to problems as they arise instead of planning for improvement. The simple truth is that if the corporate culture didn't support and favor the reactive mode of management, the reactive mode would be quickly replaced.

Participants in a TQM undertaking must become *results-oriented.* They must measure their performance and achieve their goals. Monitor the indicators that lead to success and stress the importance of meeting stated objectives on time. The test of commitment is that everyone, from the top down, should be involved; giving the Critical Business Success Factors priority over all other activities; spending time trying to improve every aspect of the business via internal customer satisfaction; and playing the roles dictated by a true desire for improvement.

A feeling of commitment and involvement should begin to grow out of the training sessions and team meetings. TQM cannot be viewed as a program that is being carried out in some other division or by some other department. It must be communicated that involvement and commitment are important and expected. The time for cultural change to take place, and the degree of change attainable, is determined by effort, the size of the gap, and the time taken to address that change.

Teams can eventually become self-directed, if management chooses, as they work through one successful project and move on to another. The TQM implementation process is a management training and development tool. Self-managed teams and their management need to be trained in a customized program to meet specific desires and requirements. Strong, assertive team leaders often lend their personality to the team, so that a group will tend to work above their personal levels of performance and achieve a group synergy. At the same time, pessimistic or intractable individuals, even those with strong technical skills, will drag down team progress. These people should be quickly identified and counseled. Something as simple as arriving at productivity/quality-improvement meetings on time, with assignments completed, should become part of the team culture and eventually the company culture.

Successful productivity/quality-improvement teams will serve as models for the other employees. Even with training in all of the recommended areas, people will generally attend their first few team meetings with little idea as to their purpose or what is expected. Groups that have been through the process should be encouraged to share their experience, including mistakes, with newly formed teams. There is a great temptation to continue on with the same approach to business, and the same projects, under the guise of a productivity/quality-improvement team. If this happens, resources may be applied in a manner inconsistent with the vision, mission, and values statements.

Techniques to Foster the Cultural Change

The organization fosters cultural change by implementing organizational values and recognizing the behaviors of the values. Failure to measure the implementation of values is like playing tennis without a net: it provides a lot of exercise, but few results. Lack of results in implementing values equals little or no cultural change.

Many times individuals balk at the idea of measuring the extent to which values are practiced. Their primary reason for this reluctance is that it seems impossible to measure values; furthermore, if a measurement method were to be established, many fear it would be arbitrary. Let's look at how a value can be measured within an organization.

Assume that a desired value is mutual trust, respect, and integrity. In order to instill this behavior in the organization, it would be necessary to define what is desired; define how to recognize the desired behavior; then determine how to instill it. The following example shows the steps involved in defining, identifying, instilling, and measuring a value. In addition, the following plan provides specific recommendations for management regarding implementing and measuring the value of mutual trust, respect, and integrity. Implementation plans for the other desired values need to be developed and executed.

Value: Mutual Trust, Respect, and Integrity

What is desired?

- Actions based upon commitments.
- Treating others as equals.
- Mutual trust. To have the same confidence in each other; to believe in each others' intentions.
- Respect. To have regard for and appreciation of worth; to treat with consideration.
- Integrity. Uprightness of character; honesty.

How to recognize the desired behavior

- Employees who are proud to work at the company.
- Employees who believe in the company's vision, mission, and values.
- Employees who are aware of their departmental charters.
- Employees who readily assist those who require information.
- Employees who consistently meet or exceed their internal customers' expectations.
- Employees who constantly strive to refine their skills.
- Employees who submit suggestions for improvement.
- Employees who are proactive.

How to instill the desired behavior

- Implement employee suggestion box.
- Listen and respond to all suggestions.
- Base decisions on facts, not feelings.
- Answer with honesty.
- Explain reasons/benefits for assignments.
- Develop, incorporate, and communicate position summaries.
- Treat all employees as equals.
- Create guidelines to demonstrate uniformity among employees.

How to measure the behavior

- Check increased employee involvement.
  (number of suggestions)
  (number of different people submitting suggestions)
  (number of suggestions implemented)
- Perform audits on knowledge of departmental charter.
  (percent of employees with knowledge of charters)
- Perform audits on knowledge of position summary.
  (percent of employees knowing job descriptions)
  (percent of position summaries implemented)
- Review employee general survey on management performance.
  (number of positive comments compared to previous survey)
- Track implementation of company guidelines.
  (number of guidelines implemented by date scheduled)

- Analyze outcome of meetings at all levels.
  (number of items on agenda completed as scheduled)
  (number of disruptions)
- Trace the amount of formal recognition made public.
- Check the number of awards given or recognition for value-enforcing role behavior.

What must management do to instill the value?

- Manage with facts only.
- Be punctual, keep appointments.
- Keep records and return calls and feedback when promised.
- Don't get involved in gossip.
- Correct other management people on their negative behavior.
- Give feedback on survey results within one week of collecting data.
- Critique meetings; improve content and process each session.

How to measure management's behavior?

- Chart meetings' performance leaders and participants.
- Make promises with due dates; chart results of keeping word.
- Fine people who are late to meetings.
- Issue daily "to-do" lists, with completed items listed.
- Make weekly plans for accomplishments, review last week, outlook future performance to avoid surprises.
- Keep weekly record of people acknowledged and rewarded for accomplishment; review the list with management.

Other Techniques to Foster Cultural Change.   It should come as no surprise that there are a variety of techniques which an organization can use to foster cultural change. Since there are a variety of techniques, some are more useful than others. The following table provides a list of other cultural change techniques; they are arranged in descending order from the most significant or useful techniques to the techniques of some or marginal usefulness.

Degree of importance	Technique
Very great	Display top management commitment and support for values and beliefs.
	Train employees to convey and develop skills related to values and beliefs.

Great	Develop a statement of values and beliefs. Communicate values and beliefs to employees. Use a management style compatible with values and beliefs. Offer rewards, incentives, and promotions to encourage behavior compatible with values and beliefs. Convey and support values and beliefs at organizational gatherings. Make the organization's structure compatible with values and beliefs. Set up systems, procedures, and processes compatible with values and beliefs. Measure and report results.
Moderate	Replace or change responsibilities of employees who do not support desired values and beliefs. Use stories, legends, or myths to convey values and beliefs. Make heroes or heroines of exemplars of values and beliefs.
Some	Recruit employees who possess or will readily accept values and beliefs. Use slogans to symbolize values and beliefs. Assign a manager or group primary responsibility for efforts to change or perpetuate culture.

One pitfall which one must avoid when discussing the implementation of cultural change is the distinction between attitude and behavior. Behavior is what we say and do. Attitude is what we feel and believe. Values drive attitudes, which drive behavior, in turn driving performance.

Behavior is what first evokes a response in others, but behaviors come from attitudes. So it is important to work on attitudes first. When people are clear on their attitudes it will show in their behavior. Changing behavior will be realized in a change in performance. Remember, role modeling is a very powerful tool for changing behavior. People like to be recognized for doing things right and for doing the right things. This can create a role model. "Catch a person doing something right."*

Tip   Negative reinforcement, telling someone what they should not do, cannot elicit desired behavior; it can only discourage unacceptable behavior.

Tip   No consequences, either positive or negative, can aid the behavior to become nonexistent. This will happen with both desirable and undesirable behaviors.

* Ken Blanchard in *One Minute Manager.*

As you have probably figured out, there is no cookbook approach to implementing cultural change. One tool which can be used to develop a plan and then monitor adherence to the plan is the following list of concrete steps which an organization can implement to foster cultural change.

- Develop guidelines for a culture-fostering team
- Select team leaders with good communication skills who are supportive of management but willing to speak their minds.
- Rotate the team leaders every six months.
- Select team members from a wide sector of the organization.
- Create and publish team charters, i.e., expectations, guidelines, authority, and resources.
- Review values and develop a questionnaire for each value, focusing on the positive examples which represent that value.
- Set a schedule for observations of the culture to see if the values are being adhered to.
- Keep score—actual versus expectations.
- Cite positive examples and role models whenever possible through the use of recognition boards or company newspapers. Don't use negative examples.
- Have culture-fostering team members at meetings to cite positive examples.
- Make it a game for anyone to cite a person for positive value reinforcing behaviors.
- Choose suitable recognition for people who are role models of positive value behavior.
- Make periodic reports (present the score) to the management via the steering or resource committee.

Crucial Management Responsibilities to Effect Cultural Change

In order to effect a cultural change, there are seven crucial responsibilities of management.

1. Assess and understand the existing cultural.
2. Determine what culture is desired. This can be ascertained by identifying the desired values in a decision-making meeting.

3. Develop implementation plans to transform the visions of the desired values to reality.

4. Become role models.

5. Provide expectations to all employees and empower them.

6. Create a culture-fostering group to monitor the progress of values and the feedback process.

7. Provide adequate recognition and reward programs.

The behavioral changes necessary to improve productivity and quality through process control cannot be made overnight. Management is faced with changing both individual and group behavior—two slightly different tasks requiring training and new attitudes. It is possible to make changes overnight through shifts in paradigms, but the more common mechanism for change is through gradual ongoing feedback, analysis, and continuous improvement.

Management's first responsibility in effecting cultural change is to assess and understand the culture and paradigms that presently exist in the organization. Insights to these value sets can be gained through:

1. Obtaining assessment by an outside organization that will evaluate the organization's dynamics as well as operational issues, customer satisfaction perception(s), strategic direction, quality policy and practices, and marketing and sales aspects.

2. Surveys of employees to characterize their perceptions of the factors that define corporate culture. Formally and informally, TQM practitioners need to get out and talk to all the employees to see what is working and what needs help.

3. Interviews with team members centered on perceived barriers to reaching objectives. Once the team accepts the commitment to reach an objective, the barriers will overshadow the planning process. Teams will need support in sorting out and removing these obstacles.

4. Observations during training classes or problem-solving sessions. What myths are being brought up to maintain the status quo? The trainer can help amplify thoughts expressed in class and suggest ways to apply the lessons to actual work situations.

5. Analysis of improvement programs that have failed. (Be honest!) Conduct "what works and what doesn't work" in the existing organization.

6. Evaluation of the company atmosphere. Does work flow well? Is there spirit and teamwork? Do people participate and are they motivated?

7. Analysis of cultural effects that will support or prevent achievement of the vision, mission, and values statements. Are restraining forces balanced by supporting forces? Where is reinforcement required?

8. Discussions with consultants who have had the opportunity to observe other business systems, and who will more easily identify cultural characteristics. Organizations may feel that internal personnel should be able to drive the program, but they will seldom have the time, training, experience, or support available from consultants specializing in TQM training and implementation.

Management's second responsibility is to determine the future culture of the company. This is done by a series of surveys and exercises which identify what values are required to make the company successful. These values, when achieved, will favorably influence the Critical Business Success Factors which were determined in the tactical planning session.

Management's third responsibility is to take action through training and implementation. This will change the attitudes and behaviors of the organization, which in turn will reshape the culture. The most direct approach is to provide new and/or additional knowledge through training. Then develop implementation plans that will make reality of the desired values and reinforce the new knowledge. For example, if people are taught how to develop control charts, then senior management should be asking questions about the charts that demonstrate understanding and commitment at all levels of the organization. Employees can usually accommodate a perceived conflict of values but may fail to be committed to the goals and aspirations of the organization in the long term. They will begin to see the role they play in the company as a means to an end and at this stage their commitment will cease. This plan must be monitored on a monthly basis.

Management's fourth responsibility is to be a role model. This is the easiest and most effective approach. People do not listen to what management *says*—they listen to what management *does*.

An example of a role model is Ralph White, vice president and general manager, of IMO WIGGINS, an aerospace fuel connector manufacturer. Ralph attended all the training sessions, was involved in the resource committee, placed all major issues under the umbrella of TQM, created time in his busy schedule for TQM, and maintained TQM as the company's number one priority. As a result, the company remained extremely profitable in a business environment that became very competitive.

Management's fifth responsibility is to give expectations. This is actually accomplished through the means of empowerment.

Management must give direct reports—clear expectations of the behavior they expect from employees, guidelines on how to achieve them, limits of their authority, adequate resources, and assurance that personnel have learned the requisite skills.

One of the pitfalls that managers should anticipate is that employees may not change. This is an especially difficult problem if the employees have been with the company for years and are loyal and trustworthy.

In this case, the definition of "loyal and trustworthy" must be changed. Loyalty is years of committed contributions that meet expectations. Trustworthiness is measured not only as personal character but as technical competence which meets expectations.

Once these new definitions of values are understood and accepted, the process of changing attitudes is facilitated. This process has a proven track record.

Til Layne, vice president of Cipher Data, had a situation where the engineers and supervisors thought that implementing SPC with event logs, corrective/preventative-action matrixes, process control procedures, as well as process control, was too much work. Til emphasized to them the expectations, gave them guidelines and limits of authority, and reassured them that they had adequate resources. He also advised them about the requirement of IBM Rochester (a Malcolm Baldrige National Quality Award winner two years later) and emphasized that if he didn't have their support they could peddle their talents to Cipher's competition.

Six months after the plant implemented SPC, it passed IBM's production verification test on its new product, and 14 executives from IBM held a plantwide celebration for Cipher Data.

Til was a believer, and his belief paid off.

Sustaining Cultural Change

When gaps have been minimized between the "existing culture state" and the "desired culture state," it is necessary to hold on to the positive changes that have occurred. Company culture must be monitored by regular management review. It may be that changing business conditions require new modifications to the corporate culture. Like the rest of TQM, the improvement process is ongoing. The following list is a series of 10 suggestions on how to sustain the desired changes once they have been achieved.

1. Maintain a sense of urgency to sustain the desired changes. (People should smell the smoke, but make sure they don't panic and run out.) Everyone has a stake in the company's success.

2. Provide a clear-cut vision of where it is you are headed and what is expected.

3. Remove rewards that promote behavior opposite to that which you want to promote. This may sound unreasonable, but these situations occasionally exist.

4. Show the people what you want by *example* and *message.*

5. Allow them to do what it is you want them to do.

6. Reinforce the desired behaviors. Recognize and reward them.

7. Expect breakdowns, and provide assistance to put things back in order.

8. Encourage people to take risks and keep trying. (Follow the *147/805 Rule;* that is, Edison tried *147* times before he perfected the light bulb, and the Wright Brothers tried *805* times before they achieved sustained flight.)

9. Reward success. Train people how to handle mistakes.

10. Continually express your commitment to the success of the people doing the work and to the job getting done.

Empowering the Workforce

When trying to communicate new concepts in business, it is often advantageous to adopt a phrase or buzz word that expresses the intended idea, stimulates creative thinking, and is easy to remember. These phrases can become battle cries in the war against lost productivity.

One such word is "empowerment." Empowerment means that all employees feel they have the responsibility and authority to participate in decision making and problem solving in their appropriate operating levels. It's obvious that a whole company of skilled and capable problem solvers will have a distinct competitive advantage over an organization with only a few key contributors and an army of drones.

Empowerment is a word coined in 1849 to refer to "the gaining of power." In the context of TQM, empower means to enable, to endow, to give permission to, or to give the ability of power to. The opposite of empowerment is impotence.

Empowerment is the authority to act independently to meet expectations. This authority is given by management for the purpose of developing a human connection with the decision-making process, which sustains improvements through the TQM program.

Exploring the concept of empowerment includes defining the reasons for management to empower the workers, needs analysis, and techniques for empowerment. When management is comfortable with the justification for an empowerment program, the company can begin to determine strategies and establish an implementation plan.

An early step in implementing empowerment is determining what will motivate people and how that motivation can be used to further the empowerment process. Barriers of insecurity and parochialism must be removed so that empowerment and change can take place. A conscious effort must be made to create an environment that encourages people to take what would normally be considered personal or professional risks. The employees should feel free to suggest solutions and act on problems they encounter.

Most people make the mistake of thinking they will feel good about empowerment. The truth is that it may involve personal sacrifice or risk taking, and many people may be uncomfortable in that position. In one example, George Andrews, president of Cherry Textron, became frustrated with the teams' reluctance to take action and offered each team the authority to spend $5000 on their project without management approval. They were "empowered" to allocate the most coveted resource, cash, and to spend it as they saw fit. Very little of the money was ever spent, revealing that it may be very difficult for people to accept empowerment until they are comfortable with risk taking and making changes. Once effort was made to make them comfortable, the monies (and even larger amounts) were spent, and the results saved millions of dollars and enabled production to ship significantly more product without additional workers or equipment.

Empowerment also requires management to take risks by turning over some control of the organization to the employees! However, this control must be planned and authorized.

Management must appreciate that relinquishing control to the effective and skilled employees will result in a more productive organization with better teamwork and faster problem resolution. An empowered organization will be able to respond more quickly to changes, improvements, and new customer requirements. This will occur as they begin to act independently in pursuit of their expectations and within the boundaries of their authority.

Management also makes a mistake in fearing that empowerment will force them to make open-ended promises with unknown consequences. They are also afraid that they have abdicated the right of vetoing bad, irrelevant, or counterproductive ideas. Managers must realize that they do not give up the responsibility to make final decisions. Neither do they give up the ability of coaching teams when appropriate. The basic concept of empowerment is management's promise to employees that they will be supported and rewarded for taking action and finding new ways to contribute. What is important is that management establishes the expectations, sets the guidelines and defines the boundaries of authority. It is not a process of giving up their responsibility.

Why Empower?

Empowerment is a crucial part of cultural change that brings the decision-making process down to the point where problems are most visible. Team members and people on the production line need to be able to take action when problems are observed, and not wait for higher-level direction.

Specifically, the primary benefits of empowering the workforce are as follows:

- It promotes employee productivity, involvement, and ownership. Employees will take responsibility for their work and become managers of the specific areas where they have authority. An employee can contribute as an individual or as part of a group or department.

- It opens more avenues of change in the business culture and improves the company's operation.

- It fosters open communication at all levels. Often managers mistakenly assume that everything they say is completely understood and faithfully implemented. Employees also often make the mistake of believing that management is aware of problems, but is consciously choosing not to correct them.

- It makes best use of the employees' talents and knowledge, and identifies areas where additional training or coaching are needed.

- It creates *careers* over *jobs* by enhancing the feeling of reward and involvement, and establishes a basis for trust. When employees feel more confident that their ideas and actions will be supported, they will begin to view work as an extension of their own personality.

- It produces action over reaction and results over excuses. Every company should adopt a "no excuses, no scapegoats, and no whiners" policy. Fix the problems by assigning cause, not blame. Turn complaints into requests for action and solutions.

- Finally, it builds the opportunity for continuing improvement by installing a network of problem solvers throughout the organization.

When empowerment begins to be effective, the company will realize the benefits of synergy. A single individual has limited knowledge and abilities. But when several individuals pool their skills, their productivity increases manyfold. When the whole company is empowered and sharing the same objectives, the result can be remarkable, and performance levels will far exceed those demonstrated by individuals. Empowerment will become a force for success that cannot be achieved in any other way.

Failing to empower the workforce will cause a division of perceptions, motivations, and priorities that will detract from the company's ability to achieve vision, mission, and values.

Empowerment versus Other Techniques

To be able to empower, managers must understand the differences between motivation, domination, and manipulation. Most people become managers because of their skill as individual contributors. Those who continue to grow are those who know how to delegate. *Motivation* encourages people to take action in concert with the vision, mission, and values state-ments. *Domination* implies influence or control by superior power or authority. *Manipulation* means to manage shrewdly or deviously without open communication. It should be obvious that people will be most pro-ductive when motivation to do their job is based on fulfillment of their own needs, rather than on intimidation.

Problems Associated with Empowerment

Empowering the employee isn't as simple as issuing an interdepartmental memo saying "all of you are now empowered." Not everyone will see this as a positive change. Some will regard the move toward empowerment with suspicion, mistrust, and insecurities. Managers will have to call on available motivational tools and skills to win employee support and coop-eration.

There are a number of pitfalls to empowering the workforce, including:

- Using empowerment without understanding; usually management is not trained in leadership and empowerment.

- Viewing empowerment as an event, not a process.

- Expecting overnight success. It takes time to build a history of success, but eventually success will become part of the company's culture. The company must stay focused and continually reinforce the modifications in behavior that allow employees to trust the system.

- Limiting information and communication about what is wanted and needed, or having unclear expectations and guidelines.

- Lack of measuring, monitoring, feedback, or appropriate consequences.

- Empowering individuals without the skills to do the job.

- Viewing empowerment as a tool, technique, or strategy instead of as a new corporate culture and mindset.

- Not asking people "what is" and "what is not" working.

- Not obtaining acknowledgment of responsibilities, i.e., understanding expectations, guidelines, authority, and resources.

- Speaking instead of listening.

How to Empower

Empowerment is a process, not an event. It is a triad of three acts: the act of empowerment, the act of process management, and the act of taking responsibility. (See Figure 4-8.) Each segment of the triad must take place effectively or empowerment cannot occur. Empowerment is like a three-legged stool; remove one leg, and the stool will fall.

Act of Empowerment

- *Expectations.* Give clear objectives, goals, deadlines, or criteria to be achieved.

- *Guidelines*

  Provide policies, procedures, or checkpoints to be adhered to.
  Furnish sources for interface, data, inputs, etc.

- *Authority.* Specify the levels or boundaries of decisions.

- *Resources.* Ensure adequate human resources, tools, information, or whatever is necessary to achieve the expectations.

- *Skills.* "How to do" the various techniques, methods, etc., to achieve the expectations.

Act of Process Management

- *Measurement.* Determine the means to measure performance.

THE EMPOWERMENT PROCESS

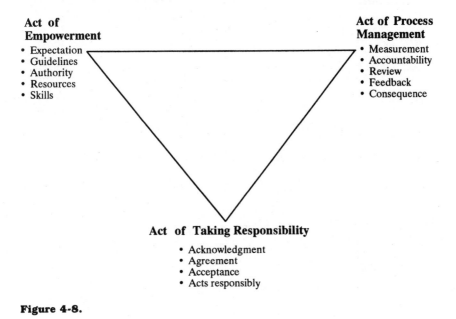

Act of
Empowerment
- Expectation
- Guidelines
- Authority
- Resources
- Skills

Act of Process
Management
- Measurement
- Accountability
- Review
- Feedback
- Consequence

Act of Taking Responsibility
- Acknowledgment
- Agreement
- Acceptance
- Acts responsibly

Figure 4-8.

- *Accountability.* Hold individuals and teams accountable for their performance.

- *Review.* Review progress or status on a periodic basis.

- *Feedback.* Provide positive as well as negative communications regarding performance.

- *Consequence.*

  Recognize and reward positive performance and behavior.

  Take corrective action measures for negative or null performances and behaviors.

An Illustration of Empowerment

A president directs the VP of sales to obtain Price Club as a customer. After six months, the VP of sales did not gain Price Club as a customer. At another meeting six months later, the topic of the Price Club arose. The president claimed he had empowered the VP of sales. When the president reviewed the chart in Figure 4-8, he jumped up and claimed, "No way did I empower the VP of sales." He did a critique of himself.

In Regard to the Act of Empowerment

Q: *Did the president give clear expectations?*

A: Yes and no. *Yes* to the criteria for success. *No* to establishing a deadline.

Q: *Did he give guidelines?*

A: No.

Q: *Did he give the authority to the VP of sales to price the product to get the order?*

A: No.

Q: *Did he give the VP of sales the resources to accomplish the expectations?*

A: Yes, the allocation of time to the VP of sales.

Q: *Did the VP of sales have the skills to accomplish the expectations?*

A: Yes.

In Regard to the Act of Process Management

Q: *Did the president have a means to measure performance of the expectations?*

A: Yes, the dollar amount of the purchase order.

Q: Did he hold the VP of sales accountable for performance?

A: No.

Q: Did he review the progress of his expectations?

A: No, he never even thought about the assignment again.

Q: Did he give feedback to the VP of sales?

A: No, he never reviewed the progress.

Q: Did he provide the consequences, positive or negative?

A: No, as there was no accountability, review, or feedback.

In Regard to the VP of Sales Taking Responsibility

Q: Did the VP of sales acknowledge his understanding of expectations, guidelines, and authority?

A: No. This would have been clearly communicated to the president if he had asked.

Q: Was there agreement on the adequacy of skills, resources, and measurement?

A: No, there was no discussion.

Q: Was there acceptance of ownership and responsibility?

A: No, there was no discussion.

Q: Did the VP of sales respond with his total ability to achieve the expectations?

A: No. He did not have the full expectations, guidelines, and authority.

Q: Did the VP of sales have the skills to accomplish the expectations?

A: Yes.

In summary, the president claimed he did not even come close to empowering his VP of sales. However, with that knowledge he quickly proceeded in turning the negative responses to his self-critique of empowerment into positive responses.

Empowerment is a process and not an event. It can work one time, and not work another time if one of the key elements does not transpire. Be aware that there are four levels of competence, as shown in Figure 4-9. One must consistently keep focused on the triad's three acts. With enough experience and analysis of what works or what doesn't work, the process will become a way of life.

Four Levels of Competence

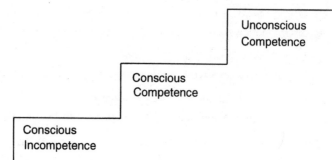

Unconscious
Competence

Conscious
Competence

Conscious
Incompetence

Unconscious
Incompetence

Figure 4-9.

Requirements for Empowerment

As noted earlier, empowerment does not just happen at the stroke of a pen. In order for empowerment to work, management must provide employees with expectations, guidelines, responsibilities, authority, and resources. These management requirements are explained and expanded upon in the following series of questions.

An empowered workforce is defined as *the extent to which people and teams feel they have the responsibility, authority, and resources to take action.* Management must provide the following:

- *Expectations*

    What results or goals are expected? Quantify, if possible, both by financial impact and/or performance.

    What criteria or requirements are necessary for successful completion of the task?

    When are the goals or tasks expected to be fulfilled or completed?

- *Guidelines*

    What steps or procedures should be taken to achieve the results?

    What data/information is required?

    What milestone should be created, and what are the criteria?

    What organizations or individuals should be interfaced with?

    What is the plan, including the preceding items?

■ *Authorities*

What decisions can be made independently?

What decisions can be recommended with management's approval?

What dollar resources and limits are available for the projects?

■ *Skills*

What skills (how-to-do techniques) are required to achieve the results?

Are the skills existent with the involved personnel?

What is the plan to acquire missing skills?

Corrective/Preventive Action Matrix

Empowerment can be a documented process as well. This is applicable when a person performing a function or process has the choices of taking preventive or corrective actions relative to the output of their process.

This is done by proving a "corrective/preventive-action matrix" at the process. It can be applied to the administrative process, production processes, or any type of process that requires decision making and action to produce conforming service or products.

The corrective/preventive-action matrix is discussed in detail in Chapter 3, "Implementation Track." However, the concept is that the operator/employee and their supervisor document the various nonconforming situations and what actions are to be routinely taken, and by whom. Example forms are shown in Figures 4-10 to 4-12.

Typical corrective and preventive actions are as follows:

Actions	*Type of action*
■ Purge stock produced since last acceptable subgroup	Corrective
■ Purge stock in-process but not yet inspected	Corrective
■ Adjust tooling	Preventive
■ Remove defective component and replace	Corrective
■ Reset cutting bit	Preventive
■ Grind excessive material	Corrective
■ Change cutting bit	Preventive
■ Change rate of feed	Preventive
■ Remove burr	Corrective
■ Replace operator	Preventive
■ Sand and repaint	Corrective
■ Make operator aware	Preventive
■ Change speed	Preventive
■ Change temperature	Preventive

CORRECTIVE/PREVENTIVE ACTION MATRIX

INSPECTION STATION: _____   PREPARED BY: _____   PAGE   OF

PRODUCT: _____   DATE EFFECTIVE: _____   REVISED BY: _____

INSPECTION INSTRUCTION: _____   APPROVED BY: _____   DATE EFFECTIVE: _____

DATE: _____   REVISION: _____   APPROVED BY: _____

DEFECT/ CONDITION	PROBABLE CAUSE	DEFECT TYPE W, D, P, C	STATION RESP.	CORRECTIVE ACTION (1) DEFECT	PREVENTIVE ACTION CAUSE (2)		
					OPER	INS/TST	OTHE

Notes: 1.   Prior to taking any corrective action, defect must be positively identified.
         2.   Prior to taking any preventive action, verify cause listed on matrix.

Figure 4-10.

CORRECTIVE/PREVENTIVE ACTION MATRIX

INSPECTION STATION: PRE-SHIP INSPECTION	PREPARED BY: _____	PAGE 1 OF 6
PRODUCT: _____	DATE EFFECTIVE: 27 APRIL 87	REVISED BY: _____
INSPECTION INSTRUCTION: _____	APPROVED BY: _____	DATE EFFECTIVE: _____
DATE: 27 APRIL 87 REV A		APPROVED BY: _____

DEFECT/CONDITION	PROBABLE CAUSE	DEFECT TYPE W, D, P, C	STATION RESP.	CORRECTIVE ACTION (1) DEFECT	PREVENTIVE ACTION CAUSE (2)		
					OPER	INS/TS	OTHE
CONFIGURATION SHEET MISSING	WORKMANSHIP	W	ALL	1, 2, 3	A,D,E		
UNIT TRAVELER							
1. MISSING	WORKMANSHIP	W	ALL	1, 2, 3	A,D,E		
2. INCOMPLETE	WORKMANSHIP	W	ALL	1, 2, 3	A,D,E		
CHASSIS, exterior							
1. RAILS & BRACKETS							
a. loose hardware	WORKMANSHIP	W	F.A.	1, 2, 3	A,D,E		
b. location	WORKMANSHIP	W	F.A.	1, 2, 3	A,D,E		
2. DAMAGE	2a. WORKMANSHIP 2b. HANDLING	W / P	ALL / M.E.	1, 2, 3 / 5, 6, 7	A,D,E		D,E,F, G
3. LABELS							
a. damage, missing not correct	WORKMANSHIP	W	ALL	1, 2, 3	A,D,E		
4. HINGES							
a. loose hardware	WORKMANSHIP	W	F.A.	1, 2, 3	A,D,E		
b. location	WORKMANSHIP	W	F.A.	1, 2, 3	A,D,E		

Notes: 1. Prior to taking any corrective action, defect must be positively identified.
2. Prior to taking any preventive action, verify cause listed on matrix.

Figure 4-11.

CORRECTIVE/PREVENTIVE ACTION MATRIX

INSPECTION STATION: PRE-SHIP INSPECTION	PREPARED BY: _____	PAGE 2 OF 6
PRODUCT: _____	DATE EFFECTIVE: 27 APRIL 87	REVISED BY: _____
INSPECTION INSTRUCTION: _____	APPROVED BY: _____	DATE EFFECTIVE: _____
DATE: 27 APRIL 87 REV A		APPROVED BY: _____

DEFECT/CONDITION	PROBABLE CAUSE	DEFECT TYPE W, D, P, C	STATION RESP.	CORRECTIVE ACTION (1) DEFECT	PREVENTIVE ACTION CAUSE (2)		
					OPER	INS/TS	OTHE
TOP COVER							
1. LABELS							
a. missing	1a. WORKMANSHIP	W	F.A.	1, 2, 3	A,D,E		C, F
b. damaged	1b. HANDLING	W	ALL	1, 2, 3	A,D,E		
2. DAMAGED, missing or wrong parts	2a. WORKMANSHIP	W	ALL	1, 2, 3	A,D,E		C,D,F
	2b. HANDLING	W	ALL	1, 2, 3	A,D,E		
3. HINGES							
a. hardware	3a1 WORKMANSHIP	W	F.A.	1, 2, 3	A,D,E		
(1) loose/missing	3a2 WORKMANSHIP	W	F.A.	1, 2, 3	A,D,E		
(2) wrong							
4. OPERATION/FIT	WORKMANSHIP	W	ALL	1, 2, 3	A,D,E		
TOP PLATE							
1. HOLD DOWN SCREWS							
2. CONTAMINATION	WORKMANSHIP	W	F.A.	1, 2, 3	A,D,E		
3. LABELS							
a. damaged	WORKMANSHIP	W	F.A.	1, 2, 3	A,D,E		
b. missing	WORKMANSHIP	W	F.A.	1, 2, 3	A,D,E		
4. REEL STOP PAD							
a. missing	WORKMANSHIP	W	F.A.	1, 2, 3	A,D,E		
5. BLOWER COVER	WORKMANSHIP	W	F.A.	1, 2, 3	A,D,E		
	BAD GLUE	C	M.E.	1, 2, 3			
6. COBRAR COVER	WORKMANSHIP	W	F.A.	1, 2, 3	A,D,E		E, F
TAKE UP HUB							
2. CONTAMINATION	WORKMANSHIP	W	ALL	1, 2, 3	A,D,E		

Notes: 1. Prior to taking any corrective action, defect must be positively identified.
      2. Prior to taking any preventive action, verify cause listed on matrix.

Figure 4-12.

The Results of Empowerment

By empowering employees, management is harnessing a powerful tool. One of the effects of this tool is that it will generate change. When the phenomenon of change is occurring, there are two primary outcomes: *danger* and *opportunity*.

Danger and opportunity can each be further subdivided into two additional phases, providing a model of four phases people commonly go through when facing change:

- *Danger* can be subdivided into
  Denial
  Resistance

- *Opportunity* can be subdivided into
  Exploration
  Commitment

Most people move through these four phases in every transition. However, some may go quickly or get bogged down in different phases. Effective leadership can help a group and each of its members move through the phases from denial to commitment. Figure 4-13 provides a graphic representation of the phases of change.

When approaching change it is important to understand what change is and is not. There are six basic principles of change which leaders need to understand.

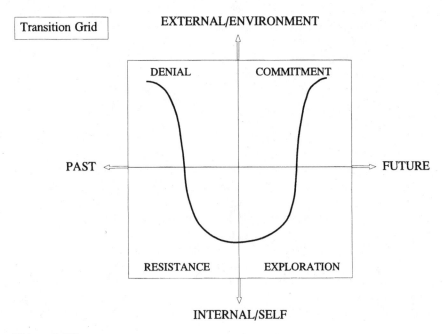

Figure 4-13.

1. Change takes time. There are no magic potions or cures to implement change.

2. Change is a process. It is not an event.

3. In order for change to occur, there must be positive incentive or negative consequences communicated to all at the start of the initial process.

4. Successful change requires experience. It requires role-playing exercises in the new paradigm to internalize the process.

5. Change should be incremental if outside pressures will allow it.

6. Change is more easily adapted if the leaders demonstrate the change in their behavior.

Common Responses to Changes Caused by Empowerment

Within the organization, different groups will react in different ways. It is important that leaders recognize that these reactions are normal. Instead of viewing certain behaviors as insubordinate, management must recognize certain reactions as part of the change process. The following paragraphs address common reactions by certain levels of the organization.

In a traditional company, senior management has a hard time coming to grips with the direct implications of the change. They often underestimate the impact that change has on their employees.

Members of senior management tend to isolate themselves. Often they engage in strategic planning sessions and gather information in survey reports. They avoid communicating or seeking "bad news," because it is difficult for them to admit they "don't know." They expect employees to "go along" when a change is announced. They often feel betrayed when employees don't respond positively.

Managers in the middle feel the pressure to "make the organization change" according to the wishes of top management. They feel pulled in different directions. Middle managers often lack information and leadership direction to focus on multiple priorities. Caught in the middle, they are often fragmented because they don't have clear instructions. They feel besieged with upset, resistant, or withdrawn employees who no longer respond to previous management approaches, while also feeling deserted, blamed, or misunderstood by their superiors.

Workers often feel attacked and betrayed by changes announced by management. They are often caught off guard, not really believing that "my company could do this to me." Many respond with resistance, anger, frustration, and confusion. Their response can solidify into a wall of "retirement on the job." They become afraid to take risks, be innovative, or try new things. They experience a loss of traditional relationships, familiar structure, and predictable career advancement patterns.

Phases of Change Generated by Empowerment

Changes in your organization will transport people through the four phases of the transition process shown in Figure 4-14. You can think of this process as descending into a valley and then climbing back out. The transition leads from the way things were done in the past to how they will be done in the future. During change, people focus on the past, and *deny* that they change. Next, everyone goes through a period of preoccupation, wondering where they stand and how they will be affected. This is normally where *resistance* occurs. As they then enter the *exploration* and *commitment* phases they start to look toward the future and the opportunities it can bring.

The different stages call for different strategies. During the change process you will probably have employees in each of the different stages. You will need to be situational in your participation of techniques to help your people through the change. Also, it is not uncommon to find an employee swinging between two stages.

The *denial* stage can be prolonged if employees are not encouraged to register their reactions, or if management acts like employees. Denial is harmful because it impedes the natural progression of movement that management expects. Workers habitually prefer to do things the old way, and even though this may lack some promised advantages, they know what to expect. They have reached a comfort level, in spite of conditions that may be less than perfect. This is what they have come to expect, and getting them to alter their behavior will be impossible unless they can talk about the issues and problems they are having. If your organization is not one which speaks freely between levels of the company, you can expect that the resistance will be great, and change will be slow and painful.

People are often blind to problems during the denial phase; the company's performance may not be acceptable or, if currently acceptable, the long-term outlook may not be favorable. However, the executives themselves do not recognize this. This is habitual arrogance based on the achievements of the past. When change occurs, all that worked yesterday no longer applies today, and it is foolish to expect that you know what will happen.

Resistance occurs when people have moved through the numbness of denial and begin to experience self-doubt, anger, depression, anxiety, frustration, fear, or uncertainty because of the change that is taking place. Some types of organizational change can be intensely painful, and similar to a death experience. No matter what the change or how subtle you may think it will be, denial and resistance should be expected. It's not a matter of will it happen, rather when and for how long. Resistance manifests itself by a lack of anything happening; progress is not made. Nobody takes action on their own, or they are allowed to take this position. While it is difficult for an organization to experience negative expressions, it is absolutely necessary that people be allowed to communicate and vent their frustrations, concerns, upsets, and dislikes. They must also be allowed to express their positive affirmations of how well things are progressing. Not expressing the negative aspects associated with change can be *a red flag* that denial and resistance are taking place. When resistance occurs, management must listen and develop action plans with the personnel, make clear expectations, and perform the three points of empowerment.

The next phase is *exploration.* Energy is released as people focus their attention on the future and toward the external environment once again. Another word for this phase is "chaos." As people try to figure out new responsibilities, search for new ways to relate to each other, learn more about their future, and wonder how the company will progress through this change, they have many questions which must be addressed. There is uncertainty during this phase, including stress among those who need a lot of structure. During exploration people tend to draw on their internal creative energy to figure out ways to capitalize on the future. This can be an exciting and exhilarating phase. It can, and often does, create powerful new bonds in work groups and between individuals who previously did not work well together. When it begins to appear that everything is being pulled together, the shift is then made to commitment.

During the *commitment* phase employees are ready to focus on a plan. They are willing to focus on their vision and mission, and build action plans to make it work. They want to work together, and have renegotiated roles and expectations. The values and actions needed to commit to a new phase of productivity are in place. This is a phase where employees are willing to solidly identify with a set of goals and be clear about how to reach them. The commitment becomes more intense as success is achieved. This phase will last until a new cycle of transition begins with another major change.

How to Diagnose Each Phase of Change Caused by Empowerment

Denial. It is common to observe the following: withdrawal, "business as usual," and focus on the past. There is activity, but often productivity is very low. Emphasis is on form or style, not results.

Resistance. You will probably see anger, blame, anxiety, depression, and even "retirement on the job." Some workers will feel, "What's the difference? This company doesn't care any more." Very little activity is observed.

Exploration. You will recognize the following: overpreparation, confusion, chaos, and hyperactivity. "Let's try this and this and what about this. . . ." Lots of energy and new ideas, but a lack of focus.

Commitment. Occurs when employees begin working together. The three points of the empowerment triangle have been satisfied. There is cooperation and a results-oriented focus: "How can we work on this?" Those who are committed are looking for the next challenge.

Strategies for Dealing with the Phases of Change Caused by Empowerment

During Denial. Confront individuals with information. Let them know that the change will happen. Explain what to expect and suggest actions they can take to adjust to the change. Give them time to let things sink in, and then schedule a planning session to talk things over. Clear expectations, guidelines, authority, resources, and skills need to be emphasized.

During Resistance. Listen, acknowledge feelings, respond empathetically, and encourage support. Don't try to talk people out of their feelings or tell them to change or pull together. If you accept their response, they will continue to tell you how they are feeling. This will help you respond to some of their concerns and develop a plan of action. After the act of empowerment is completed, the act of taking responsibility must occur.

During Exploration. Focus on priorities and provide any needed training. Follow up on projects that are under way. Set short-term goals. Conduct brainstorming, visioning, and planning sessions. Management must perform the act of process management.

During Commitment. Set long-term goals. Concentrate on team building. Validate and reward those who are responding to the change and producing results. Look ahead.

Traps and Pitfalls to Avoid During the Change Process

1. *Ignoring or resisting resistance.* Resistance is not pleasant to experience. It can feel like everyone is angry and you are to blame. This is usually temporary. Denying resistance only makes it go deeper and last longer. Invite it, seek it out through listening and good communication.

2. *Jumping to team building.* When faced with change, many managers think that what they need most is to get people back to working together. When a group is in denial, resistance, or the early phase of exploration, you are wasting your time to work on team building. The group needs a chance to complain and assess their loss before beginning to rebuild trust and cooperation.

3. *The "Drano" approach—pushing productivity too soon.* Some managers believe that if you demand performance you will get it. Employees may respond in the short run, but tend to plateau and actually decrease productivity if their feelings don't match their actions. The danger is that you will end with a "clogged" organization where everything breaks down.

Some Final Thoughts

The purpose of this chapter has been to discuss ways to manage the cultural change. In order to effect a cultural change, management must first create commitment to the goals of the organization. Without a foundation of commitment, there can be no cultural change.

Assuming that the organization is committed to the cultural change, the process of cultural change becomes precisely that—a process. As with all processes, it has a series of steps in a sequential order. The process can be managed and measured. In fact, if the process is not managed and measured, it will ultimately fizzle out and be relegated to the scrap heap of other management fads.

The purpose of effecting the cultural change is ultimately to implement the desired values in the company. Successful TQM implementation requires an empowered workforce—a workforce which has clearly defined expectations, responsibilities, limits of authority, resources, and skills. Empowerment can be a scary process for many individuals in an organization because it causes change, the harbinger of opportunity and danger.

Organizations which successfully manage the process of cultural change find themselves in an arena full of opportunity and some difficulties. The opportunities are frequently extolled in the TQM literature: proactive, driven toward excellence, adaptable to change, and continuously improving. The difficulties are not insurmountable; rather, they represent a need to develop new management skills which place an emphasis on influencing and molding behaviors rather than dictating desired behaviors.

A metaphor for the focus on empowerment is to say it is like driving your car. If you wish to reach your destination, you must focus on keeping your foot on the accelerator. If the foot is taken off the accelerator, the auto does not progress. Empowerment is identical. Stay focused!

5
Recognition and Rewards

Richard Bush

Purpose

The purpose of a recognition and reward system is to influence employees' performance by reinforcing desired values and behavior. Organizations which effectively implement TQM try to set up a system which "catches" people in the act of doing the right things from a values or behavior perspective.

Objective

The objectives of this track are to:

- Understand how behavior drives performance.
- Outline a systematic approach to employee rewards and recognition.
- Learn how to get results from your recognition and reward program.
- Identify typical rewards and recognition.
- Create guidelines for rewards and recognition.

Definitions

Recognition. A reward in the form of an acknowledgment of gratitude perceived as a commendation by the recipient.

Reward. A gift or prize considered to be of value by the recipient.

Introduction

Behavioral psychologists tell us that people seek rewards and recognition to satisfy basic psychological needs that exist in each of us. Although the theory of positive reinforcement is complex, generally speaking, rewards can be divided into two groups. *Intrinsic rewards* are the internal feelings we get that are based on satisfying our own personal values by doing a "good job," such as the feeling of pride that accompanies performing a task exceptionally well, or completing a particularly difficult job, learning a skill, etc. *Extrinsic rewards* are those such as pay increases, bonuses, prizes, awards, public or private recognition, etc., that others give us when showing appreciation for our performance or accomplishment.

Principles of Rewards

People choose to work or remain working at a particular job for a variety of reasons. Very few employees have only one overriding reason for working at a particular job, and money is rarely the only reward a person receives from a job. In some cases, money may be secondary to other considerations, such as gaining specific experience, lifestyle compatibility, travel considerations, feeling that the job we are doing is important, etc.

The things an employee seeks as rewards will change over time as the employee's life situation, needs, interests, and values change. Young parents many times seek increasing salary and bonuses as their primary rewards to pay for the new mortgage and other expenses associated with establishing a household and raising a family. More established employees might prefer deferred income (for a more generous retirement plan) or tax-exempt rewards or benefits. Employees also have different psychological needs for power, prestige, and independence as they grow older.

Our Personal Changing Needs

Early Career	Mid-Career	Mature Career
■ Safety	■ Advancement	■ Self-actualization
■ Security	■ Friendships	■ Achievement
■ Steady income	■ Increased income	■ Independence
	■ Developing a discipline or specialty	■ Use of abilities
		■ Power and prestige
		■ Recognition

Reinforcing Company Values and Driving Performance

The seasoned manager knows that behavior drives performance and values drive behavior. Therefore, the manager or leader needs effective methods to motivate employee performance toward accomplishing organizational goals. The use of rewards and recognitions that are perceived to be fair and of value when used to reinforce desired employee behavior in accordance with the company's stated values is the primary way in which managers can move these values from being displayed on the "wall" into the reality of actually being used to improve the company's daily performance.

Since perceptions can differ greatly, management should charter a "rewards and recognition team" made up of as broad a demographic representation as possible to seek out those forms of rewards and recognition that the majority of employees perceive as valuable. The team should begin its work by (1) preparing a list of those behaviors that are recognized and rewarded by management, and (2) creating a listing of the current forms of reward and recognition given formally and informally by management. Next, the team seeks input from as many employees as possible for behavior consistent with the company's stated values and those traditional behaviors considered to be worth retaining in the new "total quality" culture. Finally, these recommendations are presented to the resource committee for approval and subsequent budgeting and implementation.

Since behavior drives performance, during periods of change some initial or intermediate achievements may have to be rewarded to reinforce the continued new behavior that will generate performance improvement. Once new behavior develops into new skills and the desired level of performance is achieved, smaller rewards will usually reinforce and maintain the desired behavior.

Positive behaviors need not be of an earth-shattering or revenue-generating type. An interesting example of this principle occurred at Arral Industries in Ontario, California. As part of Arral's total quality journey, they appointed a charismatic first-line supervisor, Linda Larsen, to head up their Cultural Change-Fostering Committee (CCFC). The CCFC was made up of both hourly and salaried employees.

One of the early objectives of the CCFC was to increase the amount of time which upper management spent on the shop floor. In order to reinforce this positive behavior (increased time on the shop floor by management), the employees of Linda's solder room began keeping a tally of the amount of time which managers spent on the shop floor interacting with employees. This tally was then reported positively to the resource committee—whose managers spent the most time on the shop floor interacting with employees.

Spending time on the shop floor interacting with employees is not a behavior whose benefit can easily be quantified; however, it is a behavior which is highly valued by employees—especially those from the lower ranks. This positive behavior provides two key benefits: it reinforces the concept that everyone in the organization is part of a team and it tends to break down organizational barriers. Incidentally, as you can imagine, managers now spend considerably more time on the shop floor generating employee enthusiasm.

Recognition and Reward System

Recognition and reward team should be comprised of representatives throughout the company. The team should represent all groups in the company at all levels.

Guidelines

1. *Focus*   All participants in training
   Selection criteria   Specific minimum hours of training per employee
   Recognition   Company pins, etc.

2. *Focus*   Involvement
   Selection criteria   All team members, mentors, and resource committee members
   Recognition   Coffee cups individually printed with affiliations

3. *Focus*   Individual suggestion plan
   Selection criteria   Suggestions which provided benefits or cost savings
   Selection process   Individuals submit suggestions
   Recognition   Cash awards, a percentage of savings, or levels of benefits (15 to 20 percent of real savings)

4. *Focus*   Achievements from teams, groups, or individuals
   Selection criteria   Teams either achieved or overachieved their goals and objectives within target dates

Scope	All team projects
Type of Benefits	■ Cost savings
	■ Cycle time reduction
	■ Customer satisfaction
	■ Leadership
	■ Role model
Recognition	Inducted to "Wall of Fame," which includes pictures and statements of accomplishments
	Choice of luncheons, stock certificates, shirts, caps, jackets, or whatever is their currency

5. *Focus*   Departments

Selection criteria	Meet or exceed department targets or internal customer satisfaction agreements
Selection process	Measured monthly for quality, schedule, cost, and absenteeism. The goals increased attainment of goals 10% each time that they are achieved
Delivery method	Executives serve lunch; a percent of the positive financial variances if significant
Recognition	Luncheon

6. *Focus*   Individuals not participating in teams or group

Selection criteria	Peer recognition
Selection process	Peer vote based on contribution
Delivery method	Executives serve lunch
Recognition	"Wall of fame"

7. *Focus*   All employees knowing the *vision, mission, and values* statements

Selection criteria	Correct entry drawn
Selection process	Individuals receive tickets for drawing when they know vision, mission, and value statements.
Recognition	Participate in a drawing for a significant prize: trip to Las Vegas, Hawaii, TV, etc.

Implementation

Initiating a program of recognition and reward can best be illustrated by an example, which follows.

Recognition Awards Program Criteria

Purpose

The Recognition Awards Program consists of the *President's Award, Employee of the Month,* and *Employee of the Year* awards. It has been established for the purpose of acknowledging employees' notable ideas and accomplishments in improving the company, its processes, procedures, and products. Another purpose is to inspire and promote the involvement of others to comparable achievements.

The President's Award

This is a certificate award signed by the president and given with a discretionary monetary award for employees whose ideas have significantly benefited the company.

Employee Eligibility for the President's Award. Any employee other than the division president may be recommended for a *President's Award.*

Accomplishments Eligible for Consideration for the President's Award. Accomplishments eligible for consideration for the *President's Award* include those employees' ideas which are of significant benefit to the company, but which cannot be approved under the Product Improvement Program (PIP) because they are within the employee's job responsibility. These include cost savings, technical achievements, new products ideas, and efficiency improvements.

Submitting Recommendations. Recommendations for the President's Award may be submitted by the PIP Committee after it is determined by that Committee and/or the evaluator that the suggestion has a cost savings that could result in an award of $500 or more (this represents a savings of at least $10,000), but is ineligible because it falls within the employee's job responsibility.

Submission to the President will be made on PIP forms with cost savings completed by the PIP evaluator in accordance with PIP Policy 5.12b with the recommendation of the PIP Committee.

Those recommendations not related to cost savings, such as technical achievements and new product ideas, may be submitted by any employee on a President's Award Recommendation Form, signed by the submitter, his/her supervisor and the executive staff member responsible for his/her department.

Employee of the Month Award

The Employee of the Month shall be selected by the President and his/her executive staff from among candidates recommended during the prior month.

Employee Eligibility. Any employee at the level of Director may be recommended for Employee of the Month. Previous Employee of the Month winners may be nominated for Employee of the Month in any calendar quarter following the quarter in which they were a winner, as long as the basis for recommendations is accomplishments subsequent and different than those which led to the prior recommendations.

Achievements Eligible for Recommendation for Employee of the Month. Because this program is for the purpose of inspiring employees to become more involved with their company, achievement eligibility for consideration is deliberately broad. Further, because the award is competitive, its purposes will best be served by not limiting the award topics in order to encourage a steady flow of candidates for recommendation.

Receipt of a PIP award in the month for which an employee is recommended would be an achievement that would be appropriate. Also, leadership efforts on a departmental problem-solving team would be an appropriate basis for nomination.

Submitting Recommendations for Employee of the Month. On the last Friday of each calendar month, department supervisors or managers shall submit to their executive staff the name of one person who is the employee they believe best represents the employee of the month from among the employees under their supervision.

Where multiple organization reporting levels exist, employees recommended shall be reduced to one by each manager reporting to a member of the executive staff. Each executive staff member then may submit one name to the division president on the Recommendation Form, for the selection of one employee for that month.

Announcements for winner of the current calendar month, based on recommendations submitted from the prior month, will be made no later than the fifth working day of the current month.

The Award. Monthly award winners shall receive preferential parking for the calendar month. In addition, monthly award winners shall receive one day off with pay, to be taken during the same calendar month, subject to the employee's supervisor's approval as to the specific date.

Award winners' pictures and names shall be posted prominently in designated areas in Buildings 1, 2, and 3.

Second and Third Runners-Up. Runners-up shall receive a Certificate of Mention.

Employee of the Year Award

From the 12 Employees of the Month during the year, one will be selected by the division president for Employee of the Year.

The Employee of the Year will be named in January of the year following the award for Employee of the Month.

The Employee of the Year shall be awarded a certificate from the President and a $500 gift certificate to a retail establishment of his/her choice.

Suggestion System

Overview

The traditional United States industrial suggestion system is typically focused on the economic gains resulting from the suggestion. This leads to review committees carefully screening suggestions to identify the big winners. The process is slow, few individuals get recognized, and many bad feelings are generated since the majority of suggestions are rejected. The system creates winners and losers, does not stimulate creativity, and generally loses its momentum over time.

The suggestion or improvement system used in Japan focuses on generating and implementing as many new ideas as possible, regardless of the magnitude of the economic impact. In the 1980s, suggestions in Japanese companies averaged close to 20 per worker per year, with 75 percent being adopted. In 1986, 48 million total suggestions were generated in Japan.

The system is broken down into three stages and is based on the premises of starting with quantity and improving quality (economic impact) in the second phase. Stage one stresses participation and involvement. Stage two involves education and development of workers in better problem-solving and team skills. In stage three, more emphasis is placed on the economic impact of suggestions. The first stage breaks traditional barriers and makes improvement a normal part of the job. As skills improve, better ideas with greater impact are generated.

This system recognizes that the major difference between humans and the other animals is the ability to think and generate new ideas. To deny workers the opportunity to create and change their work environment is to deny their humanity.

Behavioral psychologists have long demonstrated that to change behavior one must reward and recognize the new, desired behavior. The key questions for management then become—what is the behavior we want to change? And then, how are we going to reward, recognize, and encourage the new behavior? The suggestion, reward, and recognition system outlined in the following sections is based on the desire to get employees involved in improving the performance of the company. Measurements (scorekeeping) are one of the key feedback and motivational tools to accomplish this objective.

Stage One. This is the simplest and perhaps the most effective approach—a good starting point and ice breaker. It is most useful in situations where employees have not been involved and have not been listened to, but where they are anxious to bring about changes. This approach puts a lot of pressure on management, and they had better be ready to respond and take action. Supervisors must encourage participation, including helping employees write up their ideas.

This can also be used on a short-term basis, at any time, to focus on a specific issue (safety, reducing inventory costs, etc.).

A. Place flip charts or white boards in each department in the plant. The primary focus is on the employee's own work area.

B. Any employee can write an idea, specific suggestion, or specific request on the board. It must be signed and dated. *No complaints allowed, only requests.* Ideas or suggestions should be focused on safety, making the job easier, removing drudgery, removing nuisances, making the job more productive, improving quality, or saving time or cost.

C. Management must respond and take action as soon as possible. A specific time frame should be established. An employee committee could be used in each department to screen and decide which ideas to implement. Timeliness is very important to build momentum. The supervisor or screening committee must respond in less than one week. If higher-level approvals must be sought, they must be completed in less than one month. If the suggestion is rejected, a satisfactory reason must be given (see D). If accepted, a plan for implementation should be given, with a date marked on the chart. Where possible, let the suggestor implement the improvement.

D. The employee who raised the issue erases it from the board when he or she is satisfied. If the employee is not satisfied, he or she can appeal to the suggestion committee and/or the resource committee.

E. CEDAC is an acronym for *Cause* and *Effect Diagram* with the *Addition* of *Cards*. A CEDAC approach could also be used as an alternative, with a CEDAC chart and cards set up in each area of the plant, targeted on a specific issue like scrap reduction.

F. Rewards and recognition must be set up for employees and departments with most ideas, best ideas, most progress to goal, etc. Plantwide awards could also be used for achieving milestones (i.e., paid days off for all employees if quality is improved by X percent). The following approach is recommended, but can be modified in reward amounts or to incorporate other ideas:

1. Monthly drawing of $100 in each department or, if plantwide, $250 for anyone *submitting* a suggestion. The initial goal would be to have everyone in the pool every month after one year.

2. Quarterly recognition of employee with most valid suggestions submitted during quarter, $50 award, letter of commendation, picture on suggestion system board.

3. Monthly recognition of department with most suggestions submitted: $100 award, traveling trophy, group picture on board, lunch with president. All department awards can be used by department as they see fit. They can have a drawing or split equally or have a dinner or some other comparable reward.

4. Quarterly recognition of department with highest number of suggestions accepted: $100 award, group picture on board, lunch served by president and staff to department. To qualify, a department must have submitted suggestions equal to more than 50 percent of the all-department average.

5. Annual awards day—lunch served by president and staff to all employees. The following awards will be given:

- All employees with five suggestions submitted during year: $25 award.

- All employees with 10 or more suggestions submitted during the year: $100 award.

- The employee with the most suggestions submitted during the year: $500 award, picture on board all year, feature article in company newsletter.

- Department with most valid suggestions submitted during the year: $1000 award, name on trophy, feature article in company newsletter.

- Department with highest percent of suggestions accepted during the year: $1000 award, feature article in newsletter.

G. A suggestion committee must be formed to coordinate the program, maintain the bulletin board, maintain the measures, resolve disputes, publish a handbook detailing significant suggestions, and publish a monthly newsletter. The majority of the committee should be operators. The bulletin board will also show honor role lists of monthly drawing winners, those employees submitting 5 suggestions, and those for 10 or more.

H. The following measures (metrics) must be maintained. This will help everyone get comfortable with using measures and show them how they work.

- Number of suggestions submitted each month, total, by department, plus the individual record.

- Number of suggestions accepted each month.

- Percent of suggestions accepted.

- Total suggestion dollars awarded each month and year—chart.

- Average number of suggestions per employee each month.

- Audit results—percent of accepted suggestions audited that are in use.

- By department and total, the number of suggestions on the white board or flip chart that are over one month old.

Stage Two. This stage is basically a training phase and overlaps stage one. Personnel would be trained in SPC, problem solving, being an effective team member, and taking responsibility. These skills are trained in the Core Techniques track.

Stage Three

A. In stage three, the flip charts and white boards are replaced with a simple suggestion form and suggestion boxes are scattered around the factory. Suggestion forms can also be used in stage one instead of the flip charts.

B. Other major changes in stage three are to encourage the use of teams, the development of additional performance measures in each department, the implementation of SPC charting, and bringing processes into statistical control. These additions could be introduced in steps instead of all at once. Stage three requires training in problem-solving and SPC methods.

C. The rules remain the same as in stage one, but additional recognitions, rewards, and measures are required. The changes to the system should be announced with great fanfare by the president of the company.

D. All project teams must be approved by the suggestion committee. They will verify that the team has established a specific and clear objective, a milestone plan, a measurement for performance, and define a review point. Project teams can be established to develop measures, introduce SPC, solve specific problems, generate suggestions, or work on performance improvement.

E. Additional rewards and recognitions: Alternative option for dollar awards is paid time off.

1. Monthly, all departments adding a new performance measure: $50 award. Old measures reintroduced within 12 months do not count.

2. Monthly, $10 award for each SPC chart initiated where appropriate and in use for the full month.

3. Department award for each process that has been brought under statistical process control, complete with events log and C&PA matrix: $50.

4. Department award for performance improvement on any measure of an average of 10 percent for any consecutive three-month period versus the previous three-month period: $100 or 20% of annual savings for the group.

5. Department with the most completed team projects for the year: $500 award.

6. Completed project team with largest estimated cost savings for the year: $1000 (savings must be in excess of $5000 to qualify).

7. Department with the most processes under statistical control: $500.

8. Departments with all process under statistical control: $1000 (each department may win only one time).

F. Additional measures of the system:

1. Number of active project teams—monthly

2. Number of measures established—average per department monthly

3. Number of SPC charts in use—monthly

4. Percent of processes in statistical control

G. Individual contributions

- Individuals or groups would receive 20% of the annual savings.
- Safety or other suggestions would have a fixed award granted.

Getting Results from Your Recognition and Reward System

Since rewards and recognition vary in value in the eyes of each individual, not all reinforcers will have the same effect on all employees. What acts to reinforce desired performance in one situation may not have the same desired effect in another. Timing is also important—the more immediate the reward given after the desired performance is achieved, the more powerful the effect of the reward or recognition becomes. Using a graph as the basis for rewarding desired behavior will also increase the positive effect of the reinforcement.

When giving recognition, the person giving the acknowledgment should:

- Describe in detail the performance being rewarded and why it deserves commendation.

- Express personal appreciation.

- Offer help in making the team or employee's work more effective, rewarding, and challenging in the future.

Social or noneconomic methods should always be tried before adopting economic rewards. They are less expensive to use, and it is much easier to withdraw or modify them if the rewards are not achieving the desired performance levels or if the rewards create an undesirable side effect.

Examples of typical recognition and rewards (noneconomic and economic) follow. (They are not exhaustive lists.)

Types of Recognition

- In person, verbal praise (private or public)
- A visit by the company president or other company leader to the employee or team at their work area
- Letter of commendation from the company president
- Recognition memo, certificate, or award

- Giving credit and praise during a management presentation
- Posting the person's name and a brief description of the reason for acknowledgment on a bulletin or "thank-you" board
- An article in the company newsletter
- A call from the president or other company leader

Types of Rewards

Typical Noneconomic Rewards

- Special parking privileges or use of company car
- Photo of the team and result
- Hats, T-shirts, belt buckles, pins
- Pen and pencil set, professional book(s)
- Coffee cups
- Decals or embroidered team or project patches
- Lunch, breakfast, or other time with the boss
- Tickets to a ball game, concert, or other social event
- Paid dinner out with spouse
- Dinner with the boss or other company leader
- Days off with pay
- Promotion
- Additional authority
- Special prizes or trips
- Invitation to sit in a meeting, task force, etc.
- Invitation to speak to other teams to share new learning (can use video)
- Technical or business books
- Opportunity to attend a business seminar or conference
- Introduction to top management
- Taking part in a seminar or special training program
- Use of new equipment

Typical Economic Rewards. Economic rewards in the order of 20 percent of the savings works very effectively. These can be provided as follows:

- Cash awards
- Savings bonds

- Shares of company stock
- Stock options
- Bonus plans
- Commissions
- Pay increases
- Low- or no-interest loans

Why Rewards Sometimes Don't Work

Since rewards and recognition are intended to reinforce desired behavior and performance in people, there can be no guarantee that the same results will be achieved with each and every person. Supervisors and managers know from experience that each of their employees is unique and may react differently from the majority. In some situations, giving a reward or recognition does not act to reinforce desired performance levels or behaviors. When this occurs, it may be because the recipient has competing behaviors which are receiving a greater reward by someone else (e.g., peer, other manager, wife and family or significant other); or the reward was, in fact, not seen as having value by the recipient, as in the case of a pay raise that moves the employee into a higher tax bracket resulting in less take-home pay. Also, if the reward does not occur frequently enough or in a timely manner, the value of the reward could be diminished for the recipient. To avoid these problems, when possible the individual(s) to be commended should be consulted on what they agree would be a reasonable reward (within preestablished limits) for their level of performance or behavior. If that is not practical, the supervisor or manager should give serious thought as to how the reward(s) might be viewed by each recipient beforehand. This is another important benefit of having a rewards and recognition team to help obtain employee feedback.

A Word on Punishment

Most behavioral research conducted on punishment (a form of negative reward or consequence to behavior) has shown that its effect often has an unpredictable result. Punishment can stop undesired performance, but does not necessarily ensure that the desired performance will occur. In the past, some managers and supervisors have attempted to use punishment as a negative reward when performance fails to reach desired levels. The primary problem with this approach is that its effects are very unpredictable, create resentment, and engender fear. Punishment can stop an undesirable behavior, but does not necessarily ensure that the correct behavior will occur, nor does it teach or reinforce new desirable behavior or levels of performance. Another problem with punishment is that it tends to eliminate not just one behavior, but many times a number of desirable associated behaviors as well.

6
Leadership/Team Building

Purpose

The purpose of the Leadership/Team-Building Track is to educate a company's employees—at all levels of the organization—in effective leadership and team-building techniques. Leadership can be executed by managers as well as individuals. It is paramount that individuals focus on effectiveness first (i.e., doing the right thing), then on efficiency (doing things right). These techniques are crucial for companies which wish to attain the TQM plateau.

Objective

The objective of this track is to achieve the following:

- Understand the difference between managing and leading.
- Understand the impact of role modeling and how to do it.
- Learn and use the five practices of successful leaders.
- Learn and use the 10 commitments of leadership.
- Learn the seven habits of highly effective people.
- Better understand your own style of dealing with people and its impact.
- Determine what makes a winning team.
- Be capable of developing teams.
- Identify normal stages of team development.
- Be capable of managing conflict.

Introduction

In this chapter, you will be exposed to effective leadership techniques. One of the premises of this chapter is that leaders are made and not born. Of course, there will always be individuals who are more effective leaders than other individuals even though they have similar training and experience; however, individuals who are willing to learn and practice the techniques presented in this chapter can produce results as leaders of teams, departments, and ultimately companies. After all, if TQM is to succeed in a company, then leaders will have to emerge from all ranks—from the mail room to the board room.

Leadership is a crucial component to the success of TQM. It supplies direction with vision, mission, and values statements, and supports the actions necessary to meet TQM goals. Leaders continually monitor progress, and make adjustments when needed, to keep the company moving toward its TQM goals. Leadership is a process by which leaders influence others to want to follow their direction. Strong leadership and commitment will bring significant success. Conversely, weak leadership may well bring partial or total failure. By providing leadership in every integral part of TQM, the probability of success will be enhanced exponentially.

Interrelationship with Other Tracks

Without guidance (leadership) from management, TQM would not be taken seriously by the workers. A company's attitude is sourced from the top down. If it appears that the people at the top don't care, then management does not care either. The organization is a mirror of the leadership.

Leadership has a direct correlation to the effectiveness of everyone in the company. Each person must feel that he or she has the authority to be a leader in their position. People need to know they can make decisions to ensure their job gets done properly. Employees who fully utilize their leadership authority will become role models for the other employees. They will provide a core of excellence and influence that other workers will strive for in an effort to achieve the company's desired culture.

Each track relies on leadership to achieve success. The Foundation Track deals with "doing the right thing" while preparing the company to implement TQM. Each employee's leadership abilities will be strengthened as their knowledge of TQM grows, as they participate in establishing the direction and vision of the company, and as they work toward aligning their department with the company's goals.

The Cultural Track is affected by leadership because the leaders are the ones who share their visions and rally the company around the vision, mission, and values statements. These same leaders plan and execute the implementation of their values and help to empower others by serving as role models.

In the Core Techniques Track, effective leadership provides motivation to the employees. They provide the expectations for their company's employees to learn, and actively use, core techniques. These leaders take the potential energy of knowledge and transform it into the kinetic energy of implementation.

The Leadership Track is the key to success of the Implementation Track. After the expectations have been communicated, the leaders become involved in the process management function necessary to meet these expectations. Leaders will ensure that the outcome of implementation is related to the performance appraisal and recognition system of the individuals involved.

The Advanced Techniques Track uses leadership in the same way as the Core Techniques Track, but is more sophisticated because the TQM "tools" applied in this track are more complex and sophisticated.

The Recognition and Reward Systems Track uses leadership to reward the positive behaviors and results of the Implementation Track in order to encourage replication of the results and behaviors. The results of this track not only establish the goals, but sustain the culture in demonstration of recognition of the values.

In the Customer Focus Track, leadership combines the techniques used in the Core and Advanced Techniques Track with the customers.

The leadership role in the Train the Trainer Track is in the selection of the trainers who will be effective in training and inspiring the personnel to continually sustain the process.

The Management Skills Track uses leadership in the selection of the skills that the company intends to use to increase its efficiency.

Leadership and Teamwork

Companies will learn to apply the SPC and problem-solving "hard" skills, and develop leadership and teamwork "soft" skills. Unlike technical skills, leadership and teamwork are not learned by studying formulas or theory. Leadership and teamwork are learned by knowing the principles and experience, i.e., role playing and practice. Trainers can set forth a concept in a classroom environment, but the lesson will not be complete until participating individuals have had the opportunity to practice the concepts.

In the classroom potential leaders will emerge. Management is encouraged to find those individuals who demonstrate the commitment and results-oriented attitudes that will work best for TQM implementation. Just because one has the position of leader, it should not be presumed that person has the leadership skills necessary for that position. Of course, this is not to be taken for granted. Some managers have been promoted because they have special technical skills, training, or experience that has nothing to do with leadership. This latter group will also benefit from the leadership skills taught in the TQM experience. Those managers who today possess leadership skills and experience can only enhance their effectiveness by training.

A successful leader must be able to create and sustain a climate of teamwork. In addition, a successful leader must become a skilled motivator, delegator, and above all, communicator. High energy and a bias toward action are also typical leadership characteristics. These traits are evaluated relative to the rest of the organization and are the most difficult to instill.

The Leadership Challenge

Definition of Leadership

TQM programs need heroes and champions. Heroes and champions are the people who see Total Quality Management as a means toward accomplishment of the organization's long-range goals, and who have the charisma and enthusiasm necessary to motivate others with their infectious attitude. They behave in a proactive, results-oriented manner, and adapt to change easily. Heroes are, to some degree, like supermen/superwomen dedicated to truth, justice, and continual improvement. They are people who make others feel safe with risk taking and committing to new objectives.

Another way of looking at the TQM heroes and champions described in the previous paragraph is to recognize that they are TQM leaders. In the previous paragraph, the TQM heroes/champions/leaders were described in some fairly general terms. Let's take some time to describe these individuals in specific terms. We will describe them in terms of the qualities which they must have in order to fulfill the three-pronged role of hero, champion, and leader.

There are seven qualities of leaders which need to be enumerated. Truly effective leaders display all seven of these characteristics in one form or another.

1. *Vision.* The ability to see an end in mind or goal in complete form. Tangible examples of this trait include:

- Seeing the big picture, conceptualizing what will be achieved, and how it will be achieved
- Positive mental image of the desired results or behaviors
- Keeping focused (when distracted, being able to redirect yourself upon your mission)
- Ability to communicate the vision to others through metaphors

2. *Confidence.* Knowing that you and your team can accomplish assigned tasks, no matter what obstacles stand in the way. The difficult we can do, the impossible takes a little longer. Examples of this quality include:

- Inner strength
- Accepting ethical goals without question; ensuring they are reasonable and committing to what is reasonable and identifying what is not reasonable
- Faith in oneself and others to achieve the goals
- Ability to make an impact on others, i.e., making a difference
- Sense of urgency to achieve the goals
- Resourcefulness to assist the organization to develop plans
- Asking the right questions to zero in on root causes
- Striving toward the desired results in spite of any obstacles

3. *Risk taking.* The willingness to try new methods. This quality is displayed by:

- Evaluating the challenge and exploring options
- Changing unacceptable external environments to desired environments
- Experimenting with new approaches
- Challenging paradigms and doing paradigm pioneering
- Accepting experiencing errors in commission; not accepting errors of omission

4. *Decision making*

- Making the right decision (doing the right thing)
- Being effective in doing the right thing
- Courage to take a stand
- Ability to make the tough decisions

5. *Development of others.* The ability to create leadership thought and action in your teams, spreading responsibility and credit for work.

- Sharing responsibility
- Decentralizing power
- Getting others to solve problems
- Patience—allowing others to develop

   6. *Influence on others*

- Inspiring others to the mission
- Being contagious with energy and enthusiasm

   7. *Communication.* The ability to channel your ideas into action.

- Being simple, not simplistic
- Listening to individuals and groups
- Picking up signals from observing
- Having a caring and respectful attitude

The Process of Leadership

Leadership is a process which influences the actions of an individual or team toward achievement of a goal in a given situation. These goals can be values or quantified objectives. There are four minidefinitions which expand on the preceding definition: (1) A process is a series of actions leading to an outcome. (2) A process is what we do to achieve desired outcomes. (3) A process is the way we work and interact with one another. (4) A process transforms resources into a product or service.

If we take these minidefinitions and combine them with the concept of leadership, then we arrive at the ideas that leadership is a series of steps leading to a desired outcome. We must keep in mind the ideas that leadership is the process by which leaders interact with the people being led, and that the process of leadership consumes resources. (The resources being consumed can be people's talents.)

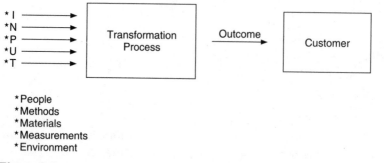

* I
* N
* P
* U
* T

Transformation
Process

Outcome

Customer

*People
*Methods
*Materials
*Measurements
*Environment

Figure 6-1.

Since leadership is a process, it is important that we understand the components of a process. A *process*, regardless of its outcome, will have the following components.

- A process has a customer. This means that the customer is the receiver of the process, and that the customer establishes quality expectations.

- A process involves an interaction between a customer and a supplier. In order for this interaction to be effective, expectations must be clear to both the customer and the supplier. Besides clearly expressed expectations, a process requires that feedback be given and received.

- A process is transformational. This means that something changes its state.

- A process consumes resources; therefore it has a cost. Because a process has costs, it is our responsibility to ensure that the processes we are involved in add value and that they are aligned with the organization's goals.

- A process has controls which can be expressed as rules, boundaries, or parameters.

- A process is improved through the analysis of feedback: from the customer, from the process itself, or from measurement of the results or outcome.

Five Common Practices of Successful Leaders

Now that we know what leadership is, we need to look at how successful leaders practice this process called leadership. There are five leadership practices which seem to show up repeatedly.

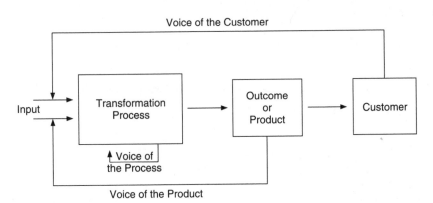

Figure 6-2.

The first common practice is that leaders will *challenge the process*. Another way to phrase this idea is to say that they search for opportunities to change the status quo and improve the organization. Searching for ways to change the status quo means that they experiment and take risks. Since risk taking involves mistakes and failure, leaders accept the inevitable disappointments as learning opportunities.

Second, leaders inspire a *shared vision*. Leaders passionately believe that they can make a difference. They envision the future, creating the ideal and unique image of what the organization can become. Through their strong appeal and quiet persuasion, leaders enlist others in the dream. They breathe life into visions and get us to see the exciting future possibilities.

Challenging the Process

Leaders are pioneers—people who search out opportunities and step into the unknown. They are willing to take risks. They innovate and experiment, and they treat mistakes as learning opportunities. Leaders also stay prepared—physically, mentally, and emotionally—to meet whatever challenges may confront them. Challenging the Process involves:

- Searching for opportunities
- Experimenting

Suggested Actions for Challenging the Process

Treat every job as an adventure.

Provide challenging assignments (beat the system).

Question the status quo.

Find something that is broken—and fix it.

Break free of daily routines.

Institutionalize processes for collecting innovative ideas.

Set up little experiments.

Honor risk takers.

Foster psychological hardiness.

Figure 6-3. From *The Leadership Challenge*, by J. M. Kouzes and B. Z. Pozner, 1987, San Francisco: Jossey-Bass.

Inspiring a Shared Vision

Leaders spend considerable effort gazing across the horizon of time, imagining what kind of future they would like to create. Through their enthusiasm and their skillful communication, leaders enlist the emotions of others and inspire others to share the vision. They show others how mutual interests can be met through commitment to a common purpose. Inspiring a Shared Vision requires:

- Envisioning the future
- Enlisting others

Suggested Actions for Inspiring a Shared Vision

Learn from the past.

Act on your intuition.

Test assumptions.

Know your followers.

Appeal to a common purpose.

Communicate expressively.

Believe in what you are saying.

Develop a stump speech.

Figure 6-4. From *The Leadership Challenge*, by J. M. Kouzes and B. Z. Posner, 1987, San Francisco: Jossey-Bass.

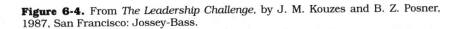

Third, leaders *enable others to act*. They foster collaboration and build spirited teams. They actively involve others. Mutual respect is what sustains extraordinary efforts, so leaders create an atmosphere of trust and human dignity. They strengthen others, making each person feel capable and powerful.

Fourth, leaders model the way. They *establish values* about how employees, colleagues, and customers ought to be treated. They create standards of excellence and they set an example for others to follow. Because complex change can overwhelm and stifle action, leaders plan small wins. They unravel bureaucracy, put up signposts, and create opportunities for victory.

Finally, leaders *encourage the heart*. Getting extraordinary things done in organizations is hard work. To keep hope and determination alive, leaders recognize contributions that individuals make to climb to the top. And

---

Enabling Others to Act

Leaders gain the support and assistance of all those who must make the project work or who must live with the results. They stress cooperative goals and build relationships of mutual trust. Leaders make others feel important, strong, and influential. Enabling Others to Act consists of:

- Strengthening others
- Fostering collaboration

Suggested Actions for Enabling Others to Act

Always say "we."

Create interactions between and among people.

Delegate.

Focus on gains, not losses.

Involve people in planning and problem solving.

Keep people informed.

Give people important work on critical tasks.

Be accessible.

Give people the opportunity to be autonomous and to use their discretion.

Figure 6-5. From *The Leadership Challenge*, by J. M. Kouzes and B. Z. Pozner, 1987, San Francisco: Jossey-Bass.

every winning team needs to share in the rewards of their efforts, so leaders celebrate accomplishments. They make everyone feel like heroes.

Seven Common Habits of Successful Leaders

In order to be effective, it is necessary that one develop certain habits which will reinforce one's leadership. In 1989, these traits were identified and written about in a book entitled *The Seven Habits of Highly Effective People* by S. Covey, published by Simon and Schuster. The importance of Mr. Covey's observations is that he managed to distill seven habits which seem to be common among leaders in all endeavors ranging from the military to academia. Those habits are:

Modeling the Way

Leaders are clear about their business values and beliefs. They keep projects on course by behaving in a way that is consistent with these values—by modeling how they expect others to behave. Leaders also make it easier for others to achieve goals by focusing on key priorities and breaking down big projects into achievable steps. They Model the Way by:

- Setting an example
- Planning small wins

Suggested Actions for Modeling the Way

Do what you say you are going to do.

Walk the halls.

Publicize your "rules of the road."

Talk with others about your values and beliefs.

Be expressive (even emotional) about your beliefs.

Spend time on your most important priorities.

Get started; build on your successes.

Build commitment by offering choices.

Make people's choices public and visible to others.

Figure 6-6. From *The Leadership Challenge*, by J. M. Kouzes and B. Z. Posner, 1987, San Francisco: Jossey-Bass.

1. *Be proactive.* Take the initiative and the responsibility to make things happen.
2. *Begin with an end in mind.* Start with a clear destination, understand where you are now, where you're going, and what you value most.
3. *Put first things first.* Manage yourself. Organize and execute around priorities.
4. *Think win/win.* See life as a cooperative, not a competitive, arena where success is not achieved at the expense or exclusion of the success of others.
5. *Seek first to understand.* Understand, then be understood, to build the skills of empathetic listening that inspires openness and trust.

Encouraging the Heart

Leaders must give encouragement and recognition if people are to persist, especially when the climb is steep and arduous. To continue to pursue the vision, people need to feel that they are part of a team. Leaders Encourage the Heart by:

- Recognizing contributions
- Celebrating accomplishments

Suggested Actions for Encouraging the Heart

Foster high expectations.

Make creative use of rewards.

Say "thank you."

Link performance with rewards.

Provide feedback about results.

Be personally involved as a cheerleader.

Create social-support networks.

Love what you are doing.

Figure 6-7. From *The Leadership Challenge,* by J. M. Kouzes and B. Z. Pozner, 1987, San Francisco: Jossey-Bass.

6. *Synergize.* Apply the principles of cooperative creativity and value differences.

7. *Renew.* Preserve and enhance your greatest asset, yourself, by renewing the physical, spiritual, mental, and social/emotional dimensions of your nature.

Your Professional Profile

To effectively lead and participate in teams, you must understand your professional strengths and weaknesses as a leader and as a team member. Once you have identified these strengths and weaknesses, you can then set about strengthening and improving them. In the following sections of this chapter, readers will find a system devoted to identifying and enhancing their unique skills. This system is called the Personal Profile System.

What Is the Personal Profile System?

The Personal Profile System reveals your behavioral patterns, your distinctive style of thinking, feeling, and acting. The Personal Profile System is not a test. You cannot pass or fail. There is not a "best" pattern. Research evidence supports the conclusion that the most effective people are those who know themselves, know the demands of the situation, and adapt strategies to meet those needs.

What Will the Personal Profile System Do for You?

In summary, the Personal Profile System enables you to:

1. Identify your work behavioral style.
2. Create the environment most conducive to your success.
3. Increase your appreciation of different work styles in order to adapt your approach to be effective with individuals of other behavior styles.
4. Create a work environment which maximizes productivity and harmony.

Identifying Your Work Behavioral Style

Figure 6-8 provides snapshots of sample behaviors which illustrate the four behavioral styles. Can you identify your category?

Create the Environment Most Conducive to Your Success

Once you have identified your own behavioral style and perhaps the styles of those around you, you can begin to identify strategies for leading and working with such individuals. Figure 6-9 contains a table of possible tactics. Figure 6-10 contains a compatibility chart. Its purpose is to rank the compatibility of individuals based on their behavioral style.

Increase Your Appreciation of Different Work Styles

As Figure 6-10 demonstrates, there are differences of compatibility between individuals of different behavioral styles. The following prose scenarios look at the compatibility of various combinations of behavioral

The Four Behavorial Styles

Figure 6-8. Personal Profile System. Source: Carlson Learning Company, Copyright 1993.

Strategies for Blending and Capitalizing

Dominance	Influence
Remember a *High D* May Want:	**Remember a *High I* May Want:**
Authority, challenges, prestige, freedom, varied activities, growth assignments, "bottom line" approach, opportunity for advancement.	Social recognition, popularity, people to talk to, freedom from control and detail, favorable working conditions, recognition of abilities, chance to motivate people, inclusion by others.
Provide direct answers, be brief and to the point. **Ask** "what" questions, not how. **Outline** possibilities for person to get results, solve problems, be in charge. **Stress** logical benefits of featured ideas, approaches. **When** in agreement, agree with facts and ideas rather than the person. **If** timeliness or sanctions exist, get these into the open as related to end results or objectives.	**Provide** favorable, friendly environment. **Provide** chance for them to verbalize about ideas, people and their intuition. **Offer** them ideas for transferring talk into action. **Provide** testimonials of others on ideas. **Provide** time for stimulating, sociable activities. **Provide** detail in writing, but don't dwell on these. **Provide** a participative relationship. **Provide** incentives for taking on tasks.
Compliance	Steadiness
Remember a *High C* May Want:	**Remember a *High S* May Want:**
Personal autonomy, planned change, personal attention, exact job descriptions, controlled work environment, reassurance, precise expectations.	Security of situation, time to adjust, appreciation, identification with group, repeated work pattern, limited territory, areas of specialization.
Take time to prepare your case in advance. **Provide** straight pros and cons of ideas. **Support** ideas with accurate data. **Provide** reassurance that no surprises will occur. **Provide** exact job description with precise explanation of how it fits the big picture. **Review** recommendations to them in a systematic and comprehensive manner. **If** agreeing to be specific. **If** disagreeing, disagree with the facts rather than the person. **Be** prepared to provide explanations in a patient, persistent, diplomatic manner.	**Provide** a sincere, personal and agreeable environment. **Provide** a sincere interest in them as a person. **Focus** on answers to "how" questions to provide them with clarification. **Be** patient in drawing out their goals. **Present** ideas or departures from current practices in a non-threatening manner, give them a chance to adjust. **Clearly** define goals, roles or procedures and their place in the overall plan. **Provide** personal assurances of follow-up support. **Emphasize** how their actions will minimize risks involved and enhance current practices.

Figure 6-9. Source: Carlson Learning Company, Minneapolis, Minn.

styles. Let us now examine the Dominant, Influential, Steady, and Compliant styles as they interact one with the other.

Hi-D and Hi-D. One of the basic tenets of the Hi-D is: "There is only one tiger to a hill—and I'm it." Therefore, some behavioral clash is inevitable when two or more Hi-Ds live or work together. The common ground on which the Hi-D can "negotiate," however, is the one founded

*Compatibility Chart

Styles	Excellent 1 2	Good 3 4	Fair 5 6	Poor 7 8
D-D		S	W	
D-1		S	W	
D-S	W		S	
D-C			W	S
I-I	S			W
I-S	W		S	
I-C		W		S
S-S	S	W		
S-C	S W			
C-C	S	W		

*KEY
S=Social Interaction   W=Work Tasks   1=Best Possible   8=Poorest Possible

Figure 6-10.

on mutual respect. Hi-Ds tend to come on strong, get to the point quickly, and generally move through issues to decisions and conclusions rapidly, and they will appreciate that "approach" in others. Hi-Ds should be careful, however, that they do not move so quickly that they "go off half-cocked" or leave themselves open to later criticism of committing errors of both omission and commission.

Hi-D and Hi-I. The Hi-D prefers to "tell" and the Hi-I to "sell," so it is usually a game of who maneuvers best on a given day. If the Hi-D is well enough prepared, he or she will not often be "sold a bill of goods." If, however, the preparation is weak, the Hi-I is at his or her best while dealing in noncommittal, broad-based generalities. Both the Hi-D and Hi-I are risk takers, good competitors, and enjoy moving quickly through or around details. If the Hi-D is going to "score" with the Hi-I, he or she should have all the facts straight, lay them out, and get a commitment as to understanding, etc. On the other hand, the Hi-I will do best with the Hi-D if he or she deals in specifics, stays away from "broadbrushing" major issues, and, in general, does not substitute charm for substance.

Hi-D and Hi-S. Give round 1 to the Hi-D, because the Hi-S will back down in the face of initial, direct confrontation. In other words, the Hi-D's knack of "winning through intimidation" will generally carry opening day. It is round 2, however, that is almost always the more important, and this can go either way. Sometimes the Hi-S will carry it by shrewdly observing a policy of "watchful waiting" and carefully timing the next move. You will recall the race between the tortoise (Hi-S) and the hare (Hi-D) and, as you know, the tortoise won that time. Since Hi-Ds tend to see Hi-S's as "moving too slowly" and having "little sense of urgency," it would behoove the Hi-S to move more enthusiastically and quickly to the bottom line when dealing with the Hi-D. The Hi-D should also exercise caution when the Hi-S "clams up," as this may be a sign of passive aggressiveness and a general policy of "getting even" by deliberately "slowing things down." The Hi-S will respond favorably, however, to a policy of being "included" and "appreciated"—both of which are hard for the Hi-D to do.

Hi-D and Hi-C. Toughness (Hi-D) and strictness (Hi-C) also present as natural protagonists. Dominant people tend to see compliant people as "nitpickers" and "bogging down in details," while the compliant ones perceive dominant people as "arrogant" and "pushy." In debating issues or arguing decisions, the Hi-D will hit fast and hard, and the final authority will be himself or herself. The Hi-C, conversely, will "quote the book," get defensive, and generally attach specific details. The Hi-D should learn to slow down in the presence of the Hi-C, deal in facts, be well prepared and normally expect to be met with plenty of doubt, questions, and contemplation before action. The Hi-C, on the other hand, has to speed up a little in the presence of the Hi-D, be less evasive, get to the point quicker, and learn to be a little less rigid, if all the "*i*'s aren't dotted."

Hi-I and Hi-I. When two Hi-Is wheel and deal, it's a question of who out-impresses whom. Hi-Is will instantly like and relate to each other.

Both will be easy to get to know and respond to the "innate" people skills of the other. They can work comfortably together, support dreams and schemes, and capitalize on the input from people they admire (admiration is the basis of Hi-I respect). Hi-Is like to have fun along with their work, which is obviously OK as long as it doesn't interfere with performance and, especially, deadlines. Prone to enthusiasm and optimism, two Hi-Is in any relationship may, on occasion, need to exercise restraint as it relates to spending, budgets, and short-term projects versus long-term objectives. They must also learn to listen instead of thinking only about what they will say next.

Hi-I and Hi-S. People skills (Hi-I) and patience (Hi-S) will get along well for openers. The Hi-I will sell, convince, and charm, while the Hi-S will listen, nod approvingly, and not openly disagree. The Hi-I and Hi-S do not, however, share the same capacity for taking risks and, while they will appear to "hit it off," they will never be entirely sure of each other. Hi-Is have to tone down their innate enthusiasm, suggest rather than direct, and be sincere above all else with the Hi-S. Anything less will be suspect, and the Hi-S, never wishing to appear foolish, will protect his or her flanks by not giving complete cooperation. Hi-S's, on the other hand, should be less placid and possessive in the presence of the Hi-I, say what they really think, and generally let the Hi-I know where they stand.

Hi-I and Hi-C. This particular relationship has a high potential for conflict, as most of the Hi-Is strengths are at odds with the preferred style of the Hi-C and vice versa. Basically, the Hi-I is comfortable with outgoing people and the Hi-C is comfortable with things. The Hi-I should go out of his or her way to impress the Hi-C with being well prepared, dealing in specifics and never letting "fun and games" interfere with getting the "job" done first. The Hi-C tends not to relax before and during, but only after, the job is completed, while the Hi-I's preference will be to blend fun with work from beginning to end. The Hi-C, conversely, should learn to speed things up in the presence of the Hi-I, have some fun, play down the facts in favor of "gut" feelings once in awhile, all while trying to be more stimulating and less "fussy."

Hi-S and Hi-S. Patience, patience, and more patience. The support given by one is only to be outdone by the support returned. Very rarely will two Hi-S's find the basis for a good, old-fashioned argument because they are both too low-key to really care. Tending to plod and persevere, one Hi-S will not push the other (Hi-S's never push but rather "motivate" by good example). Hopefully, when two or more Hi-S's work or live together, there will be goals and deadlines from some source, or else it will be a very pleasant place where very little happens.

Hi-S and Hi-C. The Hi-S and Hi-C both agree on being cooperative, reducing risk, and exercising care in decision making. The Hi-S, however, will come across as a little too "easygoing" and "lenient" to most Hi-Cs. The Hi-C, as you know, can be very demanding of facts and evidence, and the Hi-S doesn't usually feel that same "sense of urgency." To most effectively relate to a Hi-C, the Hi-S should build the "friendship" bridge slowly, deal in raw facts, and realize that the Hi-C's initial "coolness" is nothing personal. The Hi-C, at the same time, would do well to remember that to "err is human" and that the Hi-S will often make decisions and/or take action based on trusting someone rather than because the facts are all perfectly clear.

Hi-C and Hi-C. Very few philosophical differences will occur when two or more Hi-Cs get together. They will cooperate, remain cool, and exercise a high degree of self-control. Since the Hi-C deals in direct answers and volunteers evidence to support conclusions, another Hi-C is rarely frustrated in the planning and control aspects of a given endeavor. The difficulty is that the Hi-C would rather be "right" than "finished" and, therefore, deadlines are sometimes sacrificed for accuracy and that, at times, can be very costly. I would suppose that some timely advice for two Hi-Cs in a relationship would be to remember that, yes, "Rome wasn't built in a day"—but it was built!

Create a Work Environment Which Maximizes Your Productivity and Harmony

Once you can recognize and appreciate different behavioral styles, it is important that you understand how to manage and influence those styles in order to achieve the organization's stated goals. Table 6-1 contains a series of activities down the left-hand column. Opposite these activities, you will find a matrix showing how the various styles react to specific activities.

Strength Management

After taking the personal profile, you will better understand your strengths and improvement areas. The personal profile is a diagnostic

Table 6-1. Managing Styles

	High Dominance	High Influence
Delegating:	Can't wait; too impatient; tells or shows others how to do it.	May overdelegate and not follow up; trusts people too much.
Communicating:	Direct but may be blunt; may be too brief, too fast.	May tell too much; likes to talk but hates to write.
Time control:	Putting out brushfires; involved in a day-to-day production rather than long-range planning.	Overinvolvement with people may throw off schedule.
Decision making:	Can be impulsive; shoot from the hip.	Unpopular decisions difficult; may overestimate results of his or her projects.
Directing people:	Commands rather than leads.	May not be firm enough; too lenient.
	High Cautious	**High Steady**
Delegating:	Fearful and hesitant; does work oneself so he or she knows it's right; may oversupervise.	Slow and reluctant; honestly likes to do work oneself to feel important and secure.
Communicating:	Highly detailed; brief face-to-face; lengthy in writing.	Need-to-know basis only; possessive of information.
Time control:	Checking and rechecking; details.	Personal production slows this person.
Decision making:	Tentative; guarded; low risk.	Puts off.
Directing people:	Strict and by the book.	If you don't push this person, he or she is lenient. If you do, the individual is stubborn and rigid.

tool. It helps us to identify our strengths; better use the strengths we already have; and it will help us find other areas to use our strengths. Besides identifying our strengths, the professional profile allows us to use other people's strengths to cover areas where we are not strong.

There are several points which we need to keep in mind when using a tool such as the professional profile. First, don't allow yourself to always use the same strategy. This is referred to as "hardening of the strategies." Second, when working with others, keep an eye on the overall outcomes.

Finally, be flexible in using your appropriate style to help reach group outcomes.

Summary

The following table summarizes much of the information provided in this section on the professional profile. Use this table as a quick reference regarding the professional profile information.

| | D I S C Summary | | | |
	Dominance	Influence	Steadiness	Compliance
Appearance	Businesslike Functional	Fashionable Stylish	Casual Conforming	Formal Conservative
Workspace	Busy Formal Efficient Structured	Stimulating Personal Cluttered Friendly	Personal Relaxed Friendly Informal	Structured Organized Functional Formal
Pace	Fast Decisive	Fast Spontaneous	Slow Easy	Slow Systematic
Priority	The task Results	Relationships Interfacing	Maintaining Relationships	The task Process
Fears	Loss of control	Loss of prestige	Confrontation	Embarrassment
Under tension	Dictate Assert	Verbally Assert	Submit Acquiesce	Withdraw Avoid
Generally want	Productivity	Recognition	Security	Accuracy
Gain security by	Control	Flexibility	Close Relationships	Preparation
Want to maintain	Success	Popularity	Relationships	Credibility
Support their	Goals	Ideas	Feelings	Thoughts
Achieve acceptance by	Leadership Competition	Playfulness Stimulating Environment	Conformity Loyalty	Correctness Thoroughness
Like you to be	To the point	Stimulating	Pleasant	Precise
Want to be	In charge	Admired	Liked	Correct
Irritated by	Inefficiency Indecision	Boredom Routine	Insensitivity Impatience	Surprises
Measure worth by	Results Track record	Compliments Applause	Compatibility with others	Precision Accuracy
Decisions are	Decisive	Spontaneous	Considered	Deliberate

Team Building

A *team* is a group of people with different backgrounds, skills, and abilities who work together to achieve clearly defined common goals through clear communication and action. Teams are an integral component of the TQM process because they provide the leadership necessary to achieve the goals established by the company. Teams are self-maintaining through regular evaluations, which keep them focused on their goals. Once the goals or objectives have been achieved, the teams are discontinued.

Team building is a critical element of the TQM process. Selecting the right team leader and members from the beginning is easier, and certainly more effective, than replacing the team leader and members after the fact. The right combination of people on the team will bring a balance in technical and business skills and will generate synergy. An effective team can accomplish more as a group than they would as individual contributors. The effectiveness can be enhanced by training team leaders in how to build teams and how to have team members work as a team.

Formation of teams underlines management's commitment to TQM through dedication of resources and the establishment of an infrastructure which will enable the organization to begin and sustain a continual improvement process. The investment of this resource does not go unrewarded. The accomplishments of the teams are directly related to the Critical Business Success Factors, and are the recipients of the recognition and reward systems. Soon the whole organization will begin to develop a sense of ownership and responsibility directly related to the success of the company.

Associates of Vince Lombardi explain that his greatest talent as a football coach was his uncanny ability to quickly evaluate potential players. He could talk to a prospect for 15 minutes and determine, with a high rate of accuracy, whether that player had sufficient desire to become a winner and how well he would fit into the team. Coach Lombardi understood the importance of recognizing individuals with the ability to work with and lead the team. He was able to identify attributes during the interviews that contributed to a greater team effort and desire to win. It is essential that managers and team leaders learn this skill too, i.e. identify behavior that matches individuals to their position requirements to ensure successful performance.

The group approach to problem solving gives a new dimension to the organization. With people in the habit of working together, new lines of communication and cooperation are established. Organizations that fail to develop these cooperative behaviors encounter duplicated effort, frustration, and the inability to efficiently respond to challenge.

The organization which develops cooperative behaviors will develop a number of new skills and characteristics that may have been lacking, or poorly utilized, prior to TQM implementation. These behaviors will contribute to a new culture in the company.

The results which successful companies enjoy as a result of TQM are normally greater than initially expected. These results can be seen in both financial and intangible benefits. For some, this is quite surprising; others entered the program with this as a motive. In addition to creating greater awareness and ownership of quality, they find that participation in the effort provides employees the opportunity to enhance their leadership skills and put down a foundation for team building.

The Stages of Team Development

There are four main stages of team building:

1. *Forming.* This involves awareness and the task of becoming oriented. Typically, this involves goal setting, commitment, acceptance (what can and can't be done), and rules of behavior.

2. *Storming.* Includes conflict and resistance to goals and tasks. Typically, people can be heard to boast, criticize, attack others, and generally struggle for control.

3. *Norming.* Here efforts to cooperate and promote open communication are developed. Typically, this includes respecting others, listening, learning, supporting the team, trying to work effectively together, and being polite.

4. *Performing.* Involves productivity and problem solving. Typically, team members work collaboratively, use milestones and goals, seek achievement and pride, and promote interdependence.

Why Teams? The idea of forming teams is sometimes scorned because people equate forming teams with forming committees. They view teams as another excuse for blowhards to pontificate. However, a well-run team can accomplish a lot more, more quickly, and generally at less cost than a single individual. The following table provides a rationale for why teams are so much more effective than individuals or groups. (A group is a collection of individuals brought together by fiat rather than a common vision.)

Group members	vs.	Team members
Compete against each other and internal goals.		Compete against external goals.
Think they are grouped together for administrative purposes only. Work independently; sometimes at cross-purposes with others.		Recognize their interdependence and understand both personal and team goals are best accomplished with mutual support. Time is not wasted attempting personal gain at the expense of others.
Personal agenda and task focused.		Team, company, and goal focused.
Focus on themselves because they are not involved in planning or setting objectives.		Feel a sense of ownership and responsibility for achieving objectives they helped set.
Stodgy, staid thinking; do only what is assigned.		Look for innovative and imaginative ways to improve the product.
Kick starters—need to be directed to move forward.		Self-starters, feed off of team accomplishments, forward focused.
Autocratic. Do as told; fear of loss is the main motivator.		Participative management—close the gap between where decisions are made and implemented.
Tolerate each other and what goes on at work.		Enjoy each other and what they do for a living.
Independent or overdependent.		Interdependent. Offer assistance and ask for help.
Distrust the motives of colleagues because they do not understand the role of other members. Expressions of opinion or disagreement are considered divisive or nonsupportive.		Work in a climate of trust and are encouraged to openly express ideas, opinions, disagreements, and feelings. Questions are welcomed.
So cautious about what they say that real understanding is not possible. Game playing may occur and communication traps be set to catch the unwary.		Practice open and honest communication. They make an effort to understand each other's point of view.

Consensus: How Teams Operate. One of the main goals of a team is to reach consensus on a project. Consensus is achieved when a proposal is acceptable enough with all members that they can support the project. To reach consensus the team must let all members participate fully in the decision making process. This will require an investment in time, listening skills, creative thinking, and open-mindedness. Do not be misled. Consensus does not mean that everyone, or even a majority, is totally satisfied.

What it means is that all team members have worked together to develop a project proposal. This proposal meets enough of the needs of all the members that they will support it in the work environment. It does not mean the conclusion is their first choice, but they can support it. All team members share responsibility for the project and its successful completion.

Expectations of Management and Teams. In an organization, team members and management will have certain expectations. These expectations must be understood and met if a productive work environment is to be achieved and maintained.

Teams will expect management to:

- Make decisions and take on projects the team sees as important in keeping with Critical Business Success Factors and company needs.
- Provide an environment that allows creativity, empowerment, and encourages accomplishments.
- Inform them about organization/company vision, mission, goals, projects, and progress.
- Give them some measure of autonomy and authority to take on delegated projects and to make decisions free of the bureaucratic process.
- Keep communication between management and other teams open.
- Appreciate and acknowledge success, and take responsibility for mistakes and breakdowns.

Managers will expect the team to:

- Commit to achieve the desired results through a relentless pursuit of the stated objectives.
- Cooperate between themselves, management, and other teams.
- Communicate with management via regular written reports, milestone plans, and resource committee presentations.
- Contribute by achieving the expectations as stated in the formal milestone plan.

Agreements and Expectations of Team Members

Agreements

1. I will start to engage as a team member, creating an understanding of what is expected of me as a team member and employee.
2. I will gain a complete overview of what it takes to be a successful team member.

3. I will be completely informed about the training and skills that will be provided for me as a team member.

4. I will be at choice to place myself actively on the team.

5. I will communicate openly and honestly with my team.

6. I will attempt always to reach consensus with team members for the success of the team.

7. I will make my commitments.

Expectations

1. Discover just why I am here, if I don't already know.

2. Have all questions answered regarding the team's role in the bigger picture for my company.

3. Have all questions answered regarding my role on this team and the entire company team.

4. Understand what will be expected of me in my job and on the team in the future.

5. I will find out what happens if I don't want to be on the team.

6. I will find out what is in this for me.

7. Make it happen. In other words, achieve the desired goal or objective.

What Makes Strong Teams?

The key to a strong team can be summed up in a few words: two-way communication. In fact, according to many experts, combined with job satisfaction, this is the most important ingredient in effective teamwork. Identifying two-way communication as crucial to establishing a strong team is easier than accomplishing it.

Several situations may arise in the team that can ruin the lines of communications. They must be guarded against to ensure the team will achieve its goals. Be sure not to *judge* an idea. Instead, evaluate whether or not it will solve the problem. If it will, then use it. If not, then search for another idea. No member of the team should feel *superior* to others; everyone is an equal. Trying to *control* other team members will also result in a breakdown of communication. All members of the team must feel that they are free of unnecessary constraints. They must not feel like their power has been taken away. *Manipulating* team members will not only damage communication, but also build resentment. Any member who feels *indifferent* toward the team will destroy the "team" feeling and communication. All members must feel dedicated to the teams goals. *Certainty* also stalls communication since you have shut the door to discussion from other team members, assuming you already have all the answers.

Communication can be established by avoiding the previously mentioned qualities, and by integrating the following qualities into the team. *Describe* your ideas. Don't assume that other team members will automatically know all the points of your idea. Ensure understanding with clear communication of all the points. Treat all members as *equals* to ensure camaraderie in the group. You must be *open* with all team members. This helps to build trust between all members. Stay *problem-oriented* with all your efforts. Don't waste your time or energy on projects which will not lead to solving the problem. Always keep a *positive attitude.* Allowing a negative attitude to cast a shadow over the group will reduce the team's productivity and quality of work. Be *understanding* of the other team members. See a situation from their point of view, and be empathetic toward that view. *Trust* is also a vital component of communication. It allows the other team members to communicate with you openly, without holding anything back.

Functional Roles of Team Members

Team members serve many roles within a team. These roles can be both constructive and destructive to the harmonious workings of the team. In the following paragraphs you will find definitions of both constructive and destructive team member functions. Following each functional description, you will find examples of each type of behavior.

Task-Related Member Functions.   These are activities that help a group work on its task. By performing one or many of these functions, any member of a group can help the group determine the exact nature of its problem and work toward its solution. When any of these functions are omitted, the effectiveness of the group declines.

1. *Initiating.*   Helping the group get started by proposing tasks or goals; defining a group problem; suggesting a procedure or idea for solving a problem.
2. *Information or opinion seeking.*   Requesting facts, asking for clarification of statements that have been made; trying to help the group find out what people think or feel about what is being discussed; seeking suggestions or ideas.
3. *Information or opinion giving.*   Offering facts or additional useful information; expressing what one thinks or feels; giving suggestions or ideas.

4. *Clarifying or elaborating.* Interpreting or reflecting ideas and suggestions; clearing up points of confusion; offering examples to help the group imagine how a proposal would work if adopted; distinguishing alternatives of issues before the group.

5. *Summarizing.* Pulling together related ideas; restating suggestions after a group has discussed them; organizing ideas so that the group will know what it has said.

6. *Setting objectives.* Expressing objectives for the group to achieve; applying standards in evaluation; measuring accomplishments against goals.

7. *Testing workability.* Applying suggestions to real situations so that groups can examine the practicality and workability of ideas.

8. *Consensus checking.* Sending up trial balloons to see if the group is nearing conclusion; checking to see how much agreement has been reached.

Maintenance-Related Member Functions. These are activities that maintain or build the morale or spirit of a group. They help the members of the group work together so that they develop a loyalty to one another and to the group and its task. When any of these functions are omitted, the effectiveness of the group declines.

1. *Gate-keeping.* Attempting to keep communication channels open; making it possible for others to make their contributions to the group; suggesting procedures for better sharing in the discussion.

2. *Willingness to change.* When one's own idea or status is involved, offering to change one's position in the group for the sake of goals of the group; admitting error, disciplining oneself in order to maintain group unity and win/lose decisions.

3. *Harmonizing.* Attempting to reconcile disagreements; trying to provide common-ground compromises for opposing points of view so the group can continue to work; getting people to explore their alikeness as well as differences.

4. *Relieving tensions.* Trading off negative feelings by jesting or pouring oil on troubled waters; putting tense situations in wider context.

5. *Encouraging.* Being friendly, warm, responsive to others and their contributions; helping others to contribute; listening with interest and concern; reinforcing others' participation.

6. *Diagnosing.* Determining and publishing sources of difficulty; seeking appropriate steps to take next.

Disruptive-Member Roles. Normally these roles or behaviors occur in the group while it is in a developmental stage. They are attempts by members of a group to satisfy their own individual needs, in the process of which they block progress toward group goals or loyalty to the group or its task. Some of these are:

1. *Blocking.* Being negativistic and stubbornly resistant; disagreeing and opposing without or beyond "reason"; attempting to maintain or bring back an issue after the group has rejected or bypassed it.

2. *Attacking.* Deflating the status of others; expressing disapproval of the values, acts, or feelings of others; attacking the group, the leader, or the problem being worked on; joking aggressively; showing envy toward another's contribution by trying to take credit for it.

3. *Playing.* Displaying lack of involvement in a group's processes by cynicism, nonchalance, horseplay.

4. *Recognition seeking.* Boasting; reporting on personal achievements; acting in unusual ways; struggling to prevent being placed in an "inferior" position.

5. *Deserting.* Withdrawing in some way; being indifferent, silent, aloof, excessively formal; daydreaming; deliberately talking about own experiences when unrelated to discussion of group.

6. *Pleading special interests.* Speaking for the shop-floor worker, the grass-roots community, the "father," etc., usually cloaking one's own prejudices or biases in the stereotype which best fits his or her individual needs.

7. *Dominating.* Asserting power or superiority to manipulate the group or certain members of the group by flattery; asserting a superior status or right to attention; giving directions autocratically; interrupting the contribution of others.

Building Trust in the Team

Trust must be built into any relationship, and this is especially true in a team environment. At times this trust may be betrayed, and you must know how to handle this situation. The best approach is to begin the relationship with trust. Continue to trust the person until he or she betrays that trust. At that time you should quickly confront the person. You must ascertain why the trust was betrayed, then correct the situation. Once this is done, return to the original state of trusting the person. By returning to the trusting state you are reinforcing the conditions of the relationship that you want. People don't want trouble, so once they are back in your good graces they will want to do whatever is necessary to stay there.

There are several ways in which people can be influenced, reciprocity being one of them. Most people feel obligated to reciprocate. When a friend invites you over to dinner, you then feel obligated to have that person over to your house. This can also be applied to customers. If you give them great service, they will feel obligated to use your service or product in the future.

Scarcity will influence people in that they will pay more for the scarce product or service. An example of this is people at an auction. People will pay high prices for rare art and collectibles. This also represents the downside for the customer; they may be so caught up in the crowd's desire for the item that they will pay more than the item is worth. Once the customer realizes this, he or she will become resentful and angry.

Be very careful with *authority*. You can easily be carried away by it, and just as easily be hurt. Many people resent strong authority figures, especially those who do not listen to those under them. When this kind of resentment builds up in people they will not tell the authority figure when they see that person making mistakes. Instead they will stand back and watch the authority figure make the mistakes and take a major fall.

Consensus is a powerful tool, but it is also a double-edged sword. Following the crowd can be very helpful if the crowd is correct. It is usually a safe bet that if a lot of cars are parked at a restaurant, the food is good. But you should be sure the people who parked the cars are actually eating in the restaurant. In Singapore, a bus strike caused a large crowd to form at a bus stop. Unfortunately this bus stop was in front of a bank. People passing by thought the crowd was taking their money out of the bank, so they went in and took their money out. Soon there was a run on the bank. It's not necessarily a bad idea to follow the crowd—as long as you know where they are going.

It is a common belief that people who are *consistent* possess intellectual strength. This is partially true since to be consistent requires knowledge and planning. However, to be consistent for the sake of being consistent can cause you great harm. If old ways are followed consistently without taking new facts into consideration, then you will consistently make mistakes. This is not the reputation that anyone wants.

You should strive to have people *like* you. People will like you if they share things in common, whether it's hobbies, careers, or points of view. Praising a person for a job well done is also a good way to get them to like you. Both of these elements can be fostered through teamwork. It is essential that enough rapport be built between people that a strong bond is created.

Pitfalls in Teams

As you have already deduced, establishing and running effective teams is not a simple matter. There are a myriad of problems which can befall a team; however, there are some problems which seem to crop up more often than others. The most common problems are:

- Not preparing the group to be a team. It is essential the group understands its role and expectations.
- Not selecting a strong team leader—a strong leader can lift the level of performance of a team. A team cannot lift the level of performance of a team leader.
- Not holding members accountable for their assignments being complete, on time, and effective.
- Selecting a process which no one is interested in improving.
- Not having team members on the team who can benefit from the solution. The beneficiaries will become the cheerleaders.
- Not having team members on the team who can technically contribute to the solution. Problems will not be solved.
- Not having regular weekly meetings unless a valid reason exists. The regular meetings will establish a heartbeat for the team.
- Selecting a desired solution instead of studying the process.
- Not publishing weekly progress and milestone reports.
- Not putting measurements in place to determine if progress is being made.
- Not having process management conducted by higher levels of management.

Teamwork Communication Checklist

If you are heading a team, or if you are a member of a team which is not experiencing much success, you might want to consider using the following teamwork communication checklist. The intent of the checklist is to provide a simple diagnostic tool to help identify what is ailing a sick team.

1. *Is my team in sync?*

_____ Restate specific objectives.

_____ Ask task-related questions.

_____ Express feelings about staying on task.

_____ Ignore off-task issues.

_____ Reinforce on-task results.

_____ Are measurements in place to evidence success?

2. *Difficult problems*

   _____ Gather team perspective.

   _____ Make summarizing statements for clarity.

   _____ Use polarity issues to clarify.

   _____ Expect the team to own the problem.

   _____ Don't jump at the first answer.

   _____ Facilitate conversation with open-ended questions.

3. *Difficult people*

   _____ Use concrete examples of problems.

   _____ Separate people from problem.

   _____ Solicit team perspective.

   _____ Ensure that everyone gets heard.

   _____ Align with the purpose statement.

   _____ Recall commitments.

   _____ Reissue challenges.

   _____ Realign the team.

Managing Conflict

In order to achieve results with a team, we have to recognize the fact that some conflict is inevitable. Since conflict within a team is inevitable, it will be necessary to learn how to manage conflict. When approaching conflict as a situation to be managed, it is important to recognize that different individuals will have different ways of dealing with conflicts. Figure 6-11 presents a model of the various ways in which people respond to conflict. Responses to conflict are as follows:

1. *Avoiding (uncooperative and unassertive).* A person neglects his or her own concerns as well as those of the other person by not raising or addressing the conflict issue.

2. *Accommodating (cooperative and unassertive).* One who seeks to satisfy the other person's concerns at the expense of one's own.

3. *Competing (uncooperative and assertive).* The opposite of accommodating; one uses whatever seems appropriate to win one's own position.

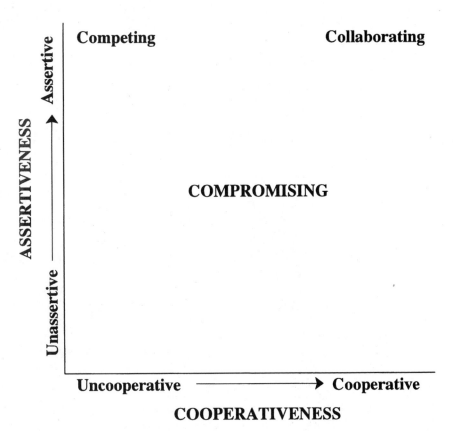

Figure 6-11. Responding to conflict.

4. *Collaborating (cooperative and assertive).* The opposite of avoiding; one works with the other person to find a solution that satisfies the concerns of both parties.

5. *Compromising (intermediate in cooperativeness and assertiveness).* One seeks an expedient middle-ground position that provides partial satisfaction for both parties.

Conflict Content. Once we recognize that different individuals respond to conflict in different ways, we need to also recognize that conflict can stem from a variety of causes. Four general causes of conflict are:

Facts	People see the same fact from distinctly different viewpoints.
Methods	People disagree on how to do something.

Goals   The goals toward which people work are different.

Values   People differ in their basic values.

In order to illustrate the point about how individuals can view the same fact from different viewpoints, we have developed a sample conflict situation. The situation pits two bureaucratic potentates: an office manager and an accounting manager.

Sample Conflict Situation

	Office manager	Accounting manager
Facts:	Automation will save the company money.	The new system will be more expensive to install and operate.
Methods:	The new system should be fully installed at once.	We need to move slower, one step at a time.
Goals:	We want accurate data rapidly, whenever we want it, and it should be managed by the people who will use it.	We need flexibility to meet our changing needs and it should be managed by the accountants who solve unexpected and complex problems.
Values:	We must be modern and efficient.	We must consider our long-term employees and be respectful of tradition.

Outcomes of Conflict. As with any situation which must be managed, conflict can have negative and positive consequences. In addition, we need to recognize that all of the responses to conflict, such as competition and accommodation, have their place. The key is to understand the consequences of these actions and to employ them when the situation warrants their use.

Negative Outcomes of Conflict

1. Results not being achieved
2. Decreased productivity
3. Relevant information not being shared
4. Unpleasant emotional experiences
5. Environmental stress
6. Excessive consumption of time
7. Decision-making process disrupted
8. Poor work relationship

9. Misallocation of resources

10. Impaired organizational commitments

Positive Outcomes of Conflict

1. Increased motivation and creativity

2. Healthy interactions/involvement stimulated

3. Number of identified alternatives increased

4. Increased understanding of others

5. People forced to clarify ideas more effectively

6. Feelings aired out

7. Opportunity to change bothersome things

When to Choose Certain Conflict Styles

When to Compete

1. When quick, decisive action is needed

2. On important issues for which unpopular courses of action need implementing

3. On issues vital to company welfare when one knows one is right

4. When protection is needed against people who take advantage of non-competitive behavior

When to Collaborate

1. When both sets of concerns are too important to be compromised

2. When the objective is to test one's own assumptions and/or better understand the views of others

3. When there is a need to merge insights from people with different perspectives on a problem

4. When commitment can be increased by incorporating others' concerns into a consensus decision

5. When working through hard feelings that have been interfering with an interpersonal relationship

When to Compromise

1. When goals are moderately important but not worth the effort of potential disruption of more assertive modes

2. When two opponents with equal power are strongly committed to mutually exclusive goals

3. When temporary settlements are needed on complex issues

4. When expedient solutions are necessary under time pressure

5. If a backup mode is needed when collaboration or competition fail

When to Avoid

1. When an issue is trivial

2. When there is no chance of getting what you want

3. When the potential damage of confrontation outweighs the benefits of resolution

4. When one needs to cool down, reduce tension, and regain perspective and composure

5. When the need is to gather more information

6. When others can resolve the conflict more effectively

7. When the issue seems symptomatic of another fundamental issue

When to Accommodate

1. When one realizes one is wrong

2. When the issue is much more important to the other person

3. When "credits" need to be accumulated for issues that are more important

4. When continued competition would only damage the cause

5. When preserving harmony and avoiding disruption are especially important

6. When subordinates need to develop and to learn from mistakes

The Potentially Negative Consequences of Certain Conflict Responses

Potentially Negative Consequences of Competing

1. Eventually being surrounded by "yes" people

2. Fear of admitting ignorance or uncertainty

3. Distorted perceptions

4. Reduced communications

5. Damaged relationships

6. No commitment from the other people

7. Having to keep "selling" or policing the solution during implementation

Potentially Negative Consequences of Collaborating

1. Too much time spent on insignificant issues
2. Ineffective decisions made from input from people unfamiliar with the situation
3. Unfounded assumptions about trust

Potentially Negative Consequences of Compromising

1. No one fully satisfied
2. Short-lived solution
3. A cynical climate through perception of a "sell out"
4. Losing sight of the larger issues, principles, long-term objectives, values, and the company welfare by focusing on short-term goals

Potentially Negative Consequences of Avoiding

1. Decisions made by default
2. Unresolved issues
3. Energy sapped by sitting on issues
4. Self-doubt created through lack of esteem
5. Creative input and improvement prevented
6. Lack of credibility

Potentially Negative Consequences of Accommodating

1. Decreased influence, respect, or recognition by too much deference
2. Laxity in discipline
3. Frustration as own needs are not met
4. Self-esteem undermined
5. Relinquished best solution

Conflict-Control Tactics

1. Use an avoidance style and ignore the issue.
2. Use an accommodating style and allow the other person to resolve the conflict in his or her favor.
3. Structure the interaction so that a triggering event is unlikely to occur.
4. Strengthen the barriers that inhibit the expression of conflict.
5. Avoid dealing with the party with whom you are in conflict.
6. Learn better coping mechanisms and refocus energy in a more productive area. Seed out support and assistance in this regard.

Steps for Confronting Conflict

1. Explain the situation the way you see it.
2. Describe how it is affecting performance.
3. Ask for the other viewpoint to be explained.
4. Agree on the problem.
5. Explore and discuss possible solutions.
6. Agree on what each person will do to solve the problem.
7. Set a date for follow-up.

Summary

With strong leadership, a company will be successful in implementing TQM and be able to meet any future business challenges. A company may not have capital; it can acquire it. A company may not have equipment; it can purchase it. But a company without leadership is one that is bankrupt.

Leadership is a process which can be learned and refined. In fact, effective leaders need to frequently hone their skills and develop habits which reinforce their effectiveness as leaders.

Leadership is not a trait which is manifested by employees only in the boardroom. Employees at all levels must exhibit leadership if TQM is to survive in an organization. One very useful vehicle for practicing leadership skills for all levels of the organization is in the framework of teams. Like leadership itself, effective team building and membership is a skill which requires training and practice. With the opportunity for personal growth which a team provides, there are also responsibilities. The responsibilities extend to management regarding the teams which they have chartered, to teams regarding management, and finally to individual team members.

Because leadership and team building is a process, one of the processes which one must be prepared to employ is the process for managing conflict. Implementors of TQM must acknowledge that conflict is a natural part of human existence and that it is inevitable when an organization undertakes a difficult tasks such as implementing TQM. Part of the process of managing conflict is the recognition that different people have different responses to conflict. All of these responses have consequences which can be either positive or negative. There is no standard response to conflict. Effective leaders school themselves in the management of conflict and possess a variety of skills which they can employ, depending upon the given situation.

7

Management
Skills Track

Purpose

The purpose of the Management Skills Track is to develop managers to become more efficient in achieving desired business results. The crucial point to keep in mind is that the use of these techniques must reinforce the vision, mission, and values of the organization and achieve the goals of the Critical Business Success Factors.

Establishing the vision, mission, and values is the responsibility of the senior leaders. Once the leadership has firmly committed to implementing TQM, then it is the responsibility of the organization's management, including the senior leaders, to effectively implement the leadership's goals or vision.

Keep in mind that there is an important contrast between leadership and management skills. *Leadership* involves influence over group or individual behavior (doing the right thing), while *management* is concerned with an organization's administration or efficiency (doing things right).

Managing a department means selection of personnel, planning, staffing (i.e., right-sizing the workforce to the activities), integration of activities with other departments, directing work within the department, coordinating projects and scheduling the work, developing measurements of performance, and monitoring performance and effectiveness. All of these skills are important in the TQM world along with the leadership qualities. Team leaders will be responsible for assigning work to team members, developing personnel, presenting progress reports to the resource committee, and maintaining the project milestone chart. Implementing and sustaining TQM will serve as practical management experience for all team leaders. Without competent personnel, any TQM undertaking is destined to fail. Therefore the significance of the development of management skills becomes a top priority.

Management skills include the techniques employed to select and coach employees, set work priorities and allocate resources, run effective meetings, and use performance appraisals in TQM organizations to link performance to the appraisals rewards. These techniques are addressed as a track in TQM for the reason that, without effective management skills, the difficult and arduous process such as implementing TQM cannot succeed.

Objectives

The objectives of the Management Skills Track are as follows:

- Gain knowledge of effective techniques to use during the hiring interview.
- Develop the ability to observe and evaluate performance behaviors, and coach to develop desired behaviors.
- Develop the ability to observe and evaluate behaviors and match the behaviors to the position, i.e., in normal functioning mode as well as during interviews with new-hire candidates.
- Be capable of effective coaching to achieve results.
- Improve individual efficiency by prioritization and time management.
- Be capable of operation in the TQM roles and responsibilities and in the reporting process.
- Be capable of managing effective and efficient meetings.

Introduction

The Management Skills Track deals with skills which are crucial for managers in a TQM organization. In this track, five major topics will be addressed. These topics are:

- Observing and interpreting behavior for hiring and coaching personnel
- Coaching for results and behavior
- Managing multiple priorities
- Results-oriented TQM meetings
- Performance appraisals in a TQM organization

The key to an organization's success is the caliber of the people within the organization. It is impossible to create a championship team with second-rate individuals. It is possible to improve the performance of less-talented individuals, but the amount of accomplishment is ultimately limited by the amount of raw talent an individual possesses.

Interviewing can cost a company money—and it can be *expensive*. In 1987, considering managerial time alone, American business spent $26 billion preparing for, conducting, and evaluating interviews. Add to this the expense of personnel to handle the staffing process, employment ads, record keeping to comply with regulations, and money spent on "hospitality" for applicants. The American Management Association estimated that it takes $50,000 to find and relocate a manager for an $80,000 job.

There are all of the associated costs of turnover if a newly hired candidate doesn't perform on the job or leaves in search of greener pastures: lost production, advertising, recruiting, training, severance, outplacement, and unemployment insurance. Add to that the hidden costs of time disciplining, documenting performance, and terminating. *Fortune* magazine found that an employee who flops and leaves after a few months can cost a company anywhere from $5000 for an hourly employee to $75,000 for a manager.

Some other statistics show that if you make a mistake in hiring it could cost the company *two times* the position's actual salary. One way to avoid these costs is to perform an effective evaluation process before bringing an employee on board. This evaluation, or cost-avoidance process begins with an evaluation of the candidate's résumé. The following section contains some rule-of-thumb tips to use when evaluating résumés.

Observing and Interpreting Behavior for Hiring and Coaching Personnel

Since competent personnel are a vital ingredient to successful TQM, one can either develop the existing staff or hire the necessary talent. Developing from existing employees is preferred as it creates personal growth in the organization and demonstrates the performance will be recognized and rewarded. In addition, the personnel do not need a learning curve to learn the business, i.e., only to learn the behavior required for their development.

It is important that the individuals to be developed fulfill two criteria: first, they desire to be developed to greater breadth, depth, and responsibility; second, they have the capacity to be developed in keeping to their objectives. The latter will show up as the individual grows; however, the former will show up during the interview.

Since individual behaviors are derived from attitudes, and attitudes from values, it is imperative that the managers and supervisors are skilled in identifying the behaviors and coaching the individuals who have a gap between observed and desired behaviors.

The same knowledge and skills are required by the managers and supervisors when hiring new personnel. If the selection is right, the new personnel can quickly start to achieve the behaviors and need only learn the organization and the company's products. Behavior models for each position should be developed. These behavior-model and job-description requirements should be used to evaluate all new applicants and to evaluate performance of current employees.

Evaluating Résumés

Things to Look For

- Signs of achievement and profit-mindedness
- Patterns of stability and career direction
- Specifics in job descriptions
- Willingness to work hard as evidenced by descriptions of major completed tasks and projects

What to Be Wary Of

- Lengthy descriptions of education
- Obvious gaps in background
- Trivia in "personal" section
- Overabundance of qualifiers
- Bitter tone in describing past employment
- Overly slick print and graphics

After the avalanche of résumés that resulted from the newspaper ads has been sifted through, it is time to begin the interview process. Keep in mind that the interview process is the single most powerful tool in the hiring process. Because it is such a powerful tool, use it wisely. Following are some tips for effective use of the interview tool.

First, use a systematic approach—don't "wing it." Second, have specific objectives when you sit down to interview someone. Third, remember, you are also the representative of the company; the candidate must sell his or her talents, and you must sell the organization to the candidate. Fourth, don't prejudge on the basis of the résumé; the résumé is simply the first screen in a multiscreen process. Fifth, don't make the mistake of assuming you are a good judge of character; this can result in costly mistakes and the possibility of law suits.

Types of Interviews for New Hires

There are two types of interviews for hiring, the first type of interview is the focused interview for meeting position's behavior model's requirements. Typically, the focused interview is job-specific. It often consists of a lengthy, structured discussion to determine if the candidate possesses the knowledge, skills, behaviors, and experience necessary to perform well in the position for which the candidate is being interviewed.

The second major type of interview is the evaluation interview for growth potential. It consists of in-depth discussion to elicit information from the candidate to determine his or her desire and ability to advance beyond the position being interviewed for. Normally, it is desirable to target for at least one level above the current position.

Basic Elements of the Interview. Once you have determined the general type of interview that you will be conducting, you need to keep in mind the basic functions of the interview:

1. Determine relevance of applicant's experience and training to the requirements of the specific position.
2. Appraise the applicant's behavior relative to the requirements of the specific position's behavior model.
3. Provide a process for judging high performers and upward potential.

Basic Principles. Behaviors are indicators or values. Since TQM is value-driven, it is essential to observe and interpret the behaviors. As the interviewer, you do only 10 percent of the talking by asking open-ended questions. The accuracy of your evaluation will depend on the amount of information you elicit from the candidate.

Fifteen Steps to a Successful Interview. In order to make the task of interviewing easier, here is a list of key interviewing tips:

1. Screen carefully.
2. Have a plan.
3. Follow a logical sequence.
4. Create a proper interview environment.
5. Put the candidate at ease.
6. Let the candidate do the talking.
7. Perfect your questioning technique.
8. Become a better listener.
9. Keep your reactions to yourself.
10. Stay in control.
11. Take notes.

12. Don't oversell the position.

13. Conclude on a proper note.

14. Write an interview summary.

15. Learn from each experience.

Questioning the Candidate. The questioning should initiate at the point of high school. Ask interviewees what they decided to do after high school and why they made the decision. Then have the interviewees chronologically discuss each of their jobs, why they changed, what their responsibilities were, what they liked about their responsibilities, and what they didn't like about the job. However, emphasis should be placed on the more recent positions.

In questioning the candidate, you might want to concentrate on the areas of work history, job-related skills and knowledge, general intelligence and aptitude, attitudes and personality, education, and life experience. One tool that you might use in order to hit upon these areas is the following list of questions.

Observing and interpreting behaviors		
Question	Performance criteria	Use?
What is your supervisor's title and what are your supervisor's functions?	_____	_____
Describe a typical day in your job.	_____	_____
Tell me about people you hired in your last job. How long did they stay with you and how did they work out?	_____	_____
Tell me your single most noteworthy accomplishment in your career.	_____	_____
If you ran into this situation [give a typical problem situation the candidate might encounter at your company], how would you handle it?	_____	_____
How do you go about making important decisions?	_____	_____
What are some things your company could have done to be more successful?	_____	_____
What do you consider are your biggest failures or frustrations in your business life?	_____	_____
Tell me risks you have taken and the result of such risks.	_____	_____

(Continued)

Observing and interpreting behaviors		
Question	Performance criteria	Use?
What do you do when you're having trouble solving a problem?	_____	_____
What could you do to make yourself more effective in your job?	_____	_____
Describe your best boss.	_____	_____
What is the most monotonous job you ever had?	_____	_____
Why did you decide (or not decide) to go to college?	_____	_____
What does a job have to have in order to give you satisfaction?	_____	_____

In order to use the list effectively, you should read each question. Decide if it is relevant to the position for which you are interviewing. If it is, check the "Use?" box. Then identify the performance criteria required by the position which it addresses.

Criteria for Performance-Oriented Individuals

As mentioned earlier, in order to effectively use the preceding list of questions, it is necessary to identify the performance criteria required by the job which each question addresses. Following are definitions of useful performance criteria along with examples of how these criteria are often displayed by individuals.

Intellectual ability. The capacity to think logically, practical analytically, and conceptually, as well as the ability to express oneself articulately.
Example: Observes, thinks, and delivers accurate impressions.

Decisiveness. Analyzes alternatives and commits to definite choices in a timely manner. Not tentative, quick on uptake, definite career plan.
Example: Makes his or her position known and understood.

Energy/enthusiasm. The capacity to work vigorously and actively without fatigue. The tendency to express positive attitudes, emotions and energy.
Example: The ability to exude energy even when negativism abounds.

Result orientation. The intrinsic desire and commitment to achieve results and complete what one starts. Answers the questions asked, gets to the point, emphasizes accomplishments and does not digress.
Example: Task-oriented and focused. Record of achievement.

Maturity.   Appropriate dress and behavior, poised and relaxed. An ability to exercise emotional control and self-discipline. The ability to behave responsibly and learn from past experiences. Accepts responsibility for own actions.
Example: Responds to social values.

Assertivness.   An ability to take charge; speaks convincingly, disagrees, expresses unpopular options and intrupts.
Example: Is convincing and persuasive.

Interpersonal skills.   Friendly, responsive, and emphasizes people in responses. The tendency to be aware of and demonstrate appreciation, understanding, and concern for the feelings of others.
Example: Communicates warmth and caring to people.

Openness.   The ability to express opinions and feelings in a frank, candid, and straightforward manner.
Example: Willing to discuss failures and personal experiences.

Courage.   An ability to make difficult business and decisions regarding people in an objective and timely manner. Evaluates events and people critically.
Example: Independent thinker.

Proactivity.   An ability to take self-initiative, anticipates problems and accepts responsibility to get things done.
Example: Doesn't wait for boss to tell them what to do next.

Empowering ability.   The ability to trust and give expectations, guidelines, and authority to others to perform their responsibilities. Shares power and gives recognition.
Example: Provides clear expectations, guidelines, and resources to others.

Technical.   Knowledge, skills, decisions, behaviors, and responsibilities.
Example: Should the position be a manufacturing engineer, the individual should have technical education or experience in the tasks of the job description of the engineer.

Figure 7-1 provides a framework for recording your evaluations of the performance criteria discussed earlier. The forms are similar; however, the second form is a more in-depth from than the first in that it provides concrete examples of how specific performance criteria might be displayed. Each position should have an evaluation for the minimum rating and this should be listed under the model column. The evaluation of the interview or subject would be to compare the evaluators rating with the model and make the appropriate judgments.

Behavior Model Performance Criteria Summary

Position
        Production Supervisor

Name   Date

Critical Behavior	Observations	Model (Norms)	Rating Low-Avg-High				
Intellectual		3	1	2	3	4	5
Decisiveness		3	1	2	3	4	5
Energy/Enthusiasm		4	1	2	3	4	5
Results/Orientation		4	1	2	3	4	5
Maturity	This can be developed over time.	3	1	2	3	4	5
Assertiveness		3	1	2	3	4	5
Interpersonal Skills		4	1	2	3	4	5
Openness		4	1	2	3	4	5
Courage		3	1	2	3	4	5
Proactivity		3	1	2	3	4	5
Empowering		4	1	2	3	4	5
Technical	Depends on the position.	3	1	2	3	4	5

Refer Comments:

Figure 7-1.

Coaching for Results and Behaviors

Just What Is Coaching?

When many people think of coaching, they think of athletes and athletic coaches. Some athletic coaches were asked "What are the characteristics (qualities, approaches, and so on) of an effective coach?"

John Erickson, president of the Fellowship of Christian Athletes (former basketball coach of the University of Wisconsin and general manager of the Milwaukee Bucks) has this to say:

I have always thought that to be effective, a coach must be an excellent teacher and a person with leadership qualities.

These two general, but important, characteristics include more specific requirements such as:

1. Solid knowledge of what one is teaching
2. Good motivation skills as well as effective communication
3. A deep personal concern for each team member
4. Ability to make decisions under pressure and live with these decisions without second-guessing oneself
5. Willingness to confess mistakes and build upon the experience
6. Complete honesty in all situations
7. Willingness to be an example for players in all areas of life
8. Ability to keep all things in perspective, recognizing priorities of life— God, family, others, and work

Tom Landry, former football coach of the Dallas Cowboys, says:

An effective athletic coach must be a teacher, a psychologist, and a motivator. He must possess leadership qualities and, at the same time, he needs the confidence and concentration to operate at maximum efficiency under great stress.

These athletic coaches speak well to the needs of an effective manager as "coach." Companies asking to teach their managers and supervisors how to coach had many different ideas about what coaching meant. Some had in mind giving employees timely information about their performance or reviewing and adjusting performance expectations. Still others seemed to equate coaching with giving encouragement and inspiration— a kind of pep talk. The most common way, however, was in terms of counseling, e.g., "performance counseling" or "performance coaching."

"Coaching" is sometimes used to describe a specific action such as encouraging, reinforcing, giving feedback, demonstrating. At other times, it is used to denote the style of managers who have a developmental orientation toward employees and who manage by giving employees greater challenges, autonomy, and power.

Basically, coaching is a conversation.

Coaching is eyeball-to-eyeball management!

Coaching is a management skill which effective managers use to improve their most valuable resource: their people. Just as with a produc-

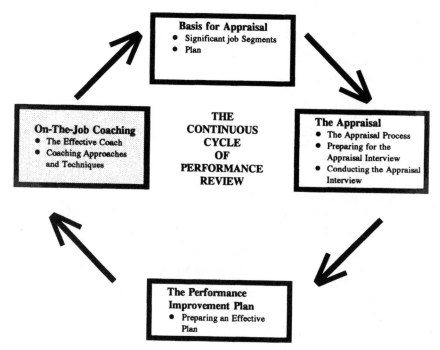

Figure 7-2.

tion process, it is possible to continuously improve employees. The model shown in Figure 7-2 is a graphic illustration of this continuous improvement process.

The Four Functions of Coaching

The frequency of coaching is a function of need. When the situation calls for a change in performance or behaviors, then coaching is appropriate.

Another situation appropriate for coaching is when it is desired to continuously and steadily improve performances and develop an action plan. In either case, a monthly meeting to review performances and improvement plans would be held.

Coaching is a management skill which allows effective managers to successfully engage in the continuous cycle of performance review illustrated previously. The following lists enumerate the four functions of counseling. Beneath each function, you will find bulleted descriptions of the specific skills which each function helps develop. The four functions of effective coaching are:

1. *Counseling*
   - Accurate descriptions of problems and their causes
   - Technical and organizational insight
   - Venting of strong feelings
   - Changes in points of view
   - Commitment to self-sufficiency
   - Deeper personal insight about one's feelings and behavior

2. *Mentoring*
   - Development of political savvy
   - Sensitivity to an organization's culture
   - Personal networking
   - Greater proactivity in managing one's career
   - Commitment to the organization's goals and values
   - Sensitivity to senior manager's likes and dislikes

3. *Tutoring*
   - Increased technical competence
   - Increased breadth of technical understanding
   - Movement to expert status
   - Increased learning pace
   - Commitment to continued learning

4. *Confronting*
   - Clarification of performance expectations
   - Identification of performance deficits
   - Acceptance of more difficult tasks
   - Strategies to improve performance
   - Commitment to continued improvement

In order for coaching to be successful, there are key characteristics which one must observe. First, effective coaching is done during a one-to-one conversation. Second, the discussion must be focused on performance rather than on nebulous factors such as "attitude." Third, the coach must communicate respect when coaching another. Finally, the emphasis of the coaching session must be on change in the future rather than dwelling on past performance. When coaching meets these criteria in a mutual conversation which focuses on problems in a disciplined way, then there will be observable outcomes such as improved performance.

Criteria for Successful Coaching

Successful coaching is characteristic of the following attributes.

- The issues are mutually a concern to both the coach and the coached.
- The focus is on the problem, i.e., performance deficiency or behavior gap.
- The process is disciplined in that the approach is systematic.
- The coaching activity is based on the employee's specific skills.
- The communication is mutually respected between coach and coached.
- The orientation of coaching is future-oriented and/or change-oriented.
- The coaching activity is a process itself.

The results of coaching are the following observable outcomes:

- A positive change in performance
- A positive work relationship between manager and subordinate
- New desired behaviors or higher levels of desired behaviors

The Two Processes of Coaching

There are two primary coaching processes. Process 1 addresses the coaching functions of counseling, mentoring, and tutoring. Process 2 addresses the coaching function of confronting. The following step-by-step processes illustrate approaches which managers can use.

Process 1: Addressing Counseling, Mentoring, and Tutoring

1. Indicate purpose and importance of the discussion.
2. Discuss and clarify details about the situation.
3. Summarize details and agree on desired outcomes.
4. Discuss possible behavior to achieve desired outcomes.
5. Agree on most effective behavior and develop a plan of action including target dates.
6. Express confidence and set a follow-up date.
7. Document the actions, target dates, and follow-up dates.

Process 2: Addressing the Coaching Function of Confronting

1. Describe the problem in a friendly manner.
2. Make a direct request to the employee to help solve the problem.

3. Discuss the causes of the problem.
4. Identify possible solutions.
5. Decide on specific actions to be taken by each of you.
6. Agree on a target date and a follow-up date.
7. Document the actions, target dates, and follow-up dates.

Earlier in this track, there was lengthy discussion about performance criteria. In addition, two forms were reproduced which could be used to evaluate a candidate's suitability for employment. These forms provided outward behaviors which could be used to judge a candidate's internalization of certain performance criteria.

The aforementioned forms can also be used to evaluate current employees as a means of identifying where an individual could use coaching. Consider using these forms in both scenarios: the hiring process as well as the identification of coaching opportunities.

Managing Multiple Priorities

Time is a precious, elusive resource requiring solid self-management for its effective use. Even with the most aggressive time management techniques, you will never be able to do everything. The management of time is a critical issue from team members through team leaders and resource committee members.

Managing time means the ability to realize which issues are significant and focus on those to their logical conclusion. Especially in the beginning of a TQM implementation program, there will be a number of activities that require attention, and choosing which to address first is fundamental to sustaining the program and meeting milestones. Of course the normal ongoing tasks involved with running the business can't be suspended while the TQM system is designed and installed. These too must be prioritized and delegated if necessary.

Managing time also means spending an appropriate amount of time on the activities necessary to accomplish your objectives. If there is something you enjoy doing, you are likely to allocate more time to that activity than would be indicated by the relative importance of the task. In other cases, low-priority tasks may take more time because of the complexity or the lack of support from other departments. It is fairly easy to allocate time spent in direct labor or in inspection activities, but with management activities such as planning, problem solving, or system development the task is more difficult. When you set priorities, also block out times to devote to each priority.

Effective time managers must accept responsibility for their own actions and inaction and must learn to make tough decisions. As soon as you accept someone else's problem and act on it, it becomes your problem. A basic ingredient to success is doing the things you should do when you should do them. The steps to normal time management are fairly simple to understand, but apparently difficult to put into action.

Clarify Your Outcomes. Managing time requires the creation of a plan for optimum use of your time. Planning requires outcomes (specific end results). Without outcomes it really doesn't make any difference how you spend your time. Most people don't think much about outcomes. They spend much of their time responding or reacting to pressures from other people or things. Success bypasses them as they shift from one activity to another without any focused or directed purpose. To set outcomes is to think about what is the best use of your time. Begin to write down your outcomes. Make sure they are specific, realistic, and measurable. Put a time schedule on your outcomes. Think in terms of results, not just activities. And, finally, establish priorities. Which outcomes are most important? Least important?

Analyze Your Time Habits. Surprisingly, most people really don't know how much time they spend in specific kinds of activities. Time is spent habitually, almost automatically, and without a single thought about where it goes. Record your time for a week or two. Where is your time being wasted? Are your time patterns consistent with your outcomes? What is the importance, or value, of each activity? How could you spend it better? Can you identify all your time wasters? Is more of your time wasted because of what other people do, or because of what you do?

Plan Your Time. To manage your time more effectively will require changing some of your time habits. Things will not take care of themselves. Things happen because people make them happen. If you don't plan things, then you settle for a rather vague, random existence. Random events are seldom as good as planned events. Plan your time daily, weekly, monthly, yearly.

Evaluate Your Progress. Are things working out according to plan? Have you reduced your wasted time? Are you achieving your outcomes? Plans must be reviewed to make sure they are still appropriate. Are your outcomes still the right ones? People and conditions change. As you and your situation change, make sure that your outcomes and your plans keep pace. And don't forget to develop new plans and outcomes as you accomplish current outcomes.

Making Managing Time a Daily Habit. At the end of the day review your day's accomplishments. What worked? What didn't work? Write out your outcomes for tomorrow and plan your activities. Look for ways to improve your time management skills. And don't get uptight. Relax. No one is perfect. If you work at it you will gain more control of your time. Keep focusing on your positive progress and your improvement will continue to mount.

Four Levels of Time Management

The introductory remarks on time management pointed out common problems that people have in managing time such as not knowing how their time is spent and not making the management of time a daily habit. Once an individual decides that he or she needs to manage time better, they frequently go through an evolution of time management methods. The four methods, in order of sophistication and effectiveness, are:

1. Notes and checklists (least effective)
2. Appointment books/calendars
3. Prioritizing/clarifying—goal setting
4. Self-management—balancing prioritizing (most effective)

In order to be an effective time manager, one needs to understand the Time Management Matrix shown in Figure 7-3. The matrix illustrates graphically that all activities have two classification descriptors: (1) urgent/not urgent and (2) important/not important.

Effective time managers use some classification system similar to that of Figure 7-3. Each of the four levels, or quadrants, address certain needs, but they also produce other consequences.

1. *Level I activities* focus on the accomplishment of short-term priorities, but they interfere with the accomplishment of long-term priorities. The consequences of level I activities are stress, burnout, crisis management, and a feeling that one is always putting out fires.

2. *Level II activities* contribute to the accomplishment of short- and long-term priorities. The consequences of level II activities are a vision of where you are headed, balance in life, discipline, control, and very few fires.

3. *Level III activities* take time but do not contribute to the accomplishment of either short- or long-term goals. The consequences of level III activities are a short-term focus, crisis management, having the reputation of a chameleon, seeing goals and plans as worthless, feeling victimized and out of control, and shallow or broken relationships.

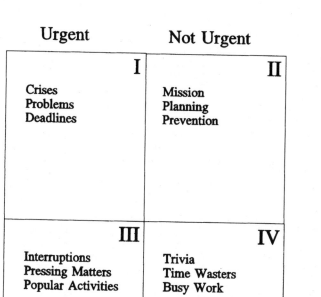

Figure 7-3. Time management matrix. Reproduced from the
book by Stephen R. Covey, *The Seven Habits of Highly Effective
People*, Simon and Schuster, New York, 1989.

4. *Level IV activities* do not contribute to the accomplishment of either
   short- or long-term priorities. The consequences of levels III and IV
   activities are total irresponsibility, being fired from jobs, and dependence on others or institutions for basics.

Priority Planning

Priorities can be accomplished only through effective *planning* and *execution*. In establishing priorities, always keep in mind the need to maximize
level II activities. *Effective plans must contain:*

- A clear goal statement. What do you want to accomplish? You must be
  able to answer the question, "How do I know when I am finished?"
- Specific steps to be taken. Be detailed, but don't get lost in minutia.

- Identified individuals responsible for accomplishing the steps. Can any of the steps (tasks) be delegated?
- A specific time table. Goals without time limits are better known as dreams.

Time Management Overview

Understanding Time

- Time management is not simple.
- Time is a precious and elusive resource requiring solid self-management.
- There is never enough time to do everything.
- There is never enough time to do it right, but there is always enough time to do it over.

Managing Time

- Develop the ability to identify significant issues, and focus on them until they are resolved.
- Spend the appropriate amount of time on the activities necessary to accomplish your objectives.
- Make tough priority decisions and act upon them.
- Accept responsibility for your own actions and inactions.
- Determine which priorities you can and cannot complete and learn to live with the difference.

Time Wasters

No discussion of time management would be complete without addressing the issue of time wasters. Time wasters are insidious traps which can wreck the most carefully laid level II planning activities. One of the secrets to effective time management is learning to control time wasters.

Everyone wastes time, but there is a difference between the people who consistently produce good results and those who don't. The producers manage to hold their wasted time to a minimum.

Exactly what is a time waster? You're wasting your time when you choose to spend it on a lower-priority task instead of a higher-priority task. Relative importance is always determined in relation to your outcomes (end results). And that's the first problem. Unless you know your outcomes, you can't accurately determine your time wasters. And if you have no outcomes, then it doesn't really make any difference what you do—anything will occupy your time. It's when you have outcomes—when you know what you want to accomplish—that wasting time is an important issue. What is a time waster to one person may not be a time waster to another. Who you are, what you do, how you approach your work, and what you are trying to accomplish all help to determine whether you're using your time well or wasting it.

As you identify your time wasters, you will discover they come from two general sources. The first source is from your environment—things that other people do, or don't do, that end up wasting your time. The second source of time wasters is you—those things that you bring on yourself that waste your time.

Working with thousands of managers has revealed a long list of time wasters. However, there are a handful that consistently appear at the top of the list. The following pages will offer comments on how you can start controlling a few of the most-often-mentioned timewasters.

Drop-In Visitors.   Realize that interruptions are a part of your job. You can't eliminate them all, and you wouldn't want to even if you could. But you can begin to manage your interruptions better. Learn to control the controllable and accept the uncontrollable. The frustration that comes with interruptions is often due to your attitude. Change your attitude and you will reduce the frustration you feel.

One of the biggest problems with drop-in visitors is the confusion surrounding the "open-door" policy. An open-door policy doesn't mean that you must literally be at the beck and call of anyone who has the strength to walk through the door. To allow others to constantly control how you spend your time is a sure way to get very little accomplished. The philosophy underlying the open-door policy means that a manager should be easily accessible to those who really need him or her. This philosophy can be easily implemented even if your office door is closed.

You can beneficially modify your open-door behavior in two ways. First, close your door occasionally, or establish a regular "quiet hour" when you will not be disturbed. But don't abuse it. This means that you must make certain that you are adequately available to your staff and others during other periods of the day. Try scheduling regular times to see staff members.

Another principle in solving the drop-in visitor problem is this: if the visitor never gets in, you don't have such a big problem. There are a number of ways you can discourage the common, time-consuming visitor. You might move your desk so you don't face the flow of traffic. Remove extra chairs from your office. Try intercepting people outside your office and you'll probably spend less time with them and still get the benefit of the contact. Remain standing and conversations won't last so long. When someone sticks his or her head in to ask if you've "got a minute or two," say "no" and suggest another time when it will be better.

Remember, too, that if you can keep interruptions short, you'll solve half the problem. Set time limits at the beginning of visits. Have your secretary interrupt you with a prearranged signal. Learn to be candid with people. Go to the other person's office. When you're the visitor it is easier to control the length of the visit. Don't contribute to small talk yourself. Get to the point and stay there.

If a few staff members are constantly coming in and out of your office, have them accumulate items and go over several things at one time. You might also analyze the situation. Should you delegate more authority? Are employees delegating their problems and decisions upward to you? Are you generally so unavailable that they feel they must grab you whenever they have a chance, or it is just a habitual pattern that should be changed?

Telephones.   Ask any group of managers about time wasters, and telephones are always mentioned in the top three. If only it wouldn't ring so often. Many managers also lament their inability to reduce social chitchat, their inability to terminate calls, their fear of offending callers, and a general lack of knowledge about how to control the telephones in their lives.

Several actions will help gain control of telephones:

1. Develop a plan for screening, delegating, and consolidating calls. Get your secretary to help.

2. Socializing helps make pleasant relationships; however, recognize that unnecessary socializing may stem from habit, ego, fear of missing something, desire to be liked, and even procrastination. Begin to sort yourself out and recognize your actions for what they are. Socializing can be reduced without becoming antisocial.

3. Plan your calls. Have information at hand; list outcomes; organize yourself before you call; be prepared to talk.

4. Establish periods for taking calls. Many callers will respect this and call then.

5. Group outgoing calls for greater efficiency.

6. Tell long-winded callers that you have another call, appointment, or emergency.

7. Stop worrying about offending people. Most people do not offend so easily as we like to believe. Don't be rude, but be firm.

Work-Area Organization.   A cluttered desk and a paperwork jungle is a prime indication of inefficiency. The purpose of the work area is to allow work to be accomplished effectively; it should be organized to make things easier, and the layout and organization should be simple. Special papers and materials should be accessible with minimal search. The phone and other office equipment should be within easy reach and placed near associated materials such as note pads or reference material.

Clutter expands to fill the available space and tends to hide important documents. It can increase stress induced by repeated searches for the right piece of paper. The key to victory over clutter is to retrieve, decide, and respond to items that pass across your desk. As an objective, handle each piece of paper only once and resist the urge to file things or to procrastinate. Respond to situations as they come up and delegate the responsibility when there is a conflict with your objectives.

Meetings.   Time wasted in meetings must be approached from two points of view: those meetings you call, and those meetings you attend. For both kinds you must consider activities before, during, and after the meeting.

Discourage and discontinue unnecessary meetings. Meetings should not be held as an excuse for failure to act individually. No meeting should be called without a definite purpose. Before calling a meeting, ask what you expect to accomplish with the meeting. Many meetings will be ended right then. (See the subsequent section on Results-Oriented TQM Meetings.)

Learn to make decisions without meetings. Never use a committee if it can be done individually. Managers are paid to make decisions and act upon them. Think through your own problems before carrying them to others.

Whenever you must hold a meeting, use an agenda and stick to it. Prepare the agenda ahead of time. Set a time limit for all meetings. Start on time and end on time. You will probably find that you can accomplish just as much in far less time than you have in the past. Don't wait for slow people to show up before starting.

Invite only those whose attendance is necessary and tell them exactly what will be expected of them. Allow people to come and go as their contribution is needed and completed. There is no point in having someone sit through an entire meeting if only a few minutes of it are significant to them.

Be prepared for the meeting. Resist interruptions. Stay on course. Insist that others be prepared for the meeting. Analyze each of your meetings from time to time. Are the right people attending? Are too many people attending? Are the people properly prepared? Is the meeting held at the proper time, in the proper place? How could meetings be improved?

Don't make people more comfortable during the meeting than is appropriate for the time you intend to keep them. In other words, don't give a person a two-hour chair if you only intend to keep them for 30 minutes. You might even try holding some stand-up meetings.

Be sure that follow-up is prompt. Summarize conclusions. Be sure that everyone knows exactly what they are supposed to do and when they should have it finished. If minutes are required, make sure they are distributed within 48 hours after the meeting. Instead of assigning a meeting participant to keep the minutes, bring in a secretary or assign one attendee as the scribe for the sole purpose of recording, word-processing, and distributing the minutes.

As a meeting participant you can encourage all these same points. Be prepared yourself. Come on time. Don't contribute to wandering or unnecessary talk. Encourage others to stick to the point. Ask the chairperson for an agenda and follow-up systems. Be the kind of participant you would like to have when you're chairing the meeting.

Crises.   Many managers believe that crises are an unavoidable part of their job. That's only partly true. Unique crises are often unavoidable. But the majority of crises in most jobs are either recurring crises, or crises brought about because of something you did or did not do. Whenever you're wasting time you are probably setting up some future crisis for yourself. When you put off a necessary task that you don't really want to do, you almost guarantee that it will become a crisis at some point.

Categorize your crises. How many of them are really unique, and how many are recurring? How many of them are your fault? Get data about your crises. Look for patterns. Anticipate problems. Develop contingency plans. Expect the unexpected. Learn to react appropriately. Don't go into crisis mode unless it really is a crisis. Keep your cool. Conquer procrastination. Learn to do it now. Stop putting things off until the last minute.

Start on projects earlier than usual. Give yourself more time to do it right in the first place and you'll spend less time having to do it over. Don't ignore deadlines.

Discuss priorities with subordinates. Check with subordinates, peers, superiors, and others to spot potential problems that may be brewing. When a crisis does strike, rest and relax for a few minutes before tackling the problem. Take time to prepare yourself for peak performance. And don't forget to practice good time management in the midst of the crisis. Don't start a second fire while trying to put out the first one.

Of course, there are some crises which are caused by events not under your control. Your boss may set unrealistic schedules, or switch priorities at the last minute. Machines break down, people make mistakes, information is distorted or delayed. Learn to live with these problems and stop fretting about them. Influence other people to change their behavior whenever possible. Solve the recurring crises so you will have more time to handle the unique crisis. But don't expect to ever have a job free of crises.

When a unique crisis occurs, turn it into an opportunity to try new ideas, develop new procedures, find better ways for doing things. A few crises may be just the impetus you need to make some positive improvement. Most crises come from our failure to use time effectively. Change your time habits and you will probably notice a decrease in your crises. And don't forget Murphy's three famous laws: (1) Nothing is as simple as it seems; (2) Everything takes longer than you think it will; and (3) If anything can go wrong, it will.

Procrastination. Procrastination was a major enemy in the development of this book. Putting things off wastes time because it allows you to avoid the things you should be doing when you should be doing them and throws you into a panic mode when deadlines arrive. Some techniques for limiting procrastination are:

- Admit that you are a procrastinator. Recognize this problem as a drain on effective use of time.

- Stop rationalizing your actions! Recognize the cost of procrastination and do something to correct the condition.

- Break tasks down into realistic steps to minimize the pain of getting into motion.

- Set deadlines for completing the subtasks and get positive momentum started. Once you get going, try to finish before you are sidetracked again.

Routine and Trivial Tasks. Ridding yourself of routine, trivial tasks becomes easier if you learn to emphasize results instead of activities. Think about your objectives. How do these trivial tasks add to your effectiveness? Work smarter, not harder. It's not how much you do that counts, but how much you get done.

Stop doing your own typing, filing, or other clerical work. Use dictating equipment instead of writing letters by hand. If you'd stop doing so many trivial tasks, you'd have more time to devote to the important things that are not getting done.

Review routine tasks. What would happen if they weren't done at all? If the answer is "nothing," stop doing them. Eliminate all unnecessary activities. Delegate all nonessential activities to the lowest level competent person.

Do important things first; do routine things last. Don't fill prime morning hours with trivial activities. And don't waste your peak performance periods on anything less than highly important activities.

Many managers have difficulty learning to delegate tasks; some never learn. Management, after all, is the ability to get things done through other people. Delegation and teamwork are critical to effective time management.

Even more serious than reluctance to delegate is the inclination to accept delegation of tasks and decisions from subordinates. You might want to be a "nice boss" or may feel that only you are fully capable of making decisions, but this is what is meant by empowerment. Everyone has to take responsibility for their own activities. Once you accept delegation from subordinates, they will gleefully assume that they are relieved of such responsibility, and your time management will be compromised. Learn to resist doing things for others that they probably could, and should, be doing for themselves.

TQM goes together perfectly with effective time management because the focus of TQM is to use objective data to set priorities for the improvement teams. Team meetings and team planning should incorporate all the concepts of time management to ensure the most cost-effective use of available resources.

Practice being efficient in performing routine tasks. Analyze them and look for shortcuts. Could they be done by machine? Could tasks be combined or modified? Don't get involved in "busy work" to fill gaps in your time. And don't waste time on routine things just because you justify it by saying you'll get them out of the way so you'll be free to tackle the bigger things later. Later may never come.

Time would be better spent coaching and training subordinates to take over your routine tasks so you can participate in higher-level decisions and activities.

One of the most important lessons of time management is how to say "no" to demands on your time that do not contribute to meeting your objectives. There are some other very simple practices that, when integrated into normal activities, will lead to more effective use of time. Elimination of the time wasters in small ways can result in substantial gains in time available during the week.

Certain tasks require higher levels of concentration and energy. College students learn that effective study habits include studying when you are at your physical and mental best. If you are a morning person, write your detailed analysis and reports in the morning. If you feel better in the afternoon, do your research and development after lunch. You probably know when your "prime time" is—use it to your best advantage. Plan times for seclusion at the best time for thought work when you can avoid interruptions.

Time Analysis Questions

1. What worked today?

2. What didn't work today?

3. What time did you start on your top-priority task? Could you have started earlier in the day?

4. What patterns and habits do you see in your time log? What tendencies?

5. Was the first hour of your day spent well, on important things?

6. What was the most productive period of your day?

7. What period of the day was the least productive?

8. Who, or what, accounted for most of your interruptions? What were the reasons for the interruptions? How could interruptions be controlled, minimized, or eliminated?

9. What were your three biggest time wasters today?

10. What could you do to solve your three biggest time wasters?

11. How much of your time was spent on high-value activities? How much on low-value activities?

12. What did you do today that could be eliminated?

13. What activities could you spend less time on and still obtain acceptable results?

14. What activities need more time?

15. What activities could be delegated? To whom?

16. Beginning tomorrow, what will you do to make better use of your time?

Finally time management means improving those things you can and learning to live with the difference. People often are drawn into or simply prefer to complain about all the things that are wrong while taking action on none of them. Sort out the most important tasks and refrain from expending energy on the others until you have worked your way down the priority list.

Results-Oriented
TQM Meetings

Work output isn't the sole province of the factory. White-collar employees and executives can make an equally important contribution to profitability by learning ways to "work smarter, not harder." Working effectively in meetings is one way to work smarter, not harder.

During the meeting each participant has some responsibility for the success of the meeting. The person who called the meeting may act as the leader or may assign a leader based on his or her familiarity with the objectives. The leader will be responsible for orchestrating the session, stating the objectives, and making sure that the group stays focused on the objectives. Each meeting should have a written agenda complete with the times allotted for each subject. The attendees should have a chance to review the agenda before the meeting begins. One member of the group can be assigned as the timekeeper, with the authority to interrupt conversation when time expires. Keeping the meeting on time will help the group arrive at the point where decisions are made.

The attendees should be encouraged to participate through presentation of information or opinions. Meeting leaders must keep the group focused on the objectives. At the same time they must not dismiss ideas quickly, even those lacking apparent relevance. Presentations should be planned to introduce a new proposal, suggestion, or course of action and then to solicit information from the other participants by asking specific questions or inviting contributions. For example, "If we improve yield at final test by 3 percent we can save the company $50,000 a year. Where would be the best place to start?"

Including all participants in the discussion will help to develop new ideas and explore the group's position on the topic. The leader might invite comment by asking more direct questions: "Bill, you haven't said anything about this. Do you think we can improve yields if we buy a new piece of equipment?" At times, the leader may find it necessary to take the opposite action and exclude comments that have drifted too far from the subject matter: "Ken, we all understand how anxious you are to replace your old computer system, but that doesn't have any bearing on our yield problem."

During the discussion the leader can restate ideas to summarize the discussion or clarify points that might not be understood by all the participants. These behaviors are particularly important at the end of the meeting so that the group can judge the effectiveness of the meeting and decide if further action is needed. One member of the team can be assigned as the scribe, who will have the responsibility of recording the points covered in the discussions, action assignments, and any decisions that were reached. It is especially important to review any assigned action in order to test the responsible member's commitment to the plan.

One behavior that should not be permitted is attacking or defending individuals instead of addressing the issues; for example, "Jim, you don't really care about the yield as long as you make your production quota." This is an emotional statement or value judgment which may evoke an argument that will be very difficult to control. Attacking or defending behaviors tend to quickly lead the group away from the original objectives. Disagreement on issues is a reasonable mode of discussion, but attack or defense of personal values should be avoided.

Effective, results-oriented meetings are the same as effective presentations. They require preparation and well-organized execution. With the addition of follow-up to action items, the team can make the most effective use possible of time spent in meetings.

Hopefully, you're convinced of the need for learning effective meeting techniques as part of one's management skills arsenal. Before continuing the discussion of meeting-related skills, let's take some time to understand the roles of various groups within a TQM organization. The reason for studying these roles is to better understand why these groups need to conduct regular meetings.

Meeting Skills

Since team concept is the media of achieving results, the effectiveness and efficiency of meetings is paramount. Now it's time to address specific techniques to make meetings more productive. First, let's start with a definition of a meeting. A *meeting* is any time three or more people get together to share information, develop plans, solve problems, or make decisions.

Some persons may question the need to make meetings more efficient. After all, aren't they simply part of one's working day? They may be part of one's workday—after all, some meetings are unavoidable; however, meetings can be costly. Some experts estimate that meetings occupy at least 35 percent of the time of middle managers and 60 percent of the time of upper managers. Another way of expressing this cost is to realize that these same experts estimate that meetings account for 7 to 15 percent of many companies' budgets.

The following checklists or guidelines help ensure effective meetings.

Musts for Results-Oriented Meetings

1. Is this meeting necessary?
2. Participants should be kept to a minimum.
3. Have an agenda.
4. Start with and stick to the agenda. The purpose of this meeting is . . .
5. Start and end the meeting on time.
6. Give important topics the most time.
7. Restate conclusion and action items.
8. Accomplish the purpose. Was the primary objective accomplished?
9. Complete and distribute progress reports.
10. Make sure decisions are made and actions are executed.

Guidelines for Making Meetings Work

1. *Frequency.*　Same time, same day, and regularly.
2. *Priority.*　Allow very few other requirements to have greater priority.
3. *Recording function.*　Record decisions, plans, future agenda items, assignments, and follow-up actions.

4. *Developing the agenda.*   The group should own the agenda and have a way to prioritize it.

5. *Kinds of problems appropriate for the group.*   Items that effect most of the group.

6. *Confidentiality.*   Have the group clear on what may be communicated.

7. *Minutes.*   Should be typed and distributed to all members within 24 hours.

8. *Continual evaluation.*   Devise a method of evaluating the effectiveness of the meeting.

In order for a meeting to be effective, there are four crucial elements. These four elements are logistics, roles/responsibilities, content, and process. Figure 7-4 shows the interlinking nature of these four elements. Note that below each element there are concrete examples. For example, the logistics element of an effective meeting ensures that the right people are on time, at the right place, and that the necessary supplies are available.

Meeting Roles and Responsibilities.   A meeting is a work process. The result of a meeting should be an identifiable product: a decision, a schedule, a plan, etc. Because it is a process, it is necessary to clearly define the roles and responsibilities of the participants. The following lists identify the four principal meeting functions with a corresponding list of roles and responsibilities.

Leader

- Sets agenda/logistics
- Opens meeting
- Guides and motivates
- Empowers
- Coaches/consults
- Participates
- Solicits consensus
- Listens
- Ensures accountability
- Pushes for closure
- Makes sure all can participate
- Focuses on process
- Defends others from attack/ridicule

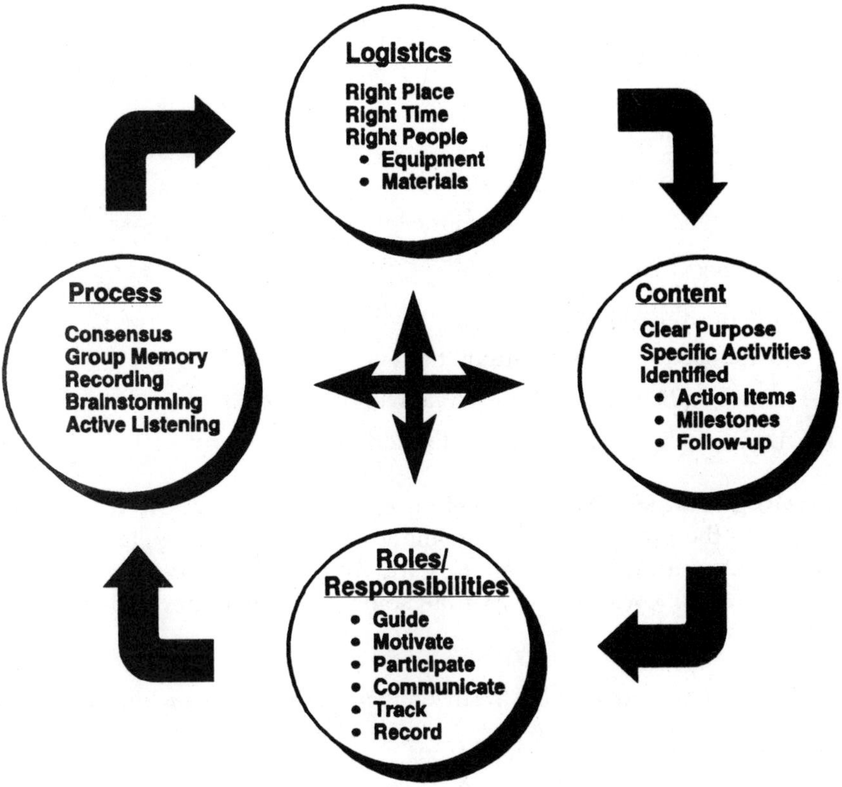

Figure 7-4. Effective meeting elements.

Participant

- Listens
- Analyzes
- Participates
- Makes/fulfills commitments
- Responds
- Helps keep meeting on course
- Keeps eye on recorder's accuracy
- Helps maintain teamwork

Recorder

- Records basic ideas
- Gets clarification when necessary
- Produces minutes
- Reviews minutes
- Updates members who missed meetings

Timekeeper

- Starts meeting on time
- Closes meeting at agreed-upon time
- Ensures sufficient time for each item on the agenda
- Ensures each item adheres to allocated time

Meeting Stages.   Since a meeting is a process, there are identifiable stages to the process. Not surprisingly, the three meeting stages are (1) before, (2) during, and (3) after. A truly effective meeting has all three stages. The following lists illustrate some of the activities which should occur at each stage.

Before the Meeting.   Plan the meeting by asking the following questions:

- Who should attend?
- What do you want to accomplish?
- When is the meeting?
- Where is the meeting?
- What materials are needed?

Plan a clear agenda:

- Set goals and objectives.
- Decide the process to be followed.
- Is any data required?

During the Meeting.   Ensure an effective meeting:

- Review/adhere to the agenda.
- Start on time.
- Appoint a recorder.
- Define participants role.

- Establish a common focus on tasks/process.
- Provide necessary data.
- Encourage participants to take responsibility for success.
- Maintain open-balance participation.
- Be positive and give reinforcement.
- Reach consensus on results.
- Review the process. Was it a productive meeting?
- Identify action items.

After the Meeting. Solidify the meeting's effectiveness:

- Follow up on action items.
- Prepare minutes.

One way to improve the effectiveness of meetings is to begin to evaluate them. A formal checklist poses objective questions which can be answered in a yes/no fashion. (See Figure 7-5 for an example.) Give it a try. Evaluating meetings has worked wonders in many organizations.

The TQM Organization

Figure 7-6 shows the organization chart for a "model" TQM organization. In this section, you will be presented with the roles/responsibilities and the reporting process of a TQM organization. Keep in mind that the TQM organization is not an attempt to overlay the existing organizational structure; rather, it is a parallel organizational structure.

The following outline of TQM teams and committees expand on and explain the various groups highlighted in Figure 7-6. The TQM teams and committees that follow have proven to be effective in many TQM implementations, in many different industries, over the past decade. This structure is recommended but can be modified. Following the outline of TQM teams and committees, you will find a definition of the roles and responsibilities of each.

I. **Executive Board**
   A. **Organization**
      1. Chairperson—Either CEO, president, or vice president/general manager
      2. Members—CFO or other so designated
   B. **Responsibilities**
      1. Ensure success of the total TQM process

Meeting Evaluation (Sample)

Name (Optional): _____ Date _____

Meeting Chairperson: _____

Meeting Purpose: _____

	Yes	No	N/A
1. Was the agenda prepared and given in advance?	____	____	____
2. Was the purpose of the meeting clear?	____	____	____
3. Did the meeting start on time? Were the expected outcomes clear?	____	____	____
4. Were participants prepared?	____	____	____
5. Was the agenda followed?	____	____	____
6. Were action items clear and assigned?	____	____	____
7. Were all past action items discussed?	____	____	____
8. Were discussions made by the proper person when consensus didn't work?	____	____	____
9. Was meeting free of interruptions?	____	____	____
10. Did meeting achieve it's expected outcome?	____	____	____
11. Did the assignees commit to their target dates?	____	____	____
12. Did the meeting end on time?	____	____	____

Figure 7-5.

    2. Encourage/reinforce culture change
    3. Empowerment of the organization
    4. Scope
       a. Champions of Critical Business Success Factors
       b. Champions of values
       c. Recognition/reward committee
       d. Training committee
       e. Culture-fostering committee
       f. Internal Customer Satisfaction Agreement

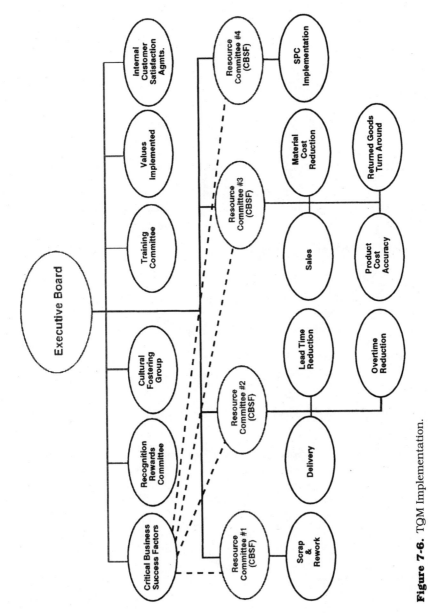

Figure 7-6. TQM Implementation.

C. Meetings
1. Purpose—Review the progress of the key elements of the TQM undertaking
2. Scope
   a. Critical Business Success Factors Champions
   b. Recognition and rewards committee
   c. Cultural-fostering group
   d. Training committee
   e. Values champion
   f. Internal Customer Satisfaction department managers
3. Frequency of meetings—monthly, 4 hours
4. Format
   a. Each reporting person presents and answers questions for a planned 10 minutes
   b. A strategy session is held to determine opportunities and provide support as needed

II. Resource Committee
A. Organization
1. Chairperson
   a. Champion
2. Members
   a. Mentors
   b. Managers with their personnel assigned to teams and/or affected by projects

B. Responsibilities
1. Ensure success of TQM team process
2. Create culture change
3. Empower team leaders
4. Identify projects (team, informal groups, individuals)
5. Monitor team performance
6. Identify/report wins and opportunities
7. Provide resources (financial, workforce, moral support)
8. Remove roadblocks
9. Secure senior management support
10. Function as role models; "walk the talk"

C. Meetings
1. Purpose
   a. Monitor team's progress
   b. Identify/provide assistance
2. Scope
   a. Team projects
3. Frequency of meetings—biweekly
4. Format

     a. Teams present progress report

       (1) 10-minute presentation

       (2) 10-minute questions/answers

     b. Strategy session

       (1) Opportunities for development

       (2) Develop action plan

          ■ Focus on assisting teams to improve performance

          ■ Identify resources required

III. Project Teams

  A. Organization

    1. Team leaders

      a. Designated 1st-level manager

    2. Team members

      a. Individuals who can contribute to solving the problems or who will benefit from solving the problems

  B. Responsibilities

    1. Improve operational performance via problem solving

    2. Monitor performance and report progress

    3. Identify roadblocks

  C. Meetings

    1. Purpose

      a. Assign action items, address status of action items

      b. Focus on solving the problems and/or achieving the expectations

    2. Scope

      a. The project being pursued

    3. Frequency of meetings—weekly

    4. Format

      a. Each reporting person presents and answers questions for a planned 30 minutes, not longer than 60 minutes

      b. Review items on progress report format

Performance Appraisals in a TQM Organization

There are several camps with varying perspectives on the application of performance appraisals. Performance appraisals are essential in establishing mutually agreed-upon goals, which initiates motivation and is a means to recognize and reward those individuals who are contributing and earning their salary.

It is a means to transform a company's environment from a point where people feel *entitled* to their salary to an environment where recognition and reward are due to employee contribution or added value to the company. Many companies are plagued with the "entitlement attitude," where employees feel that employment is their privilege and showing up is more important than results. Performance appraisals, when effective, will align the attitudes and behaviors of the employees to the company's values. Repeat performance appraisals, when effective, will reinforce the attitude of the employee to the company's values.

Human performance on the job is a complex system which involves the inherent abilities of the employee, the difficulty of the task, and the environment in which the job is to be performed. The ability of the employee to perform any given task is affected by each of these major system components. Many times, by changing only one small part of the system, a large effect on the system output can be seen.

Within each of us is a level of motivation, aptitude, training, and perception of the role we are to play in completing a given assignment that affects the outcome. These internal factors are surrounded by an external set of circumstances that further compound the difficulty of accomplishing the task.

Scientists have noted that peak performance usually occurs after long periods of training, preparation, and discipline. Fortunately, there are actions that can be taken to cause performance to improve in general and to increase those periods when peak performance will occur. By implementing a performance appraisal system and linking rewards to performance objectively, many world-class organizations have been successfully using this system to increase employee performance on the job.

Why Are Performance Appraisals Important?

A system of routinely appraising an employee's job performance provides the foundation for a system which links the accomplishment of company success factors to employee rewards. A performance appraisal system which flows down company business success factors and goals to employees' objectives and goals and routinely appraises employee performance to those objectives helps subordinates maintain focus on those significant assignments which are critical to the organization's survival.

The performance appraisal system also provides managers with a tool for rewarding employees objectively based on their individual contribution. This can be very important in arbitrating claims of bias under Title VII of the Equal Opportunities Legislation as well as demonstrating to all employees the linkage between individual rewards and job performance in the organization.

Appraising performance routinely also creates the opportunity for an ongoing two-way feedback between supervisors and subordinates regarding the entire system of factors which underlie individual job performance such as task complexity, the work environment, and roles of other team members. Since performance appraisals are written documents, they provide a system for correction of performance problems that can provide a consistent, long-term result when appraisals are prepared and reviewed with the employee on a regular basis.

Importance of the Appraisal to the Employee. A system of routine performance appraisals opens the door for an ongoing dialogue of feedback between the subordinate and supervisor to address issues which are often overlooked in the press of daily fire fighting and short-term priorities. Routine discussions with supervision also provides an opportunity for subordinate career planning.

When linked to the reward system, performance appraisals provide a system to link remuneration and other rewards to the individual's performance providing fairness and objectivity for all employees. The system also acts to document agreements on goals and performance between the employee and his or her supervisor. To maintain credibility, it is extremely important that supervisors not make any promise of rewards during the appraisal process that they cannot keep after the employee accomplishes the goal(s).

Routine performance appraisals also help the subordinate increase his or her focus on specific company success factors and bring out the best in the employee. The performance appraisal is a means for identifying needs for personal development of individuals. If it were not for performance appraisals and plans for improvement, many skills may not have been gained by employees.

Importance of the Appraisal for the Company. For a company using *Total Quality Management* for a culture of empowered employees working for continuous improvement, a system of regular performance appraisals can be the most important single device available for setting and obtaining goals, and changing and sustaining the culture critical to the ongoing success of the enterprise. The performance appraisal system provides for a systematic flow of critical success factors to each employee and follows up on the results.

The performance appraisal system gives the organization's senior management feedback about effectiveness of its employees. It also generates information about staffing and development decisions made in the placement of managers, supervisors, and employees in their current and past positions. It also can serve as an important tool in the employee problem-resolution process and protects legally both the employee and the company.

What Purposes Are Served by Appraisals? In summary, the performance appraisal system can provide input for these important business systems within a world-class organization:

- Establishes the supervisor, cares about the individuals
- Rewards results and values, i.e., merit, promotion, bonus
- Aids in promotion, separation, and transfer decisions
- Identifies under- and overachievers
- Evaluates selection and placement decisions
- Identifies training and development needs
- Evaluates relative contribution made by individuals
- Develops criteria for evaluating training and development decisions
- Provides a system for correcting performance problems

What are the Potential Problems? The benefits of performance appraisals can be realized only if the evaluations are objective and fair. If the appraisal(s) are flawed by bias or nonobjective ratings, they become a liability to management instead of a resource for performance improvement. The following list shows some of the potential problems which can occur in evaluations. Many of these problems can be prevented by an understanding of what they are and how they enter into the supervisor's evaluation.

- *Stereotyping.* The evaluator forms a theory about some group as a whole and then attributes that belief to a single member without considering that person as an individual. *("They're all like that in that department!")*
- *Contrast error.* The evaluator's judgment about a subordinate's performance is affected by the performance of an immediate predecessor. *("That's a hard act to follow!")*
- *The halo effect.* The tendency to rate a person the same way in all areas because of an overall impression. *("She sure looks the part.")*
- *Situational factors.* Irrelevant characteristics of the performer, characteristics of the organization, or the way the results are fed back. *("Too bad he spilled coffee all over his shirt before the interview.")*
- *Similar-to-me error.* An error in which the evaluator places the performer who is similar to him or her in attitude, interest, race, sex, or other characteristics in a more favorable light than those who are not. *("I'm not surprised; they went to the same school.")*
- *First-impression error.* This error occurs when the evaluator allows the first judgment he or she forms about a performer to dominate all subsequent judgments. *("He probably will be like that for the rest of his life.")*

- *Incomplete or distorted information.* Making an evaluation without having all the facts necessary to form an accurate judgment.

- *Leniency error.* Rating everyone higher than they deserve. *("Oh, they do a great job!")*

- *Central tendency.* Planning everyone in the middle of the rating, therefore having no above average performers or below average performers. *("They all seem pretty average to me.")*

- *Recency.* Only remember recent events or situations and giving; the appraisal based on recent events in lieu of the entire period. *("He didn't complete his budget last week by the deadline.")*

What Are the Steps of the Performance Appraisal System?

A performance appraisal system consists of four parts or steps. (See Figure 7-7.) Each step is an important link in the process of providing a fair and objective appraisal. The first step establishes the expectations or standards of performance for the task to be performed. Although it may appear simple at first glance, having consistent expectations or standards of performance can be a difficult problem for large organizations with many evaluators.

Once the expectations or standards of performance have been established, they must be communicated to all members of the organization as applicable for the tasks being performed (step 2). The third step is the actual performance appraisal in which the expectations and standards of performance are compared to the results and a judgment made by the evaluator as to the actual level of performance achieved by the employee. A face-to-face review of the performance is conducted between the evaluator and the performer. This step provides the vital feedback loop to both parties in the process.

Figure 7-7.

Applying the appropriate extrinsic reward to reinforce good performance or to correct poor results is the critical final step in the process. The external rewards applied here further reinforce the intrinsic rewards we receive from the evaluator's private comments and in knowing down deep inside how we performed a particular task.

Guidelines to an Effective Performance Appraisal

1. *Start out by getting other perspectives.* Utilize the Internal Customer Appraisal portion of the process. Have internal customers write up their experiences with the employee, and spend about 15 to minutes discuss their critiques before you write your comments. Remember, these internal customers probably interact with the employee more than you do.

2. *Review the past.* Make sure that you have a system for reviewing the entire appraisal period. Set up a Key Incidents file and add to this consistently. Make sure that you discuss with the employee any negative key incidents that you are adding to the file.

3. *Give employee a copy of appraisal so that he or she can complete a self-appraisal.* The employee rates him- or herself on the variables and sets objectives.

4. *Organize and outline the appraisal.*

- Give yourself hours without interruption. This may require leaving your office.

- Reread notes and key incidents. Record your thoughts and insights.

- Assign ratings at the end after you have written comments.

- Create how-to-improve lists to be added to the Performance Improvement Plan.

- Check for biases and inflammatory wording.

5. *Schedule a face-to-face discussion.*

- Schedule 1 to 1.5 hours of uninterrupted time.

- Compare sections and areas one by one. Discuss all examples.

- Listen attentively.

- Make revisions as appropriate.

- Agree upon areas for the performance improvement plan, objectives, and measurable criteria.

- Convey and justify the raise.

Components of an Effective Performance Appraisal System

- Results/contributions
- Shared values

- Performance appraisal matrix
- Summary
- Performance improvement plan

The results/contributions are objectives met with specific criteria, goals achieved which are measurable, and projects completed fulfilling objectives and time targets.

The shared values are employee behaviors that embody the spirit of the company. The employee's attitude and behavior supports the company's values. The employee advocates the values to others—thus role-modeling.

The summary is the supervisors composite rating of the results/contributions and narrative regarding the summary of the individuals performance and contribution.

The performance improvement plan denotes the areas requiring improvement, acceptable performance criteria target data requiring improvement, employee's action plan and target dates for each line item of the action plan.

Performance appraisal matrix is a composite of both results/contributions and shared values.

How Does the Appraisal Process Work? The flowchart shown in Figure 7-8 shows the general flow of the performance appraisal system.

Beginning the Cycle of Appraising Performance. The process begins when the work objectives, usually the business success factors, are defined and delegated within your organization. Each manager, supervisor, and team leader further delegates the objectives to individuals within the organization in quantitative terms that describe the details of the tasks in measurable terms. The leader then meets with the employee to negotiate details of the objectives. After reaching consensus with the employee, the major points of the agreement are documented on the performance appraisal forms (Figures 7-9 through 7-12).

A guideline common to most positions is to evaluate their tasks relative to value, delivery, quality, cost, and people management for supervisors.

Completing the Performance Appraisal—The Results/Contribution Section. Subsequent documented performance reviews are scheduled for employees every three months; for new hires, after 30, 60, and 80 days. This shorter cycle ensures that new hires are given adequate coaching during the critical first few months and, after the third review (80 days), allows a period of time for any necessary administrative action to occur for hiring-rate adjustments, job reclassification, termination, etc. (Figures 7-13 and 7-14).

A copy of the performance appraisal is given to the employee, and the master is filed for future use.

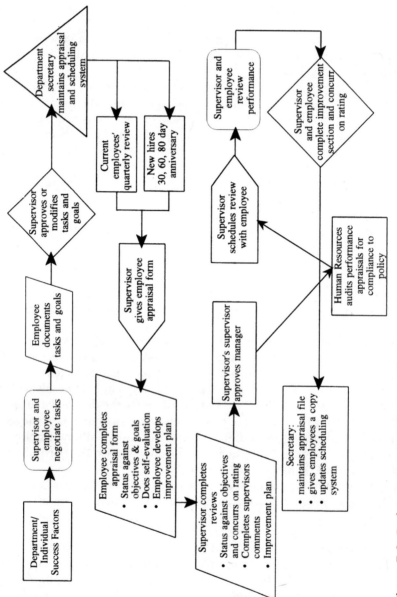

Figure 7-8. Performance appraisal process flow.

NSG

Performance Appraisal
Feedback and Development

Employee: _____

Position Title: _____

Performance Period From:_____ To:_____

Supervisor, Manager and Team Leader

Figure 7-9. This booklet is used to document objectives and evaluate performance of supervisors, managers, and team leaders (leaders).

Figure 7-10. This booklet is used to document objectives and evaluate performance of individual contributors.

Results/Contribution

OBJECTIVE/GOAL	QUARTERLY STATUS	WEIGHT (%)	MEASUREMENT CRITERIA
Status:	Status:	Final Results:	Rating:
OBJECTIVE/GOAL	QUARTERLY STATUS	WEIGHT (%)	MEASUREMENT CRITERIA
Status:	Status:	Final Results:	Rating:
OBJECTIVE/GOAL	QUARTERLY STATUS	WEIGHT (%)	MEASUREMENT CRITERIA
Status:	Status:	Final Results:	Rating:
			Overall Results Rating

Results Rating Scale

5 = Consistently Exceeds Expected Level of Performance and Standard

4 = Meets and Frequently Exceeds Expected Level of Performance and Standard

3 = Meets Expected Level of Performance and Standard

2 = Less Than Expected Level of Performance. Did not meet the Standard.

1 = Unsatisfactory Performance. Significantly below the Standard.

Objectives - discrete projects, processes, with closure (end date or completion). Concise description of key expectations.

Measurement Criteria - at least two critical measures to reach objectives which will ensure the effectiveness of the expectation.

Quarterly Status - The status against the objectives. List or describe significant accomplishments which support achievement of objectives. These correspond to the measurement criteria.

Figure 7-11. This form is utilized to record the information regarding results/contribution.

275

Results/Contribution

OBJECTIVE/GOAL	QUARTERLY STATUS	WEIGHT (%)	MEASUREMENT CRITERIA
	Status:	Status:	Final Results:	Rating:
OBJECTIVE/GOAL	QUARTERLY STATUS	WEIGHT (%)	MEASUREMENT CRITERIA
	Status:	Status:	Final Results:	Rating:
OBJECTIVE/GOAL	QUARTERLY STATUS	WEIGHT (%)	MEASUREMENT CRITERIA
	Status:	Status:	Final Results:	Rating:

Overall Results Rating

Objectives - discrete projects, processes, with closure (end date or completion). Concise description of key expectations.

Measurement Criteria - at least two critical measures to reach objectives which will ensure the effectiveness of the expectation.

Quarterly Status - The status against the objectives. List or describe significant accomplishments which support achievement of objectives. These correspond to the measurement criteria.

Results Rating Scale

5 = Consistently Exceeds Expected Level of Performance and Standard

4 = Meets and Frequently Exceeds Expected Level of Performance and Standard

3 = Meets Expected Level of Performance and Standard

2 = Less Than Expected Level of Performance. Did not meet the Standard.

1 = Unsatisfactory Performance. Significantly below the Standard.

Figure 7-12.

276

Performance Appraisal
Feedback and Development

New Hire Performance Appraisal Planner

199_

Employee Name	Hire Date	30 Days			60 Days			80 Days			Comments
		Appraisal to Employee	Appraisal to Supervisor	Appraisal Meeting	Appraisal to Employee	Appraisal to Supervisor	Appraisal Meeting	Appraisal to Employee	Appraisal to Supervisor	Appraisal Meeting	

Figure 7-13. Use this form to schedule and track performance appraisals for newly hired individual contributors or leaders.

Performance Appraisal
Feedback and Development

199__

Annual Performance Appraisal Planner

Employee Name	1st Review			2nd Review			3rd Review			4th Review		
	Appraisal to Employee	Appraisal to Supervisor	Appraisal Meeting	Appraisal to Employee	Appraisal to Supervisor	Appraisal Meeting	Appraisal to Employee	Appraisal to Superviaor	Appraisal Meeting	Appraisal to Employee	Appraisal to Supervisor	Appraisal Meeting

Figure 7-14. Use this form to schedule and track performance appraisals for individual contributors or leaders each year.

At the predetermined date, the master performance appraisal is given to the employee to appraise his or her own performance. It is recommended that a copy of the master be kept in the department files in case the original is lost. The employee completes the tasks and goals or objectives and returns the performance appraisal to the supervisor for review and completion of the Supervisor's Summary and Feedback section.

Performance Appraisals Matrix. The performance appraisal matrix (Figure 7-15) blends two factors in evaluating personal performance. The factors are results/contribution and shared values. The process for completing the results/contribution was be discussed in completing the performance appraisal results/contribution section.

Shared Values. There are five levels of shared values:

Functional—Change Agent (rating: 5)

Functional—Shaker and Mover (rating: 4)

Caretaker—Mixed Signals (rating: 3)

Nonfunctional—Status Quo Seeker (rating: 2)

Dysfunctional—The Enemy (rating: 1)

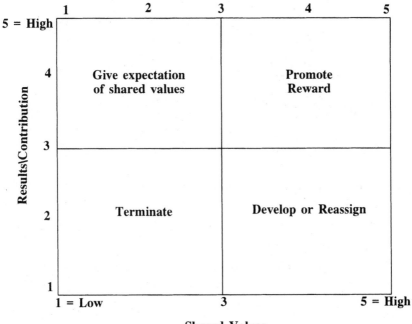

Figure 7-15.

Definitions of Classifications of Shared Values

5 = Functional → Change Agent

- Behavior overtly supports time, priority, and budget.
- Behaviors initiate changes.
- Personally involved and involves others.
- Sets the pace with a sense of urgency.
- Influences others—promotes shared values.

4 = Functional → Shaker and Mover

- Behavior overtly supports time, priority, and budget.
- Personally involved and involves others.
- Leads by supporting Change Agent—promotes shared values.
- A good role model.

3 = Care Taker → Mixed Signals

- Behavior supports budget, but does not have time or priority.
- Is not involved but supports others' involvement.
- Does not promote shared values.
- Recognizes, understands, and appreciates, but takes no initiative or leadership to make it happen.
- Is coachable.
- Cooperative when safe or costs him or her nothing to do, so "role model."

2 = Nonfunctional → Status Quo Seeker

- Behavior passively resists shared values, i.e., agrees, but does not "walk the talk."
- Behavior does not support time, priority, or budget.
- Behavior is neither involved nor supportive of anyone else's involvement.
- Allows apathy among subordinates and peers.
- Provides no leadership.
- Is a poor role model.

1 = Dysfunctional → The Enemy

- Behavior overtly resists shared values.
- Is neither involved nor supportive of anyone else's involvement.
- Provides negative leadership.
- Is a negative role model.

These shared values are rated on the evaluation form (Figure 7-16).

Evaluation Form		
1. • •	COMMENTS	Rating
2. • •	COMMENTS	Rating
3. • •	COMMENTS	Rating
4. • •	COMMENTS	Rating
5. • •	COMMENTS	Rating
6. • •	COMMENTS	Rating
	Overall Shared Values Rating	

Rating Scale

5 = Consistently Exceeds Expected Level of Performance and Standard

4 = Meets and Frequently Exceeds Expected Level of Performance and Standard

3 = Meets Expected Level of Performance and Standard

2 = Less Than Expected Level of Performance. Did not meet the Standard.

1 = Unsatisfactory Performance. Significantly below the Standard.

Objectives—discrete projects, processes, with closure (end date or completion). Concise description of key expectations.

Measurement Criteria—at least two critical measures to reach objectives which will ensure the effectiveness of the expectation.

Quarterly Status—The status against the objectives. List or describe significant accomplishments which support achievement of objectives. These correspond to the measurement criteria.

Figure 7-16. Shared values/behaviors.

Internal Customer Appraisals. Figure 7-17 shows the form utilized to acquire input from internal customers on shared values and results/contribution.

Please prepare a written critique of your internal supplier's "strengths" and "areas for improvement." One typed page would be preferred. Consider these questions and issues:

1. *Strengths.* What makes this person successful? What do you respect and admire, and what would you like to see even more of? Focus on the success factors that are ultimately your supplier's best attributes. As always, examples are helpful.

2. *Areas for improvement.* What changes in behavior and skills would make the person more effective? How would these changes improve the person's performance? Also, look at the improvement issue from a reverse perspective: consider how the lack of these behaviors and skills has held the individual back. Remember, the objective is not to dwell on weaknesses. Rather, it's to pinpoint the changes most likely to improve the individual's performance.

Meet with me for 15 minutes so we can discuss your written critique and I can ask you a few questions. At that time, you'll have the chance to comment on what you wrote, or share anything you felt uncomfortable putting in writing. Thank you for your time and feedback!

Summary and Feedback. In this section, the individual appraising performance will comment on the employee's statements regarding the task and their respective goals and objectives during the initial writing of the performance appraisal, as well as the level of performance or accomplishment achieved during the review period, and will assign an overall performance rating (refer to Figure 7-18).

After completing the Supervisor's Summary and Feedback section the supervisor schedules a meeting with the employee to discuss the employee's performance. The employee can enter any comments he or she feels pertinent in the Employee's Comments section. The booklet is signed, a copy given to the employee, and the master filed in the employee's personnel record until the next performance review cycle begins.

Figure 7-19 is an annual Appraisal Planner that is maintained on performance appraisal for new hires.

Overall Performance Rating

Evaluation

5   Consistently exceeds expected level of performance and standard

Internal Customer Appraisal Form

To:

From:

Date:

Name of Employee to be reviewed:

I'd like some candid feedback of your internal supplier listed above. Please help me appraise his or her performance fairly and accurately by preparing this internal customer review. The goal is to help this employee improve. Please review this employee on Shared Values and Results.

Shared Values

1. • •	COMMENTS	Rating
2. • •	COMMENTS	Rating
3. • •	COMMENTS	Rating
4. • •	COMMENTS	Rating

Results/Contributions

1. • •	COMMENTS	Rating
2. • •	COMMENTS	Rating
3. • •	COMMENTS	Rating
4. • •	COMMENTS	Rating

Figure 7-17.

Objectives Appraisal and Feedback Summary

Manager's/Supervisor's Summary of Individual's Performance and Contribution:

Summary of Overall Performance Rating:

Performance
Rating 1 ☐

Unsatisfactory
Performance

Performance
Rating 2 ☐

Less Than Expected
Level of Performance

Performance
Rating 3 ☐

Meets Expected Level of
Performance

Performance
Rating 4 ☐

Meets and Frequently
Exceeds Expected Level
of Performance

Performance
Rating 5 ☐

Consistently Exceeds
Expected Level of
Performance

Employee's Comments:
(optional)

I have reviewed the document and its contents and have participated in an interim review with my manager/supervisor on the status of specific objectives. My signature indicates that I have completed these discussions, but does not necessarily imply my agreement.

Employee: _____ Date: _____

Appraising Manager/Supervisor

Name: _____

Signature: _____ Date: _____

Approval by appraising managers or supervisors

Name: _____

Signature: _____ Date: _____

Figure 7-18. Use this section to comment on and rate the subordinate's performance. This section is also used for approvals, comments by the subordinate, and their acknowledgment.

Performance Improvement Plan
Feedback and Development

Date: _____

Manager: _____

Employee: _____

Area Requiring Improvement	Acceptable Performance Criteria	Target Date	Employee's Action Plan (Prepared by Employee)	Target Date
1.				

Manager's Signature | Approval: | Date: | Employee's Signature | Date:

Figure 7-19. This form is used to create a specific plan for employee performance improvement when the employee demonstrates continued less-than-expected or unsatisfactory performance.

4	Meets and frequently exceeds expected level of performance and standard
3	Meets expected level of performance and standard
2	Less than expected level of performance (did not meet the standard)
1	Unsatisfactory performance (significantly below the standard)

Example

Annual Raise Guidelines	
Rating	%
1	0
2	0
3	1–2
4	3–6
5	7–12

All employees are reviewed annually from hire date (or once a year) and receive up to a 12 percent increase. These values are determined by the company each year. Quarterly reviews of results/contributions and values do not have salary increases.

Transfers are reviewed at time of transfer. If more than six months since last review, employee receives prorated increase. Then employee is reviewed quarterly from transfer date—no increase. Next review occurs 12 months from transfer date—up to 8 percent increase.

Promotions are reviewed 3 months after promotion—no increase. Next salary review occurs 12 months from promotion date. There are quarterly reviews of progress of results and values with no salary increase.

Correction of Problem Performance. When unsatisfactory performance is encountered, a detailed plan for improvement is required. (See example in Figure 7-20.) The supervisor documents the area requiring improvement, the acceptable performance criteria (results acceptable to the supervisor), and the date the improvement is required. It may be appropriate in some cases to have the supervisor and employee negotiate the acceptable results and date by completing this section together. After recording the acceptable performance criteria and target date, the employee then documents his or her plan for improving performance and the date the action(s) will be completed. The employee signs and dates the form where indicated.

Since performance problems serious enough to require a documented Performance Improvement Plan can lead to disciplinary action for failure to

Performance Improvement Plan

Employee: John Green, Supervisor
Boss: Tom Severson, Department Head
Date: October 1
Performance to be Improved: Oriented and Training New Employees

Action to be Taken	By Whom	Due Date
1. Talk with Phil Taylor about his approach.	John Green	October 15
2. Watch Phil Taylor when he orients and trains a new employee.	John Green	The next time he does it.
3. Attend new employee orientation meeting conducted by personnel department.	John Green	The next time it is done.
4. Decide on the best time for new employee to come to the department.	John Green working with personnel department	By October 20
5. Attend a seminar on "How to Train New Employees."	John Green	November 15 University of Wisconsin
6. Read the following books: a. *Self-Development for Supervisors and Managers,* by Norman Allhiser[1] b. *No-Nonsense Communication,* by Donald Kirkpatrick[2] c. *The Supervisor and On-the-Job Training,* By Martin Broadwell[3]	John Green	by October 15 by November 10 by December 12
7. Observe John Green orienting and training a new employee.	Larry Jackson Training Director	The next time John trains a new employee.
8. Talk with John Green's next three new employees.	Tom Severson	One week after hire
9. Provide a check list to John for orienting new employees.	Larry Jackson	October 15
10. Arrange for a special office for John to use when orienting each new employee.	Tom Severson	October 15
11. Arrange for a permanent special training place for new employees.	Tom Severson	January 1

[1] Norman Allhiser, *Self-Development for Supervisors and Managers.* Madison, Wis.: University of Wisconsin—Extension. 1977.
[2] Donald Kirkpatrick, *No-Nonsense Communication,* Elm Grove, Wis.: K & M Publishers. 1979.
[3] Martin Broadwell. *The Supervisor and On-the-Job Training.* Reading, Mass.: Addison-Wesley, 1978.

Figure 7-20.

meet the agreed-to level of performance on the date(s) indicated, the supervisor is required to discuss the employee's problem performance with his or her immediate supervisor or manager and a human resources representative.

After reaching consensus with management and human resources, the Performance Improvement Plan is signed by the supervisor or managers, his or her immediate supervisor, and the human resources representative. A copy is given to the employee and human resources and the master is filed in the employee's personnel records for subsequent review and follow-up action if required.

Guidelines for Building Trust and Setting the Tone for Success

Preparing for the Performance Appraisal Interview

1. Check the appraisal to make sure it is complete.
2. Schedule an interview time when you and your subordinate will not be pressed for time by subsequent meetings, tired, or overstressed.
3. Select a private, quiet meeting place.
4. Review the appraisal and make notes for specific areas of discussion.
5. Having coffee or a soft drink available can help reduce tensions.

Communicating with Confidence

1. Be direct and to the point.
2. Show consideration, respect, and recognition of the person with whom you are communicating.
3. Focus on specifics.
4. Explain your own reactions.
5. Make it a two-way discussion.

Conducting the Performance Appraisal Interview

1. Seek the subordinate's opinion of his or her overall performance since the last appraisal.
2. Give recognition for accomplishment since the last appraisal.
3. Make no promises for rewards you cannot keep.
4. Specify one or more areas where performance might be improved, then ask for confirmation and suggestions.
5. Summarize overall performance to put things in perspective.
6. End on an encouraging note.

Counseling the Subordinate on Change

1. Describe the reasons why the change is required.
2. State the specific changes that are required.
3. Ask for subordinate's reaction.
4. Clear up misunderstandings or fill in information gaps that surface during the discussion.
5. Ask the subordinate to support the change—even if he or she disagrees.

Delegating Success Factors and Other Tasks

1. Explain the importance of the success factor or task.
2. Establish specific success measures for the assignment.
3. Set priorities for the assignments.
4. Set a date for an early progress review.

Handling Employee Complaints

1. Avoid a hostile or defensive response.
2. Get a full description of the complaint and listen carefully.
3. Recognize and acknowledge the employee's feelings.
4. State your own position calmly.
5. Set a specific follow-up date.

Listening Skills

1. Encourage dialogue with eye contact and expression.
2. Listen intently, concentrating on the individual and what is being said.
3. Seek clarification and confirmation.

Coaching Employees on Problems

1. Listen carefully.
2. Convey genuine concern.
3. Describe what you can and cannot do.
4. Keep the responsibility for solving the problem with the employee.
5. Agree on an action plan to solve the problem.

Following Up on Employee Performance

1. Review the previous discussion and define the problem as a lack of progress since the last discussion.
2. Ask for the employee's opinion of why progress hasn't been made.

3. Discuss what solutions are needed to ensure progress.

4. If discipline is to be taken, indicate your reasons for implementing it and what it will be.

5. Agree on an action plan and timetable.

6. Express confidence that your employee can do the job.

Giving Recognition

1. Describe in detail the performance you are recognizing and why it deserves recognition.

2. Give recognition by expressing your personal appreciation.

3. Offer your help in making the employee's work effective, rewarding, and challenging.

Audit of the Performance Appraisal

1. Was the correct form (Individual Contributor or Supervisor, Manager and Team Leader) used?

2. Is the employee information and performance period correct?

3. Was the appraisal and interview completed in a timely manner ($\pm X$ working days of schedule)?

4. Have the Position Accountability, Process Improvements, and Professional Development Objectives been filled in?

5. Have specific measures been specified for each objective?

6. Has Status Against Objectives been completed by the subordinate for each objective?

7. Has the manager or supervisor completed the Summary of Individual's Performance and Contribution?

8. Was the Summary of Overall Performance Rating completed?

9. Did the employee sign and date the appraisal?

10. Did the appraising supervisor/manager sign and date the form?

11. Employee comments and approval signature are optional.

12. Has the next appraisal been scheduled?

8

Core Techniques

Purpose

The purpose of the Core Techniques Track is to provide the employees at all levels in any function, be it administration, service, design, production, or sales, with the basics for measurement, statistical process control, and problem solving.

Objectives

The objectives of this track are the following:

- Be capable of using various problem-solving techniques to solve problems.
- Be capable of implementing SPC and effectively using SPC relative to the responsibility of one's position.

Introduction

The art of Total Quality Management has evolved over the last four decades. It began with the study and application of various "core techniques" to manufacturing and, more recently, to all phases of business. Originally, the intent was to contribute order and scientific analysis to the manufacturing and quality disciplines. The Core Techniques Track addresses the fundamental concepts at the foundation of Total Quality Management. Concepts in this track are designed for any member of the organization who will be part of the Total Quality Management effort. Among the topics covered are:

- Measuring success (how to tell when progress is being made)
- Statistical Process Control (SPC)
- Problem-solving techniques

Core techniques rely on two primary activities: training and implementation. Continuous improvements should be synonymous with continuous training.

Starting with measurements, problem solving and statistical process control, the training programs should be customized as much as possible to the business, the nature of the processes, and to the students' level of experience. Design engineers in a semiconductor manufacturing business will have very different needs and interests than will inspectors in a precision machine shop or line supervisors in an automobile assembly plant.

Courses should be designed with an emphasis on implementation rather than theory. The field of probability and statistics includes a rather large array of theorems, principles, and mathematical concepts that may or may not be directly related to process control. In the Core Techniques Track the focus should be on those principles that will be of immediate use. Later, instructors may suggest areas for expanded study as the participants become comfortable with the basics.

Class participants, usually engineer and management level, should be expected to learn:

- How to install meaningful measurements
- How to select the appropriate control charts, rational subgroup, subgroup size, and frequency of subgroups
- How to implement and interpret control charts
- How to design, conduct, and interpret inspection capability studies
- How to design, conduct, and interpret process capability studies
- How to determine causes of problems, using events logs and various problem-solving techniques
- How to empower the workforce with "corrective/preventive-action" matrices
- How to utilize "process control procedures" to ensure commitment and implementation
- How to perform SPC systems audits

In short, the workforce, i.e., clerks, operators, administrators, or inspectors, will learn how to collect meaningful data, understand what the data indicates, and know when to take corrective or preventive action.

Correcting Our Problems. *There's never enough time to do it right, but there's always enough time to do it again.*

Problems run our lives whether we're assemblers or engineers or senior managers. Problems are so common that we expect them. Hoards of inspectors try to find our out-of-spec product so that it doesn't get to the customer. Material review boards, customer waivers, rework, and scrap are issues which are weekly scheduled events.

Rework, scrap, inspection, and warranties cost the typical American company a significant percent of gross sales. Dollar volume is largely undiscussed, hidden from our customers, and even from ourselves. That's quite a hidden factor.

But it's so much a part of a typical factory or service system that we accept it without serious challenge. *People are human,* we think. *Mistakes will happen.*

Still, we are conscientious. At the very least, we try to determine appropriate specifications for our products, and we do first-article inspections. Our hope is that if the first item off the line is good, then the whole process will be good. Unfortunately, inspecting the first article often isn't good enough. Since all processes are variable, first-article inspection can easily miss the bad products in a lot, even if a major portion of the lot is bad. Many bad products get through to the customer.

Other factories try to solve the quality problem by inspecting every product. Even though only a small number of products are out-of-spec, they want to be sure that the customers never see any bad product at all.

But even 100 percent inspection doesn't work too well. A typical inspector will allow about 20 percent of the nonconforming products to leak through to the customer. This isn't because our inspectors are bad, but rather because our inspection systems are ineffective. Even our best systems catch no more than 90 percent of the errors. One solution to the ineffectiveness of our inspection systems is to inspect more than once. An 80 percent inspection system effectively becomes a 96 percent inspection system if we add a second inspection, and it becomes a 99.2 percent system effectively if we add a third. But these added inspections cost money and, even worse, they take time. If we inspect every part of every product, the cost of multiple inspections becomes prohibitive.

Still, we must find some way to send our customers acceptable products. We frequently tighten our specifications to improve our quality. But unless we improve our processes, tightening specifications actually increases the proportion of our product that is subject to scrap or rework.

All of these attempts to ensure quality are ineffective and expensive. Even worse, they are in natural conflict with our other two concerns: cost and schedule. When we add inspection to a product after we have detected a problem, i.e., the more quality-conscious we are, the more expensive and time-consuming the process becomes. On the other hand, if we focus on schedule, we have to be willing to pay overtime, and we have to be willing to let lower quality go out the door. And, if we focus on cost of production, i.e., the lower we try to make our costs, the more we have to sacrifice schedule and quality. The conflict seems inevitable.

And the problem seems unending. Year after year we inspect our products and services, catch the bad ones, and correct them. If we catch 5000 bad products this year, we repair or scrap all 5000. And next year, the next 5000. And 5000 the year after that. If we repair just the product, the number of problems we face each year never changes. And the cost of repairing those problems stays high.

So we see that our traditional procedures simply don't work. But we've known that all along. Then what does work? What will give our customers timely quality without breaking our bank? In order to find the answer, we have to broaden our attention to focus on our processes as well as our products.

Working on Our Systems.　Certainly, people are human and they will make mistakes. We can't stop that. But we can stop people from making expensive mistakes. Indeed, we do it all the time. We make things fail-safe; we make things idiot-proof. Planes aren't allowed to crash just because someone slips up. You have to work really hard to cut your hands on a food processor. We idiot-proof and fail-safe these systems because errors in these systems are important to us.

Of course, it is possible to cut your hand on a food processor if you try hard enough, and planes do crash occasionally. It is neither intellectually nor economically possible to eliminate all human or machine errors. But we can make it hard to make the common mistakes. It is possible to design any system so that the common mistakes either can't happen or can't hurt.

For years we have designed our workplace systems to tolerate errors, errors which have made our services erratic and our products faulty. We have acted as though fixing our systems was far too difficult and expensive. But, when we fix the process as well as the product, the system as well as the symptom, then the number of problem products goes down dramatically. If we fix one bad step in our process, then next year we may have only 4000 bad products instead of 5000. If we solve another process problem, then the next year we may go down to 3000 bad products. And if we keep solving one process problem at a time, then eventually there may be no bad product.

Fixing our process as well as our product doesn't make our product more expensive. Indeed, it makes it less expensive because, in the end, there are fewer problems to repair. And adding quality to the process as well as the product has the effect of ending the conflict between quality, productivity, and cost. The more quality there is in the system, the fewer problems will exist, and the faster and cheaper the product will be produced.

This modern management system has been given several different names by different people. Some call it *statistical process control* because the system uses counts and measurement to make our processes stable, so they can do what we want them to do. This is no motivational scheme; this is a rigorous method. It gives managers a way to understand and solve problems quickly, effectively, and with appropriate priorities. It gives workers a way to communicate with managers so that their ideas are no longer ignored.

This is no pipe dream. It can be done. It's being done in Japan. And it's being done in America by Japanese firms. And since the early 1980s, it is being done in America by American firms. Ford Motor is one of the biggest success stories.

Effective Problem-Solving Techniques

Individual and departmental performance within an organization is often judged by the individual's or department's ability to identify, prevent, and resolve problems. Of course, not all activities deal with problems. When processes are moving along consistently with only normal variation, there is no problem. Sometimes the process will change, but the change is not a cause for concern and no action is required. No problem. The real test of management ability occurs when the process changes result in undesirable output and you don't know how to fix it. Now you have a problem. A good definition of a *problem* then would be: "a deviation from a given standard or desired level of performance for which the cause is unknown and a resolution is required."

While everyone complains about problems, many don't seem to dislike them all that much. Most technical people seem eager to jump in and solve problems. It is common for newly formed quality/productivity implementation teams to tear into problems, first attacking one aspect and then another without any specific plan, objective, or justification for the activity. This hectic, undirected problem-solving behavior works to a degree, but is not generally adequate to put the problem to bed. Things get fixed, but not the right things, and problems are wounded but not terminated.

Human beings are not optimizers by nature—we are placaters. We try to make it do. When a problem arises, most will want to try the quick fix first. If a quick fix is apparent and it doesn't cost too much, give it a try. If it doesn't work, then another approach is necessary. (Sometimes if the quick fix doesn't work the tendency is to forget about it because there isn't enough time to actually solve the problem.) The objective of training in effective problem solving is to impart the means for achieving the best solution, not the one that "sort of" works. Quality/productivity implementation teams will learn how to stand back and understand the problem before pushing for the solutions, so that the solutions used will be the last solutions necessary.

There are several techniques for problem solving that help to organize the knowledge of problems and suggest probable causes. When causes are identified, solutions can be considered and tested. When all team members are familiar with these techniques and actually apply the techniques in the workplace, the teams will take on a new dimension. Team members will be motivated by their "newfound" ability to participate in a structured approach to problem analysis and solution.

Problem solving is considered by employees to be the domain of management. Everyone can tell you about problems in a company. Many people can also give you an accurate appraisal of the cause of common problems, but they don't always consider the solution to be part of their job. Here we go back to empowering the workforce. If management can facilitate and ease the path for all employees to participate in problem solving rather than complaining about the situations, the benefits will be enormous. It is much more efficient for the people closest to the problem to provide the solutions. Solving problems faster with less direct management involvement translates into bottom-line cost avoidance. Problems don't persist as long, they produce fewer defects, and fewer resources are required to implement the solution when the workforce is empowered and involved in the Total Quality Management process.

Problem-solving techniques include *events logs* to keep permanent records of the information necessary for cause identification; *diagnostic process audits* which identify non-compliance and steps to take to solve problems; *cause-and-effect diagrams* and *CEDAC* to help use your process audits to define causes of difficult problems. These will be especially effective for processes which have never worked very well. Another system called *logical decision making* will be crucial when choosing the best course of action. *Design of experiments, preventive and contingent planning,* and other techniques may be used to isolate, validate, and correct the causes of your most complex and important problems.

All of these techniques are related to the statistical process control charts you have been using. When one of your SPC charts tells you that something is wrong with your process, you will first turn to the events log to help you determine potential causes for the problem. If nothing is apparent in the events log, you will use additional methods for eliciting potential causes, i.e., diagnostic process audits, cause-and-effect diagrams, CEDAC, relations diagram, logical decision making, and preventive and contingent planning. If these tools do not enable you to solve the problem, several additional methods are available such as regression and correlation analysis, design of experiments, and the technology of asking questions.

Once having determined potential causes for your problem, you will monitor many or all of these causes with SPC charts in order to validate them as causes and to ensure that they behave in a constructive way in the future. At some important level, it is the stability induced in the system by these SPC charts that is the final and most important product of an effective problem-solving technique.

Traditionally, the problem-solving process has been broken into six steps:

1. State/define the problem.
2. Get the facts regarding the cause of the problem.
3. Restate the problem in light of the facts.
4. Analyze and form conclusions to the cause of the problem.
5. Take action on the most likely solution.
6. Follow up to assure the action was effective.

Even people as different as psychologists and engineers seem to agree on these six steps. While this six-step method provides you with an outline of what to do when a problem arises, it tells you little or nothing about how to perform each step effectively.

Events Logs

An Events Log is a book located at or near a critical work station in which involved personnel record things which are new, different, changed or otherwise significant to the process operation. Changes can be in the form of equipment, personnel, materials, suppliers, environment or anything else related to the process. It is also appropriate to record in the Events Log any special test or studies conducted on the process along with the results obtained and whether they are favorable or not.

In the event of a problem with the process, you will first turn to the Events Log in order to determine retrospectively what the cause for the trouble might have been. Since most problems are a direct result of something new, different or changed in the process, this is obviously a crucial diary, a powerful tool to provide information for problem solving and prevention. The Events Log not only documents critical events, but it also establishes the discipline to look for such changes by those who are responsible for the process. Furthermore, the Events Log is simple to implement and maintain. Entries contain only the date and time, a few words about the significant event, and the recorder's initials. (Any involved personnel can make entries.) Each entry takes less than a minute.

Still, it is very difficult to get people to use an Events Log consistently, since most of the time, nothing ever comes of the entries that are made. After all, most of the time, the monitored process doesn't have problems.

Events logs are like parachutes: clumsy, uncomfortable, usually a waste of time, and absolutely vital when you need them.

In order to keep subordinates using the Events Logs, it is essential that supervisors and managers review the Events Logs frequently and comment on the entries made. Praise is valuable here as are reminders of times that the Events Logs were an important tool in the resolution of some problem. Also, supervisors might remind workers to keep the entries explicit enough to enable reconstruction of the cause of a problem should the need arise. Whatever the supervisor does to convince the worker that he or she is paying attention to the Events Logs will help the workers to continue making entries in them. Obviously, whenever an entry actually helps to solve a problem, it is vital to emphasize the importance to the process and to the company.

An example of a structured events log is included near the end of this section. Such a form must be designed specifically for each of the areas. It is important that the structured form provide spaces for all commonly occurring entries. At the same time, structured forms must not have a lot of meaningless entry spaces, or the workers will come to treat them with disrespect. This means that some effort must be taken to be sure the structured form is up to date and relevant to current events.

Events Log Summary

- Implement Events Logs at each identified operation. If an SPC chart is there, an Events Log should be there.

- Record in the Events Log all changes or events that may affect the process.

- There are no limitations on who can make entries.

- Supervisors must actively remind workers of the importance of the Events Log.

- When a problem occurs, examine the Events Log for changes in the process which might be potential causes for the problem.

What Is New, Different, or Has Changed?

There is a tendency to approach problem solving in a way that emphasizes new and different ideas. Occasionally this can be the best way.

BUT—a very high percentage of problems can be solved by examining what was going on just before the problem occurred.

National Summit Group Daily Events Log

Process _____

Possible Events

Operator Change_____

Inspector Change_____

Material Change_____

Tooling Change_____

Other:_____

_____

_____

_____

_____

EVENTS LOG
(What's new, changed or different)

Date	Time	Events

Figure 8-1. This is a basic form which can be utilized for many processes.

Systems Daily Events Log

Operation Changes*   Inspector Changes*   Date:

_____ _____   _____ _____   _____

_____ _____   _____ _____   _____

_____ _____   _____ _____   _____

_____ _____   _____ _____   _____

Changes: Explanation:

    Process

    Material

    Tooling

    DCO Cut-in

    Other

    Machine Set-up

    Process

    Control Chart

Configuration Mix:
(Non-standard Products)

P/N: _____   Qty.: _____   Total Qty. Produced: _____

      _____   _____

      _____   _____   Yield: _____

      _____   _____

Material Shortage
New Shortage: Shortages Filled:

P/N: _____   _____   P/N: _____   _____

      _____   _____   _____   _____

      _____   _____   _____   _____

      _____   _____   _____   _____

* New Operator/Inspector
Change in O/I

Figure 8-2. This form is customized to a specific process.

Table 8-1. Application Selection

Problem Solving Method	Typical Application	Advantages	Disadvantages
Events Log	Implement as part of Process Control Procedure. Expected performance of process or product has experienced a sudden change. Implement at each specific machine or assembly operation, with or without SPC. Source of information for what's new, changed or different in process.	Provides immediate and ongoing feedback to identify causes for out-of-control points, process shifts or trends. Easy to implement and maintain.	Requires ongoing effort to maintain. Easy to omit events that may not appear important but are.
Cause and Effect Diagram	Generates possible causes of repetitive problems quickly. Generates possible corrective/preventive actions for out-of-control points on control charts. Facilitates a breakthrough to improved levels of quality and productivity. Used when potential causes already exist but require organization and prioritization.	Involves operators and inspectors. Uses synergistic resources from a group of people (PQI Team). Minimum of training needed to implement. Ideas or causes generated by brainstorming methods.	Priority of investigating causes determined by experience of group. Very little data and facts are used to generate causes. Not efficient for complex problems. Requires technical knowledge and experience with the product or process.
CEDAC	All applications included under Cause & Effect Diagram method. Can result in a documented cause and approved course of action.	Focuses on improvements rather than just determining the problem.	No knowledge or experience needed. Needs more analytical methods.

Table 8-1. Application Selection (*Continued*)

Problem Solving Method	Typical Application	Advantages	Disadvantages
Blockbusting	Whenever the advisor's solutions to problems don't work.	Enables people to tap their creative and innovative thinking. Helps to unlock your own creativity.	If course of action is not carefully planned, there is likely to be risk.
Logical Decision Making	Helps to choose between several alternatives in: process equipment course of action	Will reduce bias politics	A clever politician can manipulate the process.
Diagnostic Process Audit	Use when deviation from the process is suspected. When excessive variation in levels of performance exist.	Simple to conduct. Preventive actions are based on actually observed facts. Generates additional methods improvements.	Not efficient for complex problems, i.e., Design Components May generate too much activity with no results.
Poka-yoke	Reduce or eliminate human error.	Helps eliminate defects in making a product, or process foolproof.	
Relations Diagram	Use when problem is complex and intertwined. Problem crosses departmental boundaries.	Simple to use. Gives a graphic view of problem.	Sometimes can be vague.
Preventive and Contingent Planning	The implementation of a plan is important. Making "it" happen.	Easy to use. Focuses on the future. Will get the desired result.	None

Table 8-2. Implementation Guidelines

Problem Solving Method	Skill Level (Lo,Med,Hi)	Training Amount (Lo,Med,Hi)	Ease of Implementation (Easy,Med, Complex)	Size of Group	Usefulness of Results
Events Log	Low	Low	Easy	—	Immediate feedback. Operators, Inspectors, Supervisors make entries
Cause and Effect Diagrams	Med	Low	Med	Med to Large	Generates numerous ideas. Includes Operators and Inspectors. Priorities determined by vote.
CEDAC	Med	Low	Med	Med to Large	Generates improvement.
Blockbusting	Med	Low	Med	Med	Stimulates creative thinking.
Logical Decision Making	Low	Low	Med	Med	Good at identifying best course of action. Process very helpful for organizing and focusing thought process.
Relations Diagram	Low	Low	Easy	Small to Large	Helpful in understanding and planning complex problems and solutions.
Design of Experiments	High	High	Complex	Small	Best used to test causes. Very efficient at sorting out critical factors from among many possible factors.

Table 8-2. Implementation Guidelines (*Continued*)

Problem Solving Method	Skill Level (Lo,Med,Hi)	Training Amount (Lo,Med,Hi)	Ease of Implementation (Easy,Med, Complex)	Size of Group	Usefulness of Results
					Identifies interaction effects among causes.
Diagnostic Process Audits	Med	Med	Med	Small	Identifies deficiencies of current method (not necessarily related to the problem).
					Identifies potential preventive actions.
Poka-yoke	Med	Low	Med	Small to Large	Eliminate errors.
					Foolproofing products and/or processes.
Preventive and Contingent Planning	Med	Low	Easy, but requires some self-discipline.	Small to Large	Immediate application.
					Supports other problem solving methods.

Invariably a problem is associated with one of three occurrences. The following questions must be asked.

- What is **new**?
- What is happening that is **different**?
- What has **changed**?

These questions should be asked bearing in mind the following:

- People
- Equipment
- Procedures or process changes
- Operator
- Inspector

- Raw materials
- Machine set-up
- The environment
- Tooling
- Anything else that might impact your problem

The answers to these questions, if acted upon will lead you to the cause of the problem. Your next challenge is to decide what to do about it.

Example. An example of how an events log was used was at Komag, a manufacturer of disks for hard disk drives. An operator was instructed to record everything that was new, different, or changed. The operator recorded the relative humidity increasing as it occurred. As he did this, he noticed the yields in the process rapidly decreased. This information was shared with the engineer. The engineer verified the change and its impact. The solution was to provide an environment which controlled the humidity in the process area. This was done at an expense of several hundred thousand dollars, but the net saving exceeded a million dollars per year. Refer to Table 8-1 for Application Selection.

Cause-and-Effect Diagram

The cause-and-effect diagram is a diagram relating an effect that has been observed with possible causes. The diagram, when completed, resembles a fishbone and is often referred to as a "fishbone" diagram. The possible causes are generated using brainstorming techniques and includes all personnel involved with the problem. (See Figure 8-3.)

The cause-and-effect diagram is relatively easy to implement and requires a minimum of training. It uses the resources of all persons involved with the problem and provides a systematic approach to generating possible causes and prioritizing causes to further investigate.

The effect is usually a performance characteristic resulting from a process, whether production or administrative.

The cause is a variable or factor which may influence the variation or level of the effect. The causes may usually be grouped into the categories of people, machine, methods, materials, measurements, or movement.

Procedure for Constructing a Cause-and-Effect Diagram

Step 1: Initiate a Meeting

Invite all personnel concerned with the problem.

Invite people who have enough knowledge and experience with the problem. Include hourly people (operators and inspectors) as well as people from related departments.

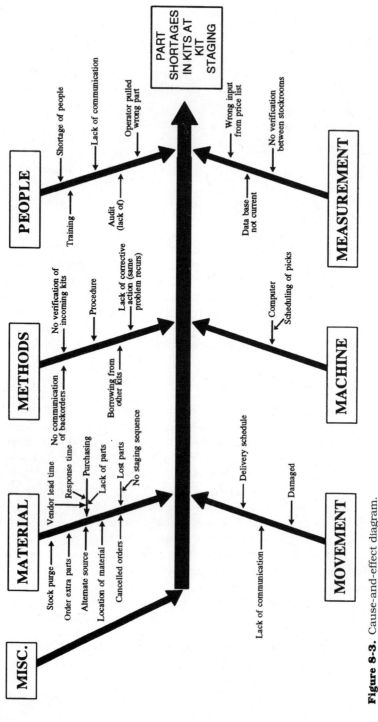

Figure 8-3. Cause-and-effect diagram.

306

Step 2: Clarify the Problem

The problem must be presented with as much definition as possible to reduce the influence of unrelated causes.

When the problem is too large or obscure, stratify the problem into several smaller ones and then do a cause and effect diagram for each one.

Write the problem effect to the right of the backbone as shown below.

$$\longrightarrow \quad \textbf{Effect}$$

The problem or effect can be a quality characteristic such as percentage of defective products, strength, size, weight, or an economic characteristic such as yield, working hours, energy consumption, response time, production rate.

Environmental characteristics such as pollution, turnover of personnel, or accident frequency can also be used.

Step 3: Generate Causes Using Brainstorming Methods

Identify the major groups of causes as branches from the main backbone. The major groups are people, material, methods, machine, measurement and movement.

The causes are obtained by going around the room in rotation with each attendee presenting one cause at a time.

As this process continues, the causes are written on the diagram on the appropriate bone or branch.

New causes may arise as new associations are developed by looking at other members' causes.

It takes several rounds to obtain all the causes. The process is continued until each attendee passes twice.

Observe the rules of brainstorming:

- Be courteous
- Encourage new ideas
- No judgments

When all causes have been presented, discussion is encouraged to clarify ideas before proceeding to the voting.

Check to determine if all causes are included in the diagram.

Step 4: Determine Importance of Causes

After all members have passed twice, a vote is taken on all causes in two rounds of voting.

In the first round, all attendees may vote for an unlimited number of causes.

The top four to six causes, based on the number of votes, are selected for round two voting.

In the second round, the attendees vote on only one of the four to six causes. The cause(s) with the largest vote count in the second round are selected for further investigation.

Step 5: Verify Cause

Test to verify that the cause(s) selected do effect the characteristic selected by using methods such as design of experiments, data collection, or control charts.

Step 6: Corrective Action

The method for verifying the cause might provide the appropriate corrective action. If it does not, or the methods used for verification are not practical under production conditions, then another brainstorming session is recommended.

Cause and effect problem analysis is a practical and effective way to arrive at the probable cause or causes. Once the causes have been identified, it is important that they are verified in a way that everyone will be comfortable with the final conclusion.

Creative Problem Solving and Conceptual Blockbusting

The problem solving methods that we are covering in this section focus mainly on **logical**, or if you prefer, **rational** approaches. These very worthwhile problem solving techniques are undoubtedly very effective. They are in some ways a variation of the "Scientific Method" taught in great detail by universities in their science and engineering schools. We can't argue with that, can we?

Let us stand back and look at the world around us. More specifically, focus on the competitive situation you face today. Superimpose upon it the rapid rate of change that is occurring in every aspect of our lives, business, knowledge and communication methods. Now consider a company or just about any organization that only utilizes logical and rational thinking. Everything they do is based upon precisely collected, statistically sound information. Facts are abundantly available and effectively used. Management is practicing what is often called "Participative Management." In other words, everyone in the whole organization is allowed to, and does, contribute to the solving of problems, the making of decisions and hence the improvements that are necessary in order to compete.

Now, compare this organization with a competitor who not only practices good sound rational thinking, but in addition, they have found a way to tap the **innovative, creative** and **conceptual** abilities of everyone.
Who do you think is likely to be the more successful?

It's obvious, isn't it? The combination of logical and creative thinking will always outclass and outperform one or the other used independently. This is why it is important to enhance the more traditional problem solving methods with alternative approaches that utilize inborn human creativity and the ability to practice conceptual thinking.

So far we have only discussed problem solving. Is that enough if we are to compete with the best in the world? Sometimes it is necessary to make a process or product better when there is no apparent problem. Another way to think of this is to state the problem in this way. "Our competitor is better than we are, and we do not know why. What can we do about it? How can we become better than they are?—when we do not know why they are better than us!"

Conceptual thinking—and the removal of the various **blocks** that inhibit it, is indubitably one of the most effective ways to **outperform** your best competitor—thus in your own unique way **becoming the best in the world**. We are now talking about continually trying to make everything a little bit better—and then making it a little bit better still—and so on—and on—and even further on—until you approach—**"Perfection"**

It is important to understand that it is not possible to tell people that they must be conceptual. You cannot order a person to be creative.

All you need to do is to create an environment that will:

- Encourage it
- Allow it
- Recognize it

Blocks That Inhibit Creative Thinking

Conceptual Blocks can occur in several forms. In order to overcome them, it is important to have an appreciation and understanding of what they are. Can creativity be learned? We believe it can, and the best way to go about this is to focus initially on the various types of Conceptual Blocks.

Perceptual Blocks are obstacles that prevent the problem-solver from clearly perceiving either the problem itself or the information needed to solve it. Examples of Perceptual Blocks are:

- **Stereotyping**—"All businessmen wear neckties." We all have a tendency to categorize and put boundaries around things. In social and professional interactions we often stick to stereotypes and generalities.
- **Difficulty in isolating the problem**—"Your car will not start." This seems to be a good problem statement. Or is it? Maybe your car has run out of gas, or the car won't crank over, or the car cranks over but will not fire. Proper problem identification is essential if we are to stand any chance of solving a problem.

- **Tendency to state the problem too narrowly**—A problem statement that is too limited will tend to inhibit creativity. If you want to encourage innovative solutions to problems your problem statement should look like this. "How best to improve the efficiency of the assembly line." Do not narrow your statement down. For example, do not say "To improve the assembly line efficiency by ten percent."

- **Inability to see the problem from various viewpoints**—An assembly line operator is likely to view a specific situation very differently to the production manager. When it is important to come up with creative solutions, it is crucial that you involve the people closest to it or at least, try to see the problem from their point of view.

- **Saturation**—For example, without looking at one, try to draw an ordinary rotary telephone dial, putting all the numbers and letters in their right place. Very few people can do this. The dangerous aspect of **Saturation** is thinking you have the data or information required to solve the problem even though you are unable to produce it when needed.

- **Failure to utilize all sensory inputs**—Problem-solvers need all the help they can get! An engineer working on an acoustics problem for a concert hall should not get so carried away with his theoretical analysis that he neglects to look at a wide variety of concert halls and the quality of sound in each. He must also bear in mind the visual aspect of the hall and, of course, not forget the comfort of the seats.

Emotional blocks can really get in the way of creative thinking and idea generation. Have you ever been at a public meeting during the time when the speaker or moderator asks for questions? What goes through your mind? "Oh, I'd better not ask my question because I might sound dumb."—and as you are thinking that, somebody else asks the exact question that you were afraid to ask.

- **Fear of taking a risk**—What happens if "it" goes wrong? When you produce and try to sell a creative and new idea you are taking a risk of making a mistake, making an ass of yourself, losing money, hurting yourself or whatever.

- **No appetite for chaos**—Very often when new creative ideas are implemented, ambiguous or confusing situations can occur. Many people have an innate desire for orderliness and this will sometimes block both the generation and implementation of new ways to do things. A way to overcome this is to think of potential negative and positive consequences and if they are important find a way to manage the negative and take advantage of the positive.

- **Judging rather than generating ideas**—If judgment is applied too early in any creative problem solving process, there is a high probability that some of the best ideas will be discarded. This is extremely common. Just take any newspaper. How much space is taken up with judgment? A high percentage—critic columns, political analysis, editorials, etc. Even in the supposed bastion of creative thinking, our universities—a high percentage of time is devoted to judgment rather than creativity.

- **Inability to incubate**—The inability to relax or sleep on it are also a common emotional blocks. Deadlines will often precipitate this. When a new idea is first thought of, it is frequently beneficial to "sleep on it." This will enable the subconscious mind to participate in the evolution of new ideas.

- **Lack of challenge and excessive zeal**—You cannot work effectively on problems unless you are motivated to do so. One of the best ways to encourage motivation is to provide a challenging environment. Conversely too much pressure for a quick results can cause an inferior solution to be implemented. This is rather like the tortoise and the hare phenomenon.

- **Reality and fantasy**—The inability to distinguish between reality and fantasy is another emotional block. It is important to be able to access the imagination if we are to be creative. However, we must be able to manage the results of this. It is important that we are able to differentiate between reality and fantasy.

Cultural and environmental blocks are acquired by exposure to a given set of cultural patterns or imposed by our immediate social and physical environment. An example is a lack of trust between colleagues. Another might be that any problem can be solved by the application of our technology.

- **Taboos**—It is not OK to make suggestions around here. Taboos are usually directed against acts that would cause displeasure to certain members of a society or organization. However it is the acts themselves that would offend. If imagined rather than carried out the acts are not harmful. Therefore, when working on problems within the privacy of your own mind, you do not have to be concerned with the violation of taboos.

- **Humor in problem solving**—A significant cultural block is "Problem solving is serious business and humor is out of place." Think of people you know that are creative. Frequently, they have a great sense of humor and can be very funny. Similarly a creative problem solving group invariably has a super time because their meetings are full of humor.

■ **Reason and intuition**—Another cultural block is that reason, logic, numbers, utility and practicality are good, whereas feelings, intuition, qualitative judgment, etc. are bad. Effective problem solving needs to utilize a balance of all of these traits and qualities. Another way to think of this is to make sure that the left and right sides of the brain are able to focus on solving the problem.

■ **Criticism**—Imagine you are discussing one of your achievements with your boss. You have just finished telling him or her all about it when you are met with, "but, what about that other project you haven't finished yet?" An atmosphere of support, trust and honesty are essential if people are to make the best of their conceptual ability. Overabundance of criticism will stifle creativity.

■ **Autocratic bosses**—If many of the ideas that get implemented come from the boss, it is highly probable that subordinates will not be forthcoming with innovative improvements. One of the best ways for a boss to be successful is to actively encourage everyone in the department to be creative and utilize their conceptual abilities to the fullest.

■ **Non-support**—A lack of physical, economic, or organizational support will also block people's creativity. New ideas are typically difficult to bring into action and therefore invariably need above average levels of support.

Intellectual and expressive blocks result in an inefficient choice of mental tactics or the use of the wrong means of communication.

■ **Choosing your problem solving language**—Sometimes a problem can be best solved mathematically. Other times a visual approach may be more effective. Brainstorming, verbalization or even inspections and other diagnostic methods may be more appropriate. It is important that you are prepared to be flexible in the selection of a problem solving strategy.

■ **Limitations**—Sometimes the elimination of various courses of action may narrow down your search for the best solution. However after you have finished eliminating you may end up with nothing. It is therefore necessary to temper elimination with caution and good judgment.

Brainstorming Rules

■ Go for Quantity of ideas, not Quality.

■ Do not apply judgment to any idea during the brainstorming meeting.

■ Hold the meeting in a neutral place—not the boss's office.

■ Write all ideas on a flip chart so everybody can see them.

■ Build on ideas already generated.

- Be as wild as possible.
- Involve all people present at the meeting.
- Turn the list of ideas into a plan of action when appropriate.

Developing a Creative Environment in Your Organization

Much has been written about creative thinking and how to develop it. Some of the most creative people in the world are young children—just about all of them. They have an inborn curiosity that enables them to explore all sorts of different things. They love to experiment and try out new approaches to the problems they come up against every day. What happens to this creativity as they grow up? It slowly disappears the older they get. The reasons for this have already been discussed. Many of the conceptual blocks start to take hold and before we know what's hit us, creativity is approaching zero.

You cannot order someone to be creative any more than you can force them to be motivated. There is only one way to tackle the issue of creativity and innovative thinking, and that is to **create a climate or environment** that will encourage it. What we are looking for is a way to stimulate a person's innate ability to be creative and feel self-motivated. Here are a few ideas that can help you achieve this.

- Accept the reality that everybody can be creative.
- Ask for it and encourage it (i.e., allow people to be creative).
- Reduce the risk associated with failure. Accept that when an innovative idea is adopted, the risk of things going wrong may go up.
- Practice positive reinforcement. Well-deserved praise is powerful.
- Recognize and reward experimentation and creative solutions to problems.
- Do not tell people how to do something. Instead, ask them what they think is the best way to reach to the goal.
- Instead of criticism make suggestions.
- Listen actively to new ideas. Do not ignore them.

Diagnostic Process Audits

Process audits comprise a thorough and comprehensive survey of every aspect of a process as measured against the process design. The diagnostic process audit, when properly applied, can be a very effective problem-solving technique as well as a tool for improving performance. Process audits are a fast and economical approach to solving problems. Only the events log is faster and more economical.

Traditionally, process audits have been used as verification tool—in a sense, to verify that what is supposed to be in place is in place and that people are complying with the established procedures. Process audits become a powerful problem-solving tool when used in a diagnostic manner. Verification is augmented by extending the audit a step further to examine if any part of the process is contributing to a known undesirable condition.

Process audits allow analysis of a process by administrative personnel or manufacturing personnel, as applicable. This provides assurance that the process documentation, tools, and material support are current and conducive to optimum producability and quality. The procedure for conducting a diagnostic process audit follows. Process audits principles can be applied to procedures and systems as well as processes. Basically ISO 9000 was developed from early process audit practices.

Planning the Audit

1. *State the problem or objective.* Provide a clear definition of the problem, preferably in the form of a variance from a standard or desired condition. The standard may be the process flowchart and associated procedures or an overall quality standard such as ISO 9000.

2. *Select the process.* Identify the process or the portion thereof where the variance is observed or originating. For the best return on investment, it is suggested to address every process via a process audit by priority of its quality history. The worst-quality history processes are the first priority.

3. *Collect all applicable documents.* Collect latest revision of all documents which are used to establish and maintain the subject process:

Assembly instructions

Workmanship standards

Inspection instructions

Test instructions

Fabrication methods

Product specifications

Calibration procedures

Tooling drawings

Station layout

Training records

4. *Select the individual(s) and stations(s) to be audited.* If the problem has been narrowed down to a specific station or set of stations, identify which ones they are. If more than one operator is assigned per station, select the

one which most closely represents the average skill level. It is recommended that a matrix be established displaying each operator and process with a date of an approved process audit. Such a matrix can be the cornerstone for an operator certification program.

5. *Select the audit team.* Include individuals who have knowledge of and experience with the product, either as internal suppliers or customers. In a manufacturing plant the team might consist of:

Production supervisor

Process engineer

Quality engineer

Inspection supervisor

Members of the audit team should be technically competent to diagnose potential problems. The audit team leader should be trained in auditing and should have previous experience serving on audit teams. The person who developed or is currently conducting the audit should be responsible for the process.

Conducting the Audit. Persons conducting the audit should have as much training and/or experience as possible in auditing techniques. They should also have some technical or administrative knowledge of the area or process to be audited.

1. *Review methods and documents.* Prior to conducting the physical audit, verify and understand the inspection and assembly methods. Also verify that the foregoing documents are current and complete. Determine the applicable standard against which the audit will be performed. Without a standard there is no basis for evaluation of the audit findings.

2. *Inform the responsible personnel.* Inform the responsible operation's line supervisor that the audit will be performed and obtain the required interface support. Agree on a convenient time for the audit when the subject station will be in normal operation. Some audit teams feel that audits should be conducted unannounced. The problem with this approach is that the key interface people may not be available or the subject process may not be in operation. Some short notification is desirable. If things are really out of control, they won't be easily corrected in a day or even a week.

3. *Hold a brief introductory meeting* with the operator/inspector and area supervisor and explain the purpose of the audit.

4. *Conduct the audit.* The following audit sequence is recommended. Use a checklist, developed for the audit, to ensure comprehensive coverage of all critical areas.

a. Review documentation at the station being audited for availability, currentness, and completeness.

b. Review tools and gages for conformance to methods and calibration.

c. Review station layout for conformance to plans.

d. Review material handling in, around, and out of the station.

e. Review human factors.

f. Review machine setups.

g. Audit the process.

Note: As the audit team proceeds through the audit, the following question must be continuously asked: "What about this element could be done differently to prevent the problem in question?"

Establishing Corrective/Preventive Action. Review audit deficiencies and opportunities with the audit team. Whenever possible, immediate corrective action shall be taken to resolve the deficiencies.

1. Develop preventive action and contingency plans for the opportunities identified.

2. Develop a plan for implementing preventive action: What, Who, When.

3. Obtain commitment from responsible individuals and their managers.

4. Follow up, follow up, follow up.

Reporting to Management. A well written concise report of an important audit will result in positive and timely action on the recommendations. If the report is vague or unclear, no action will be taken. Prepare the report using the following items as a guideline for the main sections of the report:

1. *Objective.* Why was this audit conducted?

2. *Audit summary.* A brief description of how, where, and why the audit was done, along with any noteworthy points that may have influenced the results.

3. *Findings/conclusions.* What was the diagnosis? List each major finding in detail. Finding should be referenced to the applicable portion of the audit standard.

4. *Recommendations.* Corrective/preventive-action plan.

Example. An example of an effective process audit is one conducted by a computer printer manufacturer. The problem was that there was a high frequency of service calls which required adjustment of the printer controls. The finding was that all the assembly operators were using torque wrenches. However, the high-paid test and rework operators were using screwdrivers and readjusting the machines to the best of their knowledge by feel. The machines being repaired and adjusted in the field were the problem ones that had been attended to by test and rework technicians. Once the torque wrenches were purchased and given to the test and repair personnel, the problem disappeared.

Example. Another high-impact example was at a fastening plant. The header operations were planned to be process-audited. Hence the operators prepared the instructions and the engineers published them. To every supervisor's and manager's surprise, the operators as a group were spending a total of 90 hours per shift—three shifts a day—with their machines down while they sharpened their tools. This was quickly changed so that sharpening the tools was contracted to an outside source for half the cost. The machines were then run for 7000 hours more each year, producing $10 million more in sales for the cost of materials only—equipment and labor were "free."

Preventive and Contingent Planning

Most organizations, if they really are prepared to admit it, know deep down what they need to do. That is why the cause-and-effect diagram works. When the answer to problems is known, the task is simply to surface and address them in an orderly fashion. With preventive and contingent planning, another tool for problem solving (or problem avoidance) is a means for improving the probability of success in Total Quality Management.

Preventive and contingent planning is useful in the avoidance of problems associated with new-product introduction or any task. It is a way of making sure that new process control systems are introduced properly and are accepted by all the participants. It serves to ensure proper execution and reduces the risk associated with implementation of improvement ideas.

Preventive action is the set of steps taken to prevent problems from occurring. If a particular manual processing task is known to be difficult and prone to errors, automation of the task may prevent problems arising from operator lack of attention or inadequate training. Another type of preventive action is requiring suppliers to demonstrate acceptable process capabilities which will prevent material or component failures.

Contingent action is an alternative course of action designed to take over in the unlikely event that preventive action doesn't work. Contingent action plans are actions taken to minimize the impact of a problem and require a flag or alarm to signify when preventive action has not been fully

```
┌─────────────────────────────────────────────────────────────┐
│              Process Audit Report Coversheet                  │
│                                                               │
│  Date          March 15, 1991                                 │
│  Product Line  Downconverter                                  │
│  Station       Final Assembly                                 │
│  Operator(s)   Tran Puam                     _____      │
│                Charles Odette                _____      │
│                                                               │
│                                                               │
│  Statement of Problem  High failure rate of downconverter at  │
│  post pack                                                    │
│  audit due to leak                                            │
│                                                               │
│                                                               │
│                                                               │
│                                                               │
│                                                               │
│                                                               │
│  Audit performed by:                                          │
│  ☐ Production Supervisor  _____                         │
│  ☑ Process Engineer   Ray Willis                              │
│  ☑ Quality Engineer   Harvey Rosenbloom                       │
│  ☐ Other   _____                                        │
└─────────────────────────────────────────────────────────────┘
```

Figure 8-4.

Process Audit Checklist

Plant ___*Sunnyvale*___   Date ___*March 15, 1991*___

Product/Area ___*Downconverter*___   Auditor _____

Station ___*Final Assembly*___

DOCUMENTATION
PART I

No.	DOCUMENT DESCRIPTION	AVAILABILITY YES	AVAILABILITY NO	COMMENTS
1.	Assembly Instructions		✔	*Working to "red lined" prints*
2.	Inspection Instructions	✔		
3.	Workmanship Standard		✔	*Available, but not completely applicable to this area*
4.	Calibration Procedure	✔		*Torque drivers and electronic test equipment*
5.	Test Instructions		✔	
6.	Fabrication Methods		✔	
7.	Station Layout	✔		
8.	Training Records		✔	*No formal training program*
9.	Product Specification	✔		
10.	Tooling Drawings	*N/A*		

Figure 8-4. (*Continued*)

effective. The "flag" may be indications on a control chart or detection of a particular level of error that indicates emergence of problems. A skydiver's reserve parachute is a clear example of contingent action planning. The alarm telling the skydiver to trigger the reserve chute would be failure of the main chute to open. In a factory, there could be a contingent action plan that switches a production line to building a secondary model when things go wrong with the scheduled production plan.

Preventive and contingent action planning is very similar to a design engineering function. The first step is to determine any imaginable poten-

DOCUMENTATION
PART II

ELEM. NO.	ELEMENT DESCRIPTION	VERIFICATION			DIAGNOSTIC			COMMENTS
		Acc.	Marg.	Unacc.	Acc.	Marg.	Unacc.	
1.	Are documents complete?	✔						No assembly instructions
2.	Are documents clear?				✔			Engineering review
3.	Are documents correct?				✔			Engineering review
4.	Are operations in proper sequence?	✔						
5.	Are rework/reinspect instructions included?			✔				Not available
6.	Are interactions included for defects to be reinspected?			✔				Not available
7.	What is the latest DCO revision?	✔						Revision C per doc control
8.	Are reference documents included where applicable?	✔						Test spec reference
9.	Are necessary support documents available? (e.g. variations, "workmanship standards," photographs, sketches, minimum acceptable samples, etc.)	✔						Only have workmanship standards for electronic assembly
10.	Does the manufacturing instruction adequately address all quality issues called for on the inspection instruction?	✔						Rely on experience of lead operator
11.	Does the inspection instruction adequately represent the product specification/ workmanship standard?	✔						

Figure 8-4. (*Continued*)

tial problems. Utilizing the experience and intuitive abilities of your team members, prioritize these potential problems on the basis of probability of occurrence and degree of severity. The second step is to determine likely causes for each problem, again using the experience of everyone involved or associated with the situation. Defining causes is important because it leads to more detailed preventive measures.

HUMAN FACTORS
(OPERATOR/INSPECTORS)

ELEM. NO.	ELEMENT DESCRIPTION	VERIFICATION			DIAGNOSTIC			COMMENTS
		Acc.	Marg.	Unacc.	Acc.	Marg.	Unacc.	
1.	Has the operator/ inspector been given skills training, on-the-job training, other? Has the operator been tested for his skills and knowledge? By whom? Method?			✔				*On the job training*
2.	Can the physical elements of the process be accomplished by the operator/inspector?	✔						
3.	What is the operator's/ inspector's attitude toward adequacy of the documentation?					✔		*Feel that they are responsible and get no support from engineering*
4.	Are working conditions conducive to quality workmanship?	✔						

STATION LAYOUT

ELEM. NO.	ELEMENT DESCRIPTION	VERIFICATION			DIAGNOSTIC			COMMENTS
		Acc.	Marg.	Unacc.	Acc.	Marg.	Unacc.	
1.	Does the station location conform to the area floor plan?	✔						
2.	Does the station physically conform to station layout plan?	✔						
3.	Is the station identified by number of the function/ operation performed?	✔						
4.	Is general housekeeping maintained?	✔						
5.	Are safety practices adhered to?	✔						

Figure 8-4. (*Continued*)

OPERATION (PROCESS)

ELEM. NO.	ELEMENT DESCRIPTION	VERIFICATION			DIAGNOSTIC			COMMENTS
		Acc.	Marg.	Unacc.	Acc.	Marg.	Unacc.	
1.	Are process documents present? (e.g. manufacturing method, inspection instructions, test instructions, etc.)			✔				
2.	Is document revision status correct?			✔				*Old drawing, no revision*
3.	Do the documents reflect the master copy? (No unauthorized changes)			✔				*Red lined drawing*
4.	Is set-up done according to instructions? (e.g. machine, work station, etc.)			✔				*No written instructions*
5.	Are required tools and gauges available at the station?	✔						
6.	Are tools and gauges in good working order? Are calibration stickers current where applicable?	✔						
7.	Does the operator/inspector perform the operation exactly according to the process documents?			✔				*Using wrong epoxy on isolator*
8.	Does the operator/inspector follow the sequence prescribed by the method?			✔				*Not allowing adequate cure time before test*
9.	Does the operator/inspector properly utilize his tools and equipment?			✔				*Doesn't use fixture—"Slows him down"*

Figure 8-4. (*Continued*)

Once problems and causes have been identified, develop and implement preventive action, working on the highest-priority items first. Finally, develop contingent action plans and alarms that will activate when progress to the desired objective is blocked.

The formula for preventive and contingency planning development is as follows:

DATA COLLECTION AT
TEST AND INSPECTION STATION

ELEM. NO.	ELEMENT DESCRIPTION	VERIFICATION			DIAGNOSTIC			COMMENTS
		Acc.	Marg.	Unacc.	Acc.	Marg.	Unacc.	
1.	Are the control charts posted? Are they clearly visible; properly labelled?	✔						
2.	Is the data on the control charts current to today's date?	✔						
3.	Are actions for out-of-control points indicated in the legend?			✔				
4.	Are signatures of supervisors and engineers on the control chart?	✔						
5.	Are center lines and control limits on the chart? —if X-Bar, R Chart—on both charts?	✔						
6.	Is there a "Corrective/Preventive Action Matrix?" Current?			✔				*Not available*
7.	Are the PQI teams performing to an "Inspection/Test Process Control" procedure?			✔				*Not available*

Conclusion

■ Operators were occasionally using the wrong epoxy when the correct material was not available.

■ Insufficient cure time was allowed prior to test (needs 24 hours).

■ Not using the holding fixture because it is hard to use and slows down the process.

Corrective/Preventive Action

■ Remove wrong epoxy from the plant.

■ Document assembly instructions and hold training class.

■ Build improved version of holding fixture.

Figure 8-4. *(Continued)*

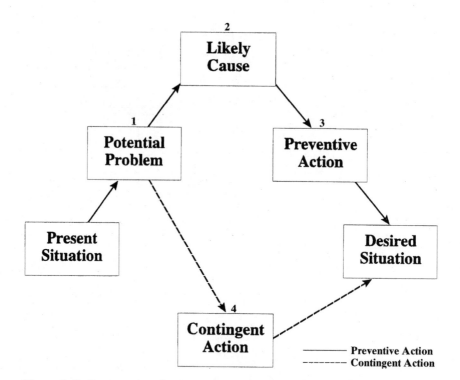

Figure 8-5. Process steps for preventive and contingent planning.

1. *Determine potential problems and set priorities.* Utilize your experience to assess those problems that are likely to impact your course of action. Prioritize them on the basis of probability and seriousness.

2. *List likely causes for each potential problem.* Use the experience of everyone involved or associated with the situation to determine "What could go wrong?" There may be more than one likely cause for a potential problem. Problems and causes could be identified in the relationship-diagram model mentioned earlier.

3. *Develop preventive actions for each likely cause.* You cannot prevent a potential problem unless you focus on the likely causes of that problem. The hard part of this exercise is to actually take the time to map out the problems, causes, and action. People will generally not view this as an important activity when they have "fires to fight," but solid planning sessions will result in fewer fires and much more efficient overall operation.

4. *Develop contingent action.* In the unlikely event of your preventive action not working, it will be important that you have an alternative

course of action planned and ready to go. A trigger, alarm, or flag needs to be built into the plan so that the need for contingent action is recognized in time.

5. Attain the desired result . . . success. Validate the results by testing to see that the desired state is achieved.

An example of a Preventive and Contingent Planning Worksheet is shown in Figure 8-6.

The Relations-Diagram Method

The relations-diagram method is a technique developed to clarify inter-twined causal relationships in a complex problem or situation in order to find an appropriate solution. Examples of use for the relations diagram are as follows:

Determine and develop quality assurance policies.

Establish promotional plans for TQM introduction.

Reengineer administrative departments.

Design steps to address market complaints.

Promote and obtain control of quality in purchased items.

Improve delivery schedules.

Plan new product, process, and/or equipment introduction

Simplify complex problems into manageable bite-size pieces.

Prioritize the most important and urgent tasks and actions associated with a problem.

Clarify interdepartmental relationships and action required.

Document various ideas and comments resulting from a discussion or meeting.

Encourage and enable free expression and the development of new and innovative ideas.

The process for use of the relations diagram is summarized as follows:

1. *Determine the broad subject or issue.*   This can be just about any type of problem. Examples are:

Successfully releasing a new product on time

Improving profitability

Reducing absenteeism

Objective Install the new computer system and have operational by end of October 1991

Potential Problems	Likely Causes	Preventive Action	Contingent Action	Alarm
Late delivery of hardware. L/M	Shortage of memory chips.	Audit supplier - focus on adequate supply of memory chips.	Plan to maintain old manual system for at least 3 months after end of October.	System does not arrive.
	Technical problems with the software.	Call other users of the software and discuss problems with supplier's Engineer.		
Internal users not able to use the system effectively. H/H	Inadequate training.	Start training in mid August. Involve supplier with the training design.	Involve the supplier, and if necessary, a consultant, to modify software prior to delivery.	During the initial stages of the user training (end August).
	System is not "user-friendly."	Make absolutely sure in the supplier selection process that it is user friendly.		
System downtime adversely impacts productivity. M/H	Internal system maintenance personnel not familiar with the system.	Have supplier's technician on site for the month of November.	Use manual system while supplier finds the cause of the problem and permanently fixes it.	Productivity control chart falls below LCL.
		Make sure that maintenance manuals are adequate and up to date.		

Figure 8-6. Example of preventive and contingent planning.

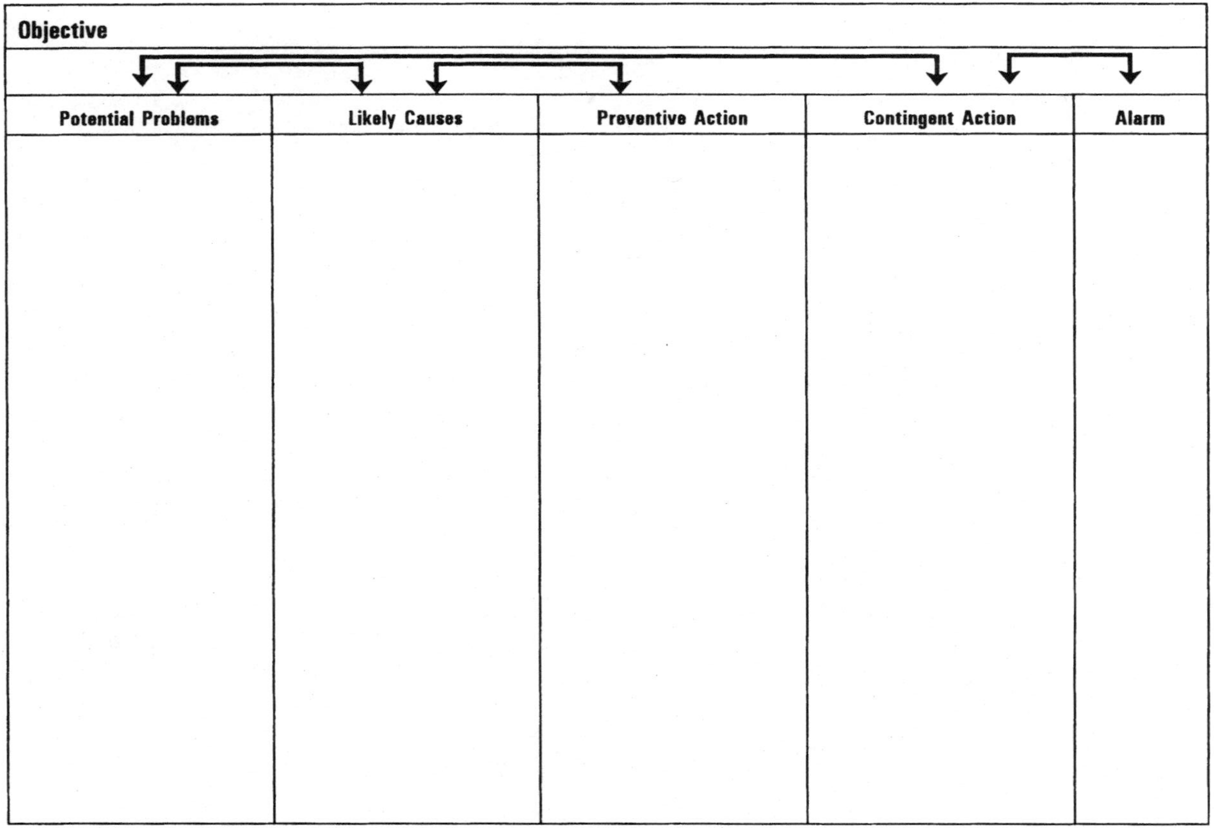

Objective					
Potential Problems	Likely Causes	Preventive Action	Contingent Action	Alarm

Figure 8-7. This is a blank form to work with for preventive and contingent action planning.

327

Improving morale

Minimizing paperwork and process efficiency

Improving communications

2. *Examine closely the factors and major issues.* This can be done by established quality/productivity implementation teams. It is important that every person who can throw some light on the problem is encouraged to participate in these meetings. Otherwise it is quite probable that a vital factor that is affecting the problem will be missed—a virtual guarantee of failure to adequately solve the problem.

3. *Create the relations diagram.* Using a large sheet of paper (flipchart size minimum) draw a circle and write the problem or issue inside it. Using arrows and more circles record on the sheet of paper all of the things, factors, and issues that impact your problem in any way. Some of these will be interrelated and should be interconnected accordingly. Please refer to Figure 8-8.

4. *Review and revise, if necessary.* Before taking any action or trying to analyze the problem further, it is a good idea to sleep on it and carefully review your relations diagram a little later. Be sure to involve the "right" people.

5. *Extract key items.* An effective way to do this is to rate each factor or issue on the basis of its impact on the problem or situation. Then it is important that you also rate each factor or issue with respect to its urgency. An effective way do this is to use *H* for high, *M* for medium, and *L* for low. This will enable you to quickly determine what needs to be addressed and acted upon first, hence getting the "biggest bang for the buck."

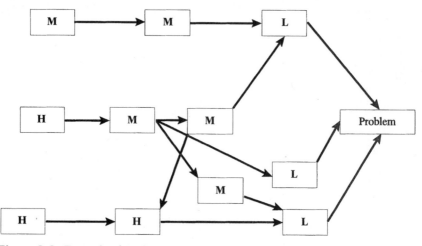

Figure 8-8. Example of a relations diagram.

6. *Plan and take action when appropriate.*   This sounds simple, doesn't it! However, many problems do not get solved because somehow or another the appropriate action does not get taken. Some or all of the other techniques covered in this workshop will be necessary to ensure success. It is vital that you determine and validate the right cause, make a logical decision to improve or correct it, and then develop a written plan that has preventive and contingent actions built into it.

See Figure 8-8 for an example of a relations diagram.

Logical Decision Making
Relative to Causes

Decision Making.   Sometimes there is no need to use one of the problem-solving processes we have discussed earlier. The person closest to a particular situation already knows the reason why something is not working as it is supposed to. Our challenge is to be sure that we listen to him or her. Then we have to do something about it. This is where decision making comes in.

Usually, there are several alternative actions that can be taken to resolve the situation. The question is, "Which is the best?" Everybody involved, directly or indirectly, with the problem has an opinion. Bias, politics, and many other reasons can and often do impact the decision. If we are to select the best way to revolve the problem, it makes sense that we use a method that will assure us that we meet the goals and objectives required.

This decision-making process is invaluable when a choice has to be made between several alternatives. It is not sufficient to come up with the cause of a problem. Invariably there will be more than one approach that may be used to overcome the cause and thus attain an improvement. Sometimes different people involved with the problem will have very diverse views of what has to be done.

You will find that this method of approaching decision making will help select the most useful course of action and will reduce the influence of biased thinking. Logical decision making is involved in all of the following examples:

Selection of the best candidate to fill an open position.

Selection of a supplier of a particular product, piece of equipment, or service that will provide the best-balanced product and/or service.

Picking the best course of action to improve a specific situation. This could be involved with cost reduction, quality improvement, better on-time shipments, improved communications—in other words, just about anything.

Of course, most functional individuals have a process for decision making, but there is always room for improvement. Outlining a logical process for decision making will help establish a more efficient approach. A great deal of time is wasted in meetings and on the telephone complaining about the problems. This decision-making process should help to achieve tangible results:

Step 1.   Determine the purpose of the decision. A good way to state this is to start with the words "to select." For example, "To select the best word processing software package for our personal computers."

Step 2.   List the specific objectives that need to be considered using a decision worksheet that lists objective and options side by side. These will usually be associated with different themes such as cost, timeliness, quality, reliability, ease of maintenance, and any other criteria that will impact your decision.

Step 3.   Split these objectives into musts and wants.

Musts are objectives that are absolute and mandatory. They have to be practical, of course.

Wants are other criteria of varying degrees of relative importance. Each one should be assigned an impact value. Use a scale of 1 to 10, with 10 being the most important.

Step 4.   Generate a few options that seem to be practical. It is important that you do not consider alternatives before you have determined the objectives that are key to the decision you are trying to make.

Step 5.   Score these options against the decision objectives.

Musts are determined on a go/no-go basis.

Wants are scored on a 1 to 10 arrangement; the best option is considered to be a 10, and the rest should be scored relative to it. Then multiply your impact value by the score and total for each option.

Step 6.   Take action to implement your best alternative. Be sure to minimize your risk of failure by utilizing "preventive and contingent planning."

"Poka-Yoke" or Opportunities for Foolproofing. Human beings tend to be very forgetful and, because of this, make a lot of mistakes. Usually, we blame the people who made the mistakes for their errors. The problem with this approach is that it will discourage the work force and lower morale. More significantly, it does not even begin to solve the problem and fosters "finger-pointing."

Decision Purpose Statement — To select the right house to buy

Musts

Musts	1 9 Green Road	Yes	No	2 33 Princes Street	Yes	No	3 2 Petit Road	Yes	No
1. At least three bedrooms	4 bedrooms	✓		3 bedrooms	✓		3 bedrooms	✓	
2. Maximum cost - $140,000	$139,500	✓		$128,000	✓		$145,750		✓
3. Available in 4 months	4 months	✓		3.5 months	✓		2 months	✓	

Wants

Wants	Impact	1 9 Green Road	Sc	xImp	2 33 Princes Street	Sc	xImp	3 2 Petit Road
A. Close to schools	10	2 blocks	10	100	Just under a mile	4	40	
B. Large back yard	6	65 x 90 ft	5	30	100 x 120 ft	10	60	
C. Fairly new	4	New	10	40	5 years old	8	32	
D. Tidy and safe neighborhood	9	A new tract on edge of town	6	54	Well established, many mature trees	10	90	
E. Large living room	5	12 x 17 ft	3	15	18 x 23 ft	10	50	
F. High appreciation potential	7	Probably slow	4	28	Good track record	10	70	
G. The lower the price the better	6	$139,500	8	48	$128,000	10	60	
				315			402	

Figure 8-9. Example of using the decision-making process.

331

Poka-yoke is a way to address the issue of reducing human errors at work. Shigeo Shingo, a Japanese manufacturing engineer, developed the idea of "foolproofing" into this formidable tool for achieving "zero defects." In this way we can move toward the elimination of quality control inspections.

The translation of *poka-yoke* in general terms is "mistake-proofing" or "fail-safing." *Yokero* means to avoid and *poka* means inadvertent errors.

Many things can go wrong in the complex environment of the workplace. Errors are extremely wasteful. It does not matter whether we are talking about the production shop floor, the front office, or the sales department. Human errors occur all the time. Just imagine your company's potential if all errors were eliminated. How successful would you be?

To become a world-class competitor, an organization must adopt not only a philosophy but a practice of producing zero defects. Poka-yoke is a way to achieve this. Conceptually, it is not difficult. In practice, however, everyone in the whole company needs to be involved, and often this concept is foreign to American business culture. Operators, assemblers, and janitors must all participate in poka-yoke, as well as engineers, scientists, and managers.

Foolproofing or Error Prevention

1. *Are errors unavoidable?* There are two different ways to think about and deal with errors:

a. Errors are unavoidable. People always have made and always will make mistakes. The only way to deal with this is to blame them and discipline them in some way. With this kind of attitude we are quite likely to overlook defects as they occur and instead rely on final inspection or tests to find them. Worse still, there is a good chance that some defects will slip through final inspection and be detected by the customer!

b. Errors can be eliminated. Any type of error that people make can be reduced or even eliminated. Proper training supporting a production system and attitude that is based upon the principle that errors can always be prevented will enable us to reduce or even avoid human error.

2. *Is sampling inspection always the best method?* One way to eliminate errors is by inspection. There are two predominant methods:

a. Sampling inspection. This is a very economical way to inspect products and processes. Properly done, sampling inspection will guarantee a predetermined quality level. For example, an Average Quality Level (AQL) of 0.1 percent means that one customer in a thousand will get a defective product—hardly zero defects!

b. Does 100 percent inspection make sense? If you wish to eliminate all defects and errors, it most certainly does. However, this can become very expensive unless the 100 percent inspection is automatic. Then the expense associated with it is extremely low. Today, even one defective product can destroy a customer's confidence in your product or service.

3. *The user is the best inspector.* This is a customer—external or internal. Whoever is either using the finished product or applying a subsequent process to a partially completed product has the highest probability of finding previously undetected errors and defects.

4. *Different kinds of errors:*
a. Forgetfulness: Sometimes we forget to do something when we are not concentrating. An example is forgetting to close the gate and letting the cattle out. *Safeguard:* Alert operator in advance or check frequently.
b. Misunderstanding: In this case we are not familiar with the situation. A person not familiar with a manual (stick shift) transmission is likely to stall the engine. *Safeguard:* Training, standardization of procedures, checking in advance.
c. Errors in identification: Sometimes we make an error because we cannot see or hear it properly. For example, thinking that a $1 bill is a $10 bill. *Safeguards:* Training, proper observation methods, vigilance.
d. Errors made by amateurs: Lack of experience or the skills required will invariably cause errors. Imagine the number of errors made by a person who is just starting to learn the game of golf. *Safeguards:* Work standardization, skill-based training, assigning the right person to the job.
e. Willful errors: Sometimes we decide that it is OK to ignore rules or regulations. An example is crossing the street when the sign tells you not to. *Safeguard:* Include the reason for a rule when training.
f. Inadvertent errors: Occasionally we can make mistakes without realizing how they happened. For example, someone lost in thought tries to cross the street without noticing that the light is red. *Safeguard:* Emphasize discipline, work standardization.
g. Surprise errors: They will sometimes occur when something unexpected happens. An example is a machine malfunctioning without warning. *Safeguard:* Focus on preventive maintenance and total productive maintenance.
h. Intentional errors: These differ from willful errors in that they are calculated and deliberate. Sabotage is an example. *Safeguard:* The creation of a loyal and supportive environment; team building.

5. *Poka-yoke or the avoidance of errors.* What is the basic function of poka-yoke? A defect occurs in one of two states. It is either about to happen or it has already occurred. Poka-yoke uses three basic functions to eliminate defects.

Detection Devices and Techniques

Control Mechanisms and Shutdown Procedures.   There are three basic functions used to eliminate errors or defects. The idea is to build into a process something that makes it absolutely impossible for human error to happen. If this is impossible, then an automatic way to prevent the error from moving on to the next step of the process must be installed to help the generation of appropriate poka-yoke actions.

1. Identify items by their characteristics:

By weight

By dimension

By shape

By any other characteristic that makes sense

Then determine a way to modify procedures such that these characteristics are always correct.

2. Detect deviations from procedures or any omitted processes. An example is the impossibility of drilling a hole in sheet metal after is has been bent or formed.

3. Use these types of approaches:

Unique location devices

Error detection and alarm systems

Limit switches

Counters

Who should be involved with and apply poka-yoke? If the ultimate goal is to reach zero defects, then everybody in the whole organization must be involved. Poka-yoke requires the ideas, thoughts, and action of whoever is the closest and most knowledgeable about the product or process. Examples of poka-yoke worksheets are shown in Figures 8-10 and 8-11.

The Eight Principles of Basic Improvement Leading to Foolproof Processes and Zero Defects

1. Build quality into all processes. Use poka-yoke to build in safeguards to all processes and associated equipment.

2. All inadvertent errors and defects can be eliminated. We must assume that mistakes are not inevitable. Where there is a will, there has to be a way to prevent them.

3. Stop doing it wrong and start doing it right—*Now!* Let's get rid of the "buts" and excuses.

Poka-yoke Worksheet

Process:	Greasing ground fixture on cassette decks	Prevent Error X	Shutdown
Problem:	Grease on pulley belt		
Solution:	Install stopper on grease brush so pulley belt can't be greased	Detect Error	Control X
Key Improvement:	Tool modified to guarantee correct processing		
Description of Process:	A ground fixture on a cassette deck mechanism greased with white grease, applied with a brush. However, if grease accidentally gets on the nearby pulley belt, the auto-stop mechanism will not work.		
Before Improvement:	Despite the vigilance of skilled workers, grease sometimes got on the pulley belt, causing defects.	After Improvement:	The brush was outfitted with a stopper so the brush cannot reach all the way to the belt. Grease no longer gets on the belt, and auto-stop failures are completely eliminated. The addition of the stopper makes it possible for experienced workers and novices to do the work equally well.

Figure 8-10. Poka-yoke worksheet.

4. Think only about how to do it right. Rather than think of excuses, focus only upon doing it right—the first time.

5. A 60 percent chance of success is good enough. Implement new ideas when you are about 60 percent sure that it will work. You can refine and perfect it as it is being implemented.

6. Mistakes and defects can be eliminated when everyone works together to eliminate them. Everybody in the entire company must be involved if you are to realize your potential to achieve zero defects.

7. Ten heads are better than one. Teamwork is the key.

Poka-yoke Worksheet

Process:	Casting		Prevent Error		Shutdown
Problem:	Unreliable pressure gauges				
Solution:	Install redundant pressure gauges for comparison				
Key Improvement:	Tool modified to make additional test		Detect Error X		Control X
Description of Process:	When casting products with a large casting machine, undesired changes in machine conditions can result in blowholes and other defects. Pressure gauges are installed at key locations to monitor the casting process and the state of the machine.				
Before Improvement:	One pressure gauge was installed at each measuring site, but if a gauge was reading a nonstandard condition, it was difficult to determine whether it was the process or the gauge at fault.	After Improvement:	Two pressure gauges are installed on the same outlet at each measuring site. The operator can quickly determine the reliability of the readings on the gauges by comparing them.		

Figure 8-11. Poka-yoke worksheet.

8. Seek out the true cause. Ask why. If you do not get a satisfactory answer, ask why again, and again, and *again*—until you really know "why." Then and only then should you ask, "How do we fix it?"

Statistical Process Control

Variability

A crucial concept in the Core Techniques Track is the idea of variability. To understand the concept of variability, we need to change some of the fundamental ways in which we think. Instead of thinking of steps in our process as potential sources of problems, let's think of them as sources of variability. The way we position our drill, the hardness of our materials, and the way we attach the spigot to the mold all vary from time to time. The temperature of our metal and the temperature of our rooms all change from one day to the next. To one extent or another, each of these variations contributes to variability in our product or service.

Natural variability comes from many sources: material (thickness or hardness, etc.), slop in the machine or the jigging (positioning or bearing wear or tool wear, for example), operator (e.g., fatigue, feed rate, or feed position), maintenance (e.g., lubrication, alignment), or environment (e.g., voltage, air pressure, humidity).

Just as there are different sources of variability, we can identify different types of variability. Among them are: lot-to-lot, stream-to-stream, piece-to-piece (jigging of each part or accuracy of each bookkeeping entry), place-to-place (when measuring a single piece), measurement equipment, measurement technique, and long-term (tool wear or procedural change).

However we choose to think about variability, its various sources add together to create the total variability in the product or service.

Whether small or large, some variability is a natural and constant feature of any of our products or services. This natural and constant variability can be considered a random result of the entire set of aforementioned causes. We call this set of causes the *common cause* of product variability, and we usually agree to live with the constant variability it produces—at least for a while.

Occasionally there is an unusual problem that causes our output to be highly variable and unpredictable. We call such a source of variability a *special cause*. Whenever a special cause occurs, it is essential that we note its occurrence, identify it, and remove it from the system. It is the energy with which special causes are removed that marks the most noticeable difference between traditional and modern management systems.

Special causes, such as broken tools or sudden power shifts, often can be removed by action of line workers or by simple intervention by local management. Special causes rarely require intensive investigation or detailed strategy sessions by top-level management. It's an everyday responsibility to keep a process stable, and this responsibility is most efficiently handled at the local level.

The constant variability resulting from the common cause, however, is a product of a stable system. It cannot be reduced without changing the basic elements of the system. Management must get involved to study and decide on which changes to make. For example, if material variability is contributing to the constant level of variability in the product, it might be appropriate to reduce the number of vendors involved. Or if the measurement system being used is not consistent, it might be better to use a new measurement system. Ultimately, management must make both of these decisions. Fortunately, actions involving the common cause are much less frequent than actions involving special causes.

The responsibility to solve commonly caused problem is management, who designs the process—not the operators.

Once we accept the idea that variability is a natural part of any process, we can use graphic tools to reinforce this idea. Two graphic tools which we can use for this reinforcement are histograms and control charts. Histograms provide a snapshot of data taken at a given point in time. Control charts graphically depict variability occurring along a period of time; they will be covered later in this chapter.

Histograms

Before we begin our discussion of histograms, we need to talk about the concept of a frequency distribution. (In fact, a histogram is a type of frequency distribution.)

The output of a process usually forms a stable pattern. For example, if we have a process which produces machined shafts with a 1.00-inch nominal diameter, we would expect to find that most of the shafts produced by this process have a diameter of approximately 1.00 inch. Because we understand the concept of variability, we know that not every shaft will be exactly 1.00 inch. Instead, some will be a little under 1.00 inch and some will be over the nominal dimension. Some may be exactly 1.00 inch in diameter.

A very frequently occurring pattern of output from a process has most of the parts near the nominal value and very few parts far away from the nominal value. In many cases, this pattern takes on the bell shape with which many of you are familiar. Another frequently occurring pattern has most of the parts near the low end of the measurement scale and fewer and fewer parts as the measurements get larger. Still another familiar pattern has equal numbers of parts at all measurement values. Whatever the exact shape, a stable pattern of variability is called a *distribution*. Distributions can differ from one another in terms of their center and spread as well as their shape. From these differences, you can tell a great deal about the causes of the variability of the part.

A *histogram* is a frequency distribution consisting of vertical rectangles whose width corresponds to a defined range of measurements and whose height corresponds to the number of readings that occur within the range. The purposes of a histogram are:

- To visually determine the central tendency
- To visually determine the variation
- To visually determine the shape of the distribution

Histograms require at least 30 data points, and preferably 75 to 100 data points. People rarely use more than 200 data points.

Histogram Construction Procedure.   To construct a grouped frequency distribution, there are three basic guidelines that should be followed:

- Five to fifteen classes are desired.
- Each class must have the same width.
- Classes should be set up so that they do not overlap. Each piece of data should belong to exactly one class.

Procedure

1. To find the approximate number of classes, count the number of pieces of data and take the square root of that number. If the square root is less than 5, adjust to 5. If the square root is larger than 15, adjust to 15.

2. Subtract the smallest reading from the largest reading to find the range of the data.

3. Divide the range by the number of classes found in step 1. This will give you the approximate width for each category.

4. Round that width to a "convenient" multiple of the measurement unit. For example, if the measurement unit is .001 inch and the width obtained in step 3 is .003264 inch, round to .003 inch. If the width found in step 3 is .017264 inch, round to .020 inch. In general, "convenient" multiples are considered to be 1, 2, 3, 4, 5, 10, 20, 30, 40, 50, 100, 200, etc. If possible, use one measurement unit as the class width. When circumstances suggest a different whole number to be "convenient," use your good judgment.

5. Lower limits of successive classes are separated by the class width, not lower and upper limits of the same class. If, for example, 3 is chosen as the class width, each class will contain 3 potential values. If one class starts at .087, the class will contain the values .087, .088, and .089. The next class will start at .090 and run to .092. The next class will start at .93, etc. Note in particular that if the class width is 3, the class .087 to .090 would be wrong because it would contain 4 values: .087, .088, .089, and .090.

The choice of class width determines the degree of "smoothing" accomplished by the histogram, so your choice of class width is very important. If you choose too small a class width, the histogram will be "choppy," and it will be hard to see groupings. If you choose too large a class width, the histogram will be too smoothed, and important details will be missed. Therefore, follow the rules in steps 1 to 3 and the suggestions in step 4 carefully.

6. Establish the lower limit for each class. There are many different rules for determining where the lower limit should be. This issue rarely makes any difference, so use whatever value you like so long as you wind up with a lowest class. Some people use whole number multiples of the class width. Some people try to put the lowest reading at or just below the

midpoint of the lowest class. The shape of the histogram will not be greatly influenced by your choice, so do as you wish.

Example: If the lowest reading is 1.018 and you have determined the class width to be .005, the lowest class could be any of the following:

1.014–1.018	Not many people choose this one.
1.015–1.019	Choose this if you want the lower limit to be a whole number multiple of the class width.
1.016–1.020	Choose this if you want the lowest reading to be at the midpoint of the lowest class.
1.017–1.021	Choose this if you want the lowest reading to be just below the midpoint of the lowest class.
1.018–1.022	Choose this if you want the lowest reading to be the lower border of the lowest class.

Any of these choices are acceptable and all are likely to give you very similar histograms.

7. Record the class limits on a worksheet. Count and tally the number of readings in each class. It is easy to make simple errors in this tally, so check your work carefully.

Construct the Histogram.   To construct a histogram:

1. *Count and tally the number of readings in each class.* It is easy to make simple errors in this tally, so check your work carefully.

2. *Determine the axes for the graph.*

- The vertical axis is the frequency of each class and should exceed the largest class frequency.

- The horizontal axis is labeled in some convenient scale corresponding to the measurements and should be at regular intervals independent of the class width. It is not necessary to label all the class limits since this will result in a cluttered-looking histogram.

3. *Draw the bars.*

- Mark off classes.

- Draw the bars or rectangles with heights corresponding to the number of readings in each class.

4. *Title the histogram.* It is very important to label the histogram completely. The labeling information should include the source of the data, sample size, and the date. Both of the axes should be labeled. The vertical axis is usually labeled "frequency."

Analysis with the Histogram. Histograms often reveal, without elaborate analysis, much information about the population under study and, because they are easily understood, can aid in making improvements. The histograms shown in Figures 8-12 and 8-13 indicate typical patterns observed and an explanation for the cause of each pattern.

The histogram does have some limitations to take into consideration:

- It requires many measurements.
- It is a "snapshot" of time.
- Time is not taken into consideration.
- Trends are not shown.

Histograms, in general, are very useful in determining the shape and location of process distributions. Comparing the histogram with the specification is especially helpful in determining if the process will meet the specification or not.

Process Flow Diagram

The next element in the project identification step is the preparation of a process flow diagram for the process being addressed. This diagram helps the team to determine the critical operations and to pinpoint where process control charts will be most applicable and effective. A flow diagram shows the processes that collectively or sequentially produce the final product. (A *process* can be defined as any combination of materials, machines, tools, methods, and people that create through specifications the desired products or services.) Processes are found in manufacturing, service, and support operations. Support processes are typically found in office or administrative operations. The flow diagram also establishes the boundaries in which the team must confine its project involvement. Steering committees, anxious to see results, may occasionally make assignments not relative to implementing SPC within the defined project. Teams also may tend to sidetrack and want to investigate an interesting but unrelated potential opportunity. The combination of a clear objective and a process flow diagram focuses the activities of the team within the originally defined project.

The team should determine if a process flow diagram exists and if so, review it for accuracy and completeness. To review the flow diagram or to prepare one, the team should visit the location(s) where the process is being performed, observe the activities, and interview the key people in the process. While reviewing the process, the team should pay particular attention to the items or characteristics checked at each inspection point.

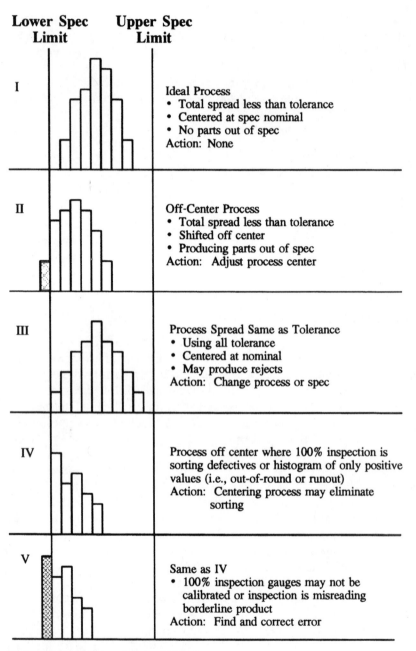

**Lower Spec   Upper Spec
   Limit   Limit**

I
Ideal Process
• Total spread less than tolerance
• Centered at spec nominal
• No parts out of spec
Action: None

II
Off-Center Process
• Total spread less than tolerance
• Shifted off center
• Producing parts out of spec
Action: Adjust process center

III
Process Spread Same as Tolerance
• Using all tolerance
• Centered at nominal
• May produce rejects
Action: Change process or spec

IV
Process off center where 100% inspection is
sorting defectives or histogram of only positive
values (i.e., out-of-round or runout)
Action: Centering process may eliminate
         sorting

V
Same as IV
• 100% inspection gauges may not be
  calibrated or inspection is misreading
  borderline product
Action: Find and correct error

Figure 8-12. Interpretation of histograms.

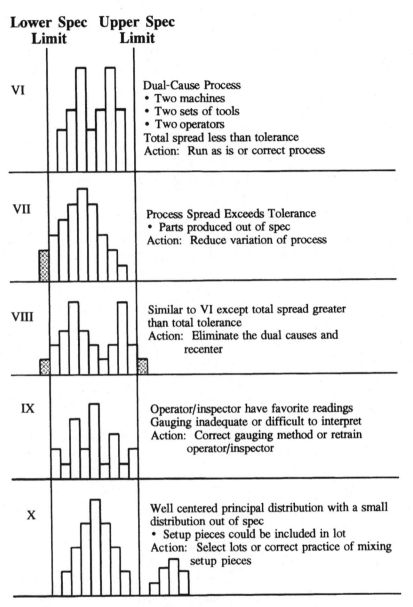

Lower Spec Limit **Upper Spec Limit**

VI — Dual-Cause Process
• Two machines
• Two sets of tools
• Two operators
Total spread less than tolerance
Action: Run as is or correct process

VII — Process Spread Exceeds Tolerance
• Parts produced out of spec
Action: Reduce variation of process

VIII — Similar to VI except total spread greater than total tolerance
Action: Eliminate the dual causes and recenter

IX — Operator/inspector have favorite readings
Gauging inadequate or difficult to interpret
Action: Correct gauging method or retrain operator/inspector

X — Well centered principal distribution with a small distribution out of spec
• Setup pieces could be included in lot
Action: Select lots or correct practice of mixing setup pieces

Figure 8-13. Interpretation of histograms.

A flow diagram is constructed with symbols that represent various operations or processes. Figure 8-14 shows suggested symbols for constructing a manufacturing flow diagram. Figure 8-15 shows how the symbols are used. Figure 8-16 is a complete flow diagram for assembling printed circuit boards.

The actual symbols in a flow diagram are not as important as ensuring that all key operations or processes are identified on the diagram and that they are understood. A standard set of symbols eliminates confusion from diagram to diagram.

Flowcharting Techniques. Effective flowcharting involves a standard process. We suggest that you follow this process in order to realize the greatest benefits from flowcharting.

■ Bring together representatives from all departments responsible for the process so they can collectively perform the analysis.

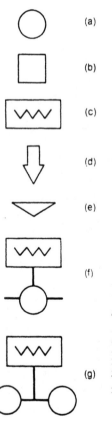

(a)

(b)

(c)

(d)

(e)

(f)

(g)

Figure 8-14. Some symbols used in constructing SPC flow diagram. (a) Operation step (production or activity). (b) Inspection step (100% inspection or appraisal). (c) Inspection or appraisal using control chart. (d) Transportation step. (e) Storage. (f) In-process inspection or appraisal with control chart performed by operator. (g) In-process inspection or appraisal with control chart performed by inspector or supervisor.

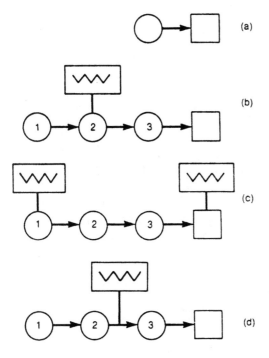

Figure 8-15. Some examples of groups of symbols in SPC flow diagrams. (*a*) Single operation followed by 100% inspection operation. (*b*) In-process inspection with control chart performed by operator at second of three operations followed by 100% inspection. (*c*) In-process inspection with control chart performed by operator at first of three operations with 100% inspection with control chart after third operation. (*d*) In-process inspection with control chart performed by inspector after second process with 100% inspection after third operation.

■ Identify and define the process in writing.

■ Define boundaries of the process.

■ Brainstorm all steps in the process. Document what actually takes place outside of the written procedure.

■ Sequence the steps in the process. Chart the flow of the steps using standard symbols. (The advantage of using standard symbols is that everyone is "speaking the same language" when analyzing a flow chart.)

You may choose to chart vertically, from the top of the page to the bottom, or you may begin charting at the upper left-hand side of the page and work left to right to the bottom of the page.

The flowchart in Figure 8-17 depicts an SPC process flow diagram.* For this example, only three symbols were used: rectangle—process, step, diamond—decision point, and circle—process start or stop point.

---

* Note: For further definition of SPC, refer to the text *Productivity/Quality Improvement: A Practical Guide to Implementing Statistical Process Control,* John L. Hradesky, McGraw-Hill Publishing, 1988.

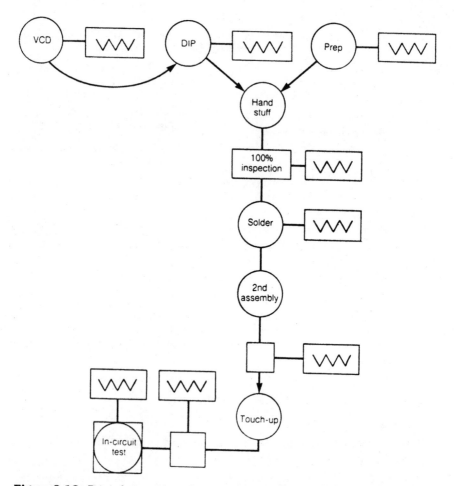

Figure 8-16. Printed circuit board assembly flow diagram. VCD stands for variable component device machine, DIP for dual in-line package machine.

Control Chart Basics

A control chart is the statistically oriented method that allows taking small samples at specified time intervals, like a series of snapshots, to record the process on a real-time basis. The control chart does not keep the process stable; the actions management takes does that.

The small samples, taken at specified time intervals, are called *sub-groups*. Samples for the subgroups are taken consecutively or one after the other from the process. The success of the control chart method depends upon obtaining samples for the subgroups where the variation between

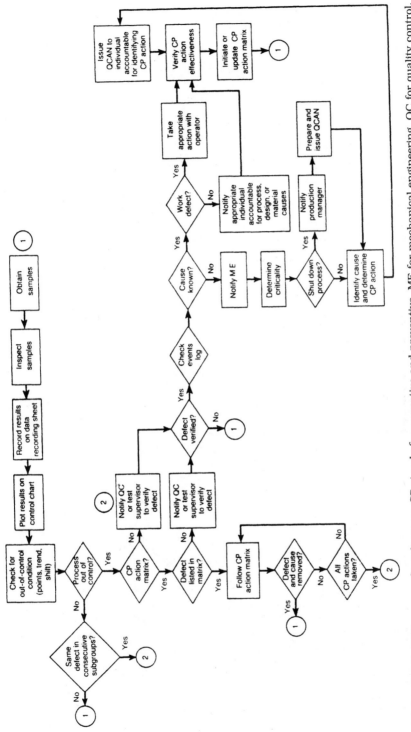

Figure 8-17. SPC process flow diagram. CP stands for corrective and preventive, ME for mechanical engineering, QC for quality control, and QCAN for quality control action notice.

the samples within the subgroup is as small as possible. And the variation between subgroups is the largest possible. This is achieved by collecting consecutive samples or parts from the process for each subgroup.

The time between the subgroups, called the *subgroup frequency,* is constant and is expressed as once per hour, once per shift, or once per some quantity, e.g., once per 200 pieces. For each subgroup, a value is computed and plotted on the control chart. Typical values computed for subgroups are the average, range, percent defective, or number of defects. These values are plotted as shown in Figure 8-18.

After a minimum of 25 subgroups are obtained, three values are calculated from the data obtained and plotted on the control chart. These values are the average or centerline value, the upper control limit (UCL) and the lower control limit (LCL) as shown in Figure 8-19.

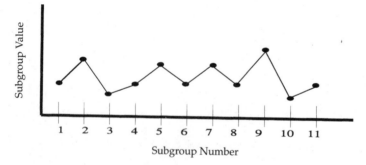

Figure 8-18. Subgroup value plotting.

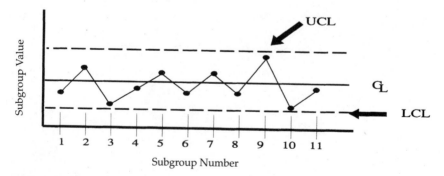

Figure 8-19. Control chart centerline and control limits.

If all the plotted points are within these limits, as shown in Figure 8-19, the process is said to be "in control." This means that the variation observed is the normal, usual, expected variation of the process. It may not meet the requirement, but it is what the process is doing. This usual normal variation is the result of the variation of all the things in the process such as machine, material, people, or environment contributing equally and at random. As long as the points fall within these limits, the process should be left alone.

If a point falls outside the control limits, the process is "out of control." This is shown in Figure 8-20. When the process is found to be out of control, it should be stopped and corrective action taken to bring it back in control. "Out of control" implies that one or more of the process variables, machine, material, method, or people is causing abnormal variation. This variable is called the *assignable cause* for the out-of-control condition; such an assignable cause must be found and removed. Detection of the out-of-control condition and the correction of it must be real-time to have effective, ongoing statistical process control.

Because subgroup averages are normally distributed (more on that later), there are only 3 chances in 1000 that a subgroup mean will fall outside the control limits if the process has not changed. Thus, if a point does fall outside the control limits, either the process has shifted or an extremely rare event has occurred. The best action, then, is to take action. The laws of probability or "odds" are in your favor that the process has changed.

The control limits on the control chart act as a warning device much like the temperature light on the dashboard of an automobile. When the temperature light comes on it is a signal or warning that the engine is over-

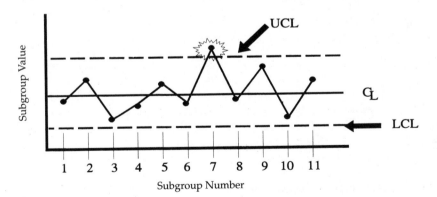

Figure 8-20. Out-of-control process.

heating. We do not know what the cause is, so the car is stopped to determine the cause or we continue driving with a potential risk that the engine may be ruined.

When the driver is checking for the cause of the overheating condition, things relating to overheating are checked, such as broken fan belt, leaking hose, or leaking radiator. The driver does not check for a flat tire.

The control chart works in the same way, when the chart indicates the process is out of control, the process is stopped and the cause is found and removed. If the control chart goes out of control due to any of the out of control conditions, causes for the increased average are investigated.

The selection of the subgroups is based on the following three items. These will be discussed for each type of chart explained in Sections 5, 6, and 7.

- *Subgroup size.* Function of the type of chart used, typically, 4 or 5 are used for X/R charts and one for chart of individuals.

- *Subgroups.* Frequency is a function of the time and cost to inspect and the cost of impact if defects are found.

- *How often are the subgroups taken?* Primarily a function of your goals, the production rate, and the cost of inspection.

The major steps for obtaining and plotting data on a control chart are as follows:

Step 1. Select consecutive pieces for the subgroup.

Step 2. Measure or inspect the parts.

Step 3. Record the data on the proper form.

Step 4. Compute the subgroup value.

Step 5. Plot the subgroup value on the control chart.

Step 6. Interpret the results.

If the point is within the control limits, continue running the process. If the point is outside the control limits, take the appropriate action to find and remove the assignable cause.

Once the centerline and control limit values have been established from the first 25 subgroups, these values are used for the ongoing control charts if the process is in control. If the process is found to be out of control in the first 25 subgroups, the assignable causes must be removed and 25 new subgroups are obtained.

Variable/Attribute Data. Several different types of control charts exist. The type of control chart you will use depends on the type of data that is being collected.

Data collected for control charts can be classified into one of the two following types:

- *Variable data.* Data generated when measurements of parts are obtained. Typical examples are measuring length, thickness, width, diameter, hardness, and resistance.
- *Attribute data.* Data generated by counting defects or detectives. Typical examples are from visual inspection, go/no-go gages, accept/reject decisions, drawing errors, invoice errors, and line-item errors.

Defects are defined as any single nonconformity on a unit. *Defectives* are units with one or more defects.

Variable Control Charts (Measurement Data). The common control charts used for variable data are the X-bar and R charts. These two charts appear on the same form.

$\overline{X}$ chart—Plots the average (X) of each subgroup.

R chart—Plots the range (R) of each subgroup.

Attribute Control Charts (Attribute Data). Attribute data results from the following:

- Visual inspection
- Dimensions checked by go/no-go gages
- Measurement data where the result is recorded as accept or reject

Attribute control charts are useful when inspecting many characteristics on a part. The results of inspection are identification of defectives (nonconforming part) or defects (nonconformities).

The distinction between a nonconforming part (defective) and a nonconformity (defect) is:

- A nonconforming part (defective) is a part or unit that in some way fails to conform to one or more specifications.
- A nonconformity (defect) is the lack of conformance to a specification of any single item on the part.
- A nonconforming part is defective and will have one or more nonconformities (defects).

The type of attribute control chart to use depends upon whether defectives or defects are being plotted. The types of attribute control charts are as follows.

"p" chart	Plots the fraction defective (*p*) for each subgroup.
"c" chart	Plots the total number of defects (*c*) in each subgroup. The subgroup size may be one unit or any equal quantity of units or opportunities of error.
"u" chart	Plots the average number of defects per unit (*u*) for the subgroup.

Basic Statistical Concepts

The statistics needed to understand the statistical methods used in *Implementing Statistical Process Control* will now be described. These statistics are concerned with computing measures of the location and variation of the process distribution. These can then be used to estimate the process distribution and determine if the process is meeting the requirement.

Population—Sample. When implementing statistical process control, it must be determined if the process being studied (population) is meeting the requirement. The requirement is a specification or target. However, not much is known about the process until pieces or parts produced by the process (samples) are obtained and inspected or measured. Once these parts are inspected or measured, the resulting data or numbers (statistics) are used to determine if the process is meeting the requirement.

The population is the process to be studied; however, the parts produced by the process over the next 6 months or year is the actual population of interest. Instead of measuring all the parts from the process to see if the requirement is being met, a portion or *sample* of the parts is measured.

The concepts of a population and sample are illustrated further by the following examples.

Tasting of stew. The whole pot of stew is the population. The spoonful tasted is the sample or subgroup.

Bead box. A bead box with 1000 beads is the population. Withdrawing 100 beads from the bead box is the sample or subgroup.

Parameter—Statistic. When studying a process, there is usually a characteristic of the parts (length, diameter, performance, or appearance) that has a requirement to meet. All the values of these characteristics have a distribution. This distribution is described by values called *parameters* that measure the location and variation of the distribution. The parameters used to measure the location and variation are the *average* and *standard deviation*. These are needed to determine if the process is meeting the requirement. In order to determine these values, the data from the sample is used to calculate statistics so that the values of the parameters can be

Population   ⇒   Parameter

⇓   ⇓   **Figure 8-21.** Parameters of a
population are estimated by
Sample   ⇒   Statistic   the statistics in the sample.

estimated. This concept of a parameter of a population being estimated by the statistics in the sample is a very fundamental one used in statistical methods. (See Figure 8-21.)

The concepts of a parameter and statistic are illustrated further by the following examples.

Tasting of stew.   The taste of the stew is the parameter of interest. The taste of the spoonful is the statistic. In this example, the taste cannot be expressed as a number but only as being "good" or "bad."

Bead box.   The percent of purple beads in the 1000 beads may be the parameter of interest. The percent of purple beads in the 100 beads sampled is the statistic.

In these examples, it may be noted that the values of the statistics will not be the same for each sample. The values will vary around some average value. The quantifying of this variation is a very fundamental concept of statistical methods. The parts from a process will not all be the same. Statistical process control is concerned with establishing whether the variation is in statistical control or not.

The following definitions summarize the terms just described:

Population.   A collection or set of individual objects whose characteristics are to be determined.

Sample.   A subset of the population.

Parameter.   A characteristic of the population.

Statistic.   A characteristic of the sample.

The following notation is used to distinguish between parameters and statistics:

Parameters, being unknown values in the population, are denoted by Greek letters.

Statistics, the values computed from the data in the sample, are denoted by Roman letters.

The *average value of a parameter* is denoted by the greek letter μ (mu). The *average of the sample* is denoted by $\overline{X}$ (read as "X-bar").

Mean or Average. The average or mean of the process is an unknown value (parameter) and is estimated using the average or mean of the sample (statistic).

The average or mean of the sample is denoted by $\overline{X}$ and is computed by the following formula:

$$\overline{X} = \text{sum of } X's/n = \text{sum of all readings/sample size}$$

Example: A subgroup for a control chart produces the following data: 0.3, 0.4, 0.5, 0.3, 0.5. Find the average.

$$\overline{X} = \frac{0.3 + 0.4 + 0.5 + 0.3 + 0.5}{5} = \frac{2.0}{5} = .04$$

The value of the average indicates where the center or average of the process is located.

Measures of Variation. In any kind of process, variation is observed. All things vary to some degree. People are not all the same size—they vary. Parts that are manufactured are not all the same size—they vary. Although these statements are true, the variation of individual items is not predictable. The adult size of an individual cannot be predicted at birth. The size of the next part from a process cannot be predicted. A tolerance is applied on drawings to indicate the allowable variation that is acceptable. If a part has a dimension of 2.500 ± .050, the allowable variation is from 2.450 to 2.550.

The amount of variation, spread, dispersion, or variability of the values in a distribution must be expressed by a numerical value. If the values are grouped closely to the average, the value of the measure of variation should be small. If the values are widely dispersed or spread out from the average, the value of the measure of variation should be large.

Two commonly used terms to quantify or measure variation are the *range* and *standard deviation*. The range is used extensively in the average and range chart to be described later.

Range. The range is the simplest measure of variation to compute and is combined together with ranges of several small samples from a process to estimate the variation of the process distribution. The X-bar and R control charts use the average range to estimate the process variation.

Average ($\overline{X}$) and Range (R) Chart Example. Data was taken in subgroups of size 5 every 100 pieces produced from a brazing process. The shear force was determined for each sample of the subgroup. The specification for the shear force is 60 to 260 lbs. Fifteen subgroups were taken and it is desired to find out what the process is doing and if it meets the specification.

The average and range of the 15 subgroups are as follows.

Subgroup	Average	Range
1	130	30
2	117	20
3	135	40
4	109	30
5	127	50
6	110	30
7	122	50
8	13	20
9	140	50
10	145	35
11	140	50
12	120	40
13	140	70
14	125	20
15	150	60
	Sum $\overline{X} = 1823$	Sum $R = 595$

Step 1. Find the initial values of the centerline and control limits.

Step 2. Plot the subgroup averages and ranges along with the centerline and control limit values on a control chart.

Step 3. If the process is in control, determine if it meets the specification.

The range is the largest value in the sample minus the smallest value, and is denoted by R. The formula is:

$$R = \text{(largest reading)} - \text{(smallest reading)}$$

Example: Find the range of the numbers 3, 3, 5, 6, 8.

The largest reading is 8.

The smallest reading is 3.

The range is $R = 8 - 3 = 5$.

Standard Deviation. The standard deviation is denoted by S and is computed by the following formula:

$$S = \sqrt{\text{sum } (X - \overline{X})^2/(n - 1)}$$

Example 1: Compute the standard deviation for the numbers 6, 3, 8, 5, 3. Set up a table with three columns as shown. Compute the values in each column and the sum of each column as shown.

X	$(X - \overline{X})$	$(X - \overline{X})^2$
6	$(6 - 5) = 1$	1
3	$(3 - 5) = -2$	4
8	$(8 - 5) = 3$	9
5	$(5 - 5) = 0$	0
3	$(3 - 5) = -2$	4
sum $X = 25$	sum $(X - \overline{X}) = 0$	sum $(X - \overline{X})^2 = 18$

$$\overline{X} = (\text{sum } X)/n = 25/5 = 5$$

$$S = \sqrt{\text{sum } (X - \overline{X})^2/(n - 1)} = \sqrt{18/4} = \sqrt{4.5} = 2.12$$

Notice that the sum of the deviations (sum $(X - \overline{X})$) is always zero. Also note that this method of calculating the standard deviation is easy if the average is a whole number and the sample size is small. If the average is not a whole number, the values of $(X - \overline{X})$ are not always whole numbers.

A formula for computing the standard deviation that is often used in computer programs is the following computational formula:

$$S = \sqrt{[(n)(\text{sum}(X^2)) - (\text{sum } X)^2]/(n)(n - 1)}$$

Where $(\text{Sum } X)^2$ = sum of readings squared
Sum (X^2) = sum of squares of readings

Example 2: Compute the standard deviation for the numbers 6, 3, 8, 5, and 3, using the computational formula. Set up two columns and compute the values for each and find the totals as shown. These totals are then used to compute the standard deviation.

X	X^2
6	36
3	9
5	25
3	9
$(\text{Sum } X) = 25$	Sum $(X^2) = 143$

Since the sample size n is 5, the standard deviation is computed as follows.

$$S = \sqrt{[(n)(\text{sum } X^2) - (\text{sum } X)^2]/(n)(n - 1)}$$

$$S = \sqrt{[(5)(143) - (25)(25)]/(5)(4)}$$

$$S = \sqrt{[715 - 625]/20} = \sqrt{90/20} = \sqrt{4.5} = 2.12$$

Variance.   The square of the standard deviation is called the *variance.*
This value is used in advanced statistical methods. The variance of sample
values is denoted by S^2.
The variance of the numbers in examples 1 and 2 is:

$$S^2 = (S)^2 = 2.12^2 = 4.5$$

Using a computer to compute the average and standard deviation is
becoming the accepted practice and is recommended for large sets of data.
The only errors encountered with computers are those arising from enter-
ing the wrong data values.

Estimating the Process Distribution.   When the values of the average and standard deviation are obtained, the location and limits of the
process can be reasonably estimated using the following empirical rule.
This rule is very useful since almost all (90 to 95 percent) of the process
distributions approximate a normal distribution.
The normal distribution is a symmetrical bell-shaped distribution as
shown in Figure 8-22. The empirical rule which follows is used to estimate
the process distribution.

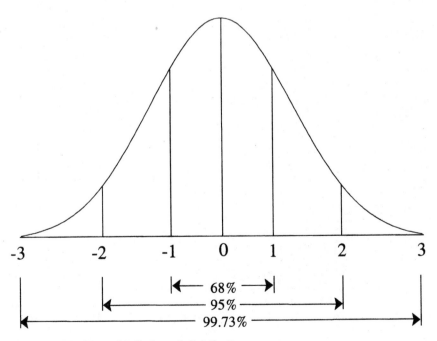

Figure 8-22. Normal bell-shaped distribution.

Empirical Rule. If a characteristic is approximately normally distributed, there will be approximately 68 percent of the values within one standard deviation of the mean, 95 percent of the values within two standard deviations of the mean, and 99.7 percent of the values within three standard deviations of the mean.

The procedure for using the empirical rule to estimate the process distribution is as follows:

Sample size must be at least 25.

Compute the average and standard deviation of the sample.

The average is used to estimate the process average.

Compute $(X + 3S)$ and $(\overline{X} - 3S)$ to find the interval where approximately 99.7 percent of the process distribution values are contained. This interval can be used to determine if a specification is being met by comparing the two computed values with the specification limits.

The following examples illustrate how a picture of a process can be drawn and compared to the specification using the empirical rule.

Suppose a sample of 30 pieces is obtained from a process and the average is found to be 0.102 and the standard deviation is 0.0015. Suppose that the specification for the process is 0.100 ± 0.010. The upper specification limit (USL) is 0.110 and the lower specification limit (LSL) is 0.090.

The process can be pictured as shown in Figure 8-23 with the average at 0.102 and −3σ value at 0.102 − (3)(.0015) = 0.0975 and the +3σ value at 0.102 + (3)(.0015) = 0.1065. This process meets specification since it is well centered and the total variation (6σ = .009) is much less than the total specification spread of 0.020.

The example indicates that an ideal process is one where the average is at the specification nominal and the 3σ values do not exceed the upper and lower specification limits.

Consider a process that is producing a part where one of the dimensions has a specification of 0.625 to 0.665. The ideal process would have an aver-

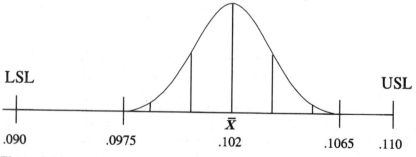

Figure 8-23. Sample process.

age of 0.645 and the +3σ value would be at 0.665 and the –3σ value would be at 0.625, as shown in Figure 8-24.

Control Chart Centerline and Control Limits
Computation Concepts

The centerline is the average of the values plotted and is denoted by $\overline{X}$.

Upper control limit (UCL) = $\overline{X} + 3\sigma_x$.

Lower control limit (LCL) = $\overline{X} - 3\sigma_x$ where σ = estimate of variation of plotted values.

It is assumed that the plotted values have approximately a normal distribution. This implies that 99.7 percent of the values plotted will be within plus and minus three standard deviations of the average. The control chart and basic computation formula is shown in Figure 8-25.

Finding the Standard Deviation

Step 1.   Find the average of the scores.

Step 2.   Subtract the average from the first score and square the difference.

Step 3.   Repeat for all scores.

Step 4.   Add the squared differences.

Step 5.   Divide that sum by (the number of scores – 1).

Step 6.   Take the square root of the result of step 5. This is the standard deviation.

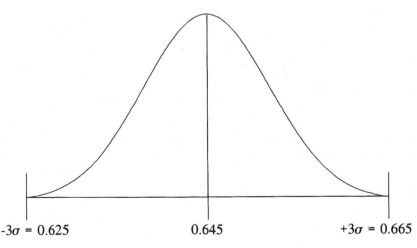

-3σ = 0.625   0.645   +3σ = 0.665

Figure 8-24. Sample of an ideal process.

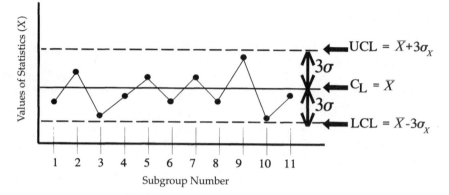

Figure 8-25. Example of plotted values with upper and lower control limits.

Average ($\overline{X}$) and Range (R) Charts. The average and range chart is commonly referred to as an X-bar and R chart. For each subgroup, two values are computed and plotted, the average ($\overline{X}$) and range (R).

The X-bar chart plots the subgroup averages, while the R chart plots the subgroup ranges. The average of the X-bar is an estimate of the process average and the subgroup ranges are used to estimate the process variation or 6σ of the process.

The following guidelines for selecting the subgroups discuss the size of the subgroup, how many subgroups, and the frequency or how often they are selected.

Subgroup Size. The size of the subgroup is denoted by n and is composed of consecutive pieces from the process. For the X-bar and R chart the recommended subgroup size is 5, which is considered to be the most effective.

The minimum subgroup size to compute a range is 2, and no more than 10 is recommended. The larger the subgroup size, the more chance there is for shifts in the average being included in the range of a subgroup. This results in an estimate of process variation that is too large.

Number of Subgroups. The number of subgroups is denoted by K for realistic ongoing control limits; K must be 25. The minimum number of subgroups for preliminary control limits is 10, but not recommended.

The number of subgroups required to establish realistic ongoing limits is a compromise between having reliable data and establishing limits quickly.

Subgroup Frequency. The frequency or time between the subgroups is a function of the production rate, cost and time of obtaining the measurements, the failure rate, and the cost of failures.

Production rates between 50 and 100 per hour should have subgroups taken at least once per hour. Production rates over 100 per hour should have subgroups obtained every 15 minutes. If the production rate is less than 50, then the time between the subgroups would be longer, possibly two to four subgroups per shift.

When obtaining data from the process for the first time, the subgroups are taken closer together until the process is in control. When the process is in control and stable, the time between subgroups can be extended. Frequently, one would start with 15 minute intervals and increase to 30–60 minute intervals as the process demonstrated it to be in control.

Data Collection and Plotting

The steps for computing and plotting data on the average and range chart, with the centerline and control limits already established, are as follows.

Step 1. Select consecutive pieces from the subgroup for the process. Select them "as is" before any touch-up or rework by the operator.

Step 2. Measure the parts and record the measurements on the appropriate data recording form.

Step 3. Compute the range of the subgroup.

Step 4. Compute the average of the subgroup. Use the shortcut method if the subgroup size is 5.

Step 5. Plot the range on the range (R) chart.

Step 6. Plot the average on the X-bar chart.

Step 7. Interpret the results. If the point is within the control limits, continue running the process. If the point is outside the control limits, take the appropriate action.

Example: Suppose subgroups of size 5 are obtained.

Step 1. Five consecutive parts are obtained from the process.

Step 2. The parts are measured and the data are as follows: 0.821, 0.819, 0.818, 0.822, 0.820.

Step 3. Compute the range. $R = 0.822 - 0.818 = 0.004$.

Step 4. Compute the average. The total of the readings is 4.100. Divide 4.100 by 5 = 0.8200, which is the average.

Step 5. Plot the range 0.004 on the R chart (see Figure 8-23).

Step 6. Plot the average 0.8200 on the X-bar chart (see Figure 8-23).

Step 7. Interpret the chart. The point is within the control limits—continue running the process.

Initial Control Limits. The initial control limits are determined after the first 25 subgroups are completed. The limits for the range chart are computed first, since the average range is used to compute the limits for the average chart.

If the process has a slow production rate, then preliminary limits may be computed after 10 subgroups. This will allow evaluation of the process in about half the time required for 25 subgroups.

Range Chart. The average range, R-bar, is the centerline value for the range chart and is computed as follows:

$$\overline{R} = \frac{R_1 + R_2 + ... + R_K}{K} = \frac{\text{sum of subgroup ranges}}{\text{number of subgroups}}$$

The control limits UCLR and LCLR are computed as follows (the factors D_3 and D_4, found in Table 8-3, are a function of the subgroup size):

$$\text{UCLR} = (D_4) \times \overline{R}$$

$$\text{LCLR} = (D_3) \times \overline{R}$$

Table 8-3. Factors for the Average and Range Chart

Subgroup/ sample size n	Averages A_2	Range D_3	Range D_4	Sigma d_2	Subgroup/ sample size n
2	1.880	.0	3.268	1.128	2
3	1.023	.0	2.574	1.693	3
4	.729	.0	2.282	2.059	4
5	.577	.0	2.114	2.326	5
6	.483	.0	2.004	2.534	6
7	.419	.076	1.924	2.704	7
8	.373	.136	1.864	2.847	8
9	.337	.184	1.816	2.970	9
10	.308	.223	1.777	3.078	10
11	.285	.256	1.744	3.173	11
12	.266	.284	1.717	3.258	12
13	.249	.308	1.692	3.336	13
14	.235	.329	1.671	3.407	14
15	.223	.348	1.652	3.472	15

Formulas for Computing Control Limits

For averages	For range
$\text{UCL} = \overline{\overline{X}} + A_2 \overline{R}$	$\text{UCL}_R = D_4 \overline{R}$
$\text{LCL} = \overline{\overline{X}} - A_2 \overline{R}$	$\text{LCL}_R = D_3 \overline{R}$

Standard deviation or σ

$$\sigma = \overline{R}/d_2$$

The factor D_3 is 0 for subgroup sizes of 6 or less so the LCLR is 0 for subgroups sizes of 6 or less.

The range chart must be in control before computing limits for the average chart. The limits for the average chart will be invalid if the range chart is out of control. The cause of the out-of-control variation must be corrected and new data obtained.

Average Chart. The average of the subgroup averages is the centerline value for the average chart, denoted X (read as "X-double-bar") is computed by the following formula:

$$\overline{\overline{X}} = \frac{X_1 + X_2 + ... + X_K}{K} = \frac{\text{sum of subgroup averages}}{\text{number of subgroups}}$$

The control limits, UCL_X and LCL_X are computed by the following formula: (The factor A_2, found in Table 1, is a function of the subgroup size.)

$$UCL_{\overline{X}} = \overline{\overline{X}} + [A_2 \times \overline{R}]$$

$$LCL_{\overline{X}} = \overline{\overline{X}} - [A_2 \times \overline{R}]$$

Plotting the Centerline and Control Limits. The centerline values are plotted as a solid horizontal line and the control limits are plotted as dashed horizontal lines.

Process Evaluation

If the control chart is in control, then it can be determined what the process is doing by estimating the process average and variation.

Once the process average and variation are estimated, then it can be determined if the process is meeting the requirement or the specification. It is important to understand that the process may be in control but not meet the specification. The variation may be too big or the process may not be centered properly. This cannot be determined by looking at the control chart. Separate calculations must be made as shown following.

The process average is estimated by the centerline value of the average chart or X-double-bar.

The process variation is determined by first estimating the standard deviation and then multiplying this by 6 to obtain the total process variation (6σ or $\pm3\sigma$).

The process standard deviation is estimated by the following formula:

$$\sigma = \frac{\overline{R}}{d_2}$$

The process spread, 6σ, is computed and compared to the specification spread or total tolerance. The total tolerance is the upper specification limit U minus the lower specification limit L.

If 6σ is equal to or less than $(U - L)$, the process is capable of meeting the specification. If the process is not centered, adjust the center to the specification nominal.

If 6σ is greater than $(U - L)$ the process is not capable of meeting the specification. The process may have to be changed or the specification may have to be changed. If none of these are possible, then the process output will have to be inspected 100 percent.

An example of an X-bar and R chart with process evaluation will now be examined.

Example Solution

Step 1.   Determine the initial control limits.

- Range chart

  Centerline:

$$\overline{R} = \frac{595}{15} = 39.67$$

  Control limits:

$$\text{UCLR} = D_4 \times \overline{R} = (2.114) \times (39.67) = 83.86$$

$$\text{LCLR} = D_3 \times \overline{R} = 0 \times (39.67) = 0$$

- Average chart

  Centerline:

$$\overline{X} = \frac{1948}{15} = 129.87$$

  Control limits:

$$UCL\overline{X} = \overline{X} + [(A_2) \times \overline{R}]$$

$$UCL\overline{X} = 129.87 + [(5.77) \times (39.67)] = 152.76$$

$$LCL\overline{X} = \overline{X} - [(A_2) \times \overline{R}]$$

$$LCL\overline{X} = 129.87 - [(.577) \times (39.67)] = 106.98$$

Step 2.   The subgroup values, centerline and control limits are plotted on the chart shown in Figure 8-26.

Step 3.   Process evaluation.

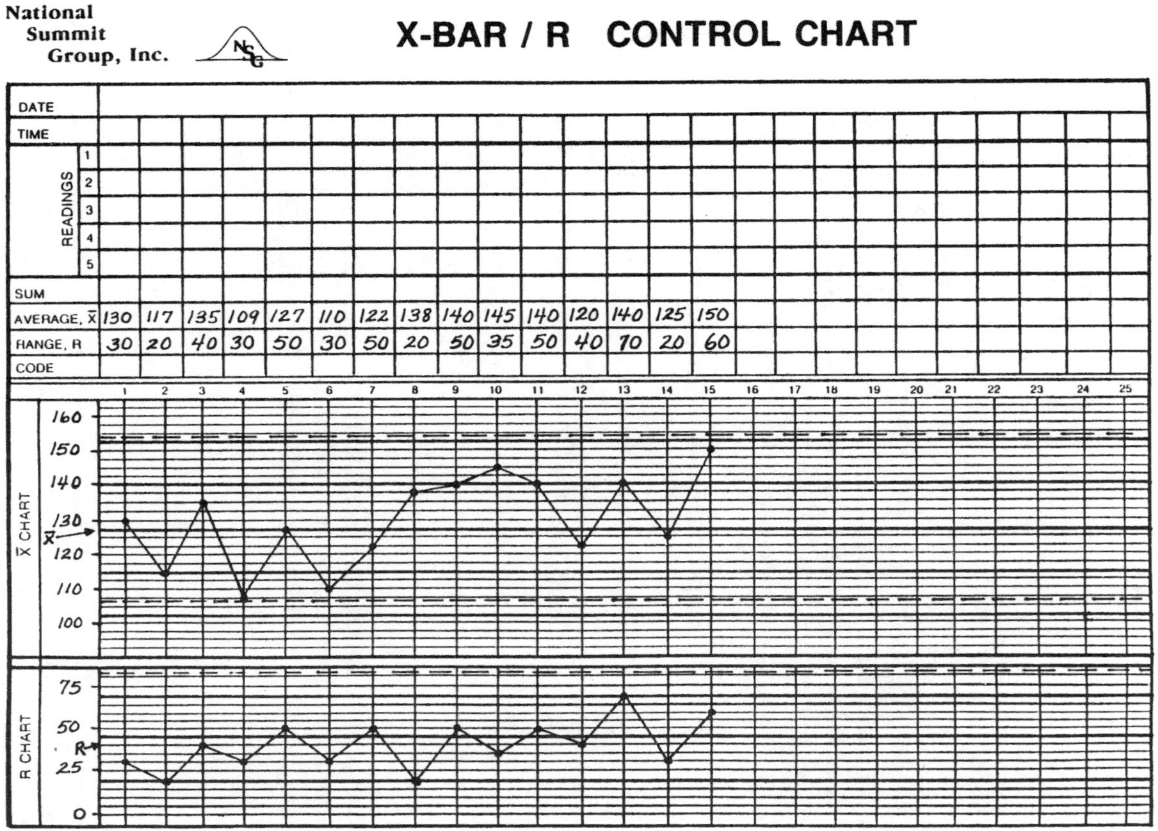

National
Summit
Group, Inc.

X-BAR / R   CONTROL CHART

		1	2	3	4	5	6	7	8	9	10	11	12	13	14	15	16	17	18	19	20	21	22	23	24	25
DATE																										
TIME																										
READINGS	1																									
	2																									
	3																									
	4																									
	5																									
SUM																										
AVERAGE, $\bar{X}$		130	117	135	109	127	110	122	138	140	145	140	120	140	125	150										
RANGE, R		30	20	40	30	50	30	50	20	50	35	50	40	70	20	60										
CODE																										

$\bar{X}$ CHART

160
150
140
130 $\bar{\bar{X}}$
120
110
100

R CHART

75
50 $\bar{R}$
25
0

Figure 8-26. X-bar/R control chart.

The control chart is in control.

The process average is estimated to be 129.87, the centerline value of the average chart.

The process standard deviation is estimated as follows:

$$\sigma = \frac{\overline{R}}{d_2} = \frac{39.67}{2.326} = 17.055$$

The process spread is:

$$6\sigma = 6 \times (17.055) = 102.33$$

The total tolerance is $260 - 60 = 200$. The process spread is less than the total tolerance so the process is capable of meeting the specification. The center of the process and the process limits place the process well within the specification as shown in Figure 8-80C.

The process limits are determined as follows:

$$\text{Upper process limit} = \overline{X} + \left[3 \times \frac{\overline{R}}{d_2} \right]$$

$$\text{Upper process limit} = 129.87 + [3 \times 17.055]$$

$$\text{Upper process limit} = 129.87 - 51.17 = 78.78$$

$$\text{Lower process limit} = \overline{X} - \left[3 \times \frac{\overline{R}}{d_2} \right]$$

$$\text{Lower process limit} = 129.87 - 51.17 = 78.70$$

X-Bar and R Chart Control Limit Computation Rationale. The control limit computations for both the X-bar and range charts are based on the general control limit formula presented in the preceding example. This formula is:

$$\text{UCL} = \overline{X} + 3\sigma_x$$

$$\text{LCL} = \overline{X} - 3\sigma_x$$

X-Bar Chart Limits. The value plotted on the $\overline{X}$ chart is the subgroup average $\overline{X}$. This is substituted for the statistic X in the above formula to give the following:

$$\text{UCL} = \overline{\overline{X}} + 3\sigma_{\bar{x}}$$

$$\text{LCL} = \overline{\overline{X}} - 3\sigma_{\bar{x}}$$

The variation of the X-bar values $\overline{X}$ has the following formula based on statistical principles:

$$\sigma_{\bar{x}} = \frac{\sigma_x}{\sqrt{N}}$$

Thus, the control limit formulas are:

$$UCL = \overline{\overline{X}} + 3\left(\frac{\sigma_x}{\sqrt{N}}\right)$$

$$LCL = \overline{\overline{X}} - 3\left(\frac{\sigma_x}{\sqrt{N}}\right)$$

which is estimated by $\overline{R}/d_2$ using the values from the range chart. Substituting $\overline{R}/d_2$ for σ_x gives the following control formulas:

$$UCL = \overline{\overline{X}} + 3\frac{\overline{R}/d_2}{\sqrt{n}} = \overline{\overline{X}} + \frac{3\overline{R}}{(d_2)\sqrt{n}}$$

$$LCL = \overline{\overline{X}} - 3\frac{\overline{R}/d_2}{\sqrt{n}} = \overline{\overline{X}} - \frac{3\overline{R}}{(d_2)\sqrt{n}}$$

These formulas can be rewritten as:

$$UCL = \overline{\overline{X}} + \left[\left(\frac{3R}{(d_2)\sqrt{n}}\right)(\overline{R})\right]$$

$$LCL = \overline{\overline{X}} - \left[\left(\frac{3R}{(d_2)\sqrt{n}}\right)(\overline{R})\right]$$

$$\text{Let } A_2 = \left(\frac{3R}{(d_2)\sqrt{n}}\right)(\overline{R}) \qquad \text{and}$$

$$UCL = \overline{\overline{X}} + [(A_2)(\overline{R})]$$

$$LCL = \overline{\overline{X}} - [(A_2)(\overline{R})]$$

R Chart Limits.   The formulas for the control limits are:

$$UCL = \overline{R} + 3\sigma_R$$

$$LCL = \overline{R} - 3\sigma_R$$

The variation of the distribution for the range σ_R is estimated by the formula:

$$\overline{R}\left(\frac{d_3}{d_2}\right)$$

Substituting this in the preceding formula gives:

$$UCL = \overline{R} + 3\left((\overline{R})\left(\frac{d_3}{d_2}\right)\right)$$

$$LCl = \overline{R} - 3\left((\overline{R})\left(\frac{d_3}{d_2}\right)\right)$$

This can be rewritten as:

$$UCL = \overline{R}\left[1 + 3\left(\frac{d_3}{d_2}\right)\right]$$

$$\text{Let } D_4 = \left[1 + 3\left(\frac{d_3}{d_2}\right)\right]$$

$$UCL = (D_4)(\overline{R})$$

Similarly,

$$LCL = \overline{R}\left[1 - 3\left(\frac{d_3}{d_2}\right)\right]$$

$$\text{Let } D_3 = \left[1 - 3\left(\frac{d_3}{d_2}\right)\right]$$

$$LCL = (D_3)(\overline{R})$$

Individuals Control Chart—Variable Data Subgroups of Single Measurements. In some applications of variable control charts, it may be difficult to obtain subgroups of size 5. The production rate may be so slow that it takes several hours to obtain the samples. In applications of this type, a control chart with subgroups containing one reading may be useful. This type of control chart is called a *control chart for individual measurements.*

A ribbon-inking process is a good example to illustrate this type of variable control chart. A spool of cloth has ink applied to it by running the cloth through a bath of ink so the cloth can absorb the ink. The amount of ink on the spool is controlled by weighing the spool after the inking process. The spools are the samples to be used for the subgroups and the production rate is slow (one spool every half hour). The amount of ink on the spool is determined by the weighing process and this value is to be plotted on a control chart.

Initially, a subgroup of size 4 was considered, but it would take 2 hours to complete a subgroup. If a subgroup value were out of control, any resulting actions would be delayed 2 hours. This chart would not be "real

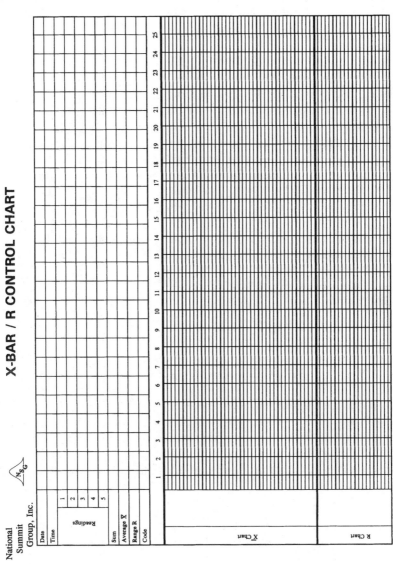

Figure 8-27. X-bar/R control chart.

time." Plotting the results of each spool as it is made would be more real time, so the individuals control chart was implemented.

In other applications there is only one number to represent the process at a given time. Examples of this are in accounting data (orders, efficiencies, productivity, earnings per share, costs of quality, costs as percent of sales) or certain production process records (temperature, humidity, voltage, pressure).

The data for the individuals control chart can be plotted on the X-bar and R chart. The labels of the charts should be changed to reflect the individual values. Change the X-bar scale to X for plotting the individual values. Change the R chart scale to the moving range (MR) for plotting the moving range.

The range chart plots the difference between two consecutive points on the X chart, without regard to the sign. Since a subgroup of size 1 is used, the range value cannot be computed within the subgroup. This difference is called the *moving range* (MR). The label on the range chart should be changed to an MR chart.

Example of Moving Range Computation. Suppose the first three subgroup values are:

$$X1 = .502 \qquad X2 = .501 \qquad \text{and} \qquad X3 = .503$$

The first moving range is

$$MR1 = X1 - X2 = .502 - .501 = .001$$

The second moving range is

$$MR2 = X2 - X3 = .501 - .503 = .002$$

Centerline and Control Limits. The control limits for the individuals control chart use factors based on estimates of the process variation from the moving range. Thus the factors for a subgroup of size 2 is used only for finding an estimate of σ and the total process variation (6σ). The formulas for the centerlines and control limits are calculated as follows after 25 values (subgroups) are obtained.

Moving Range (MR) Chart

$$\text{Centerline } \overline{MR} = \frac{\Sigma(MR)}{K - 1}$$

Note: The number of moving ranges is 1 less than the number of points plotted. For K points there are $(K - 1)$ moving ranges.

$$UCLMR = (3.268)(\overline{MR})$$

$$LCLMR = 0$$

Note: If any single moving range value is bigger than 3.5 times the average moving range in the initial capability study, the value should be removed and the average moving range recalculated.

The variation or σ is estimated by

$$\frac{\overline{MR}}{1.128}$$

Individual (X) Chart.   The centerline is $\overline{X}$ found in the usual manner. Total all readings and divide by the total number of readings.

The control limits for the X chart are the usual ±3σ limits. The formulas for the control limits are as follows:

$$UCLX = \overline{X} + 3\left(\frac{\overline{MR}}{1.128}\right) = \overline{X} + (2.66)(\overline{MR})$$

$$LCLX = \overline{X} - (2.66)(\overline{MR})$$

In some cases ±2σ limits may be used. This increases the chance of error from .3 to 5 percent and may result in limits that are more useful in practice. The control limits would be calculated as follows:

$$UCLX = \overline{X} + 2\left(\frac{\overline{MR}}{1.128}\right) = \overline{X} + (1.77)(\overline{MR})$$

$$LCLX = \overline{X} - (1.77)(\overline{MR})$$

Process Capability.   The capability (6σ) of the process is computed as follows:

$$6\sigma \text{ of process} = 6\left(\frac{\overline{MR}}{1.128}\right) = 5.32(MR)$$

To calculate control limits with modified limits being considered:

1. Find the averages.

   Sum the averages, then divide by the quantity you added. This will find X-bar-bar.

   Sum the ranges, then divide by the quantity you added. This will find R-bar.

2. Find the UCL for *R*.

$$UCLR = D_4\,(\overline{R})$$

3. Find the LCL for *R* if the subgroup size is greater than 6.

4. Find the UCL for X-bar.

$$UCLX = \overline{X} + A_2\,\overline{R}$$

5. Find LCL for the X-bar.

$$LCLX = \overline{X} - A_2 \overline{R}$$

6. If the control limits for X-bar are less than ½ the print, then recalculate the control limits using the modified method. Choose the limits which will give you the widest spread.

7. Calculate modified control limits.

Find the standard deviation for the individuals (STD$_i$).

$$STD_i = \frac{\overline{R}}{d_2}$$

Determine the customer requirements.

CPK	STD
2.00	6
1.34	4
1.00	3

Calculate the distance.

Distance = customer requirement X (STD$_i$)

Calculate UCLX.

LCLX = upper tolerance limit (UTL) – distance

Calculate LCLX.

LCLX = lower tolerance limit (LTL) + distance

Note: The range limits must remain the same.

Delta Charts. Some job shops produce small runs of similar but not identical parts. For example, they may produce a dozen 1-inch cylinders, then four dozen cylinders with a nominal diameter of 1.2 inches, and finally 20 cylinders with a nominal diameter of .75 inch. If means and ranges of the diameters of these parts were plotted on a single X-bar/R chart, the chart would look strange indeed. (See Figure 8-28.)

National
Summit
Group, Inc.

X-BAR / R CONTROL CHART

DATE																				
TIME																				
READINGS	1	2	3	4	5	6	7	8	9	10	11	12	13	14	15	16	17	18	19	20

READINGS																				
1	1.007	.983	1.002	1.214	1.178	1.220	1.206	1.182	1.200	1.232	1.062	1.215	1.088	1.226	1.212	1.243	1.715	.736	.743	.757
2	.974	1.000	.976	1.207	1.219	1.196	1.190	1.193	1.207	1.174	1.185	1.187	1.221	1.188	1.764	1.745	.725	.748	.724	.758
3	.985	.992	.977	1.176	1.190	1.187	1.184	1.217	1.181	1.106	1.219	1.129	1.200	1.108	1.194	.745	.774	.767	.728	.743
4	.990	1.011	.978	1.215	1.108	.992	1.186	1.215	1.921	1.211	1.211	1.191	1.687	1.207	.731	.752	.726	.757	.776	
5																				

SUM	3.943	3.981	3.935	4.824	4.775	4.537	4.744	4.821	4.744	4.844	4.838	4.744	4.829	4.901	3.003	3.024	2.977	2.952	3.034	
AVERAGE X̄	.991	.997	.983	1.206	1.194	1.207	1.192	1.205	1.192	1.212	1.210	1.192	1.207	1.200	1.752	1.757	.744	.738	.759	
RANGE R	.030	.051	.026	.030	.041	.028	.026	.035	.019	.016	.078	.094	.013	.035	.024	.036	.050	.091	.083	.033
CODE																				

X̄ CHART

1.2
UCL → 1.1
LCL → 1.0
.9
.8

R CHART

.075 UCL̄R →
.050 R̄
.025

Figure 8-28. Sample X-bar/R control chart.

373

Diameter	Diameter	Diameter	Diameter
1.009	1.182	1.220	0.763
0.979	1.174	1.196	0.764
0.985	1.219	1.219	0.745
0.990	1.171	1.192	0.734
0.983	1.213	1.208	0.775
1.000	1.185	1.190	0.725
0.992	1.229	1.182	0.774
1.014	1.211	1.186	0.752
1.002	1.188	1.182	0.736
0.976	1.187	1.209	0.748
0.977	1.200	1.217	0.767
0.978	1.191	1.213	0.726
1.216	1.226	1.200	0.743
1.207	1.221	1.193	0.724
1.186	1.188	1.181	0.728
1.215	1.194	1.192	0.757
1.178	1.212	1.222	0.757
1.219	1.188	1.207	0.758
1.190	1.194	1.206	0.743
1.188	1.207	1.211	0.776

One solution to this problem is to plot separate $\overline{X}$ - R charts for each job. This is what is done in typical large-run applications. This must be done even for small runs if both the nominal and the spec range are different for the different jobs, e.g., 1.2 ± .02, 1.0 ± .005, and .75 ± .001. But, for small runs, not enough data is generated to make these charts useful.

Often, therefore, job shops turn to delta charts. On these charts, the actual diameters of individual cylinders are not used. Instead, the nominal diameter is subtracted from each actual diameter, and the difference ("delta") is recorded. Delta charts can be used whenever the processes (drilling, milling, etc.) for the various job runs are the same and the tolerance ranges are similar (e.g., 1 ± .030, 1.2 ± .030, and .75 ± .030.) See Table 8-24 for an example of the computations. The delta chart itself is Figure 8-29.

Data Collection. In order to identify when a process needs attention, information needs to be transmitted from the process to us. In order to convince management that their intervention is required, information needs to be transmitted from us to our management. In either case, information starts with data, so let's next consider how we gather the data.

Data collection includes the following important steps:

1. Define the objectives. Data collected without a clear objective is often worthless. No matter how carefully we analyze poorly collected data, the results will be meaningless.

Table 8-4. Example of Delta Chart Computations

Diameter	Nominal	Delta	Diameter	Nominal	Delta
1.009	1.000	0.009	1.182	1.200	−0.018
.979	1.000	−0.021	1.174	1.200	−0.026
.985	1.000	−0.015	1.219	1.200	0.019
.990	1.000	−0.010	1.171	1.200	−0.029
.983	1.000	−0.017	1.213	1.200	0.013
1.000	1.000	0.000	1.185	1.20	−0.015
.992	1.000	−0.008	1.229	1.200	0.029
1.014	1.000	0.014	1.211	1.200	0.011
1.002	1.000	0.002	1.188	1.200	−0.012
.976	1.000	−0.024	1.187	1.200	−0.013
.977	1.000	−0.023	1.200	1.200	0.000
.978	1.000	−0.022	1.191	1.200	−0.009
1.216	1.200	0.016	1.226	1.200	0.026
1.207	1.200	0.007	1.221	1.200	0.021
1.186	1.200	−0.014	1.188	1.200	−0.012
1.215	1.200	0.015	1.194	1.200	−0.006
1.178	1.200	−0.022	1.212	1.200	0.012
1.219	1.200	0.019	1.188	1.200	−0.012
1.190	1.200	−0.010	1.194	1.200	−0.006
1.188	1.200	−0.012	1.207	1.200	0.007
1.220	1.200	0.020	.763	.75	0.013
1.196	1.200	−0.004	.764	.75	0.014
1.219	1.200	0.019	.745	.750	−0.005
1.192	1.200	−0.008	.734	.750	−0.016
1.208	1.200	0.008	.775	.750	0.025
1.190	1.200	−0.010	.725	.750	−0.025
1.182	1.200	−0.018	.774	.750	0.024
1.186	1.200	−0.014	.752	.750	0.002
1.182	1.200	−0.018	.736	.750	−0.014
1.209	1.200	0.009	.748	.750	−0.002
1.217	1.200	0.017	.767	.750	0.017
1.213	1.200	0.013	.726	.750	−0.024
1.200	1.200	0.000	.743	.750	−0.007
1.193	1.200	−0.007	.724	.750	−0.026
1.181	1.200	−0.019	.728	.750	−0.022
1.192	1.200	−0.008	.757	.750	0.007
1.222	1.200	0.022	.757	.750	0.007
1.207	1.200	0.007	.758	.750	0.008
1.206	1.200	0.006	.743	.750	−0.007
1.211	1.200	0.011	.776	.750	0.026

National Summit Group, Inc.

X-BAR / R CONTROL CHART

		1	2	3	4	5	6	7	8	9	10	11	12	13	14	15	16	17	18	19	20	21	22	23	24	25
DATE																										
TIME																										
READINGS	1	.009	-.017	.002	.016	-.022	.020	.008	-.018	.000	.022	-.018	.013	-.012	.026	.012	.013	.025	-.014	-.007	.007					
	2	-.021	.000	-.024	.007	.019	-.004	-.010	.009	-.007	.007	-.026	-.015	-.013	.021	-.012	.014	-.025	-.002	-.026	.008					
	3	-.015	-.008	-.023	-.014	-.010	.019	-.018	.017	-.019	.006	.019	.029	.008	-.017	-.006	-.005	.024	.017	-.022	-.007					
	4	-.010	.014	-.022	.015	-.012	-.009	-.014	.013	-.008	.011	-.029	.011	-.009	-.006	.007	-.016	.002	-.024	.007	.026					
	5																									
SUM		-.037	-.011	-.067	.024	-.025	.027	-.034	.021	-.034	.046	-.054	.038	-.034	.029	.001	.006	.026	-.023	-.048	.034					
AVERAGE, $\bar{X}$		-.009	-.003	-.017	.006	-.006	.007	-.009	.005	-.009	.012	-.014	.010	-.009	.007	.000	.002	.007	-.006	-.012	.009					
RANGE, R		.030	.031	.026	.030	.041	.028	.026	.035	.019	.016	.048	.044	.013	.038	.024	.030	.060	.041	.033	.033					
CODE																										

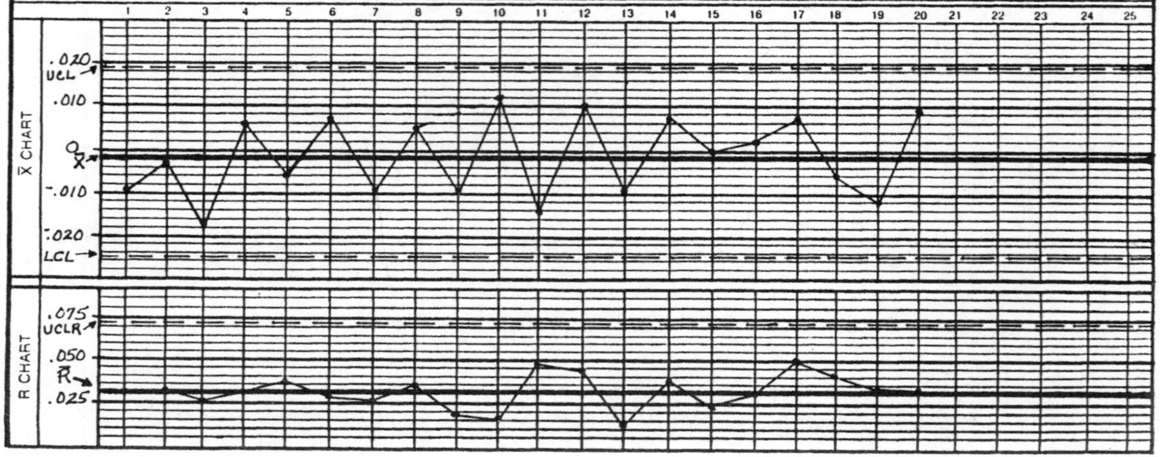

$\bar{X}$ CHART
.020
UCL
.010
0
$\bar{\bar{X}}$
-.010
-.020
LCL
R CHART
.075
UCLR
.050
$\bar{R}$
.025

Figure 8-29.

2. Define the population of interest and the variable to be measured or inspected.

3. Define the data collection and measurement/inspection methods. Defining data collection includes sampling procedures, sample sizes, and the data measuring device (questionnaire, equipment, telephone, files, gage, inspection method, etc.). The data must be obtained under typical conditions. The sampling methods are often influenced by the objective of the data collection. The following examples indicate how the objective influences the data collection.

Example 1: There is a problem with variation of a product, and data is to be collected to determine the amount of variation. Instead of one sample, a sample size of the product that is statistically valid would be required to get a good picture of the variation.

Example 2: In comparing the variation in the output of several workers, it would be necessary to look at the output from each worker over a period of time. The data should be collected and recorded to simplify analysis. Data sheets should be designed to facilitate calculating grand totals, averages and ranges, or data entry into a computer. When recording the data, consider the source, date, lot/serial numbers, measuring instruments, person recording the data, and the procedure used. Workers should be compared to each other as well as their composite average.

4. Determine the appropriate data analysis techniques. In the next several sections, we will introduce many data analysis techniques. Be sure the analysis chosen is appropriate to the information you want to transmit, and that the data collection procedures make analysis easy.

5. Determine the appropriate sampling procedures. This will include considerations of size, makeup and frequency of subgroups. The size will be determined by the type of analysis we are doing. It will be discussed when we consider each of the analysis techniques. Let's now consider the makeup and frequency of subgroups.

Outliers. Occasionally, you will note one or two data points that seem too extreme, too unusual to believe. If this occurs, you have an outlier problem. On the one hand, outliers may be caused by errors in observing or recording. On the other hand, outliers may be extreme but genuine reflections of the process you are examining. If outliers are reflections of errors, they obviously should be discarded. However, if outliers are reflections of genuine phenomena, they are extremely valuable and should be studied extensively.

If you seem to have outliers, first determine if an obvious error has occurred. Has a "1" been recorded as a "7"? Does the technician report strange meter behavior when making the measurement? (Do you observe strange technician behavior when he or she was making the measurement?) If something is obviously wrong, discard the data. Either recompute all related values without the missing data, or remeasure to fill in the missing data and then recompute.

If nothing is obviously wrong with the data, do not discard it just because it is extreme! First you must determine if it is extreme enough to warrant special treatment. To do this, compile all your data into a frequency distribution. From this distribution, find the approximate 25th percentile score and the approximate 75th percentile score. Multiply the difference between these two scores by 3 to compute the critical difference. If the extreme data is further from the next highest (or lowest) score than the critical difference, it is a true outlier. If it is not a true outlier, include it in the data and proceed as usual.*

If you have one or more true outliers, we have no further statistical guidance for you. Examine the data very carefully. If you are convinced that it is bad data, discard it and proceed as above. If you are convinced that it is real data, keep it in. Most of the time, you will not be sure. In that case, do the analysis both ways (with the data in and with the data out). If the two analyses give you roughly the same answers, no problem. If they give you very different answers, good luck!

Process Analysis

Step 1 Project Objective.   The objective of this step is to identify all of the components or elements of the process and to develop a means of improving performance.

Purpose

- To characterize the various aspects of the process so that all important factors are recognized and understood.

- To optimize the effectiveness of the process by clearly defining flow and responsibilities.

- To measure the impact of improvements as viewed by the internal customer and ultimate customer, if applicable, and to provide feedback to management to facilitate control of the process.

Methodology.   The procedure for process analysis is as follows:

- Define the objectives of the prime process by determining needs of subsequent internal customers and the ultimate customer. Find out, precisely, in the customers' words, what the process is expected to provide.

* (Tukey, J.W. 1977. Exploratory data analysis. Addison-Wesley.)

- Divide the prime process into component processes by analyzing existing systems and procedures by using functional flowcharts. As the flowchart is developed, potential improvements will be easier to visualize.
- Determine the component process boundaries, i.e., subsets of the prime process. Identify responsibilities and areas where interdepartment coordination will be necessary. Identify necessary resources (e.g., computer systems). Where does the component process begin and end?
- Define specific tasks for the component process:

Definition of task.
Document forms properly designed?
Who performs the tasks?

- Determine value added within the process. List benefits provided by the process.
- Determine location of process control and audit points. Determine who will collect and analyze data, who will perform audits, and how the results will be reported. Specify the procedure and frequency of measurement.

It is critical to identify a person or organization with the overall responsibility for maintaining the process once the Quality Producing Improvement team activities have concluded. Identify any requirements for training and resources that will be required for training.

- Document the process.

Closure Requirements. Completion of this step is satisfied with these requirements:

- Flowchart the component processes.
- Define responsibilities for maintaining the process in control.
- Define value added within the process.

Closure Criteria. The team may close this step when the resource committee has verified satisfaction of the following criteria:

- Document flowchart with functional responsibilities, measurement locations, and characteristics to be measured.
- Submit flowchart to resource committee for approval.

Resource Committee Approval

Project Objective. Define internal/external customers, respective boundaries, value added, description of measurement system, data collection, and reporting methods with related responsibilities (when appropriate, attach copies of reports).

Develop an Improvement Methodology
Frequency of reports

APPROVAL:

Prepared by _____   Date _____

Approved _____   Date _____

Approved _____   Date _____

Process Analysis
for Nonproduction Areas

Step 1. Define the objectives or requirements of the prime process by determining needs of subsequent internal customers and the ultimate customer.

Step 2. Divide the prime process into component processes by analyzing existing systems and procedures by using functional flowcharts.

Step 3. Determine the component processes boundaries, i.e., subsets of the prime process.

Step 4. Define tasks for the component process:

- Definition
- Document
- Performed by

Step 5. Determine value added.

Step 6. Determine what is important to measure. Define the parameters to be measured and what indicators to use.

Step 7. Define the criteria—what is acceptable and what is unacceptable.

Step 8. Determine location of measurement or audit, who will conduct it, how, and with what frequency.

Step 9. Document the process.

Example Project: ESR Process

Objective. Provide justifiable and adequate information for an "engineering change request" to design engineering.

Component Processes—Flowchart

- Review "ESR" (Engineering Support Request) process
- Conduct "Internal Department Corrective Action" process
- Draft "ECR" (Engineering Change Request) process
- "Implement and Approved Deviation" process

Boundaries

- Engineers review ESR process

  Start: Department clerk sign-in ESR

  End: Department clerk received ESR from engineer with a decision

- Internal department corrective action

  Start: Engineer decides to implement internal department corrective action.

  End: Department clerk receives ESR from engineer annotated "corrective action implemented."

- Draft ECR process

  Start: Engineer decides to draft an ECR.

  End: Engineering change coordinator logs in the ECR and receives the written request.

- Implement "approved deviation"

  Start: Engineer decides to implement an approved deviation.

  End: Engineering department clerk receives copy of an approved deviation which has been implemented.

Define Task

- Engineer reviews ESR
- Engineer conducts internal department corrective action
- Draft ECR
- Prepare, approve, and implement deviation

Identify Value Added

- Qualifies ESR
- Solves problems
- Defines and documents the change
- Immediate implementation of change

Review Processes Parameters

- Review ESR

  Elapsed time average working days per three ESRs (date received versus date completed)

  Quality by error or omission type from submitter (number of errors per three ESRs)

  Input and output (quantity per week)

Productivity = average worker-hours per ESR processed (decision) completed

- Internal department corrective action process

    Elapsed time (average days per three ESRs)

    Effectiveness of action taken measured by process audit

    Input and output quantity by week

    Productivity—worker-hours to plan, prepare, and implement corrective/preventive action.

- Draft ECR

    Elapsed time

    Quality (accuracy)—number errors and omissions

    Input and output quantity

    Productivity—worker-hours to draft an ECR

    Number of recycles

- Implement and approved deviation

    Elapsed time

    Input and output

Criteria—Acceptable and Nonacceptable

Elapsed Time: Measured in days

Quality: Definition of each error and space to be completed

Input: Number ESRs received at each component process

Output: Number ESRs processed complete at each component process

Location Design Engineering

Document Process

Has a procurement to be Documented

Example Project: ECR Process
1. Component process: ECR preliminary approval
2. Task: Preliminary approval ECR
3. Document: ECR
4. Performed by: Engineering manager
5. Boundaries
    a. *Start:* ECC receives ECR (physical date) from requesting manager
    b. *End:* ECC receives ECR from engineering manager with preliminary approval and assigned engineer with dates (physical date)
6. Value added: Avoids unnecessary engineering investigation time

7. Process parameters
   a. Elapsed time (ECC receipt date and engineering managers approval/ reject date) $\_$
      (1) Charts: $\bar{X}$,R by product
   b. Quality
      (1) Errors/opportunities
         (*a*) Format
         (*b*) Technical content
      (2) Charts—"u" chart by engineering group and composite
   c. Quantity/input
      (1) Quantity
      (2) Chart "c" chart by product
8. Component process: Developing the ECO
9. Task: Develop specifications, etc., for ECO
10. Document: ECO
11. Performed by: Design engineer
12. Boundaries
   a. Start: The design engineering manager assigns the ECR
   b. End: The ECO is received at the check group
13. Value added: Develop the change content, justification, and prepare an ECO form
14. Process parameters
   a. Elapsed time (date assigned to the design engineer and the date received at the clerk)
      (1) Chart $\bar{X}$,R by product
   b. Quality
      (1) Errors/opportunities
         (*a*) Format
         (*b*) Technical content
      (2) Charts—"u" chart by group
   c. Delinquent
      (1) Quantity delinquent
      (2) Days (variance from planned date)
      (3) Chart—"c" chart by product and composite
   d. Productivity
      (1) Hours/ECO
      (2) Chart—$\bar{X}$,R by engineering group and composite
15. Component process: Engineering check of ECO
16. Task Review ECO for completeness and Accuracy
17. Document: ECO
18. Performed by: Checkers
19. Boundaries
   a. Start: Receive ECO package from design engineer
   b. End: Accept/reject date from checker

20. Value added
    a. Verifies changes are made to engineering standards
    b. Avoids missing information, incorrect information, future changes
21. Process parameters
    a. Elapsed time (date ECO package received and date ECC receives package)
       (1) Chart—$\overline{X}$,R by product and composite
    b. Quality (errors/opportunities)
       (1) Changes
       (2) Format
       (3) Technical content
       (4) P.C. charts—"u" chart by product, by engineer, by organization
    c. Productivity
       (1) Hours/ECO
       (2) P.C. chart—$\overline{X}$,R by person
    d. Backlog
       (1) Queue for ECO checking
       (2) P.C. chart—"c" chart by product and composite
22. Component process: Technical review board
23. Task: Review ECO for technical content
24. Document: ECO
25. Performed by: Checkers
26. Boundaries
    a. *Start:* Input date to ECC
    b. *End:* Output date the day the tech review signs
27. Value added: Verifies technical merit of the change
28. Process parameters
29. Elapsed time
    a. Date to ECC and date tech review board accepts or rejects
    b. P.C. chart $\overline{X}$,R chart by product and composite
30. Quality
    a. Effectiveness
       (1) Solve the problem
       (2) Inspectability
       (3) Manufacturability
       (4) Best solution
    b. P.C. chart—"u" chart
31. Component process: CCB
32. Task members of the CCB will concurrently review and approve the ECO for effectiveness, cut-in date, and coordination of implementation
33. Document: ECO

34. Performed by: Checkers
35. Boundaries
    a. *Start:* Acceptance date by technical review
    b. *End:* Date CCB approval of ECO
36. Value added
    a. Commitment to implement per the effective date
    b. Disposition parts
    c. Obtain customer approval as applicable
37. Process parameters
    a. Elapsed time
       (1) Date to CCB and date CCB approved
       (2) $\overline{X}$,R chart by product and composite
    b. Quality
       (1) Effectiveness
           (*a*) Number committed versus number submitted
           (*b*) Unacceptable commitments per opportunities
38. Component process: "F" implementation coordination
39. Task:
40. Document: Coordination plan
41. Performed by: Checkers
42. Boundaries
    a. *Start:* CCB date
    b. *End:* Implementation date
43. Value added
    a. On-time implementation
    b. Happy customers
44. Process parameters
    a. Elapsed time
       (1) Date of actual implementation per the effective date
       (2) By $\overline{X}$,R chart by product and composite
    b. Quality
       (1) Effectiveness
       (2) Number problem—number implementation
    c. Delinquent

Defect (c and u) Control Charts

This section describes the control charts that are used to implement statistical process control on processes involving the number of defects or nonconformities. The control charts used are the number of defects per subgroup, called the *c chart*, and the average number of defects per unit, called the *u chart*.

Number of Defects (c) Control Charts. The c chart is used to control the number of defects or nonconformities in a process. The c chart plots the number of defects in a subgroup. It is used when there are a large number of characteristics being inspected and the total number of defects may be large. The count of defects or nonconformities, c, in each subgroup is recorded and plotted on the control chart. It is important to have the opportunities for error to be equal to each subgroup.

Since it is possible to have more than one defect in each unit, it is also possible that the count of defects, c, per subgroup will be larger than the size of the subgroup. Subgroup sizes are usually small (1 to 5). The subgroup size must be constant; if not, another type of chart (u chart) should be used.

The following guidelines are provided for selecting the subgroups.

Subgroup Size. The size of the subgroup is denoted by n. The subgroup is composed of consecutive pieces from the process. For the c chart the subgroup size is usually from 1 to 5.

Number of Subgroups. The number of subgroups is denoted by K. The number required for realistic ongoing control limits is 25, the same as for the average and range chart. The minimum number of subgroups for initial control limits is 20.

The number of subgroups required to establish realistic ongoing limits is a compromise between having reliable data and establishing limits quickly.

Subgroup Frequency. As with any control chart, the frequency or time between the subgroups is a function of the production rate, cost of inspection, the failure rate, and the cost of failures.

When obtaining data from the process for the first time, the subgroups are taken closer together until the process is in control. When the process is in control and stable, the time between subgroups can be extended.

Data Collection. The steps for computing and plotting data on a c chart with established centerline and control limit values are as follows:

Step 1. Select consecutive pieces for the subgroup from the process. Select them "as is" before any touch-up or rework by the operator.

Step 2. Inspect the parts; count and record the number of defects on the appropriate data recording form.

Step 3. Plot the number of defects in the subgroup (c) on the chart.

Step 4. Interpret the results. If the point is within the control limits, continue running the process. If the point is outside the control limits, take the appropriate action.

c Chart Example. Suppose subgroups of size 5 are obtained once per hour from a process. The centerline and control limit values have been established as shown on the chart in Figure 8-30.

Step 1.   Five consecutive parts are obtained from the process.

Step 2.   The parts are inspected and three defects are found.

Step 3.   Plot the number of defects; three on the chart.

Step 4.   Interpret the chart. The point is within the control limits; continue running the process.

Initial Control Limits.   The initial control limits are computed after 25 subgroups are obtained. The average number of defects per subgroup is the centerline value and is computed by the following formula:

$$\bar{c} = \frac{\text{sum of all defects found in all subgroups}}{\text{total number of subgroups } (K)}$$

The control limits, UCL_c and LCL_c, are computed using the following formula:

$$UCL_c = \bar{c} + 3\sqrt{\bar{c}}$$

$$LCL_c = \bar{c} - 3\sqrt{\bar{c}}$$

Often the LCL_c is computed as a negative or minus number. When this occurs, use $LCL_c = 0$.

The centerline and control limits are plotted on the chart the same as the X-bar and R chart described earlier in Control Chart Basics.

Process Evaluation.   The process evaluation will answer the questions of what the process is doing and what the process is supposed to do.

If the process is in control, the natural or usual variation of the process has been quantified. The value of the process average indicates what the process is doing. If the process average is 2.3 defects per subgroup, then the process provided is capable of making 2.3 defects per subgroup.

If the process is out of control, the assignable cause must be removed and new data obtained to determine the new process average.

The process target or goal must be known in order to determine if the process is doing what it is supposed to be doing. The process average is then compared to the target or goal. This is described in more detail later in the process capability section.

c Chart Example.   Suppose 18 subgroups of size 5 are obtained from a process and the number of defects for each subgroup are plotted. The detail data is shown in the chart in Figure 8-31.

$$\bar{c} = \frac{\text{sum of all defects found in all subgroups}}{\text{total number of subgroups } (K)}$$

$$\bar{c} = \frac{20}{18} = 1.11$$

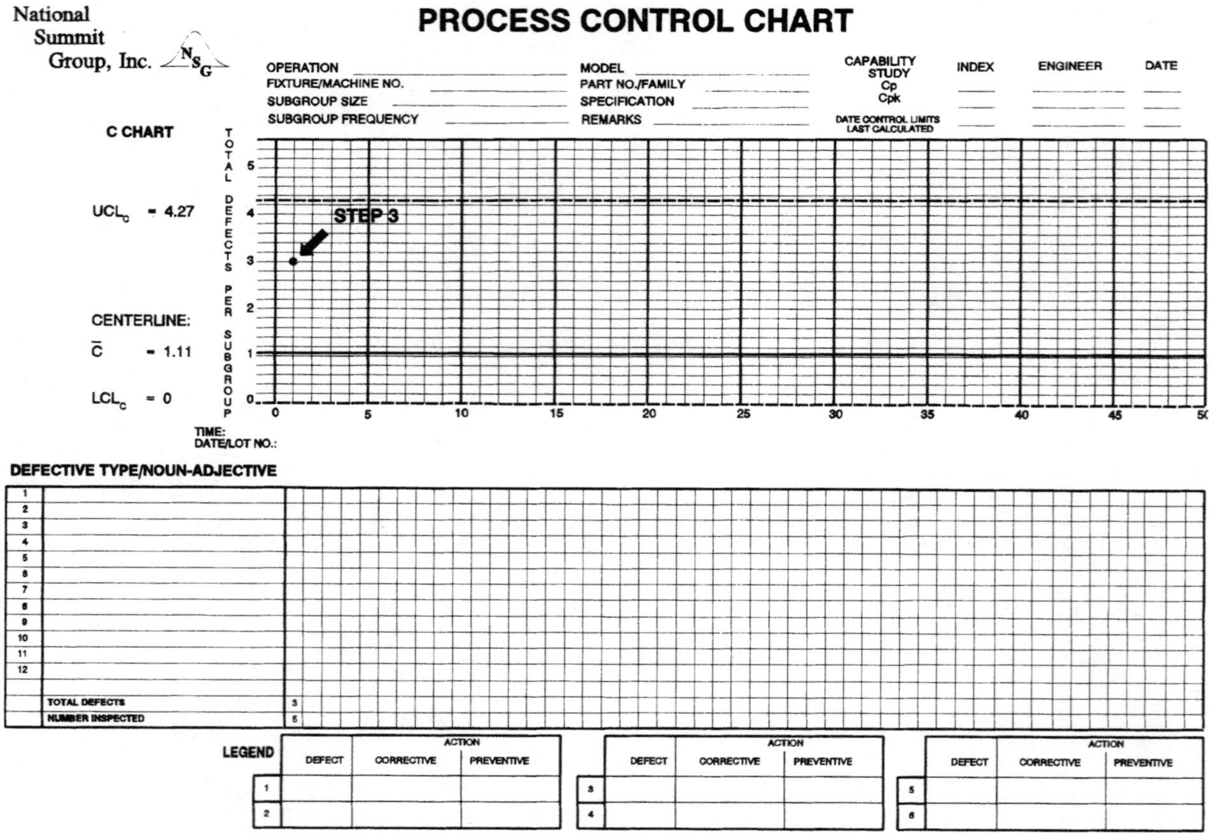

PROCESS CONTROL CHART

National
Summit
Group, Inc.

OPERATION
FIXTURE/MACHINE NO.
SUBGROUP SIZE
SUBGROUP FREQUENCY

MODEL
PART NO./FAMILY
SPECIFICATION
REMARKS

CAPABILITY
STUDY
Cp
Cpk

INDEX

ENGINEER

DATE

DATE CONTROL LIMITS
LAST CALCULATED

C CHART

UCL_c = 4.27

CENTERLINE:
$\bar{C}$ = 1.11

LCL_c = 0

STEP 3

TIME:
DATE/LOT NO.:

DEFECTIVE TYPE/NOUN-ADJECTIVE

TOTAL DEFECTS
NUMBER INSPECTED

LEGEND

	DEFECT	ACTION	
		CORRECTIVE	PREVENTIVE
1			
2			

	DEFECT	ACTION	
		CORRECTIVE	PREVENTIVE
3			
4			

	DEFECT	ACTION	
		CORRECTIVE	PREVENTIVE
5			
6			

Figure 8-30.

PROCESS CONTROL CHART

National Summit Group, Inc.

OPERATION
FIXTURE/MACHINE NO.
SUBGROUP SIZE
SUBGROUP FREQUENCY

MODEL
PART NO./F-FAMILY
SPECIFICATION
REMARKS

CAPABILITY STUDY
Cp
Cpk
DATE CONTROL LIMITS LAST CALCULATED

INDEX   ENGINEER   DATE

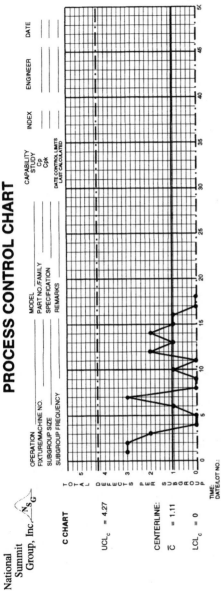

C CHART

$UCL_c = 4.27$

CENTERLINE:
$\bar{C} = 1.11$

$LCL_c = 0$

TIME:
DATE/LOT NO.:

DEFECTIVE TYPE/NOUN-ADJECTIVE

#	Defective type															
1	MISSING COMPONENT	2	1			1			1	1					1	1
2	REVERSED COMPONENT		2			1		2								
3	WRONG COMPONENT				1		2		1							
4	DAMAGED COMPONENT							1								
5	IDENT. ILLEGIBLE	2														
6	LEADS BENT UNDER	1														
7																
8																
9																
10																
11																
12																
	TOTAL DEFECTS	3	3	2	0	1	3	0	0	1	0	2	1	2	1	1	0	0
	NUMBER INSPECTED	5	5	5	5	5	5	5	5	5	5	5	5	5	5	5	5

LEGEND

	DEFECT	ACTION	
		CORRECTIVE	PREVENTIVE
1			
2			
3			
4			
5			
6			

Figure 8-31. c chart example.

389

The control limits, UCL_c and LCL_c, are computed using the following formula:

$$UCL_c = \bar{c} + [3\sqrt{\bar{c}}] = 1.11 + [3\sqrt{1.11}]$$

$$UCL_c = 1.11 + [3 \cdot 1.054] = 1.11 + 3.16 = 4.27$$

$$LCL_c = \bar{c} - [3\sqrt{\bar{c}}] = 1.11 - [3\sqrt{1.11}]$$

$$LCL_c = 1.11 - [3 \cdot 1.054] = 1.11 - 3.16 = -2.05 \qquad \text{use } 0$$

Average Number of Defects per Unit (u) Control Chart. The u chart can be used on the same processes that a c chart can. The u chart must be used whenever the subgroup size is not constant. The c chart is preferred when controlling the number of defects in a process. The value plotted for the u chart is the average number of defects per unit in the subgroup.

The guidelines for selecting subgroups is the same as for the c chart.

Data Collection. The steps for computing and plotting data on the u chart are as follows:

Step 1. Select consecutive pieces for the subgroup from the process. Select them as is before any touch-up or rework by the operator.

Step 2. Inspect the parts; count and record the number of defects on the appropriate data recording form.

Step 3. Compute the average number of defects per unit.

$$u = \frac{\text{total number of defects in the subgroup}}{\text{size of subgroup } (n)}$$

Step 4. Plot the value of u on the chart.

Step 5. Interpret the results. If the point is within the control limits, continue running the process. If the point is outside the control limits, take the appropriate action.

Initial Control Limits. The initial control limits are computed after 25 subgroups are obtained.

The average number of defects per unit is the centerline value and is computed by the following formula:

$$\bar{u} = \frac{\text{sum of all defects found in all subgroups}}{\text{total number of units inspected}}$$

The control limits are computed using the following formula:

$$UCL_u = \bar{u} + 3\sqrt{\bar{u}/n}$$

$$LCL_u = \bar{u} - 3\sqrt{\bar{u}/n}$$

If the subgroup size is changing, the control limits must be calculated for each subgroup using the following formula. This is similar to the p chart with changing subgroup sizes.

$$UCL_u = \bar{u} + \frac{3\sqrt{\bar{u}}}{\sqrt{n}}$$

$$LCL_u = \bar{u} - \frac{3\sqrt{\bar{u}}}{\sqrt{n}}$$

Plot the centerline and control limit values on the control chart the same as the c chart.

If the subgroup size varies, plot the control limits the same as the control limits are plotted for the p chart with changing sample size as described in Section 9.

The process evaluation is the same as for the c chart.

Pareto Chart.   Pareto analysis is a technique to identify the most significant items ("vital few") among many ("trivial many"). This method was coined the "Pareto principle" by Dr. J. Juran. It is based upon studies of wealth distribution by Vilfredo Pareto, an Italian economist.

Pareto analysis illustrates the "80–20" rule. This rule says that 80 percent of the variation of a characteristic is caused by 20 percent of the possible variables. Applying this to attribute date involving defectives means that 80 percent of the total defects are caused by only 20 percent of the defect types.

Constructing a Pareto Chart.   The Pareto chart facilitates the Pareto analysis and will provide information to determine what item among many to attack first. An example is shown subsequently in the Pareto chart illustration.

Step 1.   Decide on the classification of the items to be plotted. For example, classify causes of defectives and defects, kinds of product, size, machines, workers, and processes, etc. Develop a checklist for ongoing data recording.

Step 2.   Determine the period of time the chart will represent. Is it one day, one week, one quarter, etc.? There is no rule—it depends on the purpose of the data.

Step 3.   Calculate the percentage of occurrence of each item and the cumulative percentage. Use a worksheet with the following headings:

Item	Quantity	Cumulative
A	10	10
B	7	17
C	3	20
Total	20	

Step 4. Label the vertical axis of graph. The left side represents the number of occurrences for each item. This scale is to be labeled from 0 to the total quantity. The right side represents the cumulative percent of the total (0 to 100 percent) and is related to the left scale. This scale is optional but is very useful in making comparisons before and after improvements or corrective actions have occurred. If the total number of items is 50, label the left scale from 0 to 50. The right scale is labeled with percent corresponding to the left scale. Where 10 is on the left scale, 10/50 or 20 percent is on the right scale.

Step 5. Label the horizontal axis. The horizontal axis is used to identify the classified items. The first item is the one corresponding to the largest percent of the total. The second item has the next largest percent of the total. Continue until the last item is labeled. Many items with small frequencies can be grouped together as "other" on the extreme right of the horizontal axis.

Step 6. Draw the bars. The height of the bars corresponds to the number of occurrences of each item and uses the left vertical scale. The width of all the bars is the same.

Step 7. Draw the cumulative line graph (optional). The cumulative line starts at the lower left corner of the bar corresponding to the first item and is drawn to the upper right corner of the bar. The next item is plotted at the number on the frequency scale corresponding to the cumulative frequency for the second item above the top right corner of the bar for the second item. Continue until the cumulative line is completed at the 100 percent point on the cumulative percent scale.

Step 8. Label the chart completely. Include the data source, data, and conditions (inspection method, inspector, operator, process conditions, etc.).

Step 9. Determine the vital few items and develop the corrective action required to reduce or eliminate the item.

Pareto Chart Illustration. Suppose the following error data has been obtained from one week's output of a particular operation:

Error type	Qty. of error
A	2
B	12
C	2
D	18
E	10
F	2
Others	4

The computations are as follows:

Error	Qty.	Cum.	Cum. % of total
D	18	18	36
B	12	30	60
E	10	40	80
A	2	42	84
C	2	44	88
F	2	46	92
Others	4	50	100
Total	50		

The Pareto graph is shown in Figure 8-32.

Figures 8-33 and 8-34 indicate how a Pareto chart can be used to provide corrective action direction and to graphically show improvements made after corrective action is complete.

Figure 8-33 shows data on the number of accidents in a factory. The vital few accidents injure fingers, but it is not known what to do to reduce the number of finger-injuring accidents. The Pareto graph of the type of accidents resulting in injuries to fingers indicates that the vital few finger accidents result from fingers being struck or crushed. Armed with this additional information, we can begin to reduce the number of accidents causing finger injuries. Figure 8-34 shows the Pareto graphs before and

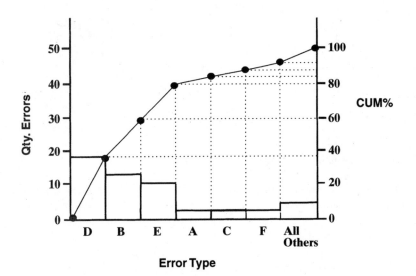

Figure 8-32. Pareto graph.

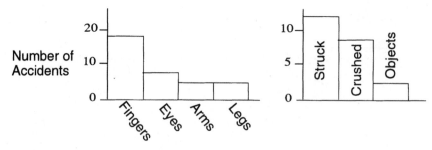

Figure 8-33.

after improvements were implemented. To indicate the improvement it is important to maintain the same scale for both graphs. The fewer number of total defects in the Pareto graph after the improvement is demonstrated graphically in Figure 8-34.

Fraction Defective (p) Control Charts.   This section describes control charts that plot attribute data where defectives or nonconformities are to be placed under statistical process control. The control charts that plot this type of data are the fraction defective and number defective control charts. The fraction defective chart is called a p chart and the number defective chart is called an np chart.

The p chart is used to control the fraction or proportion defective of a process. The p chart plots the proportion or fraction defective, p, of a subgroup. The proportion defective, p, is usually shown as a decimal number and is computed by dividing the number of defective units in the subgroup by the number of units inspected in the subgroup.

Example:

One defective unit is found in a subgroup of 50 units.

The fraction or proportion defective is $1/50 = .02$.

The percent defective would be $(.02) \times 100 = 2\%$.

For convenience in computation and ease of interpretation, the size of each subgroup should be the same. Otherwise, a separate pair of control limits must be computed for each subgroup size used. Unequal subgroup sizes occur when results of 100 percent inspection are plotted where the number of parts produced in one day or one shift is the size of the subgroup and the number of parts produced each shift or day varies. Following are guidelines for selecting the subgroups.

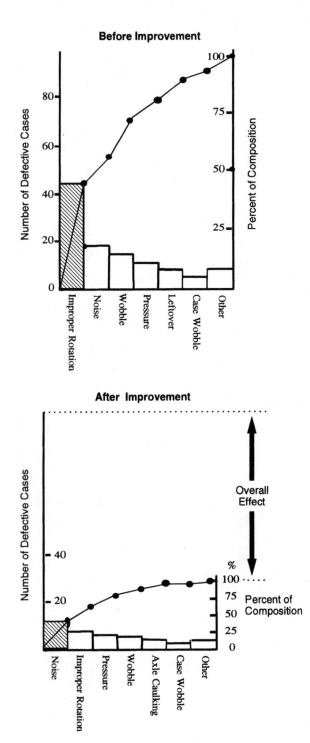

Before Improvement

Number of Defective Cases (left axis): 0, 20, 40, 60, 80

Percent of Composition (right axis): 25, 50, 75, 100

Categories: Improper Rotation, Noise, Wobble, Pressure, Leftover, Case Wobble, Other

After Improvement

Number of Defective Cases (left axis): 20, 40

Overall Effect

% — 100, 75, 50, 25, 0 — Percent of Composition

Categories: Noise, Improper Rotation, Pressure, Wobble, Axle Caulking, Case Wobble, Other

Figure 8-34.

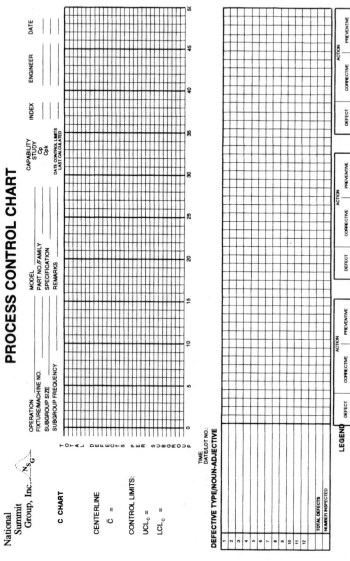

Figure 8-35.

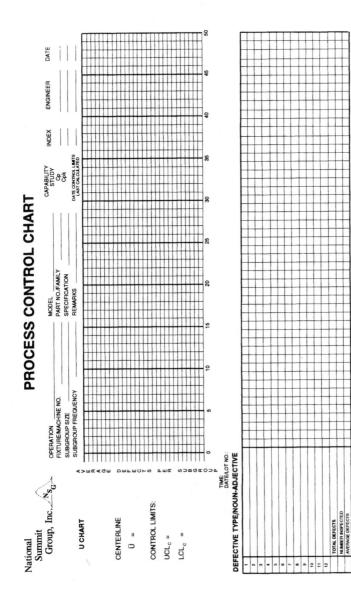

Figure 8-36.

397

Subgroup Size. The size of the subgroup is denoted by n and is composed of consecutive pieces from the process. For the p chart, the subgroup size is larger than the average and range chart. The minimum size is 25, with 50 to 100 recommended.

The results of 100 percent inspection are often plotted on a p chart.

Number of Subgroups. Variance in lot size will result in the subgroup size changing from one subgroup to the next.

The number of subgroups required for realistic ongoing control limits is 25, the same as for the average and range chart. The minimum number of subgroups for initial control limits is 20.

The number of subgroups required to establish realistic ongoing limits is a compromise between having reliable data and establishing limits quickly.

Subgroup Frequency. The frequency or time between the subgroups is a function of the production rate, cost of inspection, failure rate, and cost of failures.

The subgroups for a p or np chart are often obtained by taking consecutive parts from the process in the following manner. The first 50 parts are subgroup 1, parts 51 to 100 are subgroup 2, etc.

Data Collection for p Chart. The steps for computing and plotting data on the p chart, with the centerline and control limit values already computed, are as follows.

Step 1. Select consecutive pieces from the process for the subgroup. Select them "as is" before any touch-up or rework by the operator.

Step 2. Inspect the parts, then count and record the number of defective parts on the appropriate data recording form.

Step 3. Compute the subgroup fraction defective p.

$$p = \frac{\text{total number of defectives in the subgroup}}{\text{subgroup size}}$$

Step 4. Plot the fraction defective (p) on the chart.

Step 5. Interpret the results.

If the point is within the control limits, continue running the process. If the point is outside the control limits, take the appropriate action.
Example: Suppose subgroups of size 200 are obtained.

Step 1. 200 consecutive parts are obtained from the process.

Step 2. The parts are inspected and six defectives are found.

Step 3.   Compute the subgroup fraction defective (p).

$$p = \frac{6}{200} = .03\ (3\%)$$

Step 4.   Plot the fraction defective .03 on the chart.

Step 5.   Interpret the chart. The point is within the control limits, continue running the process.
Refer to Figure 8-37.

Initial Control Limits.   The initial control limits are computed after 25 subgroups are obtained using the following formula:
The average fraction defective is the centerline value for the p chart and is computed using the following formula:

$$\overline{p} = \frac{\text{sum of all defectives found}}{\text{total items inspected in all samples}}$$

The control limits, UCL_P and LCL_P, are computed using the following formulas:

$$UCL_p = \overline{p} + 3\sqrt{\frac{\overline{p}(1-\overline{p})}{n}}$$

$$LCL_P = \overline{p} - 3\sqrt{\frac{\overline{p}(1-\overline{p})}{n}}$$

Often, the LCL_p is computed as a negative or minus number. When this occurs, use $LCL_p = 0$.
The centerline is plotted on the control chart as a solid horizontal line and the control limits are plotted as dashed horizontal lines.

Process Evaluation.   Evaluating the process will answer the questions of what the process is doing and what the process is supposed to do.
If the process is in control, then the natural variation is being observed. The quality of the process is indicated by the centerline value. A centerline value of 5 percent says the process is capable of making 5 percent defective parts. The materials, tools, documentation, and training provided produces 5 percent defective parts.
If the process is out of control, the assignable causes must be removed and new data obtained to determine what the process is doing.
The target or goal of the process must be known in order to know what the process is supposed to be doing. The average value for an in-control process would be compared to the target or goal. This will be explained later in the section on process capability.

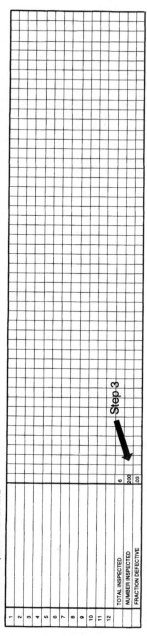

Figure 8-37. Sample process control chart.

p Chart Example. Suppose 25 subgroups of size 200 are obtained from a process and the number of defectives for each subgroup is recorded.

Compute the centerline, $\bar{p}$.

$$\bar{p} = \frac{\text{sum of defectives}}{\text{total inspected}} = \frac{70}{(200)(25)} = \frac{70}{5000} = 0.014$$

Compute the control limits.

$$\text{UCL} = \bar{p} + 3\sqrt{\frac{\bar{p}(1-\bar{p})}{n}} = 0.014 + 3\,\frac{(0.014)\,(0.986)}{200}$$

$$\text{UCL} = 0.014 + 3(0.0083) = 0.014 + 0.025 = 0.039$$

$$\text{LCL} = \bar{p} - 3\sqrt{\frac{\bar{p}(1-\bar{p})}{n}} = 0.014 - 3(0.0083)$$

$$\text{LCL} = 0.014 - 0.025 = -0.011$$

Use LCL = 0 since LCL is less than 0

Refer to Figure 8-38.

Fraction Defective (p) Chart—Unequal Subgroup Sizes. Unequal subgroup sizes occur when the results of 100 percent inspection are used and the subgroup size is the total production for the shift or day. When this occurs, control limits must be calculated for each subgroup since the number of parts for each shift or day will vary.

The selection of the parts for the subgroup and data collection is the same as described in the previously. When the data for a subgroup is computed and plotted, the control limits for that subgroup must be computed and plotted.

Initial Control Limits. The initial control limits are computed after 25 subgroups are obtained.

■ Average or centerline $\bar{p}$:

$$\bar{p} = \frac{\text{sum of all defectives found}}{\text{total items inspected in all samples}}$$

■ Control limits for each subgroup—compute:

$$3\sqrt{\bar{p}(1-\bar{p})}$$

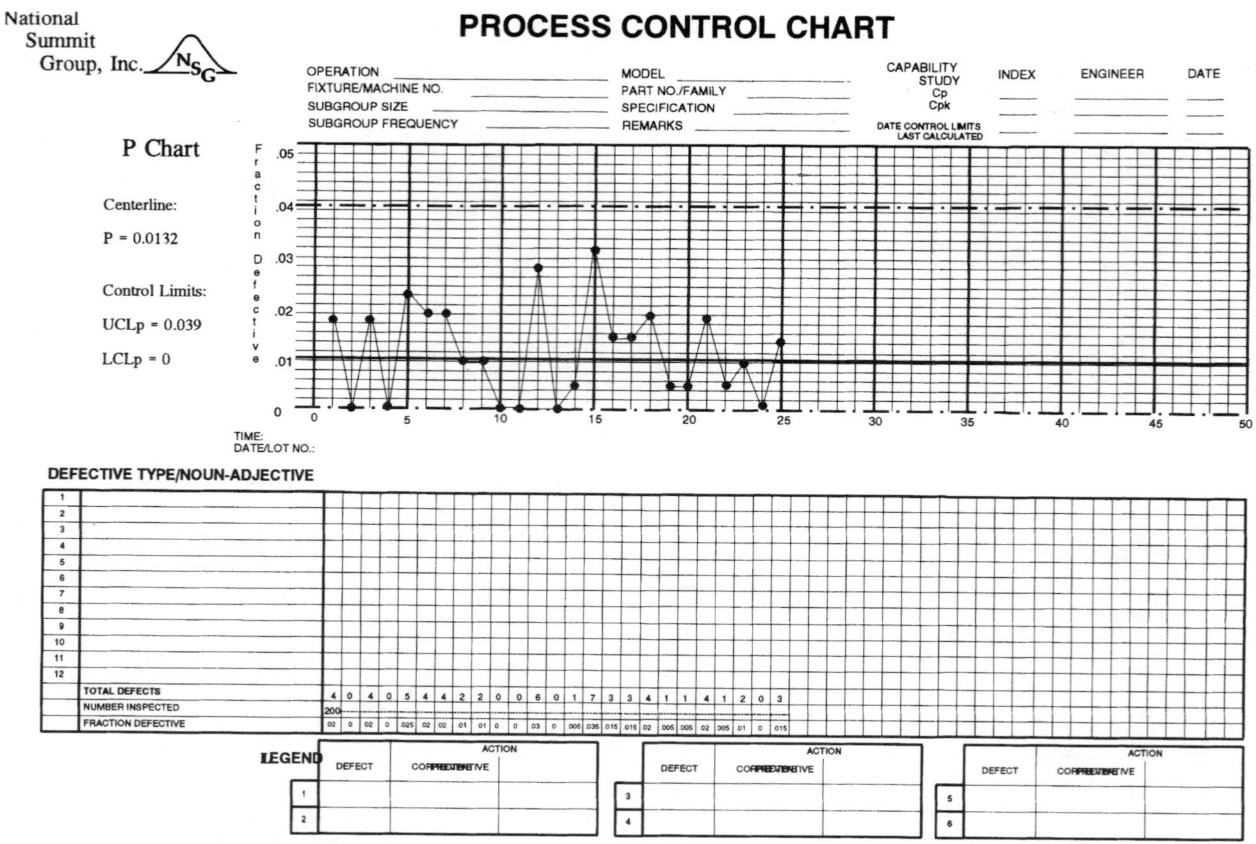

Figure 8-38. Sample process control chart.

- For each subgroup of size n compute:

$$UCL_P = \bar{p} + \frac{3\sqrt{\bar{p}(1 - \bar{p})}}{\sqrt{n}}$$

$$LCL_P = \bar{p} - \frac{3\sqrt{\bar{p}(1 - \bar{p})}}{\sqrt{n}}$$

Plot the average or centerline as a solid line on the p chart. Plot the control limits for each subgroup. The control limits will not be a straight line across the chart. The limits, however, will be centered around the average or centerline value. The limits will have the appearance of a city skyline. The process evaluation is the same as before.

Interpretation of Control Charts

This section describes patterns observed on control charts that may indicate an assignable cause is present even when all points are within the control limits. These are provided for knowing when to take action as a result of these patterns. The two basic patterns to be described are *shifts* and *trends*.

The methods presented here are useful not only for ongoing control, but also for revision of ongoing centerlines and control limit values on completed charts. In reviewing control chart data over a long period of time, this will assist in identifying additional ways to improve the process.

Shifts. The first pattern to be discussed is one that occurs when there is a shift in the process average. A shift in the process average is indicated by several points in a row on one side of the centerline, but not outside of the control limits as shown in Figure 8-39. This indicates an assignable cause has caused the process average to shift or jump to a new level. The actions taken to remove the cause are often the same as when a point goes out of the control limits.

The simplest test for a shift is seven points in a row on the same side of the centerline. This is shown in Figures 8-40 and 8-41.

Seven points in a row is used to test for a shift based upon probabilities. The chance or probability of 7 points in a row, on one side of the centerline, is the same as getting 7 heads in a row when flipping a coin. This probability is ($\frac{1}{2}$ to 7th power) or 0.008. In other words, this says the chance of seven points in a row is 0.008 if nothing changed in the process. If seven in a row occur, the best action is to stop and remove the cause.

Additional tests for shifts are shown in Figures 8-42 to 8-45. These tests are 10 out of 11 on one side of the centerline and 12 out of 14 on the other side of the centerline.

SEVERAL POINTS ABOVE
THE CENTERLINE

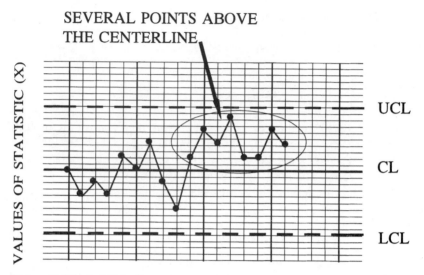

Figure 8-39. A shift in the process average.

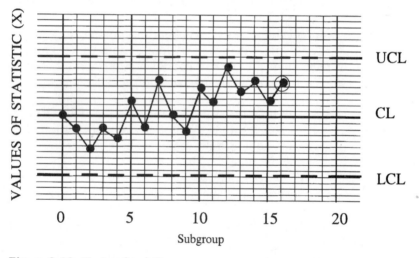

Figure 8-40. Testing for shifts.

Trends. A second pattern that is encountered on a control chart is a trend. This is indicated by several points going in the same direction, as shown in Figure 8-46. A trend pattern is most often encountered on an X-bar chart. A similar pattern is a trend in the downward direction.

The minimum number of points in a direction (that indicates a trend is occurring) is seven in a row without a change in direction. An upward

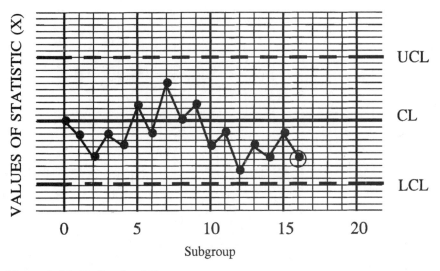

Figure 8-41. Testing for shifts.

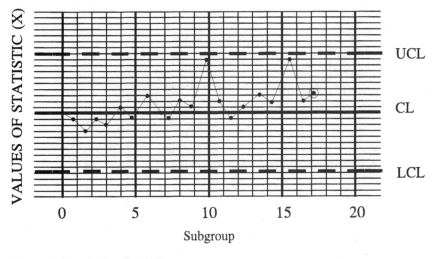

Figure 8-42. Testing for shifts.

trend is indicated by a run of seven points in an upward direction without a change in direction, as shown in Figure 8-36.

A downward trend is indicated by a run of seven points in a downward direction without a change in direction, as shown in Figure 8-47.

A trend is also present if there are eight or more points in one direction, even though there may be changes in direction. This is shown in Figure 8-48, where eight points are shown in one direction.

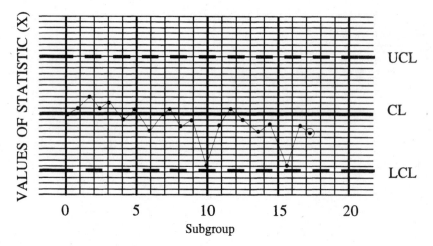

Figure 8-43. Testing for shifts.

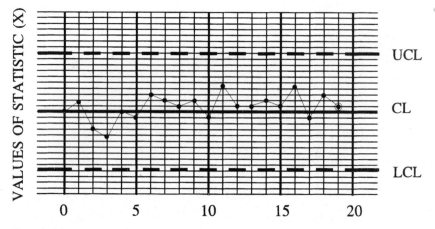

Figure 8-44. Testing for shifts.

Special Patterns.   The material in Figures 8-49 to 8-53 contains patterns that may be observed on X-bar and R charts when reviewing completed charts. Some of these patterns may be observed over several charts. The following causes, listed for each figure, will help make continuous improvements to the process that will result in less variation of the process average.

For Figure 8-49

Causes if jumps occur on X-bar chart

- Change in proportions of materials or subassemblies coming from different sources

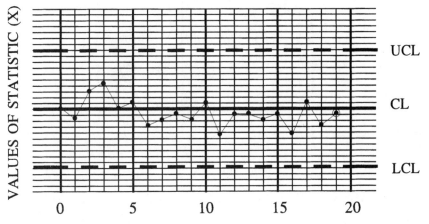

Figure 8-45. Testing for shifts.

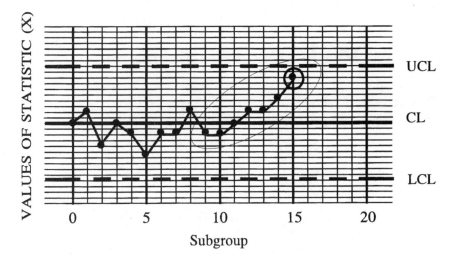

Subgroup

Figure 8-46. Trends.

- New worker or machine
- Modification of production method or process
- Change in inspection device or method

Causes if jumps occur on R chart

- Change in material
- Change in method
- Change in worker

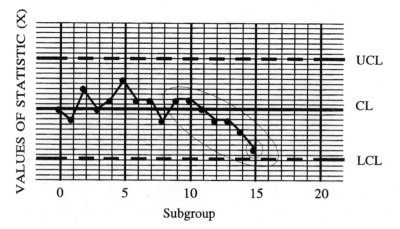

Figure 8-47. Downward trend.

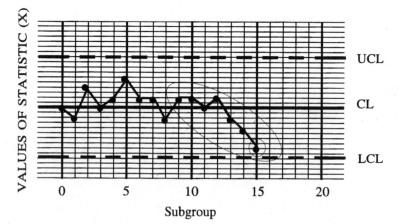

Figure 8-48. Sample trend.

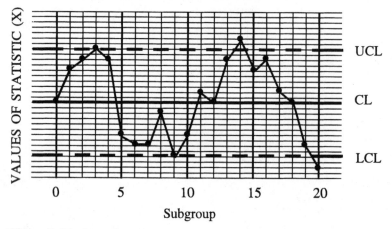

Figure 8-49. Special patterns.

For Figure 8-50

Causes if trend occurs on X-bar chart

- Gradual deterioration of equipment, which can affect all items
- Worker fatigue
- Accumulation of waste products
- Deterioration of environmental conditions

Causes if trend occurs on R chart

- Improvement or deterioration of operator skill
- Worker fatigue
- Change in proportions of subprocess feeding an assembly line
- Gradual change in homogeneity of incoming material quality

For Figure 8-51

Causes if recurring cycles occur on X-bar chart

- Temperature or other recurring changes in the physical environment
- Worker fatigue
- Differences in measuring or testing devices which are used in order
- Regular rotation of machines or operators
- Merging of subassemblies or other processes

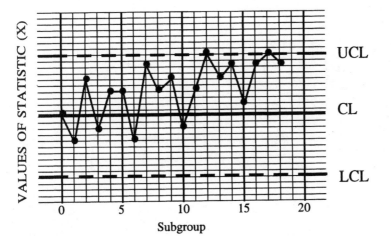

Figure 8-50. Special patterns.

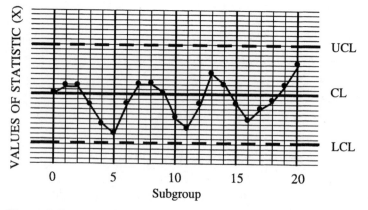

Figure 8-51. Special patterns.

Causes if recurring cycles occur on R chart
- Scheduled preventive maintenance
- Worker fatigue
- Worn tools

For Figure 8-52
Causes of lack of variability on X-bar chart
- Incorrect calculation of the control limits
- Not taking consecutive parts

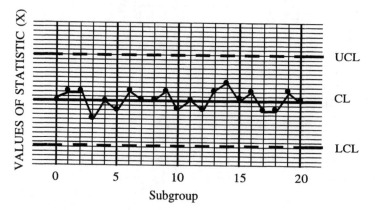

Figure 8-52. Special patterns.

Causes of lack of variability on R chart

- Collecting a sample which includes a number of measurements from widely differing universes (populations)

For Figure 8-53

Causes of points near limits on X-bar chart

- Overcontrol
- Large systematic differences in test method or equipment
- Large systematic differences in material quality
- Control of two or more processes on the same chart

Causes of points near limits on R-chart

- Mixture of materials of distinctly different quality
- Different workers using a single R chart
- Data from processes under different conditions plotted on the same chart

Refer to Figures 8-54 and 8-55 for examples of process control charts.

Inspection Capability Studies

Often, it is taken for granted that inspection results are true values with no error. It is simply assumed that measurements taken from manufacturing operations are correct. However, inspection methods are subject to varia-

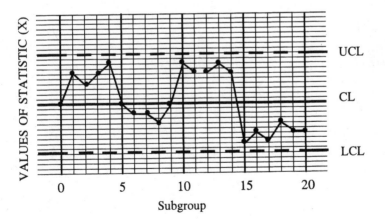

Figure 8-53. Special patterns.

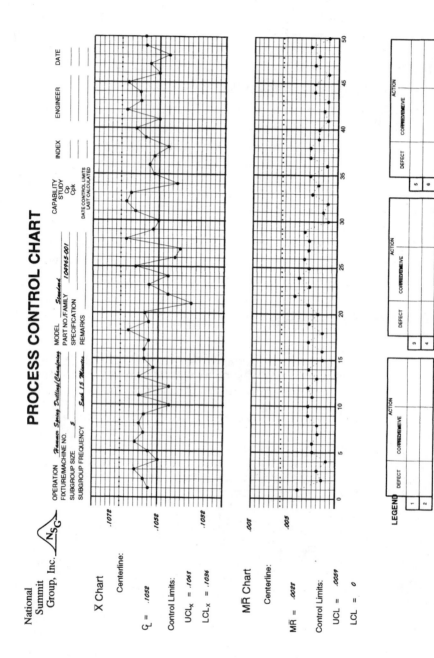

Figure 8-54. Sample chart.

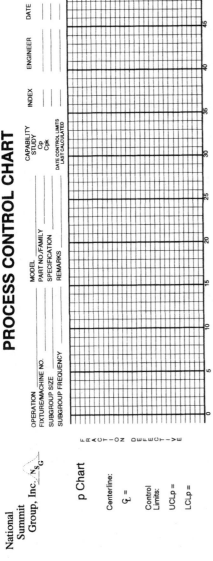
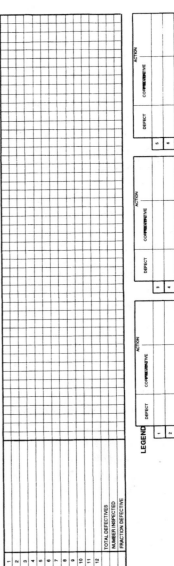

Figure 8-55. Sample chart.

tion. This is true whether the results are derived from a basic mechanical measuring device, complex electronic gear, or a simple go/no-go gage.

On the other hand, the fact that inspection results are subject to variation may be well known within the manufacturing function. Unfortunately, an evaluation of the consequences is very seldom made. All too often the possibility of fluctuating results is completely neglected, brushed off as just another variable, or included as a fudge factor in the phrase, "There are too many variables in my process."

An output-oriented production person may take advantage of the situation by sending the rejected part back for retest or reinspection. The part may pass the second time—or the third time or the fourth. In such a situation, others in manufacturing will lose faith in statistical process control, which will never achieve its savings potential.

Inspection capability is a method of evaluating and quantifying how good an inspection method is. The key result of this method is to ensure that accurate data is obtained for ongoing statistical process control. Inspection capability, then, can be considered the "missing link" in the successful implementation of statistical process control.

Inspection capability studies will determine if an inspection method or piece of equipment produces results which are acceptable, marginal, or unacceptable. Such studies can be used to:

- Evaluate new measuring equipment or inspection methods.
- Compare one or more of the same type of measuring equipment.
- Compare equipment before and after repair or adjustment.

Methods are described in this section for determining inspection capability for both measurement data (variable data) and visual data (attribute data). The method for variable data is described first and the analysis is based on control chart methods.

Attribute data methodology involves completely different terms and calculations. The end result of the study is to determine if the measurement/inspection method is acceptable, marginal, or unacceptable. Criteria for determining acceptable, marginal, and unacceptable is provided. An acceptable inspection capability study is required before performing a process capability study.

The major steps of an inspection capability study are as follows:

- Describe the study by specifying the inspection method, type of data, and the purpose of the study.
- Prepare the data collection method for the appropriate type of data (variable or attribute).

- Continually refer to the checklists.
- Inform appraisers about purpose of study.
- Collect and record the data on the appropriate form.
- Personally observe the study or have a responsible designate observe the study.
- Follow the appropriate data computation procedure.
- Evaluate the results and determine if the study is acceptable, marginal, or unacceptable using the criteria provided.
- Establish the required corrective action and follow-up for marginal or unacceptable studies.
- Report the results.
- All gages and inspectors must be studied and found acceptable as a requisite of declaring the inspection/test process as acceptable.
- Repeat the study after any corrective action is complete.

Inspection capability studies measure and quantify the repeatability and reproducibility of the measurement/inspection method. These two concepts will be defined and illustrated before describing the details of the inspection capability study.

Repeatability: Variable Data.   Repeatability is the variation resulting from the inability of the measuring instrument to obtain the same result over and over. The inability to obtain the same result is due to the numerous little things that make up the measuring system (such as friction, springs, etc.) combined with the inability of the checker to operate and read the instrument in exactly the same manner each time. This is more properly called lack of repeatability since no variation would occur if the measuring instrument were repeatable.

Repeatability may be determined by measuring the same part several times. The resulting distribution of the measurements is shown in Figure 8-56. As can be seen from this figure, the spread or "6 sigma" of this distribution should be small when compared to the total process tolerance: upper spec limit (USL) minus lower spec limit (LSL).

Repeatability for attribute data is defined as the variation in classifying parts as conforming or nonconforming when one person inspects the same part several times using the same inspection method, criteria, or equipment.

Reproducibility.   Reproducibility is the variation among the persons doing the measurements or inspection using the same methods or equipment. This is more properly called lack of reproducibility. The variation

Specification

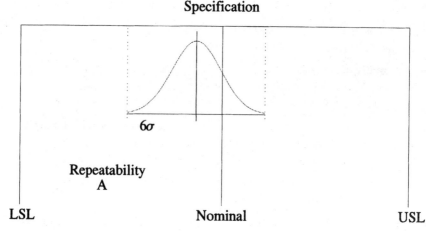

Figure 8-56. Repeatability distribution.

among identical measuring instruments used by the same person is another source of lack of reproducibility.

Reproducibility may be determined by having another person measure the same parts with the same measuring instrument. The distribution of the results of the second person is shown in Figure 8-47 as distribution B. Distribution A in Figure 8-57 represents the results of the first person. The difference between the averages of the readings in distributions A and B is the basis for determining reproducibility.

Specification

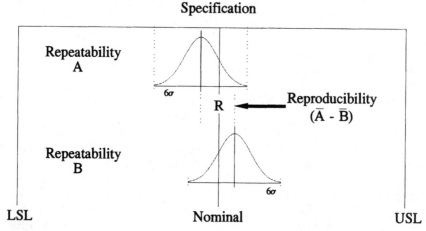

Figure 8-57. Reproducibility.

Reproducibility for attribute data is defined as the variation among people inspecting the same parts using the same inspection method, criteria, or equipment.

Inspection Capability. Repeatability and reproducibility are quantified and combined to determine the inspection capability. For variable data the percent of the total tolerance consumed by the capability (PTCC) is computed. For attribute data, the capability cannot be expressed as a percent of a tolerance. The emphasis for attribute data is on how effective the person is at detecting conforming or nonconforming parts and how biased the person is toward rejecting conforming parts or accepting nonconforming parts.

Variable Data

This section contains information on collecting, analyzing, and evaluating inspection capability studies involving variable data.

Data Collection. In order to determine the repeatability and reproducibility of a measuring method involving variable data, the parts or samples must be obtained and repeat measurements taken on all samples by every appraiser.

The following guidelines are presented to aid in the selection of samples and the collection of data.

- Select the parts from production. The selected parts should cover the entire tolerance range. Each part should be labeled with a number or code to maintain a unique identity and facilitate data collection. The minimum number of samples and repeat measurements required are indicated in Table 8-5.

- Review the inspection method or instruction and verify that it is the correct one.

- Prepare the appraisers for data collection. Review the inspection method with the appraisers. Explain the purpose of the study, the method of data collection and the role of the appraisers.

- Measure each part once in a random order and record the readings on a data sheet. Appraisers should take a normal amount of time to make each measurement.

- Measure each part again in a random order and record the readings on another data sheet. It is important to keep the readings separate so that the appraisers are not biased by the previous readings.

Table 8-5. Sample Sizes—Variable Data

Quantity of appraisers	Quantity of gauges	Minimum number of parts	Minimum number of measurements per part
1	1	10	5
1	2	15	3
2	1		
2	2	10	2
1 or 2	3 or more		
3 or more	1 or 2		
3 or more	3 or more		

- Repeat taking measurements on each part, one at a time, until the desired number of readings per part are obtained. Record each set of readings on different data sheets.

Data Analysis. The analysis of data from an inspection capability study with variable data will be illustrated with an example.

Suppose an inspection capability study is conducted with two persons measuring parts using the same gage. The specification is 1.000 ± 0.010 inch. Readings are recorded to the nearest 0.001 inch. The measurements obtained are shown in Table 8-6. The range of the repeat readings for each part is computed and recorded on the data recording sheet in the row identified as the range R.

Compute the average range R across all parts for each person and record on the data recording sheet in the appropriate space.

Compute the average of all the readings for each person and record the result on the data recording sheet. (See Figure 8-58.)

Range Evaluation. The average range of each person is recorded as shown on the worksheet in Figure 8- If more than one person or gage is used, compute the overall average range by the following formula:

$$R = \frac{\overline{R}_1 + \overline{R}_2 + ... + \overline{R}_K}{K}$$

where $K =$ number of person/gauge combinations

$\overline{R}_1, \overline{R}_2, ..., \overline{R}_K =$ average range of each person/gage combinations

Compute the upper control limit for the range (UCLR).

$$UCLR = D4 \times \overline{\overline{R}}$$

Table 8-6. Example—Variable Data

Gauge/appr. test equip.	Trial	1	2	3	4	5	6	7	8	9	10
	1	1.004	1.005	1.002	1.002	1.004	1.003	1.007	1.000	.999	.998
	2	1.004	1.005	1.001	1.000	1.004	1.003	1.007	1.000	.999	.998
$\Sigma X = 30.072$	3	1.005	1.005	1.003	1.002	1.004	1.002	1.007	1.001	.999	.998
$X = 1.0024$	4										
$\Sigma R = .007$	5										
$R = .0007$	R	.001	0	.002	.002	0	.001	0	.001	0	0
	1	1.004	1.005	1.001	1.001	1.004	1.002	1.006	1.000	1.000	.998
	2	1.004	1.006	1.002	1.002	1.004	1.002	1.004	1.000	.999	.997
$\Sigma X = 30.060$	3	1.003	1.005	1.001	1.002	1.004	1.002	1.005	1.000	.999	.997
$\overline{X} = 1.0020$	4										
$\Sigma R = .008$	5										
$R = .0008$	R	.001	.001	.001	.001	0	0	.002	0	.001	.001

where $D4$ is a factor found in exhibit 1 with n equal to the number of repeat readings on each part. The $D4$ factor is the same factor used in control charts for the range. In the example, $n = 3$, $D4 = 2.574$, and

$$\text{UCLR} = 2.574 \times 0.00075 = 0.0019$$

If any of the ranges exceed the UCLR, do the following:

Remeasure the part to determine if the out-of-control range was due to a recording or a measurement error.

If a data recording error is suspected, remove the affected data from the computations and recalculate the average range and UCLR. When more than one data recording error is found, correct the measurement process and repeat the study.

If there is more than one range exceeding the UCLR due to a measurement error, the method must be revised. No further computations need be done, and the study should be repeated when the method is revised.

When only one range exceeds the UCLR due to a measurement error, remove the affected data from the computations and recompute the average range and UCLR.

In the example, three of the ranges are 0.002, which exceeds the UCLR of 0.0019. Due to the measuring increment being 0.001, these are not considered measurement or data recording errors and are not removed.

Repeatability Evaluation. The repeatability is computed as shown in section 2 on the worksheet. The repeatability is found by computing the

Part number _____ Date _____

Dimension/characteristic _____

Specification/tolerance ___1.000 + .010_____

1. Range Evaluation

ORIGINAL	REVISED	UPPER CONTROL LIMIT RANGE (UCLR)
$\overline{R}_a$ = .0007	_____	n = __3__ $D4$ = __2.574__
$\overline{R}_b$ = .0008	_____	UCLR = ($D4$) ($\overline{R}$) = __.0019__
$\overline{R}_c$ = _____	_____	NO. POINTS ABOVE UCLR = __3__
$\Sigma\overline{R}$ = _____	_____	NO. POINTS DISCARDED = __0__
$\overline{R}$ = .00075	_____	REVISED $\overline{R}$ = _____

2. Repeatability Evaluation

SDR = $(1/d2)$ $(\overline{R})$ n = __3__ $(1/d2)$ = __.592__

SDR = __.00044__

REPEATABILITY = 6 (SDR) = __.00264__

PTCR = $[(6 \times \text{SDR})/(\text{TOL})] \times 100$ = __13.2%__

3. Reproducibility Evaluation (Appraisers or Gauges)

RM = $(\overline{X}_L - \overline{X}_S)$ $\overline{X}_L$ = __1.0024__ $\overline{X}_s$ = __1.0020__

RM = __.0004__

SDM = (D) (RM) K = __2__ D = __.709__

SDM = __.00028__

REPRODUCIBILITY = $6 \times$ (SDM) = __.00168__

PTCM = $[(6 \times (\text{SDM})/(\text{TOL})] \times (100)$ = __8.4%__

4. Inspection Capability Evaluation

SDC = $\sqrt{\text{SDR}^2 + \text{SDM}^2}$ = $\sqrt{.00044^2 + .00028^2}$

SDC = __.00052__

PTCC = $[(6 \times (\text{SDC}))/(\text{TOL})] \times (100)$ = __15.6%__

Figure 8-58. Example of worksheet computations.

Standard Deviation of Repeatability (SDR), multiplying by 6, and dividing by the total tolerance. The SDR is found by the following formula:

$$SDR = \frac{1}{d_2} \times \overline{\overline{R}}$$

where $(1/d_2)$ is a factor found in exhibit 1, with n equal to the number of repeat readings on each part.

In the example, $(1/d_2) = 0.592$ and

$$SDR = 0.592 \times 0.00075 = 0.00044$$

The percent tolerance consumed by repeatability (PTCR) is found by the following formula:

$$PTCR = \frac{6 \times SDR}{tolerance} \times 100$$

In the example:

$$PTCR = \frac{6 \times 0.00044}{0.020} \times 100 = 13.2\%$$

Reproducibility Evaluation. The reproducibility is computed as follows, using the average of all the readings for each person. Compute the difference, denoted RM, between the largest and the smallest average.

$$RM = \overline{X_L} - \overline{X_S}$$

The standard deviation of reproducibility is computed by the following formula:

$$SDM = D \times RM$$

where D is a factor found in exhibit 1, with $K =$ number of appraisers or gages. Compute the percent tolerance consumed by reproducibility using the following formula:

$$PTCM = \frac{6 \times SDM}{tolerance} \times 100$$

In the example, the reproducibility is computed as follows:

$$\overline{X_L} = 1.0024 \quad \text{and} \quad \overline{X_S} = 1.0020$$

$$\text{Thus, } RM = 1.0024 - 1.0020 = 0.0004$$

$$SDM = 0.709 \times 0.0004 = 0.00028$$

$$PTCM = \frac{6 \times 0.00028}{0.020} \times 100 = 8.4\%$$

Inspection Capability Evaluation. The percent tolerance consumed by inspection capability is computed by the following formula:

$$PTCC = \frac{6 \times SDC}{tolerance} \times 100$$

where

$$SDC = \sqrt{SDR^2 + SDM^2}$$

$$= \sqrt{.00044^2 + .00028^2}$$

$$= .00052$$

$$PTCC = \frac{6 \times SDC}{tolerance} = \frac{6 \times .00052}{.020} \times 100 = 15.6\%$$

Note: If both appraisers and gages are included in the study, the variance of the reproducibility must be computed for both appraisers and gages and added to find the overall inspection capability.

The inspection capability for the example is computed as shown on the worksheet and the value of PTCC is 15.6 percent.

Data Evaluation. The results of the study are evaluated to determine if the measurement method is acceptable, marginal, or unacceptable using the following criteria for PTCC:

PTCC value	Study result
10% or less	Acceptable
Between 10 and 25%	Marginal
Greater than 25%	Unacceptable

If the measurement method is unacceptable or marginal, corrective action is required. The study should be repeated when the corrective action is completed.

In the example, the inspection capability is marginal since the PTCC is between 10 and 25 percent.

Attribute Data

As discussed earlier, the concepts of repeatability and reproducibility are the same for attribute data as for variable data, but the measurement of these is entirely different. The emphasis is on how capable or effective the appraiser is in detecting conforming or nonconforming parts repeatedly and how biased the appraiser is toward rejecting conforming parts or accepting nonconforming parts. The effectiveness of different appraisers can be compared when assessing reproducibility.

The measures used in the inspection capability study for attribute data are defined as follows:

Effectiveness (E). The ability to accurately detect conforming and nonconforming parts. This is expressed as a number between 0 and 1, where 1 is perfect, and is computed by the following formula:

$$E = \frac{\text{number of parts correctly identified}}{\text{total opportunities to be correct}}$$

The total opportunities to be correct are a function of the number of parts used and how many times each part is inspected. If 10 parts are selected and each is inspected three times, there are a total of $3 \times 10 = 30$ opportunities to be correct.

Probability of a Miss—P(Miss). The probability of a miss, denoted by P(miss), is the chance of not rejecting a nonconforming part. This is a serious type of error since a nonconforming part is accepted. The probability of a miss is computed by the following formula:

$$P(\text{miss}) = \frac{\text{number of misses}}{\text{number of opportunities for a miss}}$$

The number of opportunities for a miss is a function of the number of nonconforming parts used in the study and the number of times each part is inspected. If five nonconforming parts are used and each part is inspected three times, then there are $3 \times 5 = 15$ opportunities for a miss.

Probability of a False Alarm—P(FA). The probability of a false alarm, denoted by P(FA), is the chance of rejecting a conforming part. This type of error is not as serious as a miss, since a conforming part is rejected. However, rejecting a conforming part causes rework and reinspection to be performed when it is not necessary. If the P(FA) gets too large, large sums of money are wasted on rework and reinspection. The probability of a false alarm is computed by the following formula:

$$P(\text{FA}) = \frac{\text{number of false alarms}}{\text{number of opportunities for a false alarm}}$$

The number of opportunities for a false alarm is a function of the number of conforming parts used in the study and the number of times each part is inspected. If six conforming parts are used and each part is inspected three times, then there are $3 \times 6 = 18$ opportunities for a false alarm.

Bias—B. Bias is a measure of a person's tendency to classify an item as conforming or nonconforming. Bias is denoted by the letter B and is a function of P(miss) and P(FA), as will be described later. Bias values are equal to or greater than 0 and have the following interpretation:

$B = 1$ implies no bias.

$B > 1$ implies bias towards rejecting parts.

$B < 1$ implies bias towards accepting parts.

The value of bias is computed by the following formula:

$$B = \frac{P(FA)}{P(miss)}$$

Data Collection.   The collection of samples for evaluating an inspection capability study with attribute data is quite different from collecting samples for variable data.

The parts are not selected at random. Parts are selected by appropriate personnel (supervisor/engineer) and must be determined as conforming or nonconforming. The number of parts to be selected is shown in Table 8-7. The parts are selected so that there will be ⅓ conforming, ⅓ nonconforming, and ⅓ marginal. Marginal parts are further divided so that they are ½ marginally conforming and ½ marginally nonconforming. This results in the total sample being ½ conforming parts and ½ nonconforming parts.

Once the parts are selected, they are inspected once in a random order by each inspector, and the results are recorded on data sheets. An inspection is repeated by each inspector and the results are recorded on separate data sheets to eliminate any unintentional bias. This is repeated until the required number of inspections is completed. Inspectors should take a normal amount of time for each inspection.

Table 8-7. Sample Sizes—Attribute Data

Quantity of appraisers	Quantity of gages	Minimum number of parts	Minimum number of measurements per part
1	0	24	5
1	1		
2	0	18	4
2 or more	1		
2	2 or more		
3 or more	0	12	3
2 or more	2 or more		

Data Analysis. Analysis of the data is performed using the appropriate worksheets to compute P(miss), P(FA), effectiveness (E) and bias value (B). The analysis will be illustrated by an example.

The example is concerned with a plating operation on a printer part. The visual inspection detects stains and deposits on the part after plating. Three persons were involved in the study: the plating operator, inspector, and lead inspector. Seventeen parts were selected initially and, after evaluation of the samples by the quality engineer, manufacturing engineer, and inspection supervisor, 14 parts (8 conforming and 6 nonconforming) were actually used in the study. Each part was inspected three times. The data obtained is shown in Table 8-8.

The column marked A/R contains the true condition of the part, where A means acceptable or conforming and R means rejected or nonconforming.

The analysis consists mainly of counting and division. The details of the computations are shown on the Attribute Data Worksheet (Figure 8.59).

Inspection/Test Capability Plan

■ *Objective:* To determine inspection capability of _____ inspection process.

■ *Expected use of information (value added):* _____

■ *Scope (test/inspection station(s)):* _____

■ *Type of data:* Attribute or variable. Special notes on data.

Number of inspection devices: _____

Number of characteristics: _____

Table 8-8. Attribute Example Data

Assembly	A/R	Appraiser								
		A			B			C		
		1	2	3	1	2	3	1	2	3
1	A	A	A	A	A	A	A	A	A	A
2	R	R	R	R	R	R	R	R	R	R
3	A	A	A	A	A	A	A	A	A	A
4	R	R	R	R	R	R	R	R	R	R
5	R	R	R	R	R	A	R	R	R	R
6	A	R	R	R	A	A	A	A	A	A
7	A	R	A	R	A	A	A	A	R	A
8	A	A	A	A	A	A	A	A	A	A
9	R	R	R	R	A	A	A	A	A	A
10	A	A	A	A	A	A	A	A	A	A
11	A	A	A	A	A	A	A	A	A	A
12	R	R	R	R	R	R	R	R	R	R
13	A	A	A	A	A	A	A	A	A	A
14	R	R	R	R	R	R	R	R	R	R

Part number _____ Date _____

Inspection instruction no. _____ Rev. _____ Date _____

Characteristics inspected: _____

Stains/deposits _____

Inspection Results

Appraiser	Number good correct (1)	Number bad correct (2)	Number correct (3)	Number FA (4)	Number miss (5)	Number total (6)
A	19	18	37	5	0	42
B	24	14	38	0	4	42
C	23	15	38	1	3	42

Calculations

Appraiser	E $\dfrac{(3)}{(6)}$	P(FA) $\dfrac{(4)}{(1)+(4)}$	P(miss) $\dfrac{(5)}{(2)+(5)}$
A	37/42 = .88	5/24 = .21	0/18 = 0
B	38/42 = .90	0/24 = 0	4/18 = .22
C	38/42 = .90	1/24 = .04	3/18 = .17

Figure 8-59. Attribute data worksheet

Number of production pieces used: _____
Number of appraisers: _____
Redundant machine: _____
Number of shifts: _____
Number of operators: _____

- *Methodology:* Characteristics to be measured, randomization plan, data collection forms, etc. Describe in enough detail to allow those collecting the data to execute from this plan. Use the inspection capability check sheet to guide this description.

- *Capability criteria*

  Variable date: Cr _____ CPK _____
  Attribute: P(M) _____ P(FA) _____ EFF _____ BIAS _____

Column (1)—Number Good Correct.   This is the number of conforming or good parts identified correctly by the person inspecting the parts. Since there are eight conforming parts each inspected three times, 24 opportunities exist for correct identification of the conforming parts for each person.

Person A in the example correctly identified the conforming parts 19 times. Assemblies 1, 4, 10, 13, 14, and 16 were all accepts (18 total), while assembly 8 had zero accepts and assembly 9 had one accept.

Column (2)—Number Bad Correct.   This is the number of nonconforming or bad parts identified correctly by the person inspecting these parts. Since there are six nonconforming parts inspected three times, 18 opportunities exist for correct identification of the nonconforming parts for each person.

Person A in the example correctly identified the nonconforming parts all 18 times.

Column (3)—Number Correct.   This is the total of columns (1) and (2) and is the numerator of the formula for computing the effectiveness *E*. This is 37 for person A.

Column (4)—Number FA.   This is the number of false alarms (FA) for each person.

Person A had five false alarms, three on assembly 8 and two on assembly 9.

Column (5)—Number Miss.   This is the number of misses for each person. Person A had zero misses.

Column (6)—Number Total.   This is the total of columns (3), (4), and (5) and should equal the number of parts inspected times the number of inspections per part. In the example, 14 parts were inspected three times, for a total of 42 for all persons.

$$E = \frac{\text{value in column 3}}{\text{value in column 6}}$$

This is the computation for the effectiveness *E*, and is the value in column (3) divided by the value in column (6) for each person.

For person A, the effectiveness $E = 37/42 = 0.88$.

$$P(FA) = \frac{\text{value in column 4}}{\text{value in column 1} + \text{value in column 4}}$$

This is the computation for the probability of a false alarm, P(FA), and is the value in column (4) divided by the sum of the values in columns (1) and (4). The sum of columns (1) and (4) is the number of opportunities for false alarms which are the same as the number of opportunities for correctly identifying the conforming parts (24 in this example). Person A has $P(FA) = 5 / (19 + 5) = 5 / 24 = 0.21$.

$$P(\text{miss}) = \frac{\text{value in column 5}}{\text{value in column 2+ value in column 5}}$$

This is the computation for the probability of a miss, P(miss), and is the value in column (5) divided by the sum of the values in columns (2) and (5). The sum of columns (2) and (5) is the number of opportunities for misses, which are the same as the number of opportunities for correctly identifying the nonconforming parts (18 in this example).

For person A, $P(\text{miss}) = 0/18 = 0$.

Data Evaluation.   The inspection capability study is evaluated using Table 8-9 that contains the criteria for effectiveness, P(FA), P(miss), and bias. For any marginally acceptable or unacceptable gages or appraisers, corrective action is required and, when corrective action is completed, the inspection capability study must be redone.

Reporting Results of Inspection Capability Study.   Reporting the results of an inspection capability study provides a means of recording the details of the study as well as a record of the results. The guidelines following are provided to assist in writing the report.

Prior to conducting the study, a "plan" should be prepared for review that contains the purpose and methodology sections only. This plan also provides the details of the study as well as the resources (time/material/equipment) required. This plan should be approved by the resource committee or supervisor as applicable.

The final report consists of the following seven sections.

1. *Purpose/scope.*   A brief statement of why the study is being done and the scope. This is the objective of the study. From this statement the number of gages, operators, or machines involved should be determinable.

Table 8-9. Attribute Data Criteria

Parameter	Acceptable	Marginal	Unacceptable
E	0.9 or more	0.8 to 0.9	less than 0.8
P(FA)	0.05 or less	0.05 to 0.10	more than 0.10
P(miss)	0.02 or less	0.02 to 0.05	more than 0.05

2. *Findings/conclusions.* This is a summary of the results. The capability should be stated as acceptable, marginally acceptable, or unacceptable. The values of percent tolerance consumed, P(miss), B, and other appropriate measures should be stated in tabular form.

3. *Recommendations.* These indicate what should be done about a marginal or unacceptable study. The recommendations should be supported by the findings/conclusions. Any resources required to implement the recommendations are identified in this section along with who is responsible for implementation and the expected completion dates. The corrective/preventive-action worksheet shown in Figure 8-60 should be used to summarize these items.

Inspection/Test-Capability Study Corrective/Preventive-Action Sheet

Below to be Completed by Responsible Individual

Corrective/Preventive Action	Responsibility	Target Date	Completion Date

Figure 8-60. Corrective/preventive-action worksheet.

Table 8-10. Inspection
Capability Factors
Repeatability Factors

N	$1/d_2$	D_4
2	0.885	3.268
3	0.592	2.574
4	0.485	2.282
5	0.429	2.114

Table 8-11. Reproducibility
Factors

K	D
2	0.709
3	0.524
4	0.446
5	0.403
6	0.375
7	0.353
8	0.338
9	0.325
10	0.314

4. *Methodology.* This section should explain what was done and how the data was collected. All data collection details, assumptions, and limitations should be stated in this section.

5. *Analysis/data.* This section contains the detail data and computation details. Any worksheets used should be in this section. Summary, date, recording, and data worksheets must be used to present data.

6. *Discussion.* This section is used for further supporting information and for descriptions of any unusual occurrences.

7. *Distribution.* Copies to team members, team leader, resource committee, supervisor, and quality engineering archive files.

Checklists. Figures 8-61 to 8-65 are checklists and worksheets provided to aid in conducting an inspection capability study. These lists provide assistance in designing, conducting, analyzing, and reporting inspection capability studies. It is essential that they be completed and attached to the inspection capability study report.

Summary. The inspection capability study is a method for evaluating how good an inspection or measurement method is for both variable and attribute data. The inspection capability measurement for variable data is a percent of total tolerance. The measurement of inspection capability for visual or attribute data involves four values: P(miss), P(FA), E (effectiveness), and B (bias).

An inspection capability study must be performed prior to conducting a process capability study to ensure that an accurate measurement or inspection method is in place. This is a very important step in implementing statistical process control.

ITEM	DESCRIPTION	DISPOSITION (A,M,U)*(Y/N)	ACTION REQUIRED
1.0	**STUDY PREPARATION AND DESIGN:**		
1.1	Is the purpose or objective stated?		
1.2	Is the type of data (Variable/Attribute) identified?		
1.3	Does an inspection instruction exist? Number Rev.		
1.3.1	Does the method reflect current practice?		
1.4	How many appraisers are to be involved?		
1.5	How many gages or pieces of test equipment are involved?		
1.6	Is a data collection method/procedure prepared?		
1.6.1	What is the sample size?		
1.6.2	How many inspections/measurements per sample are required?		
1.6.3	Does the order for inspection/measurement of the samples follow the method described in previous sections?		
1.6.4	Are there any special conditions (test method restrictions, etc.) that will effect the data collection?		
1.6.5	Is the study to be conducted in the same location as it is performed every day?		
1.6.6	Are the inspection measurement conditions (lighting, inspection time, inspection location) the same as those in current use?		
1.6.7	If there are any exceptions, how are they being compensated for in the analysis and interpretation?		
1.6.8	Is there a data recording form?		
1.6.9	Is the form acceptable?		
1.7	Has a plan with purpose and data collection method been issued?		
1.7.1	Has the plan been approved by the appropriate manager?		
1.7.2	Has the plan been approved by the resource committee?		

* A = Acceptable, M = Marginal, U = Unacceptable

Figure 8-61. Inspection capability checklist.

ITEM	DESCRIPTION	DISPOSITION (A,M,U)*(Y/N)	ACTION REQUIRED
2.0	**DATA COLLECTION:**		
2.1	Have the samples been obtained and identified with a number or code?		
2.2	Were samples for variable data selected at random? Include out of spec samples.		
2.3	Is the sample for attribute data composed of ⅓ acceptable, ⅓ unacceptable and ⅓ marginal (½ acceptable, ½ unacceptable parts)? This results in about ½ of samples being acceptable.		
2.3.1	Were the samples for attribute data selected by Quality Engineering and the Quality Control Supervisor?		
2.3.2	Are the defects on the samples selected, an element on the inspection instructions?		
2.3.3	Are any defects omitted? Why were they omitted?		
2.4	Has the purpose of the study and data collection method been reviewed with participants?		
2.5	Has the inspection method been reviewed with the appraisers?		
2.6	Have the roles of the participants been explained?		
3.0	**DATA ANALYSIS:**		
3.1	Variable Data:		
3.1.1	Is the average range equal to or greater than 0?		
3.1.2	Was the correct D4 factor used for UCLR?		
3.1.3	Any ranges above UCLR?		
3.1.4	What action is being done if any ranges were above UCLR?		
3.1.5	Is the repeatability computed correctly?		
3.1.6	Is reproducibility computed correctly? Note: The computation of reproducibility of appraisers is required if more than one appraiser is used. Computation of reproducibility of gauges is required if more than one gauge is used.		
3.1.7	Is the overall Inspection Capability computed correctly?		

Figure 8-61. (*Continued*)

ITEM	DESCRIPTION	DISPOSITION (A,M,U)*(Y/N)	ACTION REQUIRED
3.1.8	Are both appraisers and gages included, if used, in the capability computation?		
3.1.9	Is the result acceptable, marginal, or unacceptable?		
3.2	Attribute data:		
3.2.1	Are the number of conforming parts accepted counted correctly?		
3.2.2	Are the number of nonconforming parts rejected counted correctly?		
3.2.3	Are the number of conforming parts rejected (false alarms) counted correctly?		
3.2.4	Are the number of nonconforming parts accepted (misses) counted correctly?		
3.2.5	Check the computations for effectiveness, E, P(miss), P(FA), and bias, B.		
3.2.6	Is the result acceptable, marginal, or unacceptable?		
4.0	**REPORTING RESULTS:**		
4.1	Has the report containing the results of the study been written and issued?		
4.2	Does the report follow the report guidelines?		
4.3	Does the report contain purpose, findings/ conclusions, recommendations, methodology, and data analysis sections?		
4.4	Has the report been reviewed and approved by the appropriate manager?		
5.0	**PREVENTIVE ACTION:**		
5.1	Is any preventive action required? (marginal and unacceptable studies)		
5.2	Has the preventive action been assigned with responsibility using the preventive action sheet?		
5.3	If preventive action is required, when will the study be repeated?		

Figure 8-61. (*Continued*)

Variable Data—Data Recording Sheet

Part no. _____ Date _____

Dimension/characteristic _____

Gage/test equip. no. _____ Gage increment _____

Inspector _____ Quality engineer _____

Gage/Appr. Test Equip.	Trial	Sample									
		1	2	3	4	5	6	7	8	9	10
	1										
	2										
$\Sigma X =$ _____	3										
$\bar{X} =$ _____	4										
$\Sigma R =$ _____	5										
$\bar{R} =$ _____	R										
	1										
	2										
$\Sigma X =$ _____	3										
$\bar{X} =$ _____	4										
$\Sigma R =$ _____	5										
$\bar{R} =$ _____	R										
	1										
	2										
$\Sigma X =$ _____	3										
$\bar{X} =$ _____	4										
$\Sigma R =$ _____	5										
$\bar{R} =$ _____	R										

Figure 8-62. Variable data recording sheet.

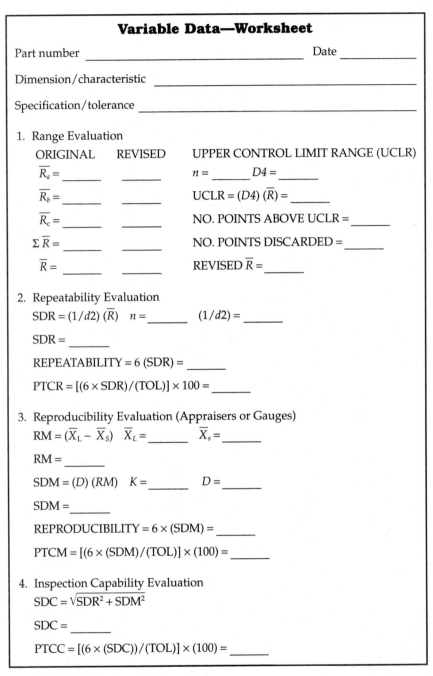

Figure 8-63. Variable data worksheet.

Attribute Data Worksheet

Part number _____ Date _____

Inspection instruction no. _____ Rev. _____ Date _____

Characteristics inspected: _____

_____

Inspection Results

Appraiser	Number good correct (1)	Number bad correct (2)	Number correct (3)	Number FA (4)	Number miss (5)	Number total (6)

Figure 8-64. Attribute data worksheet.

Attribute Data Worksheet

Part number _____ Date _____

Inspection instruction no. _____ Rev. _____ Date _____

Calculations

Appraiser	E (col. 3) $\overline{\text{(col. 6)}}$	P(FA) (col. 4) $\overline{\text{(col. 1) + (col. 4)}}$	P(miss) (col. 5) $\overline{\text{(col. 2) + (col. 5)}}$	Bias—B P(FA) $\overline{\text{P(miss)}}$

Figure 8-65. Attribute data worksheet.

Inspection Capability—Attribute Data Multiple Characteristics.
The example described here deals with a common situation found in final
inspection stations of major subassemblies and final product assembly.
The stations have several characteristics being checked by visual inspection and it is desired to determine the capability of the inspection method.
The general method for conducting an inspection capability study for
attribute data describes how to work with one or two characteristics. This
example describes how to adapt the general method to a situation involving multiple characteristics.

The P300 shuttle assembly final inspection station, where 23 characteristics are checked, will be used to illustrate a multiple characteristics
study. With 23 characteristics to check, it is unrealistic to evaluate all of
them with one set of samples. With one set of samples, the parts will have
many more defects than usual, so the inspectors may take more time in the
study than in their normal work, biasing the results.

The recommended method is to work with six to eight characteristics at
a time. First identify the criticality of the characteristics and use this as a
priority list for completing the study in groups of six to eight characteristics. The samples will be selected and defects selected to satisfy the ⅓
good, ⅓ bad, and ⅓ marginal for each characteristic. A matrix is used to
identify which parts have what defects.

The study is conducted so that the inspector checks all characteristics,
but the analysis is done on only the characteristics selected. The results on
all the other characteristics are omitted in the analysis.

Select Samples. Two inspectors were used in the study. The number
of samples required is 18, with 6 having no defects, 6 with defects, and 6
with marginal (3—marginal good; 3—marginal bad) for each characteristic. This results in 9 samples with defects and 9 without defects.

The inspectors will check each shuttle four times, resulting in a total of
72 inspections for each characteristic.

The matrix in Figure 8-66 shows the allocation of defects, for the first
eight characteristics studies, among the 18 shuttles.

This method is repeated until all characteristics have been studied; eight
in the first set of shuttles, eight in the second set of shuttles, and seven in
the third set. The same 18 shuttles can be used for the three sets of characteristics inspected.

Collect the Data. Each inspector inspects each shuttle for all 23 characteristics and records any defects found. A new data recording sheet is
provided each time the inspection is done.

The inspectors will check all shuttles once and repeat until four inspections on each shuttle are completed by both inspectors. The inspection

Matrix of Defects

Unit

CHARACTERISTIC		1	2	3	4	5	6	7	8	9	10	11	12	13	14	15	16	17	18
	1	F	F	P	F	P	F	P	F	F	P	P	P	F	P	F	P	F	P
	2	P	F	P	F	P	F	F	P	F	F	P	P	F	P	F	F	P	P
	3	F	P	F	P	P	F	P	P	F	F	F	P	F	P	P	F	F	P
	4	P	F	F	P	P	P	F	F	F	P	F	P	P	F	P	P	F	F
	5	F	P	F	P	P	F	P	F	F	F	F	P	P	F	P	F	P	P
	6	P	F	P	F	P	P	F	P	F	P	F	P	F	P	F	F	P	F
	7	F	P	F	P	P	P	F	P	P	F	P	P	F	F	F	F	F	P
	8	P	F	P	F	P	F	P	F	F	P	F	P	P	F	P	P	F	F

Figure 8-66. Allocation of defects.

results on the characteristics of interest are translated to the attribute data worksheet for analysis.

Analysis of the Data. The detail analysis will not be presented; instead, an outline of the analysis will be discussed.

The usual computations for determining effectiveness (E), P(FA), P(miss) and bias (B) are computed on each characteristic for each inspector.

The criteria for determining if the inspection capability is acceptable, marginal, or rejectable is applied to each characteristic and is the same as presented in the attribute data section.

Determining the necessary corrective actions is facilitated by developing a characteristic by inspector table, as shown in Table 8-12, for all inspectors and characteristics. This can identify if the criteria needs improvement or if the inspectors need additional training.

Referring to Table 8-12, the characteristics 1, 2, 15, and 22 have misses for both inspectors. This indicates that the criteria needs improvement.

Characteristic 4 has only one inspector with difficulties (false alarms). This indicates the inspector needs additional training.

Characteristic 8 has both misses and false alarms for both inspectors. The criteria for the item needs major work since the current criteria is not understood by either inspector. Retraining is also necessary after the criteria is corrected.

The study is repeated after all corrective actions are complete. The defects studies will be only the ones corrected.

Table 8-12. Characteristic vs. Inspector Results Table

Characteristic	Number of misses Insp. A	Insp. B	Number of F/A Insp. A	Insp. B	Action
1—S/N, P/N, W/O match	1	1	0	0	B
2—Paperwork signed	2	2	0	0	B
3—Missing label	1	0	0	0	
4—Bent, loose S/pin	1	0	7	0	A
5—Magnet separation	0	0	0	0	
6—Miss/wrong S/N C/P	0	0	0	0	
7—Damaged/miss C/plte	0	0	0	0	
8—P.C.B.A. soldering	2	1	5	6	C
9—Exposed wire	1	0	0	0	
10—Cut wire	0	0	0	0	
11—Loose cable clamp	0	0	1	0	
12—Loose A/R lug arm	1	0	0	0	
13—Tamper-proof label	0	0	0	0	
14—Miss/loose yoke	0	0	0	0	
15—Damaged H/spring	1	2	0	0	B
16—Lasermarking	0	0	0	0	
17—Loose clamp plate	1	0	0	0	
18—Clamp/plt orientate	1	1	0	0	
19—Roll-pin height	0	0	0	1	
20—Roll-pin position	0	0	0	0	
21—Missing roll pin	0	0	0	0	
22—Tip protrusion	2	2	0	0	B
23—Ins/date stamp	0	0	0	1	
OVERALL	14	9	13	8	

Process Capability Studies

A *process capability study* is formally defined as a systematic procedure for studying a process by means of control charts to determine if the process is behaving naturally or unnaturally along with investigation of any unnatural behavior to determine its cause; plus action to eliminate any of the unnatural behavior which it is desirable to eliminate for economic or quality reasons. The natural behavior of the process after unnatural disturbances are removed is called the *process capability.*

Normally, it is not feasible to determine the process capability by direct measurements on the process or machine-operating conditions. What is done instead is to measure the output or products from the process in order to quantify, measure, or determine the natural process variation.

Process capability studies can be applied in situations involving quality, cost, need for information, establishment of standards or estimates, new development, and research. Some of the benefits of a process capability study in manufacturing applications are:

Improve Productivity and Quality

- Determine the design tolerances that can be met with the current processes.

- Determine if new equipment is capable of meeting the requirements.

- Compare capability of alternative equipment/machines.

A process capability study may involve a single process (simple or complex) or a sequence of processes (production sequence). In its simplest form, the process capability study involves a single operation (machine) and one person (operator). This is referred to as a *machine capability study.* A machine capability study is conducted under normal operating conditions, but while holding as many variables constant as possible. It is usually accomplished by obtaining consecutive pieces from the process for a short time with no adjustments or material changes. This allows a measure of the inherent or natural process variation with a minimum of samples and time.

A process capability study may be conducted to determine the total process variability and stability over time under normal operating conditions and would include tool wear, operator machine adjustments, material variation, equipment wear, shift changes, and all other sources of variability.

As described earlier, the process capability is the inherent or natural variation of a process after undesirable disturbances are eliminated. The capability for a process involving variable data is expressed as 6 times the value of the standard deviation or 6σ. This number is compared to the requirement by a capability ratio number, Cr, which is 6σ divided by the total tolerance. When this number is multiplied by 100, it is the percent of the tolerance consumed by the capability, which will be referred to as PTCC. The capability for a process involving attribute data is simply the average value or centerline of the control chart.

Several indexes for quantifying the capability of a process involving variable data have been developed and are being used as a part of the requirement to become a qualified supplier. Some of the more commonly used ones are Cpk, Cp, Zmin, Cpu and Cpl. The Zmin, Cpu and Cpl indexes also include the location of the process average to the specification nominal. These are described in detail elsewhere; They are being used by many large corporations such as Ford, GM, and Xerox as a requirement for the supplier's processes.

This section contains information to assist in conducting a process capability study. We will describe how to design, conduct, analyze, and evaluate the study. A checklist will be included to ensure that all necessary items are considered when conducting a process capability study.

Design of the Study

The design of a capability study includes planning, defining the scope, and developing data collection methods.

Planning the Capability Study. Planning the study consists of reviewing the process, the manufacturing method, and the method of measurement.

Reviewing the Manufacturing Method. The following items should be considered when reviewing the manufacturing method:

- Become familiar with the part and how it is made or assembled.
- Determine the part features or characteristics that will be used in the study. Identify these characteristics as *variable* or *attribute*.
- Check the following if a fabrication process is to be studied.

  Determine the type of fabrication being used. Is it cut, grind, drill, ream, punch, or form?

  Determine if a single or multiple spindle or fixture machine is being used.

  Determine if the adjustments are automatic or manual.

  Determine the operation cycle time.

  Determine the machine or process settings such as speed, feed, pressure, or temperature.

  Determine the number of machines and operators used and how many shifts the process runs on.

- Check the following if an assembly process is being studied.

  Review the assembly method to determine if it is current and reflects current practice.

  If a fixture is used in the assembly process, determine if it is the latest configuration and if it locates the part properly.

  Determine the assembly operation cycle time.

  Determine how many operators and fixtures are used and the number of shifts.

Reviewing the Inspection Method. The inspection method should be reviewed to become familiar with the method that is used to measure, inspect, or test the product.

- Review the inspection method or instruction.
- Determine how the part is measured/inspected or tested.

- Review the inspection capability study. If the results were marginal or unacceptable, determine if the corrective actions have been completed and if the study has been redone and acceptable. If the inspection capability study has not been performed or is still marginal or unacceptable, the process capability study should not be performed.

Defining the Scope. The scope of the capability study includes identifying what exactly is to be included in the study. The following items should be considered in determining the scope of the study.

- Select which characteristics are to be studied. Study as many as possible, since the process capability can be determined for each characteristic studied from the same set of parts. If six characteristics are measured on the part, six process capability studies can be completed at the same time on the same parts.

- Determine which characteristics to be studied will provide variable data and which will provide attribute data. Separate capability studies may be required for the variable and attribute characteristics.

- Determine the number of machines or assembly operations to be studied. Usually a study will be performed on each machine or assembly operation.

- Determine the number of persons (operators/inspectors) to be used. Usually one person is studied at a time with additional studies done on each person.

If a study will include two machines and two operators, then four studies will be required, one for each machine/operator combination, to fully evaluate the process if the operators have an influence on the operation. If the machine has automatic settings, then one operator can be selected at random and used to do a study on each machine.

One person should be selected to do the inspecting or measuring. This should be a trained person and may be the best person if the inspection method has marginal capability.

- Determine how a multiple-shift process will be handled. Usually a separate study is performed on each shift or notes are made on the events log or control chart when shift changes occur.

- Determine what adjustments will be allowed to be made by the operator. Remember, the goal is to minimize adjustments. Avoid the trap of allowing all adjustments currently done. Sometimes adjustments are made "because we've always done it that way," especially if the process involves tool wear such as a cutting or grinding operation.

A good example of this occurred during the study of a grinding operation. The manufacturing method called for adjusting the grinding wheel every 200 parts. This adjustment took about 15 to 20 minutes, resulting in significant loss of production. When asked why this was done, the answer was "we've always done it that way." During the study it was decided to make no adjustments to the grinding wheel so that the "capability" of the grinding wheel could be determined. As a result of the capability study, it was found that 1200 parts could be machined before adjusting the wheel. In addition, six 15- to 20-minute delays were omitted, resulting in increased productivity.

To summarize the scope of a capability study, remember that a process capability study should be conducted under normal operating conditions using a single batch of raw material, a single operator, and a single inspector during the time the data is being collected. The operator should avoid correcting the process or making adjustments during the study. Recalibration of the measuring equipment should be avoided unless it is normally calibrated frequently.

Develop Data Collection Methods. Developing the data collection methods includes determining the number of samples to be collected, how the samples are to be obtained, and designing the data collection form. The following items should be considered in developing the data collection methods.

Number of Samples— $\overline{X}/R$ Charts

- A minimum of 25 subgroups of size 4 or 5 are required (100 to 125 total samples) for a process capability study.

- More than 25 subgroups are normally required in order to determine the effects of material, operators, shifts, and machine wear over time.

- The guidelines for subgroup selection (size and frequency) are as follows and are a function of the production volume.

  High volume: 50 parts or more per hour. Select a subgroup from a range of 15 minutes or every 25 to 30 pieces produced until it appears reasonable that the variability of the process is relatively low, then select groups every 30 minutes for a number of subgroups until it appears reasonable that the process is in control, then every hour, etc.

  Low volume: Less than 50 parts per hour. Use consecutive pieces and443 every four or five pieces constitute a subgroup.

If subgroups of size 5 are to be used and consecutive pieces are to be grouped into subgroups, the first five pieces are subgroup 1, pieces 6 to 10 are subgroup 2, etc.

Attribute Charts. Attribute data is collected and analyzed utilizing the appropriate attribute control chart. A minimum of 25 subgroups is required and the frequency is a function of the production rate similar to that of the variable data study. The following guidelines are provided for selecting subgroup size and frequency.

High volume: 50 or more parts per hour. Collect subgroups every 15 to 30 minutes or every 25 or 50 pieces. If a p or np chart is used, consecutive pieces may have to be used with every 25 or 50 pieces constituting a subgroup.

Low volume: Less than 50 parts per hour. Use consecutive pieces and break into subgroups.

A process capability study with attribute data can be performed using the results of 100 percent inspection, provided an acceptable inspection capability study has been performed. The subgroup size is often the number of parts inspected during a shift or day. This will often result in unequal subgroup sizes.

Sample Selection. The samples are to be selected as is from the process before any rework or sorting occurs. Do not let the operator select the parts. This is a very critical item in order to determine the true capability of the process.

Data Collection Forms. Design forms to record the data. These forms should allow data to be recorded quickly and easily. If computer programs are to be used, design the forms to allow for ease in data entry.

Conducting the Study

Conducting the study consists primarily of collecting the data. In order to collect the data properly, prepare both the process and the people.

Prepare Process

- Review the inspection capability study. An acceptable inspection capability study is required before proceeding with the process capability study.
- Calibrate the measuring equipment.
- Set up the machine or process under normal conditions. Ensure that the normal environment and material are used. New materials or methods should not be used. Additional studies would be required to evaluate the new materials or methods.

- Have the first samples inspected and accepted by inspection. Do not run the study until the first pieces are accepted.

- Initiate an events log at the process to record anything that is new, different, or changed during the study.

Prepare People

- Select a trained operator and inspector to be used. Normally use one operator unless process is operator-dependent. One inspector who has acceptable inspection capability is normally used throughout the study.

- Discuss the study with the operator(s) and inspector(s). Describe the purpose of the study and provide a detailed explanation of the data collection and data recording methods.

- Explain the role of the operator. Explain the adjustments that the operator is allowed to make during the study. One of the main functions of the operator is to record any changes in the process in the events log.

- Explain the role of the inspector. The inspector identifies the parts, measures or inspects the parts, and records the data on the data collection form.

It is very important that the order in which the parts are produced is maintained. This is achieved by identifying the parts with a number. Typically in a process capability study, all parts are collected before any measurements are made. Thus the identification of the parts is very important if subgroups are to be created or in analysis of the results. The measurement of the parts can be in a random order or in the same order as the parts were made, whichever is easier.

When the data is being collected, it is very important that the persons who designed the study are present to audit the data collection procedure to ensure that the data collection follows the plan. The following items are suggested for auditing the data collection.

- Record all pertinent information at the start of the study: date, operator, inspector, machine or operation number, process conditions (speed, pressures, temperature), and characteristics being checked in the events log.

- Ensure that any changes in the process that occur during the study are recorded in the events log along with the time the event occurred.

- Save all parts measured or inspected until the data evaluation is completed. This is done in case measurements or inspections must be repeated due to wild or freak readings.

Analyzing and Evaluating
the Results

The analysis and evaluation of the data depends upon which control chart was used to collect the data. The following sections describe the analysis and evaluation for the variable control chart ($\overline{X}$/R) and the attribute control charts.

Variable Data. This section contains the details of the analysis and evaluation of a process capability study using an $\overline{X}$ and R chart.

Analysis. Plot the subgroup data on an $\overline{X}$ and R chart. Compute the centerline ($\overline{R}$) and control limits for the range (R) chart. If the range chart is "in control," compute the limits for the average ($\overline{X}$) chart. If the range chart is not in control, use the following guidelines to evaluate the results.

- For any subgroup range exceeding the upper control limit (UCLR), remeasure the part to determine if there is an error in measurement of the part or recording of data. If there is a measurement or data recording error, replace the reading and redo the calculations. When more than one range exceeds the UCLR due to measurement errors, then review the inspection capability again and redo the study.

- Subgroup ranges that are more than 10 times the average range are considered "wild" or "freak" readings. If only one subgroup is affected, remove the reading; otherwise there is a major problem affecting the variation of the process. Develop and initiate the appropriate corrective action and redo the study after corrective action is completed.

Evaluation. The evaluation of the process capability study is concerned with the variation, stability, and centering.

The variation of the process is compared to the specification by computing the capability ratio, Cr, as follows:

$$Cr = 6\sigma / \text{total tolerance}$$

where: Total tolerance = upper spec limit − lower spec limit

$$6\sigma = 6 \times \overline{R}/d_2$$

The capability ratio, Cr, is multiplied by 100 to give the percent tolerance consumed by the capability that is denoted PTCC.

The stability of the process is concerned with the process average. If the $\overline{X}$ chart is in control, the process average is considered stable.

The centering of the process is concerned with where the process average is with respect to the specification nominal. If the specification nominal is within the control limits for the averages, the process average is considered not significantly different from the specification nominal.

The criteria for a process capability study being acceptable, marginal, or unacceptable are shown in Table 8-13. The criteria evaluate the three items of variability, stability, and centering.

Determine and initiate the appropriate corrective action if the process capability is marginal or unacceptable. The corrective action form shown in Figure 8-67 is to be used to define what actions are to be taken, who is to do them, and when they are to be done. When the actions are completed, the process capability study is repeated to determine if the actions were effective.

Example. Suppose a process capability study was conducted on a punch-press operation. One of the characteristics of major interest is the width of a slot having a specification of 0.4040 ± 0.007. Twenty-five subgroups of size 5 were obtained during a shift. Determine if the process is capable of meeting the specification.

Suppose the range chart and the average chart are in control, with the average range = 0.004, $X = 0.4039$, $UCL_X = 0.4062$ and $LCL_X = 0.4016$.

The 6σ of the process can be determined as follows:

$$6\sigma = 6 \times \overline{R}/d_2$$

$$6\sigma = 6 \times 0.004/2.326 = 6 \times 0.00172$$

$$6\sigma = 0.01032$$

Table 8-13. Process Capability Study Evaluation Table—Variable Data

Cr	$\overline{X}$ chart in control	Spec. nominal within control limits on $\overline{X}$ chart	Decision
80% or less	Yes	Yes	Accept
	Yes/no	No/yes	Marginal
	No	No	Unacceptable
80 to 100%	Yes/yes/no	Yes/no/yes	Marginal
	No	No	Unacceptable
100% or more	N/A	N/A	Unacceptable

Process Capability Study
Corrective/Preventive-Action Sheet

To: _____ Date: _____

Part number: _____ Quality engr.: _____

Specification/tolerance: _____ Inspector: _____

Study Disposition:
Information below to be completed by responsible individual:

Corrective/Preventive Action	Responsibility	Target Date	Completion Date

Figure 8-67. Corrective action form.

The capability ratio Cr = 6σ/tot. tol., where the total tolerance (tot. tol.) is 0.014. The total tolerance is the upper spec limit minus the lower spec limit or 0.411 − 0.397 = 0.014. The value of Cr is:

$$Cr = 0.01032/0.014 = 0.737 \text{ or } 73.7\%$$

Referring back to Table 8-3, the evaluation of the process capability can now be determined. Cr is less than 80 percent, the $\overline{X}$ chart is in control, and the spec nominal value, 0.4040, is within the control limits on the $\overline{X}$ chart, so the process is acceptable.

This formula for σ provides a reasonable estimate for the process capability.

Process Capability Indexes. The following index numbers are the common ones used throughout industry. Cp and Cpk index numbers are the most widely used.

- *Cp.* Cp is the inherent or natural variation of the process compared to the print tolerance. The range of the values of Cp are from 0 to +infinity. The target is 1.33 for Cp = $T/6\sigma$, where T = upper spec limit − lower spec limit. The Cr value used in the previous section is 1/Cp.

- *Zmin.* Zmin is the difference between the mean and the nearest specification limit divided by σ. This is used by Ford to qualify top suppliers for their Q1 rating. The formula is:

$$Zmin = \frac{(USL - X)}{\sigma} ; \quad \frac{(X - LSL)}{\sigma}$$

  Large values of Zmin indicate small process variation and process average close to the specification nominal. If $6\sigma = T$ and X = nominal, then Zmin = 3.

- *Cpk.* The Cpk index is the position of the 6σ in relation to the spec mean. The formula is:

$$Cpk = \frac{Zmin}{3}$$

  If $6\sigma = T$ and X = nominal, then Zmin = 3 and Cpk = 1.

- *Cpu.* The Cpu index is the position of 6σ in relation to the upper spec limit. If a process is centered, Cpu = Cp. If Cpu is less than 1, some portion of product exceeds the upper spec limit. If Cpu is greater than or equal to 1, all of the product is below the upper spec limit. The formula is:

$$Cpu = \frac{USL - X}{3\sigma}$$

If $6\sigma = T$ and X = nominal, then Cpu = 1. If 6σ is greater than T and $X =$ nominal, then 3σ is greater than (USL − X) and Cpu is less than 1, an unacceptable condition.

- *Cpl.*   The Cpl index is the position of 6σ in relation to the lower spec limit. If a process is centered, Cpl = Cp. If Cpl is less than 1, some portion of product is below the lower spec limit. If Cpl is less than or equal to 1, all of the product is below the lower spec limit. The formula is:

$$Cpl = \frac{X - LSL}{3\sigma}$$

Application of Cr and Cpk.   Figure 8-68 defines a typical application of Cr and Cpk combined. From this figure, Zmin = 4 and Cpk = ⁴⁄₃ = 1.33; Cr = 0.75.

Cr Value	Decision
Less than or equal to .75	Acceptable
Greater than 1.00	Unacceptable

Cpk Value	Decision
Greater than or equal to 1.33	Acceptable
1 to 1.33	Marginal
Less than 1.0	Unacceptable

Process Capability Plan

- *Objective:* To determine process capability of _____ process.
- *Expected use of information:* _____
- *Scope (process boundaries):* _____

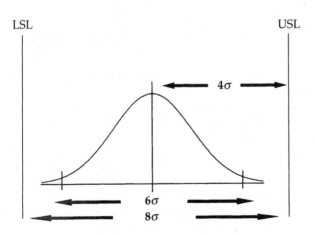

Figure 8-68. Typical application of Cr and Cpk.

- *Data:* Attribute or variable. Do the defects originate from the same cause system? Any special notes on data?

- *Number of subgroups:* (Minimum of 50 data points required)

- *Size of subgroups:* _____

- *Subgroup rational:* Why does the subgroup sampling rational make sense? Are there any problems with control limits? Do they appear unreasonably tight or wide? Could there be any biases introduced by sampling? (Examples: All subgroup samples taken at one time might yield ranges which are small but vary greatly between subgroups.)

- *Capability criteria:* Variable data—quality targets from inspection capability planning results. Inspection/test planning targets.

- *Methodology:* Characteristics to be measured, randomization plan, data collection forms, etc. Describe in enough detail to allow those collecting the data to execute from this plan. Ranges must be in control and means near control.

Use the process capability check sheet to guide this description. The process capability check sheet should be completed up to data collection and reviewed in the team meeting prior to data collection. Consider: redundant machines, number of shifts, number of operators, number of parameters, etc.

Attribute Data. The capability of an attribute data process is the value of the centerline or average. This value can be used only if the chart is in control. Suppose a process capability study is done with attribute data and a p chart is used to plot the data. Suppose the chart is in control and the value of the centerline, p-bar, is 5 percent. The capability of the process studied is 5 percent and the process is capable of making 5 percent defective product. To determine if this is an acceptable value for the average percent defective requires a target or goal to compare this to. If the requirement is 10 percent, then the process would be considered acceptable. If the requirement were 2 percent, then the process would be unacceptable.

Analysis. Using the appropriate control chart, plot the subgroup values and determine the centerline and control limits using the appropriate formulas.

Evaluation. Whether the process is acceptable, marginal, or unacceptable is determined by the relation of the process average to the target or goal. The criteria are defined as shown in Figure 8-69. If the requirement or goal is in zone A, the process capability is acceptable. If the requirement or goal is in zone B, the process is marginal. If the requirement is in zone C, the process is unacceptable.

Determine and initiate the appropriate corrective action if the process capability is marginal or unacceptable using the Corrective Action Form in Figure 8-70. When the actions are completed, the process capability study is repeated to determine if the actions were effective.

Example. Suppose a process capability study is conducted on a printed circuit board wave solder process. One of the important characteristics is the percent of boards with solder bridges. The target for the characteristic was set at 1 percent.

A subgroup of 50 boards were selected every 2 hours from the wave solder machine and the number of boards with solder bridges were recorded. This was repeated until 25 subgroups were obtained.

When the study was completed, the data was plotted on the control chart for fraction defective (p chart) and was found to be in control with p = 0.0296 (2.96%), UCL_P = 0.1031, and LCL_P = 0.

Since the control chart is in control and the process average is 2.96 percent, the process capability is 2.96 percent. We say the process is capable of making 2.96 percent defective product.

The target of 1 percent is below the process average, so the minus 2σ value is computed as follows:

$$3\sigma = UCL_P - \overline{P} = 0.1031 - 0.0296 = 0.0735$$

$$1\sigma = \frac{0.0735}{3} = 0.0245$$

$$-2\sigma \text{ value} = P - (2 \times 0.0245) = 0.0296 - 0.049$$

$$-2\sigma \text{ value} = -0.0194 \qquad \text{use } 0$$

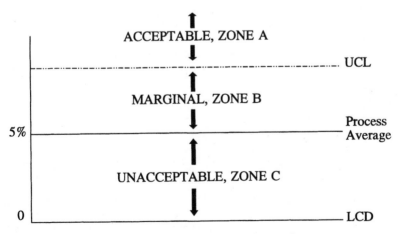

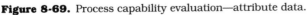

Figure 8-69. Process capability evaluation—attribute data.

Process Capability Study
Corrective/Preventive Action Sheet

To: _____ Date: _____

Part number: _____ Quality engr.: _____

Specification/tolerance: _____ Inspector: _____

Study Disposition:
Information below to be completed by responsible individual:

Corrective/Preventive Action	Responsibility	Target Date	Completion Date

Figure 8-70.

ITEM	DESCRIPTION	DISPOSITION (A,M,U)*(Y/N)	ACTION REQUIRED
1.0	**DESIGNING THE STUDY:**		
1.1	What dimensions/characteristics are to be studied?		
1.2	What type of data will be obtained (variable or attribute)?		
1.3	How many machines are included?		
1.4	How many operators are included?		
1.5	Does this process run on two shifts?		
1.6	Are both shifts to be included in the study?		
1.7	What is the production rate (min/pc; pcs/hr)?		
1.8	What are the number and revision of the manufacturing method?		
1.9	Does the manufacturing method reflect current practice?		
1.10	Fabrication Process Review:		
	What are the machine settings (speeds, feeds, temperature)?		
	Type of process cut, grind, drill, ream or other process?		
	Is the work done on a single or multiple spindle/fixture operation?		
	How is the part located?		
	Are adjustments manual or automatic?		
	How long is the operation cycle?		
1.11	Assembly Process Review:		
	How is the part located?		
	Is the assembly manual or automatic?		
	Are adjustments manual or automatic?		
	How long is the operation cycle?		
1.12	What are the number and revision of the inspection instruction?		
1.13	Has the Inspection Capability Study been performed?		
1.14	Were the results acceptable?		
1.15	If the Inspection Capability was marginal or unacceptable, is corrective action complete?		

Figure 8-71.

ITEM	DESCRIPTION	DISPOSITION (A,M,U)*(Y/N)	ACTION REQUIRED
1.16	Was the Inspection Capability Study redone?		
1.17	Was the redone study accepted?		
1.18	Have any noted exceptions been resolved?		
1.19	Inspection Instruction Review:		
	How are the dimensions or characteristics to be studied, measured/inspected?		
	Does the Inspection Instruction reflect current practice?		
1.20	Data Collection Method:		
	Has the data collection method been written?		
	Is it approved (Q.E./Resource Committee)?		
	Have the size, number, and frequency of subgroups been specified?		
	Are consecutive pieces being grouped into subgroups or are subgroups being collected periodically?		
	How are the parts being marked to determine the order of production?		
	Are parts to be measured in the same order as they are made?		
	What will be done to assure that the parts are collected prior to any screening or corrections by the operator or inspector?		
	What will be done to assure that the parts are collected properly during the study?		
	Who will monitor the data collection?		
	Are parts to be produced under normal operating conditions (lighting, temperature, pressures, speeds, feeds, material)?		
	If there are any exceptions to normal conditions, how are they being compensated for in the results?		
	Have data recording forms been developed?		
	Are the forms complete?		
	Are the forms easy to use?		
	Have the data collection responsibilities of the participants been defined?		

Figure 8-71. (*Continued*)

ITEM	DESCRIPTION	DISPOSITION (A,M,U)*(Y/N)	ACTION REQUIRED
	Has a preliminary report containing the purpose, objective, scope, and data collection method been issued?		
	Is the report approved?		
2.0	**CONDUCTING THE STUDY:**		
2.1	Is the measuring equipment calibrated?		
2.2	Are trained operators selected?		
2.3	Did the selected inspector have acceptable results in the inspection capability study?		
2.4	Has the study been described to the operators, inspectors, and supervisors?		
2.5	Were the first pieces submitted to inspection and approved?		
2.6	Is adequate material on hand to perform the study?		
2.7	Has an events log been posted to record any changes that may occur during the study?		
2.8	Will the parts be saved?		
2.9	Will the Q.E. and M.E. be present during the study to monitor the data collection?		
2.10	Are the data recording forms being used?		
2.11	Did a major process or machine breakdown occur during the study? If so, when is the study rescheduled?		
3.0	**ANALYZING/INTERPRETING DATA:**		
3.1	Were the data plotted on the appropriate control chart?		
3.2	Were the centerlines and control limits computed?		
3.4	Were the capability values computed? (PTCC - variable data) (centerline - attribute data)		
3.5	Were the capability values computed for a control chart with out-of-control points?		
3.6	For variable data, were any ranges out-of-control? What was done?		

Figure 8-71. (*Continued*)

ITEM	DESCRIPTION	DISPOSITION (A,M,U)*(Y/N)	ACTION REQUIRED
4.0	**EVALUATION OF RESULTS:**		
4.1	Has the study been evaluated as acceptable, marginal or unacceptable?		
4.2	For marginal or unacceptable studies, have the corrective actions been identified and assigned completion dates?		
4.3	Is the study to be repeated when the corrective actions are complete?		
5.0	**REPORTING RESULTS:**		
5.1	Is a report being prepared containing the results of the study?		
5.2	When will the report be issued?		
5.3	Does the report contain the purpose, findings/conclusions, recommendations, methodology, and data analysis details?		

Figure 8-71. (*Continued*)

The process is determined to be marginal since the target is 1 percent, which is between the process average and the -2σ value. The process is close to the target, but additional corrective action must be performed to get the process average reduced to 1 percent or less.

Summary

The results of the process capability study are used mainly to determine the centerlines and control limits for the ongoing real-time process control charts.

The following points summarize the main points to consider when conducting a process capability study.

- The inspection capability study must be completed and acceptable before doing the process capability study.
- If any corrective actions are required, the study will be redone when the actions are completed.

9
Advanced Techniques

Purpose

The purpose of the Advance Techniques Track is to provide the technical professionals and their managers with the knowledge and skills to utilize more advanced techniques and broader concepts.

The scope of topics includes world-class quality planning, design of experiments (DOE), ISO 9000, quality function deployment (QFD), design for manufacturability, and world-class manufacturing concepts.

Objectives

The objectives of the Advanced Techniques Track are as follows:

- Provide a quality management system.
- Establish a supplier quality assurance program.
- Develop the ability to use statistical tolerancing.
- Understand the fundamentals of quality system auditing and the knowledge and techniques in order to become effective internal auditors.
- Develop the ability to use benchmarking techniques.
- Provide effective and efficient measurements of quality parameters.
- Develop ability to perform inspection and test planning.
- Develop methods to conduct design reviews.
- Provide for quality targets for those inspection stations.

- Implement customer satisfaction measurements and feedback.
- Learn how to analyze processes in different departments and recommend measures for improvement.
- Provide advanced techniques to solve complex problems; establish set points and tolerances.
- Understand design of experiments as a problem-solving technique.
- Develop a closed-loop system for identifying, analyzing, and implementing changes in the company's operations.
- Provide knowledge to design, conduct, and analyze experiments.
- Learn how to lead and manage your organization in transforming itself into a team of QFD users.
- Gain a thorough understanding of the relevant issues that impact QFD implementation, i.e., understand customer requirements and ultimately transfer them to production specifications.
- Reduce cycle times, work-in-process, just-in-time.
- Learn the seven management and planning tools (7-M) needed for continuous improvement of quality, productivity, cycle time, and profit.
- Implement quality function deployment.
- Provide designs for manufacturability.

Quality function deployment (QFD) brings together all of the disciplines in the organization through careful analysis of customer needs and product characteristics, and ensures that every aspect of design, manufacturing, and delivery is aligned with the customer requirements. These requirements are subdivided into satisfiers, nonsatisfiers, and exciters.

World-class manufacturing focuses on cycle-time reduction, just-in-time, cells for processes, employee involvement, suggestion system, self-directed work teams, flow technology, and benchmarking.

Introduction and Overview

Any well-managed quality system is the result of thorough planning. This may again seem like a statement of the obvious, but the quality systems planning function is often ignored or understaffed in real life. Larger corporations that have had to deal with government regulations or that have historically based quality departments stemming from long experience in manufacturing apply quality planning. However, the majority of commercial firms or nonmanufacturing firms have no formal approach to establishing policy, setting targets, or formulating a scheme for inspection strategy.

Activities of the quality department are frequently determined when a test or inspection is installed in reaction to a problem already identified by the customer. Water leaks into the product; parts are rusting; parts don't fit; the product is "dead on arrival"; the service isn't satisfactory—these are the kinds of problems fed back from customers that result in attempts to bandage the process with more inspection. This is not world-class quality planning.

Of course, corrective/preventive action must be taken in response to customer complaints, but the idea of quality planning is to install a system that identifies any critical problems before they are shipped and does so in the most efficient manner possible. It is usually not effective, for example, to hire enough inspectors to verify every characteristic of every part shipped, and it should not be necessary.

World-Class Quality-Planning Policy

The first step in creating a comprehensive quality plan that will allow your company to compete in world markets that are demanding quality and service is to establish a quality policy. ISO 9000 provides an adequate guideline for policies. However, we will focus on Inspection and Test. This policy addresses the inspection and test function in detail and requires those who interpret customer requirements, those who design or define the product or service, and those who produce the product or service to share information.

A resource committee and appropriate teams can be organized to develop a quality planning policy. The policy should be customized to fit the existing organization and the business requirements and, once consensus is reached on the policy, ensure that it is communicated and understood by all personnel involved in the quality planning process.

Major elements of an inspection and test policy include the following (which supplement ISO 9000 requirements):

1. General

2. Planning requirements

3. Inspection and test operations requirements

4. Test function requirements

5. Classification of characteristic requirements

6. Repair requirements

7. Calibration requirements

8. Tryout requirements

9. Corrective/preventive-action requirements

10. Quality target requirements

11. Defect reporting and analysis requirements

12. Supplier quality requirements

By exploring the contents of each element of the policy and adapting it to your own case, it should be fairly easy to come up with the start of a quality plan. A discussion of each follows.

General Inspection and Test Requirements

- All critical and functional product specifications shall be functionally tested before the product leaves its assembly area, unless this is determined to be economically unfeasible and other feasible options are available.

- All assemblies will be function tested prior to release for shipment from their points of origin unless determined to be economically unfeasible.

- Customer configuration will be tested or inspected prior to shipment.

- Critical components will be obtained from certified or approved suppliers wherever possible. Critical components obtained from noncertified suppliers will be 100 percent inspected/tested unless it is determined to be more cost effective to detect and correct component failures in the production process.

Planning Requirements

- Flow diagrams shall be used for production, inspection, and test planning. Flow diagrams create a picture of the entire process, complete with decision points. The specific format used isn't critical but it should be consistent throughout your organization.

- A precedence diagram shall be utilized for all inspection and test requirements to determine the last opportunity to perform the inspection check and the optimum point at which to design the inspection check in the process.

- Part drawings, specifications, and other documentation used to define the product and the quality requirements shall be referenced for inspection and tested prior to shipment.

- All customer installation and operational requirements shall be inspected and tested prior to shipment.

- Inspection check (tally) sheet and test check (tally) sheets (or an electronic equivalent) which contain the respective inspection and test elements will be used at inspection/test stations.

- Detailed instructions or procedures shall be provided for each element of the checklists.

- Each instruction or procedure shall specify measurement techniques, tools, gages, workmanship standards, and related engineering specifications. Information provided to the inspectors should provide for repeatable and reproducible performance of all inspection and test steps. This information should be as complete as possible so that the user does not have to waste time investigating the actual requirements. These instructions should be the basis for training new employees.

- Instructions should include rework, repair, and special operations that may cause deviation from the normal process.

- Inspection/test elements should be based on marketing definition of customer requirements, engineering determines critical characteristics and the potential impact on customer's satisfaction. This is based on product knowledge and historical data from production and QA.

- The next-higher assembly drawing must be used in developing the inspection instructions.

Inspection and Test Operations Requirements

- Test procedures and inspection instruction, including inspection aids, print quality standards, workmanship standards, and/or minimum acceptance samples will be available at each inspection station and appropriate test operation. Documentation should be maintained and controlled so that only up-to-date information is used.

- Critical elements, or program elements, will be inspected 100 percent of the time. These are characteristics that will result in failure or perceived defects as viewed by the customer.

- Noncritical elements, or random elements, will be audited or inspected randomly, according to a planned strategy. This strategy will include which random elements will be inspected, from what sources, how frequently, and when this should be revised.

- Evidence of completion of inspection or test will be recorded on shop travelers or by similar means to indicate the inspection/test status of all products in the process. This will be done for new build as well as repairs or rework, as applicable.

- Inspectors and testers will not be permitted to inspect at an applicable inspection station unless they have been certified. Certification means they have passed an appropriate written test, have been approved by a process audit, and have passed an inspection capability study. The inspector/tester is responsible for all program elements and respective random elements assigned to the inspection station.

- All gages and instruments used for measuring critical characteristics shall be calibrated to national standards and shall have been determined acceptable by an inspection capability study. The procedure for inspection capability studies is described in the core techniques chapter, Inspection capability which is intended to ensure that measurement system error is minimized.

- No inspection or test shall be performed without complete, updated instructions, gage procedures, and specifications. All operators shall be certified. A certification system should include as requirements a written test, a process audit, a product audit, inspection capability study, and process capability study.

- All critical characteristics shall have a process control chart, an events log, a corrective/preventive-action matrix and a process control procedure. The most commonly asked question regarding SPC is "Do we have to have control charts for the right measurements?" In general, the news is good—chart only those things that affect critical characteristics or influence the process variation or critical input variables to the process.

- All critical characteristics shall have a capability index (Cpk) of a minimum of 1.3. This means that the product and process have been designed to stay in control.

Test Functions

- Test engineering or the equivalent organizational unit needs to understand the product design and should conduct an ongoing effort to evaluate and refine testing based on yield and defect data. This may not apply to nonmanufacturing or to service organizations. Testing should be automated to the greatest extent possible and computers used in automated test should be self-testing. The schedule for self-test should be called out in the test instructions.

- Media used for testing (cassettes, diskettes, or firmware) should be identified and controlled through a formal release system. All test software must be tried out and have evidence displayed that it performed successfully. Approvals of software validation shall be documented. A formal system shall be maintained for issuance and training for new or revised software.

- All test equipment must have a test capability study performed and must be found acceptable. All unplanned adjustments, repair or replacement parts, maintenance, or changes done at a workstation shall be recorded in the station's events log.

- Any defect observed during the running of a test shall be recorded and the test continued until it is unfeasible to continue.

■ Tests, e.g., functional test or bed of nails, should have an evaluation of the competency of the test. A competency of the test is to determine if each component or function is tested for. For example, if a printed circuit board assembly has 200 components, each of the components should have an adequate test applied to their specifications and the functions of the board.

■ If a product is tested and a failure occurs, the cause is determined, hence a repair, adjustment or component is replaced, the test starts from ground zero in lieu of continuing in order to pick up interactions which could cause failure on the modes tested prior to the failure.

Classification of Characteristics. The classification of characteristics should be provided on all inspection instructions.

Not all attributes or characteristics addressed during inspection or test have the same importance in terms of impact on performance or likelihood of escaping detection. It is, therefore, a common practice to classify defects (checklist elements) using the following criteria:

1. Seriousness of impact

   Critical—Will cause the product to fail may cause personal injury

   Major—May cause the product to fail or require a repair service

   Minor—Not likely to cause product failure, but indicates process problems when found in high frequency

   Incidental—Cosmetic or nonfunctional defect

2. Cause or source of defect

   Process deficiency—Fixtures/tools/equipment/methods

   Workmanship—Operator induced

   Component or material—Supplier induced

   Design—Design induced, form, fit, or function

Repair Requirements

■ All defects or errors shall be placed either in a bin or location labeled "nonconforming material," or tagged as such.

■ All nonconforming material found in inspection and test stations shall be repaired, adjusted, or replaced.

■ The repairs shall be reinspected along with any interactions. Interactions shall be identified on the inspection instructions.

■ Evidence of accepted reinspection is required to pass through the subject inspection station.

- All repairs will be performed by qualified personnel (trained and process audited) using formal methods and proper tools.

- All repairs, adjustments, and parts replaced will be documented, i.e., parts replaced, details of the adjustments, other repair actions, date, and repair person.

Calibration

- All gaging tools and instruments will have a schedule frequency for recalibration.

- Gaging, tools, and instruments are required to be calibrated and shall be maintained within calibrations and identified as such. Product tested or inspected with equipment that is out of calibration shall be purged, placed on hold, and retested or inspected after the equipment calibration has been verified.

- Damaged equipment or equipment that is out of calibration shall be submitted to the appropriate department for verification.

- An audit shall be conducted monthly to assure compliance to the audit schedule.

Tryout Requirements

- All new manufacturing methods, test procedures, and inspection instructions or revisions to documentation will have an acceptable tryout of a minimum of 30 pieces and a precontrol inspection sample prior to start of production. The tryout will be performed by a production operator, tester, or inspector, as appropriate. The manufacturing engineer, quality engineer, production supervisor, and inspection supervisor will approve and sign off on the methods and instructions.

- All new inspection stations and production operations shall have a successful tryout prior to start-up. Inspection and process capability studies are recommended.

- Tryouts, first article inspections, and capability studies shall be documented and approved by the appropriate team.

- All tryouts will have precontrol chart or SPC chart established.

Corrective/Preventive-Action Requirements

- A corrective/preventive-action matrix (discussed in Implementation Track) will be available at each inspection or test station or process control point. This serves as a training aid for new operators and ensures that mistakes aren't repeated. It will empower the operators to take the appropriate action to produced an acceptable product.

■ The production station causing defects shall be identified on the corrective/preventive-action matrix.

■ Workmanship defects should be addressed to the operators, inspectors, and testers within eight working hours. Operators, inspectors, and testers should have authority to take corrective or preventive action whenever possible and should assume "ownership" of product quality. When appropriate, defect cause and corrective/preventive action should be identified to specific individuals.

■ Any defect or inspection leakage that repeats its occurrences when reviewed on the corrective/preventive-action matrix shall have a diagnostic process audit performed on the operator and inspector, respectively.

Quality Target Requirements

■ All production operations, test operations, and inspection stations will have quality targets which will be reviewed weekly. Quality targets will be determined according to requirements for outgoing quality levels and results of initial process capability studies.

■ Achievement of a target for two or three weeks will result in a 10 percent challenge (improvement) to the target. Missing the target will result in application of appropriate corrective/preventive action, but will not result in relaxation of the target.

■ All production operations, test operations, and inspection stations exceeding their targets will have a process audit conducted to verify conformance with documented procedures.

■ A recognition and rewards program should be developed and implemented in conjunction with the quality targets.

**Defect (Nonconforming Product) Reporting
and Analysis Requirements**

■ Quality records will be maintained for all process operations, tests, and inspections. These quality records should include a defect history with the defect description and frequency. The defects will be listed first as a noun, then as an adjective, i.e., nuts, wrong; nuts, loose; nuts, missing.

■ Defect records should also be separated according to two categories: first pass and reinspection.

■ All defects should be analyzed by source using Pareto analysis methods, as follows:

By station

By element

By operator responsible

By nature of defect, i.e., visual, loose, etc.

By inspector observing the defect

■ All defects should be analyzed by source by type of defect, as follows:

Workmanship—Related to the operator

Process—Related to the inputs variables of the process, i.e., set points and tolerances (unintentional errors, methods errors, and intentional errors)

Component/material—Related to the material of the product itself

Design—Related to the design of the product and its reliability

Analysis of field data for defects will be performed on a regular basis and reported to management.

Selection of Characteristics

A second consideration for quality planning is selection of characteristics that are to be inspected, measured, or tested. Not every physical or functional characteristic will require testing inspected 100 percent of the time, but some will. The weight of a television set, for example, would probably not be important once the design had been qualified and the first article had been weighed. If, however, the TV set was for use in the space shuttle, weight might be very important and might have to be measured down to the milligram.

The general approach in selecting characteristics should take the following into consideration:

Product and process specifications from design engineering

Test environments that simulate customers' installation and use

Customer modifications and unique configurations

Applicable industry or regulatory standards

Characteristics which are important to the assembly function and are visible during the assembly process

Characteristics or components which could be damaged through improper handling and processing

Identification of characteristics with critical dimensions or tight tolerances and consideration of any special gaging that may be required

Possible safety hazards

Characteristics affecting interchangeability and function at the next-higher assembly level

Interaction with other characteristics when defects are repaired

Specification Interaction. Selecting characteristics starts with reviewing the specification to clarify the notes and terminology between the designer and manufacturing personnel.

Supplemental information may be required if the specification is vague or has incomplete or missing items.

Requirements will be vague if there is a lack of a product or design specification. When new technology is used, not all requirements may be understood. A formal design review will greatly increase the understanding of the requirements and minimize vagueness.

Vague requirements can be clarified by incorporating workmanship standards, photographs, sketches of minimal acceptable samples in the inspection instruction, and manufacturing method.

Criticality Classification. Once the specification is reviewed and interpreted, the characteristics are classified as critical, major, minor, or incidental to reflect their criticality or severity. A defective characteristic that causes the product to fail or not operate at all requires more attention in the quality planning process. Other defectives have no affect at all on the function of the product but still may cause a customer to not accept the product as fit for use. With these concepts in mind the following definitions are offered for severity of defects:

Critical Defects

Will cause personal injury or illness

Will render the product totally unfit for use

Will cause the product to malfunction at subsequent assembly, final line, or customer, and is not self-purging

Will result in high repair cost at customer's site

Will require product replacement

Will result in losses to the business greater than the value of the product

Will not fit in next assembly

Fails to conform to regulations for purity, toxicity, and identification

Note that in many cases only critical defects are considered when establishing statistical process controls.

Major Defects

Unlikely to cause personal injury or illness

May render the product unfit for use and result in rejection by user

May result in significant cost to repair

Likely to be noticed by the customer and may reduce product salability

May require product replacement

May result in losses to the business greater than the value of the product

Will cause fit problems in next assembly

Will substantially reduce production yields

Critical defects that will self-purge at next point of assembly

Fails to conform to government regulations or classified as major defect by customer

Minor Defects

Will not cause personal injury or illness

Unlikely to render product fit for use

Will make the product more difficult to use, e.g., requires removal from the package or requires improvisation by the user to make work

Unlikely to require product replacement

Unlikely to result in significant cost to the business or loss of customers

Will not effect next assembly

Minor nonconformance to government regulations or classified by customer as defect

Incidental Defect

Will not cause injury or illness

Will not affect usability of the product, may affect appearance or neatness

Unlikely to be noticed by the customer and of little concern if noticed

Unlikely to result in loss

Conforms fully to all regulations

Consider parts history when selecting the classification for each defect. Any defects that are known symptoms of nonconformance during the manufacturing process should be given special attention. Also consider field failures that occur during installation and customer use.

A specification on a characteristic such as a diameter yields two defects: oversize and undersize. The amount by which the diameter is over or under may affect whether it is a critical, major, or minor defect.

The process of classifying characteristics may be a long and tedious task, but it will identify misconceptions or confusion between various departments. Most companies will be able to generate a set of standard or generic defect classifications that apply to some or all of their product lines so that the engineering time required is minimized. A miswired or damaged power cord, for example, is almost always a critical defect for appliance or electronic device manufacturers because it represents a safety hazard. Over time, it may be possible to generate a "defect dictionary" that classifies the majority of defects regardless of the specific product or part. Some industry workmanship standards are available that will aid in this task.

For unique requirements, the classification process should be prepared by an cross-functional team. This team will determine how many classes will be used, define each class in more detail to fit the specific product, and classify each characteristic.

Some of the problems or pitfalls encountered when implementing the classification of characteristics are:

- Designers and engineers are reluctant to be involved, and express the opinion that defect classification is the exclusive responsibility of the quality control department.

- Participants are unaware of the benefits to their departments and the company. Dimensional tolerances are assigned without regard to actual use or manufacturability (too tight or too loose), or classification of defects is based solely on tolerance.

- Oversimplifying defect classification with the meaningless, timeworn bromide, "form, fit, or function."

- The company takes the default position that all characteristics are critical. This can lead to wildly excessive costs of quality and overinspection. Even the notion that all defects are critical can create problems.

Manufacturing Method.   Once the specification has been reviewed and the characteristics are classified, the manufacturing method should be analyzed to ensure that factors affecting manufacturability and productivity have been addressed. Defects not thought of as critical to a tool designer may cause excessive rework and repair in production. Evaluation of the manufacturing method should include the following questions:

- Is the process capable of meeting the specification?

- Does the process contain the necessary control in ensure conformance to quality standards?

- Is the sequence of assembly operations logical?

- Have critical parts or dimensional settings been protected properly during the manufacturing operations?

- Does the assembly process allow the opportunity for inspection at strategic points?

- Are tools and equipment adequate for the process?
- Can assembly operations be automated to minimize defects?
- Are packaging, handling, and storage methods adequate to prevent product damage or deterioration through the production process?

Defect Interaction and Reinspection.   When defects are found during the assembly process, they are reworked or repaired. An item often overlooked is the interaction of the original defect and the rework on other parts or subassemblies. Rework action may result in new defects in the reworked component or may precipitate new defects elsewhere in the product. Excessive or unnecessary rework may reduce the reliability of the products by weakening another component.

For example, if you were building a house and discovered a loose floor board you might decide to repair it by hammering it down with a bigger nail. The larger nail (or improper workmanship) might cause the board to split or the nail might break through into the ceiling of the room below.

Rework should be done in a manner that brings the product back into strict conformance with the original requirements and steps should be taken to verify conformance after rework. Verification should include all characteristics that could have been impacted by the original defect.

In some cases, risk of damage due to the rework action should be considered during the defect classification process.

Program Elements.   When characteristics have been selected and classified, the amount of inspection should be determined. Identifying each characteristic as a *program* or *random* element will assist in this determination. The program elements are defined as follows:

- Program elements are characteristics that are tested or inspected on 100 percent of the products or parts going through a station.
- Program elements usually include critical characteristics; high frequency of defects, defects identified by customers; defects that occurred in subsequent operations (internal customers); or key dimensions which correlated to one or more other dimensions which are not measured.

Random Elements.   Random elements are characteristics that are tested or inspected on a portion or sample of the parts of product processed through the station. Typical sample rates might be 1 in 3 or 1 in 5.

Random elements usually include major and minor characteristics, low frequency of defects, or defects with minimal impact on subsequent operations or customers.

Characteristics may change from program to random or from random to program depending on new data collected from the process or from the field. A change from a program to random may occur if a previously high frequency of defects should become a low defect rate. The opposite would occur if a low-frequency defect rate suddenly shows a high defect rate.

Defects identified by customers or detected in subsequent operations should be changed to program elements.

Characteristics that are program elements may be changed to random elements if the item has been proven from verifiable data to have minimal impact on subsequent operations or on customers.

Direct/Indirect Defects. The elements or characteristics being checked on an inspection checklist are either related to the cause of a defect or a symptom resulting from the cause. For example, belt tension may cause poor print quality in a printer, but poor print quality observed in the printer may result from many causes, with loose belt tension being only one of the causes.

Causes often result in one symptom, while a symptom may have more than one cause associated with it.

If an inspection checklist calls for verification of an element that is not checked again as a symptom at another test or inspection point, any undetected defects will pass directly to the customer. This is referred to as a *direct* element. Failure to catch direct elements is called *escape* or *leakage of defects.*

If an inspection element is verified again as a symptom at another station (or reverified as a cause), then any defect which is undetected and escapes from the first inspections point should be found at another station; it does not go to the customer. These defects result in rework and are referred to as *indirect* elements.

When selecting characteristics to check in major subassemblies or final assemblies, the direct and indirect elements must be identified. Characterization of direct or indirect elements will have a major impact on determination of quality targets at each station.

Configuration Control. Configuration control is concerned with meeting customer requirements, interchangeability, and revision-level control of the product. The quality planning process must make provision to ensure that the product will function properly in its delivered or installed configuration.

It may be necessary to test customer configuration functionally, rather than just to visually inspect components or subassemblies. Some companies perform "installability audits" which require complete installation and activation of the product by personnel not directly involved in manufacturing. Findings from these audits are fed back to engineering and manufacturing for corrective/preventive action through a corrective action system.

The implementation of configuration control deals with the coordination of purchasing, stores, production, manufacturing engineering, quality engineering, and inspection to ensure the effective date of the cut-in of the new configuration.

Type and Amount of Inspection

Quality planning is also concerned with determination of how much inspection should be done at each inspection station. Subjects that will be discussed include:

Inspection strategies

Amount of inspection

Inspection economics

Sampling inspection concepts

Continuous sampling

Inspection Strategies

Inspection strategies involve deciding which type of inspection is appropriate at each stage of the process. Various inspection strategies include operator inspection, product audit, tollgate inspection, automated inspection, and 100 percent on-line inspection. The new paradigms of quality assurance in Japan and the United States focus more on design for manufacturability, process control, and product audits and less on traditional tollgate inspection where 100 percent of the product is subject to verification before it moves to the next station. Automated inspection holds the most promise because of speed and potentially high effectiveness, but automation of any process requires some capital investment.

Operator Inspection. Operator inspection is performed by an operator at the machine or workstation. The operator is fixed, and is therefore in the best position to control quality, provided:

There is sufficient gaging.

The operator is trained on the quality standards required.

There is sufficient time allowed within the time standard to perform the inspection.

Operator inspection may also include verification of work done at previous stations. With this strategy, all product is inspected for specified characteristics.

In-Process Inspection.   An in-process inspector patrols from machine to machine or station to station at prescribed units of time to evaluate the current status of quality using a given inspection procedure.
In-process inspection is desirable when:

- The process has a low defect rate and is in good control with minimal operator inspection.
- The process has a highly erratic defect rate and there is a need to supplement the operator's inspection.
- There are new or untrained operators on the job.

In-process inspection is less desirable when:

- It is difficult to achieve desired quality levels unless large sample sizes are taken.
- The inspector must carry gages from place to place.
- The inspector is diverted to a troublesome process and other processes are ignored.

Tollgate Inspection.   Tollgate inspection is an in-process sampling station established in a fixed location for the purpose of monitoring the quality level of products produced in a given department prior to processing to the next.
Tollgate inspection is desirable when:

- The process has a high defect rate.
- There is a need to enforce specific quality levels between departments.
- Multiple operations requiring inspection are in the same department.
- Records depicting the quality performance of a given department are needed.

Tollgate inspection is less desirable when:

- It duplicates inspection and increases costs for a process that is operating at a low defect rate.

- It results in an increase in production flow time.
- The process is a continuous flow of product.
- The rejection rate is low.

Automated Inspection

- Automated inspection is designed to perform inspection operations where human inspection is not accurate or sufficient for a rapid production flow.

- Automated inspection is used extensively in the mechanical, chemical processing, and electronic industries where human inspection would not be possible.

- Automated inspection is desirable when:

  Human inspection is not feasible, such as for electronic parameters of electronic components produced in high volume or in strict clean-room environments.

  It is desirable to improve precision and reduce inspection monotony and automatic recording of data is desired. This can or cannot be linked to SPC charts.

- Automated inspection is less desirable when:

  Cost of equipment exceeds the current operation costs.

  Extensive product redesign is necessary to use the automatic test equipment.

  The current method of inspection has an 'acceptable' inspection capability study among all inspectors and instruments.

100 Percent On-Line Inspection. 100 percent on-line inspection is performed in the production line to sort the good from the bad product and allow only the good parts to proceed to further processing operations. Certain inefficiencies will be built into any inspection method, and therefore 100 percent inspection may not catch all of the defects. For this reason some companies will resort to 200 or 300 percent inspection and so on, meaning that all product is inspected multiple times.

100 percent on-line inspection is desirable when:

There is a need for quick feedback of defect information for corrective action in processes with high defect rates.

It is desired to prevent defects from subsequent operations where the repair of the defect is very costly.

100 percent on-line inspection is less desirable when:

It becomes redundant and routine.

When inspection capability studies reflect marginal or unacceptable findings.

Selection of the best strategy to use for inspection will depend on each specific application. Engineers responsible for quality planning should evaluate costs and effectiveness of the various methods in their particular situations and should conduct inspection capability studies to test the adequacy of their plans. Factors influencing inspection strategy include cost, production volume, defect rates, complexity of the inspection operation, and the type of characteristics to be evaluated.

The strategy considered best today may not be best a year from now and frequent evaluation is highly recommended.

Amount of Inspection. The amount of inspection to be performed at any station can vary from no inspection to sampling inspection to 100 percent inspection. Sampling inspection can be further divided into large and small samples.

No inspection means exactly that—inspection may be suspended or waived when evidence indicates that product quality targets have been met or exceeded.

Inspection/Test Instructions

Inspection/test instructions are to inspectors as manufacturing methods are to production operators. Inspection/test instructions provide a vehicle to implement repeatable inspection procedures and provide traceability, continuity, and communication.

The achievement goals of inspection/test instructions are a function of:

Clearly communicated Accept criteria

The identification of nonconformaties and malfunctions in the earliest stages of production in lieu of the field

Level of inspection training and experience required to utilize them

The contents of inspection/test instructions should include the following:

List of characteristics to be checked

Method of inspection (visual, gage, etc.)

Instruments to be used

Seriousness classification of characteristics

Tolerances and other piece-part criteria. (In some cases information on drawings will be complete or will be obscured by noncritical details. Inspection instructions should call out the specific requirements for critical characteristics.)

List of applicable standards

Sequence of inspection operations

Sample sizes and acceptance criteria

Data recording forms and directions for filling out form

Any visual characteristics should have pictures or samples available of acceptable and rejectable parts to provide consistent inspection. Reference to standards, such as workmanship standards for printed circuit boards, simplify the writing of instructions for visual characteristics.

When developing an assembly inspection instruction, understand the function of the assembly and consider the following questions:

What is the purpose of each part making up the assembly?

How do these parts interact to perform the required function?

What internal adjustments are required so that the assembly performs its function?

How does the assembly interface with the next-higher assembly?

Inspection/Test Instruction Tryout. Inspection/test instructions should be implemented using a "tryout" procedure. This tryout procedure is similar to a first-piece inspection on a part. The purpose of the tryout is to ensure the instructions, sequence, instruments, gates, etc. are appropriate for the purpose of the inspection/test instruction. The tryout is done by the inspector and is audited by the quality or manufacturing engineer and the inspection supervisor.

Immediate feedback from the inspectors and the inspection supervisor will help integrate the new procedure into the work flow and operating methods in the production area. Immediate corrective action should be taken to resolve deficiencies at the operating level.

Results of the tryout should be reported to affected management. The reports should summarize observations made during the tryout and should list all deficiencies noted with recommended corrective/preventive action.

It is strongly recommended (and mandatory for compliance with quality standards such as ISO 9000) that the completed inspection/test instructions be controlled by revision, date, or other method defined by the company that will ensure consistency and proper response to change in the product or requests for corrective action. All obsolete documents should be removed from the inspection station.

Receiving Inspection Instruction Procedure

1. Purpose
    a. To provide instructions for preparation of receiving inspection instructions.
    b. To ensure a uniform approach to the preparation of receiving inspection instructions in an effort to minimize problems of interpretation and to provide the most effective inspection possible.
    c. To provide appropriate information to the supplier for clarity, classification, and technique of inspection.
2. Scope
    a. All assemblies and parts purchased at suppliers and fabrication plants.
3. Definitions
    a. *Receiving Inspection Instruction (RII).* A document containing explicit instructions on which characteristics of a parts and/or assembly are inspected and how the inspection is to be accomplished. Form RIIs (see Exhibit I, which follows).
    b. *Drawing Change Order (DCO).* A document used to originate or change assembly drawing which may result in initiating or revising manufacturing methods.
4. General
    a. The RII document fits into the overall fabrication and processes and reflects the action to be taken at each inspection station. Once the document is completed, the RII master is kept on file in the document center and is available on request. This document is located at each receiving inspection location.
    b. If sampling plans are applicable, they will be developed in accordance with and included as backup sheets to the RII and reference in the appropriate inspection elements. If no sampling plan is referenced, the inspection level will be 100 percent or patrol as noted in Item 7 of the RII.
5. Requirements and Responsibilities: Preparation and Revision of Receiving Inspection Instructions
    a. Receipt of documentation
        (1) Issues DCO to supplier quality engineering supervisor who distributes DCO to appropriate supplier quality engineer (SQE).
        (a) *Responsibility:* Document center

    (2) Provides copy of drawings affected by the DCO and the drawing of the next-higher assembly number to the supplier quality engineer upon request.

        (*a*) *Responsibility:* Document center

  b. Prepares RII for new issues and revisions

    (3) Reviews specification and drawings for part or assembly affected by DCO and the next-higher assembly drawing.

        (*a*) *Responsibility:* Supplier quality engineer

    (4) Determines inspection characteristic, classification of characteristics, and inspection requirements with quality engineer of affected part assembly.

        (*a*) *Responsibility:* Supplier quality engineer

    (5) Completes the header information on the RII in accordance with the instructions in Exhibit I (shown subsequently) of this procedure.

        (*a*) *Responsibility:* Supplier quality engineer

    (6) Prepares the text of the RII in accordance with the guidelines in Exhibit II (shown subsequently) of this procedure.

        (*a*) *Responsibility:* Supplier quality engineer

    (7) Incorporates revisions to existing RIIs by either marking up a copy in *red*, or initiating a new RII. Changes made in red on receiving inspection area copy are to be initialed or signed by responsible engineer only. These changes are valid for 30 days.

        (*a*) *Responsibility:* Supplier quality engineer

    (8) Initiates tool authorization if required in accordance with the following.

        (*a*) Forward one copy to tool control.

        (*b*) File one copy in follow-up file.

        (*c*) *Responsibility:* Supplier quality engineer

    (9) Maintains a central file of process documentation, including RII.

        (*a*) *Responsibility:* Document center

    (10) Maintains a central file of RIIs.

        (*a*) *Responsibility:* Receiving-inspection supervisor

6. Release and Issue Cycle of Receiving Inspection Instruction (RII)

  a. Completion of RII

    (1) Complete the RII within the time frame stipulated.

        (*a*) Revisions and method improvements—3 days or as required.

        (*b*) New issues—5 days.

    (2) Date and sign the RII.

    (3) Make and retain copy of the draft of RII until typed version has been issued.

(4) Return the RII to document control.

    (*a*) *Responsibility:* Supplier quality engineer

  b. Completion of RII cycle

    (1) Process the RII through the documentation center to the appropriate receiving inspection location.

    (*a*) *Responsibility:* Document control

7. Verification of RII

  a. Provide interpretation and training to ensure understanding of RII by using personnel.

    (1) *Responsibility:* Supplier quality engineer

  b. Evaluate RII by participating in "tryout" according to procedure.

    (1) *Responsibility:* Supplier quality engineer

  c. Evaluate RII by conducting process audit studies according to procedure.

    (1) *Responsibility:* Supplier quality engineer

  d. Ensure correct utilization of RIIs by receiving inspectors. Notify supplier quality engineer when problems or incompatibilities occur on the RII.

    (1) *Responsibility:* Receiving inspection supervisor

8. Interim Update of RII in Receiving Inspection

  a. Identification of condition which required departure from RII pending completion of APCN. Usually this is when a supplier submits a variation and it is approved by MRB.

    (1) Prepare a variation in accordance with procedure number and attach to floor copy of RII.

    (2) Mark up (*in red*) instruction section of RII with the variation number to document reference to the variation log.

    (3) Forward one copy of variation to the quality control supervisor of affected area.

    (*a*) *Responsibility:* Quality engineer

9. Revision-Level Control of Inspection Visual Aids

  a. Make necessary changes to visual aid.

  b. Identify the visual aid with the following information.

    (1) Part number (same as RII or variation)

    (2) Originator (responsible engineer's name)

    (3) Date (same as RII or variation)

    (4) Revision level (same as RII or variation)

  c. Visual aids will be referred to on the inspection sketch sheet, (item 32 of Exhibit I).

Pictures and sketches, if possible, would be afixed or added as an addendum to the inspection sketch sheet.

    (1) *Responsibility:* Quality engineer

Exhibit I: Instructions for Completing RII Form

1. *Designed area.* In the upper left-hand corner of the II mark an (X) in the receiving box. This indicates the II is designated for receiving inspection use.

2. *Part number.* Self-explanatory.

3. *Revision letter.* Latest.

4. *Part name.* In full as it appears on the drawing.

5. *DCO number.* Latest drawing change order number.

6. *Product code.* Self-explanatory.

7. *Operation number.* This will be determined by reviewing the corresponding assembly process and deciding the point at which inspection will occur. All inspection operations end with the number "5," e.g., 5, 15, 25, 35. . . . Adjacent to the operation number a "P" or an "I" will be placed designating the type of inspection—either patrol (P) or 100 percent inspection (I).

8. *Using department no.* This represents the department number of the inspection area in which the AII will be used.

9. *Package code.* N/A. Package specs will be included in the body of the RII.

10. *Subsystem.* Identifies the particular area of the machine where the assembly is used (next-higher assembly number).

11. *Page.* Pages to be numbered consecutively and indicate total pages (e.g., page 1 of 10).

12. *Source code.* Make/buy code.

13. *Originator.* Name of engineer originating the RII.

14. *Approval.* The appropriate supplier quality engineering supervisor will approve and sign every RII originated or revised. This encompasses reviewing for correctness of format, completeness, and the best possible modes of selected standard inspection methods and gaging requirements.

15. *Revised by.* Name of engineer revising the RII.

16. *Date.* Date of latest revision of RII.

17. *Special gage.* All special-number inspection gages will be listed in the spaces provided beginning with reference letter "A." The reference letter adjacent to the gage number will be written in the tool/gage column ref., adjacent to the corresponding instruction in which it is used.

18. *Material.* N/A.

19. *Finish.* N/A.

20. *Other material service.* N/A.

21. *Reliability elements.* N/A.

22. *Sample modifiers.* N/A.

23. *Classification of characteristics:* The engineer is responsible for assigning the classification to each item—major, minor, or incidental. Refer to Corporate Procedure as a guideline.

24. *Sequence number.* The numerical sequence assigned to each item in the inspection procedure starting with the number 1.

25. *Element no.* This is the work element number assigned to each inspection item by industrial engineering.

26. *Instruction.* This consists of the identification of which inspection checks to make and the details of how to perform the check.

27. *Print ref.* The location on the drawing of the dimension being checked. Notes on the drawing are also referenced in this space, along with the sheet number of the dimension or note for multiple-page drawings.

28. *Specification.* List the dimensions, tolerances, and requirements of the item being inspected.

29. *Tool/gage ref.* Expense manual tooling (number tools) and methods of inspection (visual, manual, etc.) are listed in this column. Also, include the reference letter of any special-purpose gages required to inspect each sequence.

30. *Interaction.* List the sequence numbers which must be rechecked if the particular item under inspection is defective.

31. *Scale.* Size of sketch in comparison to full size, stated in fractions.

32. *Sketch.* Drawing to outline or give principal features of items inspected.

Exhibit II: Guidelines for Writing an RII

1. Plan the receiving inspection instruction.
   a. Understand the function of the assembly.
      (1) What is the purpose of each part making up the assembly?
      (2) How do these parts interact to perform the required function?
      (3) What internal adjustments are required so the assembly performs its function?
      (4) How does the assembly interface with the next-higher assembly?
   b. Determine the characteristics which are to be inspected.

(1) Characteristics which are important to the assembly function and subject to visibility in the assembly process.
(2) Important characteristics which could be damaged through improper handling and processing.
(3) Critical dimensions and tolerances which would require special gaging to inspect.
(4) Characteristics which could present a possible safety hazard.
(5) Characteristics which could affect interchangeability and function at the next-higher assembly level.

c. Arrange the inspection in a logical sequence (e.g., combine all settings with same total; do critical checks first).
d. Classify each characteristic for criticality.
   (1) Critical (level 1)
      (a) Certain to cause the product to nonfunction at subsequent assembly final line or customer.
      (b) Will not fit in next assembly.
      (c) Possibly results in an unsafe condition.
      (d) Will result in high repair cost at customer's site.
      (e) Not self-purging.
   (2) Critical (level 2)
      (a) Specific electronic parts that are certain to cause the product to nonfunction (i.e., ICs, PCBs, memory devices).
      (b) Based on complexity of product (number of devices in product that are critical).
   (3) Major
      (a) Likely to reduce the usability of the product for its intended function.
      (b) May cause fit problems in next assembly.
      (c) Defects that are critical but are self-purging at next point of assembly.
      (d) May require significant cost to repair.
      (e) Customer may have reasonable cause to complain about appearance.
   (4) Minor
      (a) Not likely to have an important effect on the product's function or useful life.
      (b) Will not affect next assembly.
      (c) Low cost to repair.
      (d) Slight appearance defects.
      (e) Substandard practice.

e. Determine the interaction of characteristics and determine how each item in the assembly process would affect the other items, dimensions, etc., it if were defective and repaired.

f. Determine the gaging requirements.
 (1) Is gaging required? Evaluate the time, complexity, cost, and accuracy of alternate methods to make this decision.
 (2) Can one gage be used to check more than one dimension?
 (3) Keep gages as simple as possible in order to perform a maximum number of checks. Remember that weight, size, and complexity will result in the gage being reluctantly accepted by assembly inspectors.
 (4) If assembly tooling and gaging are required for the same assembly specifications, tool and gage design should be compatible.
 (5) Provide direction for indicating acceptance or rejection of the parts/assemblies, e.g., inspection stamp/location and/or tags.
 (6) Surface plate setup.
2. The RII should be printed in a clear, simplified language; use sketches wherever possible to clarify instructions. Reference Quality Workmanship Standards (QWS) where applicable.
3. Guidelines for completing RII format. Refer to the preceding Exhibit I.

Location of Inspection/Test Stations

This section provides guidelines for selecting the location of inspection or test stations. The following topics will be covered:

- Typical locations
- Basic tools
- Location considerations

Typical Locations. The typical locations for inspection are as follows.

Receiving Inspection. Inspection occurs at receipt of material or parts from a supplier or supporting plants.

Setup. Inspection occurs when the process is initially set up. This is referred to as "setup approval" and occurs following the setup of a production process with final or "hard" tooling.

Fixed Point. In fixed-point inspection the product is brought to a central inspection point. This central point may or may not be near the production area. This is illustrated in Figure 9-3.

Patrol Inspection. Patrol inspection occurs when inspectors roam around or patrol an area covering several production operations. This may be called "inspection by walking around."

INSPECTION INSTRUCTION TRYOUT
CHECKLIST

PLANT: _____   DATE: _____

PRODUCT: _____   OPERATOR: _____

STATION: _____   AUDITOR: _____

DOCUMENTATION

ELEM. NO.	ELEMENT DESCRIPTION	ACCEPT.	MARG.	UN-ACCEPT-ABLE	FINDINGS (Pos./Neg. Comments)
1	Are necessary support documents (e.g. Variations, "workmanship manual", photo's sketches or minimum acceptable samples, etc.) Present?				
2	Are the documents complete, clear, correct and in proper sequence?				
3	What is the latest DCO revision?				
4	Is document revision status correct?				
5	Do the documents reflect the master copy? (No unauthorized changes)				
6	Is required documentation of work accomplishments maintained i.e. Traveler, defect reports, operator sign off, data sheet, etc.?				
7	Are all documents legible? Easily understood?				
8	Are interactions included for defects to be reinspected?				
9	Does the operator comply to the inspection instruction?				

GAGES & MEASURING/TEST EQUIPMENT

ELEM. NO.	ELEMENT DESCRIPTION	ACCEPT.	MARG.	UN-ACCEPT-ABLE	FINDINGS (Pos./Neg. Comments)
1	Are the required gages and measuring/test equipment available at the station?				
2	Are gages and measuring/test equipment properly utilized as described in the instruction?				
3	Is calibration required? What is last date of calibration? Is this date current? Is equipment properly labeled?				

Figure 9-1.

INSPECTION INSTRUCTION TRYOUT

CORRECTIVE/PREVENTIVE ACTION SHEET

TO: _____   DATE:_____

PRODUCT: _____

STATION/ASSY: _____

STATEMENT OF PROBLEM:

QUALITY ENGINEER: _____

REMARKS:

BELOW TO BE COMPLETED BY RESPONSIBLE INDIVIDUAL

Corrective/Preventive Action	Responsibility	Target Date	Completion Date

Figure 9-2.

In-Process. The in-process inspection is performed between operations. This may vary from "after every operation" to "at the end of all operations." The various arrangements are shown in Figures 9-4, 9-5, and 9-6. (The most typical configuration is the combination shown in Figure 9-6.) Patrol in-process inspections are normally sampling inspections in lieu of 100 percent inspections.

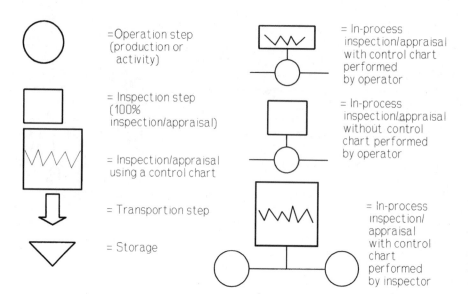

Figure 9-3. Standard symbols used in flowcharting.

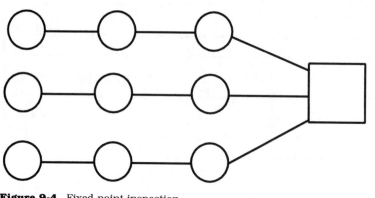

Figure 9-4. Fixed-point inspection.

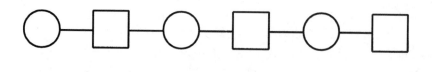

⭕ Production process   ⬜ Inspection

Figure 9-5. In-process inspection performed by inspector after every operation.

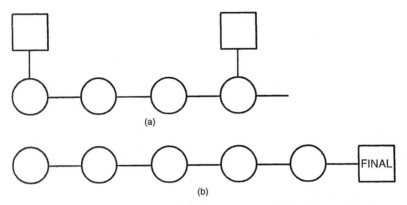

Figure 9-6. (*a*) In-process inspection by operators; (*b*) final inspection is performed at the end of all operations.

In-line or in-process inspection points are usually used with process control. A central inspection point may be dictated by material flow or size restrictions. In-process points should be used for optimum process control. In-process inspections can be performed by operators or inspectors.

Tollgate. Tollgate inspection is the inspection of product that is usually collected in lots before it moves to the next department or plant. Tollgate inspection is usually in a fixed location and all lots pass through the inspection area before being transferred to their next department.

Final Inspection. Final inspection is the inspection of completed product prior to shipping to stores or customers.

Basic Tools. To select the inspection station locations, a flowchart or a precedence diagram should be drawn. Once these are drawn, the manufacturing method and drawings of the parts/components assembly and next-higher assembly are valuable aids for selecting the location of inspection or test stations. The flowchart and the precedence diagram will be described here.

Process Flowchart. The flowchart shows the processes that collectively or sequentially produce the final product. A process is any combination of machines, tools, methods, and people that attain the qualities desired for the products. The usual symbols used to construct flowcharts are those shown in Figure 9-7.

The principles of flowcharting are illustrated in Figures 9-8 to 9-11.

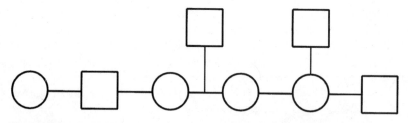

Figure 9-7. Combination in-process inspection, patrol inspection by inspector, in process inspection by operator, and final inspection.

Figure 9-8. Flowchart of a single operation followed by a 100 percent inspection operation.

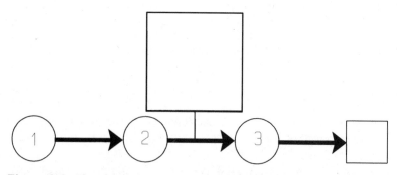

Figure 9-9. Flowchart showing an in-process inspection by inspector at the second of three operations followed by 100 percent inspection.

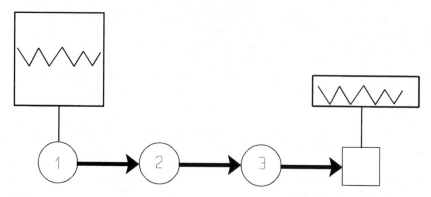

Figure 9-10. Flowchart of an in-process inspection by operator using a control chart at the first of three operations with 100 percent inspection using a control chart after the third operation.

Precedence Diagram.   The precedence diagram can be used to indicate the last possible operation at which a characteristic is able to be inspected. Figure 9-12 shows such a diagram for a process with three operations. Characteristic A1 is observable or testable only at operation 10; characteristics A2, A5, and A6 are last observable or testable at operation 20; and characteristics A3, A4, and A7 to A10 are accessible until the last station.

Figure 9-13 is a precedence diagram for a printer ribbon inking process, with an accompanying explanation of the operation performed by each operator.

Location Considerations.   The following are factors to consider when selecting the location of inspection stations.

- Locate before performing a costly or "irreversible" operation.

- Locate at natural "windows" in the process.

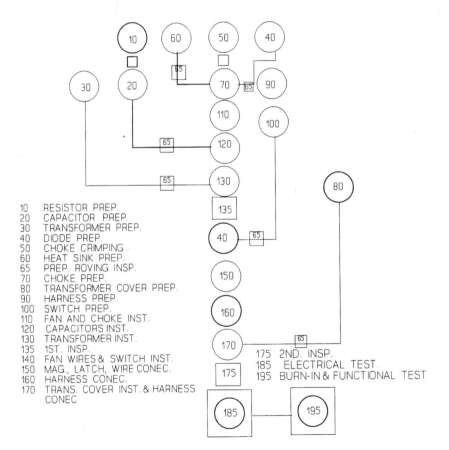

10	RESISTOR PREP.
20	CAPACITOR PREP.
30	TRANSFORMER PREP.
40	DIODE PREP.
50	CHOKE CRIMPING
60	HEAT SINK PREP.
65	PREP. ROVING INSP.
70	CHOKE PREP.
80	TRANSFORMER COVER PREP.
90	HARNESS PREP.
100	SWITCH PREP.
110	FAN AND CHOKE INST.
120	CAPACITORS INST.
130	TRANSFORMER INST.
135	1ST. INSP.
140	FAN WIRES & SWITCH INST.
150	MAG., LATCH, WIRE CONEC.
160	HARNESS CONEC.
170	TRANS. COVER INST. & HARNESS CONEC

175	2ND. INSP.
185	ELECTRICAL TEST
195	BURN-IN & FUNCTIONAL TEST

Figure 9-11. Flowchart for a PCBA power supply assembly.

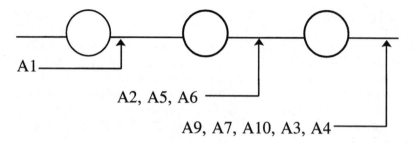

Figure 9-12. Precedence diagram for a three-operation process.

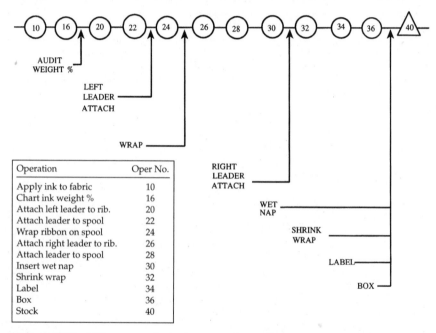

Operation	Oper No.
Apply ink to fabric	10
Chart ink weight %	16
Attach left leader to rib.	20
Attach leader to spool	22
Wrap ribbon on spool	24
Attach right leader to rib.	26
Attach leader to spool	28
Insert wet nap	30
Shrink wrap	32
Label	34
Box	36
Stock	40

Figure 9-13. Ribbon manufacturing precedence diagram.

- Locate prior to covering up a characteristic to be inspected.
- Inspect a characteristic for conformance prior to subsequent operations which may introduce interactions.
- Assembly-line inspection station work load must correspond to the cycle time of the line. If more than one inspection station is required, the work load must be distributed evenly among each inspection station.
- Inspection operations must be able to support the designed production rates of the assembly line.

The following questions will assist in making the station locations more optimum.

- Can this inspection be performed at a later point in the assembly sequence for easier accessibility and more economical gaging?
- Where will the inspection require the least amount of time to perform?
- Is it advantageous for all inspection requirements to be combined into a single operation?
- Will inspection gaging cost increase significantly if done at a later point in assembly?
- At what station can the highest degree of confidence in acceptability be attained?
- Will the cost of rework or scrap increase significantly if inspection is performed later in the process?

Production Verification Testing

Production verification testing is operating the product under actual or simulated conditions to determine if the unit will function properly when delivered to the customer.

The types of production verification tests are:

- Functional tests
- Performance tests
- Burn-in/run-in tests
- Audits

The production verification test can be automatic or manual. The basic requirements of any production test are twofold:

- Test for acceptance or rejection of the product or parts.
- Provide data on defects for process control and process improvement.

The data on defects should allow identification of the defect types and associated quantities. Providing Pareto graphs of defects and quantities periodically or on-line requests greatly improves ability to continually improve the process by preventing defects.

The major steps in development and implementation of production tests are:

- Developing test requirements
- Equipment design and planning
- Equipment procurement and construction
- Equipment installation and checkout
- Debugging equipment and operating instructions
- Inspection capability study to determine if test equipment is capable of detecting acceptable product or parts

Functional Tests

A functional test is an operational test where the product's main functions are monitored. It may be done on a 100 percent or sample basis. It should include, as a minimum, customer configuration, customer-compatible equipment, and customer modes of start-up/ongoing operation.

This type of test is often used for testing electrical items such as PCBAs or power supply assemblies and is performed on 100 percent of the product.

Often, electrical functional tests involve software to run the test equipment. When using test equipment that utilizes software and computers, consider the following:

- Computers in test functions will be self-tested and the test instructions shall call out the schedule for self-testing.

- A test competency evaluation must be conducted to verify all functions and components are capable of being tested.

- Compromises may be made on how many functions to test so that reasonable production schedules are met and testing costs are controlled. However, compromises should occur after adequate test history is recorded.

- Any cassettes/disks/tapes utilized in test functions shall be identified and controlled through a formal release system.

- All software must be tried out, have evidence displayed that it was run successfully, and show documented approvals on each revision.

- Any component or assembly replaced will require that the test sequence be reset to zero and restarted to complete the total test sequence. An alternative is to run the test sequence to the point where the risk of further failure is at a minimum.

- A formal system shall exist for issuance and training for new or revised software.

- The software should provide for accumulation of defect/error data for Pareto analysis either automatically or manually.

Performance Tests

A performance test is a complete performance check of the product's documented requirements, including:

- Product performance MTBF, MTTR, lines/minute
- Product quality
- Inspection on a 100 percent or sample basis
- Using customer configuration, customer-compatible equipment, and customer modes of start-up and operation

These tests are usually performed by a reliability test function and are referred to as *reliability demonstration test* or *life test*.

The following items must be followed for any production tests:

- All unplanned adjustments, repairs and replacement of parts will be reported on tally sheets at test.
- Any defects observed during the running of a test shall be recorded, specifying the specific time or frequency that the defect was observed; continue the test until it is unfeasible to continue.
- Any component or assembly replaced will require that the test sequence be reset to zero and restarted to complete the total test sequence. An alternative is to run the test sequence to the point where the risk of further failure is at a minimum.

Burn-in/Run-in Tests

Burn-in is a production screening technique where equipment is operated in a simulated environment at stress conditions (i.e., temperature, vibration, cycling electrical voltage) designed to cause defective units to fail within the burn-in period without causing damage to good units.

Run-in is similar to burn-in but without stress conditions of temperature, vibration, or cycling. The equipment is run continuously with possibly on and off cycling. This method is most commonly used to "burn-in" final assemblies. Due to the continuous running of the product, temperature stress is induced on the component parts.

The purposes of these tests are to:

- Detect inherent deficiencies requiring corrective action in circuit design, parts, materials, processes, or workmanship.
- Achieve inherent MTBF levels at an earlier point in time during field operation.
- Supplement other design/test activities.

The inherent deficiencies detected by burn-in are of three types:

1. *Infant mortality (early) failures.* Parts that fail very early and have failure mechanisms different from the main population failures. These early failures may be caused by various production and handling processes during assembly of the product. Such failures may be the result of "gross" mechanisms such as cracked dies or open bonds on electrical components (ICs, transistors).

2. *Freak failures.* Parts that fail very early with failure mechanisms similar to the main population of failures. These failures are from the weak parts. Very few of these occur during the burn-in or run-in test; hence the term *freak failures* is used.

3. *Design failures.* Parts that fail very early as a result of not allowing for reasonable safety margins in the design for operating environments. Certain components may operate at a higher temperature than designed for as a result of heat generated by parts nearby in the assembly.

To totally eliminate these failures, a combination of part and equipment burn-in is necessary.

Freak Failures. Freak failures may be screened out by burning in components for a long time (100 to 200 hours) in a stress environment. No amount of component screening can eliminate failures introduced by equipment used during assembly operations.

Freak failures are usually caused by a few components called the *critical components*. Once these components are identified, using a burn-in test on those components will essentially eliminate the freak failures. These critical components may be determined from prior experience, preliminary tests, or the equipment burn-in/run-in test.

The critical parts requiring burn-in tests can be burned-in by one of the following ways.

- Purchase the parts already burned-in.
- Send the parts to a test place for burn-in.
- Develop the capability to burn-in the parts within the manufacturing plant.

Infant Mortality and Design Failures. Infant mortality and design failures may be screened out by a short-term (4 to 96 hours) equipment burn-in in a normal or slightly stressed working environment. Often a run-in test is used for this.

Design of Burn-in/Run-in Tests. One of the first considerations in designing a burn-in/run-in test is the amount of time required for the test. This can be determined by the following formula.

$$T = Mf(1 + 1/2 + 1/3 + 1/4 + ... + 1/Nd)$$

where: T = Expected run-in time in hours
Mf = Total MTBF of the failed or critical components
Nd = Number of failed or critical components.

The MTBF is the Mean Time Between Failures of the failed or critical components as determined by preliminary tests. MTBF is computed from the part failure rate (FR) using the formula $1/FR$.

Example. Suppose a power supply is to be run-in before being shipped to final assembly. Early failures have occurred on five components. The MTBF of these early failed components are as follows:

Component	MTBF (hrs)
1	4
2	4
3	3
4	6
5	8

Compute the time required for a run-in test.

From the information given, $Nd = 5$ and $Mf = 25$.

$$T = 25 (1 + 1/2 + 1/3 + 1/4 + 1/5)$$

$$T = 25 (1 + 0.5 + 0.33 + 0.25 + 0.20)$$

$$T = 25 (2.28) = 57 \text{ hrs}$$

Initially, the number of critical components can be estimated along with the MTBF of each critical part. These values can be used to estimate the run-in time required at the start of testing. This will usually result in a large number of hours required. If this is larger than 96 hours, start with 96 hours. A computed initial run-in time longer than 96 hours indicates design problems that run-in cannot improve.

Following are items to consider when implementing and conducting ongoing run-in tests.

- Initiate test using the initial time calculated from the formula (96 hours maximum).
- Identify failed components along with the number of hours to failure.
- Initiate corrective action on the failed components.
- Compute the MTBF of the failed components and use in recomputing the run-in time.
- Recompute the run-in time periodically based on results of the ongoing run-in test. This should be done weekly/monthly at the start and quarterly after initial goals are achieved.
- Reduce run-in time systematically based on the ongoing run-in test results. The time can be reduced as the identified deficiencies are eliminated. If corrective action is effective, the run-in time will decrease to the range of 4 to 24 hours.
- If run-in can be eliminated, an audit type of run-in should be maintained as a minimum. This audit consists of selecting units at random and performing the run-in test.

Product Audits

Product audits are performed on finished product and are usually performed by quality or product assurance functions. The typical types of product audits are prepack or postpack, depending upon the nature of the product and the economics of repackaging.

Prepack Product Audit. A prepack audit is usually performed at the end of the product assembly line prior to packaging. The audit consists of a functional test and may be done on a 100 percent or normally on a sample basis with customer configuration, customer compatible equipment, and customer modes of start-up and operation. A chart of performance should be installed, and a written report should be issued weekly.

Postpack Product Audit. A postpack audit is performed on product selected at random from the final shipping area. The audit consists of an install, start-up, and functional test using customer configuration, customer compatible equipment, and customer modes of start-up and operation. A chart of performance should be installed, and a written report should be issued weekly. The dermit control chart is an excellent application for outout box audits, as it weighs critical defects, major defects, minor defects, and incidental defects.

Customer Data Collection

This section describes methods for analyzing customer data as well as a discussion of customer complaints.

Customer Complaints

Customer complaints are any quality-related dissatisfaction expressed by a customer when they return product, request service, or submit a claim for damages. They are measures of production service performance or customer satisfaction.

The purpose of customer service is to restore service, pay just claim, and maintain customer satisfaction. These complaints must be converted into corrective action or solutions. The types of field complaints are:

- *Isolated*—Random in nature. Troubleshoot, search for commonality of causes, and prevent recurrence.
- *Repetitive*—Common cause. Need study to determine causes.
- *Underreported*—Typical with low unit cost or when buyer has a shortage of goods.
- *Overreported*—Replacement of parts that are not defective. Part returned as defective. Returning surplus product as defectives.

Whether or not a complaint is made depends on:

- Economic climate
- Experience and skill of customer
- Seriousness of defect as seen by customer
- Unit price
  Low price—Underreported; product audit measure outgoing quality.
  High price—Use complaint rate.
- Time—Depends on nature of failure. Early failures, accidents, wearout. Measure complaints at specified intervals after delivery.

The following table shows commonly used complaint indices.

Control subject	Units of measure in use
Complaints	Total number of complaints/number of service and products
	Total number of complaints/number of service and products reported by month of origin
	Number of complaints per $1 million of sales
	Number of complaints per million units of product
	Value of material under complaint per $100 of sales for such products
Returns	Value of material returned per $100 of sales
Claims	Cost of claims paid
	Cost of claims per $1 million of sales
Failures	Mean time between failures (MTBF)
	Mean use between failures, e.g., cycles, miles
	Mean time between repair calls
	Failures per 1000 units under warranty
Maintainability	Mean time to repair (MTTR)
	Mean downtime
Service cost	Ratio maintenance hours to operating hours
	Repair cost per unit under warranty
	Cost per service call

Compare lists of principal field failures with principle factory rejects. Working to improve factory rejects may not solve field problems.

Customer information on quality can be obtained by obtaining information on all units or a sample. The system to track all units may be prohibitive if volume is large. Usually a sample, studied in depth, will yield more useful results. The process inadequately reports on all units.

Analyzing Customer Data

- *Defect matrix*—Defect/error types by product types.
- *Pareto analysis*—Order of importance.
  Failure rate by defect/error type
  Unit repair cost by product/service type
  Total repair cost by product/service type
  Complaint rate by customer
- *Cost analysis*—Justify improvement programs; usually customer satisfaction is adequate.
- *Spare parts use*
  Record of actual use from service reports
  Sale of parts to distribution chain

- *Cumulative complaint analysis*—This requires product be "dated" to show when it was made, sold, or installed. Helps in predicting failure rate of product designs.

- Data should be analyzed on the date of manufacture, not on the date it was reported. Historical charts are live and are adjusted monthly as new field input is received.

- *Predicting future performance*—Growth curves. Product is undergoing continuous improvements in design and refinements in operating and maintenance procedures. Product performance will improve ("grow") over time.

A commonly used method is Duane plots developed by J. T. Duane from General Electric. It applies any learning-curve approach to reliability monitoring, and uses log-log paper. Figure 9-14 shows a Duane plot from Duane's original article.

Data must be collected in terms of the nature of the defect, usually designated by a noun referring to what the service or product is, and then described by an adjective for type of nonconformity.

Data should also be collected in terms of when the problem occurred and whether it concerns a product or service. The date of manufacture, or date the service was rendered, must be included. Without the date of origin, data can cause confusion in the analysis.

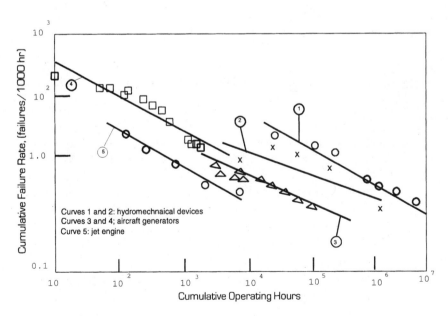

Curves 1 and 2: hydromechnaical devices
Curves 3 and 4: aircraft generators
Curve 5: jet engine

Figure 9-14. Duane plot.

Error Accountability

Error accountability deals with the area of committing people to be accountable for their errors for any process, be it software design, order entry, production planning, invoicing, etc. The purpose is to effect a reduction in the errors that lead to poor performance and unsatisfactory internal and external customer satisfaction.

Reduction of people errors is one of the most challenging areas to supervision and management. People errors are often thought to be the result of poor motivation. This is often not the case. Approximately 80 percent of the people errors are the result of management. Management accountability includes the support functions such as engineering, sales, and administration. A person is accountable for an error if the following three conditions are met:

1. Knowing what performance (goal/target) is expected
2. Knowing if they are accomplishing what is expected
3. Having the means to control the process if their performance is not meeting expectation

If any one of these conditions is not provided, management is accountable for the error.

The errors people make can be classified into one of three types:

1. Unintentional errors
2. Methods or procedural errors
3. Intentional errors

The analytical methods that can be used to determine the types of errors include:

1. *Person/error type matrix.* A table or matrix displaying errors of several people by defect type. This is the most useful in analyzing all error types. There is often reluctance on the part of management to directly associate names with errors. However, it has been demonstrated that if error accountability is presented with a professional and supportive attitude, employees will respond with positive improvement.

2. *Pareto analysis.* Separate the "vital few from the trivial many" for each person. In accord with Pareto's theory, approximately 80 percent of the problems will come from 20 percent of the people, or 80 percent of the problems will be associated with 20 percent of the known causes of errors.

3. *Trend analysis.* Examine the presence of "consistency" over time for each operator. Patterns may arise relative to time on the working environ-

ment which may lead to various frequencies of errors. Personnel performance is impacted by a complex set of factors that may be difficult to analyze, but obvious trends should be investigated.

Unintentional Errors

Unintentional errors occur because of human inability to maintain attention when they desire to be error-free. If not paying attention is deliberate, the resulting errors are intentional. Some tasks require more attention than others, and some people are better at paying attention than others over sustained periods of time and through distractions.

Characteristics

Unintended	People do not want to make them.
Unaware	At the time of making the error, person is unaware of having made the error.
Unpredictable	Nothing systematic to the errors as to when, type, or who. The random pattern of these errors creates a large number of error types with low frequencies of occurrence.

Preventive Action. The preventive action for unintentional errors is to provide aids to the person to make it easier to remain attentive. Foolproofing the operation to reduce the extent of dependence on a person's attention or training is also an approach, but at times it is more expensive.

Measure the frequency of occurrence daily; make the low-frequency errors the "problems of the day," and solve them one by one. Recognition and reward will sustain the entire process.

Methods or Procedural Errors

Errors occur because the person lacks some essential technique, skill, or knowledge. The knowledge should be documented into a procedure or work instruction.

Characteristics

Unintended	Person does not want to make error.
Consistent	Person(s) lacking the skill or knowledge will consistently make more errors than person(s) with skill or knowledge.
Specific	Unique to certain error types where the skill or knowledge is missing.

Unavoidable Persons(s) lacking the skill or knowledge are unable to perform to the level of the person(s) with the skill or knowledge.

Preventive Action. Management must be able to answer the following question: "What should I do differently than I am doing now?" The following ideas are offered to help answer this question.

- Conduct a process audit to determine if the person is complying with the methods or procedure provided.

- Obtain data which can disclose person-to-person differences using person/error-type matrix.

- Analyze data on a time-to-time basis to discover if consistency is present.

- Ask the people what they believe caused the error. Do they believe they are adequately trained and have the resources to do the job correctly?

- Identify the persons who are consistently best and those who are consistently unacceptable. Investigate the differences between the persons.

- Study the methods used by best and worst persons to identify the differences in skill or knowledge.

- Bring all persons up to the level of the most proficient:

Train persons in the necessary skills and knowledge.

Change process to include the skills and knowledge to the methods.

Foolproof the process.

Intentional Errors

These are errors that are being produced "on purpose." The errors exhibit consistency and cover a wider spectrum of error types.

Characteristics

Intended Deliberate or willful act on the part of the person.

Aware   Person is aware of making the error.

Consistent Person usually intends to keep making it.

Causes

Management Induced. Management can cause errors by shifting priorities between cost, delivery, productivity, and quality. This is the effect of either lack of planning or lack of accountability.

Person Induced. Employees may have real or imaginary grievances against supervisor or company. Sometimes the error may be the result of

Table 9-1. Operator Accountable Errors: Characteristics
and Preventive Action

	Type of error		
	Unintentional	Methods	Intentional
Description	Exhibits randomness by error type, time, or operator	Operators exhibit consistency with respect to error type	Exhibits consistency and covers a wider spectrum of error type
Characteristics	Unintended Unpredictable Unaware	Unintended Consistent Specific Unavoidable	Intended Persistent Aware
How identified	Pareto Analysis	Trend Analysis	Matrix Pareto Analysis Trend Analysis
Preventive action	Provide aids to help attentiveness	Ask people what caused the error	Make people aware of impact of error on others
	Foolproof operation	Collect data on people differences	Establish accountability
		Is person using method provided	Enhance supervisor's impact
		Analyze for presence of consistency	Reassign work
		Identify consistently best and worst people	Provide assistance to people
		Study work methods on best/ worst to identify difference	Improve communication
		Bring all operators up to best level	Conduct audits
		*train people without "touch"	
		*add "touch" to process	
		*Foolproof process	

"malicious obedience," where the employee continues with an incorrect procedure or performs a directive which will knowingly cause an error, while management is not aware of the negative impact of the directive.

Inadequate Communication. Product with errors in one department is fed to another department. A person receiving product assumes it's OK to use if it arrived at his or her operation. The subsequent departments use the service, product, or information without awareness of the errors in the incoming service, product, or information.

Supervisors post charts showing progress on meeting schedule, but do not post similar charts measuring performance against quality standards.

Supervisors launch poster campaign on doing better work. While the bulk of errors are management-controllable, the campaign makes no provision to improve quality of purchased parts, subassemblies, process capability, and machine maintenance. The message comes across as "Do as we say, not as we do."

Preventive Action for Intentional Errors. Intentional errors are oriented to the person rather than the system. The preventive action is to make the person(s) aware of the impact the error(s) have on customer satisfaction, be they internal or external customers.

Establish a system for accountability: All people behave more responsibly when their identity is known.

Provide assistance to people to identify their person-controllable errors from management's responsibility, and provide special techniques and training to the new techniques.

Empower the employees by setting expectations and guidelines, and encourage them to take action to eliminate the causes of errors.

Enhance supervisor's leadership. People mirror their supervisor's behavior pattern and will place emphasis on quality, schedule, or productivity as the supervisor does.

Conduct process audits on operators and inspectors to determine if they know the process. Once accountability has been established, and the operations and inspections have demonstrated they know the process, then a disciplinary policy needs to be enforced which addresses individuals who are causing intentional errors.

An example of such a disciplinary procedure has steps.

Steps	Action
Step 1: Initial occurrence	Awareness of error impact. Supervisor makes the responsible individual aware of the error and its impact.

Step 2: Second occurrence	Establish accountability. Conduct process audit to validate that operator has the skills and resources, hence understands they are responsible for the error.
Step 3: Third occurrence	Supervisor issues a verbal warning.
Step 4: Fourth occurrence	Supervisor issues a written warning.
Step 5: Fifth occurrence	Three-day suspension.
Step 6: Sixth occurrence	Termination.

Quality Targets

We have thus discussed selecting characteristics; locating the inspection stations; determined the type/quality of inspection/test; developed inspection instructions, and selected the type of testing, we will now establish quality targets.

Establishing quality targets is the process of determining the maximum outgoing quality level required at each inspection/test point in the process (receiving to shipping) in order to meet the outgoing quality level required by the customer.

The quality level values to be used in this section are expressed as percent or fraction defective and are 1 minus the yield values. A quality level of 2 percent is the same as a 98 percent yield. Fraction defective values are used in the calculations.

The process starts with a model representing a 100 percent inspection or test station. The basic terms are defined and principles are described and illustrated. The principles developed provide the methods for determining how to establish or calculate outgoing quality levels from an inspection or test station. These principles are used to change the computation emphasis from computing outgoing quality to computing the quality targets required at each inspection or test station to meet the outgoing quality desired.

The contents are:

- Target principles

- Sequences of inspection-test stations

- Target computations

- Determining effectiveness

- Sampling/control charts/audits

- Target economics

- Effects of rework and reinspection

- Target implementation methods

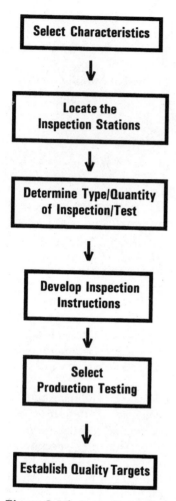

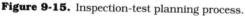

Figure 9-15. Inspection-test planning process.

Target Principles

The basic inspection-test station model is developed around a 100 percent inspection-test station as shown in Figure 9-16.

The terms associated with this model are defined as follows.

- Incoming quality denoted by p(in): The defective rate coming into the station from previous inspection station(s) or manufacturing operation(s).

- Outgoing quality denoted by p(out): The defective rate going out of the station after the inspection or testing is complete.

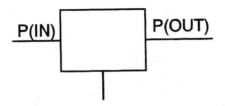

TARGET = P(OUT)[E/(1-E)]

E: EFFECTIVENESS = .80 (80%) ASSUMED

Figure 9-16. Basic station model (100% inspection).

- Effectiveness denoted by E: A measure of how good the inspection or test is in detecting the nonconforming parts. This measure is from 0 to 1 (1 is perfect effectiveness).
- Observed quality denoted by p(obs): The defective rate calculated from the inspection and test results and reported as the quality level in reports to management.

The principles of the basic model follow.

Principle 1. The incoming quality is the sum of the outgoing quality and the observed quality. This is expressed by the following formula:

$$p(in) = p(obs) + p(out)$$

Principle 2. The effectiveness can be found using p(in) and p(obs) in the following formula:

$$E = p(obs)/p(in)$$

If E = 1 then p(obs) = p(in) and, since p(in) = p(obs) + p(out), then p(out) = p(in) − p(obs) = 0.

If E = 0 then p(obs) = 0 and p(out) = 0.

Basic Computations. At any inspection-test station, the observed quality and effectiveness are either known or can be determined.

The observed quality is always known. The effectiveness may be known; but if not, it can be determined by using principle 2, by an inspection capability study, or by a special audit.

The values of p(in) or p(out) can be determined by audits of product coming into the station or product as it leaves the station.

With the values of p(obs) and E determined, the outgoing quality of the inspection-test station can be determined as follows:

From principle 1, p(in) = p(obs) + p(out);
thus, p(out) = p(in) – p(obs).

From principle 2, E = p(obs) / p(in);
thus, p(in) = p(obs) / E.

Substituting p(in) = p(obs) / E into the formula p(out) = p(in) – p(obs) gives p(out) = (p(obs) / E) – p(obs).

Simplifying this result gives Equation (9-1) for determining the outgoing quality from an inspection-test station when the effectiveness and observed quality are known:

$$p(out) = p(obs) [(1 – E)/E] \tag{9-1}$$

This result will be used to determine the formula for computing targets for each station. Once p(out) is computed from the values of E and p(obs), p(in) can be computed using principle 1, p(in) = p(obs) + p(out), directly or the formula p(in) = [p(obs) / E] derived from principle 2, as follows:

Example for Computing p(out) and p(in). Suppose p(obs) = 0.08 (8%) and E = 0.8 are given.

$$p(out) = p(obs) [(1 – E)/E] = (.08) [(1 – .8)/.8]$$

$$p(out) = (.08) [(.2/.8)] = (.08) [1/4]$$

$$p(out) = 0.02 (2\%)$$

$$p(in) = [p(obs) / E] = 0.08 / 0.8 = 0.10 (10\%)$$

Comments about the basic model:

- The outgoing quality of an inspection-test station is never observed in practice. It can only be determined by a special audit or from feedback from the next station (the customer).

- The outgoing quality of an inspection-test station is the incoming quality of the following inspection-test station.

- Any rework or reinspection test that is performed on defective material is assumed to be 100 percent effective. This is usually not true in practice and the modifications to the model are described later under Effects of Rework and Reinspection.

- If the effectiveness is 1 (or perfect) the outgoing quality is 0 (or perfect) from a 100 percent inspection or test station and the observed quality is the same as the incoming quality. This is an ideal situation that very seldom occurs.

- If the effectiveness is 0, the outgoing quality is the same as the incoming quality and the observed quality is 0 also.

- When this occurs, consider using an audit to determine if the outgoing quality is really 0 or if the effectiveness might be 0.

Sequences of Inspection-Test Stations

In order to establish targets for all inspection-test stations, the methods for determining outgoing quality levels for a sequence of inspection-test stations must be developed. The methods will depend upon whether the stations inspect-test completely different characteristics (case 1), the same characteristics (case 2), or a combination (case 3). The methods described consider only two stations to simplify the resulting formulas.

Case 1.   This involves two inspection-test stations inspecting different characteristics. When this occurs, they are referred to as *mutually exclusive stations*. The outgoing quality from these stations is the sum of the outgoing quality from each station, as shown in the following illustration.
    The value of p(out) is computed by the following formula:

$$p(out) = p(out\ 1) + p(out\ 2)$$

where p(out 1) is the outgoing quality of station 1 and p(out 2) is the outgoing quality of station 2.
    Equation (9-1) is used to find p(out) for each station and the results are added.

Quality Flow for Two Inspection-Test Stations with Different Characteristics

Case 1 Example (See Figure 9-17)

Suppose station 1 has E = .80 and p(obs) = 4%.

Station 2 has E = .90 and p(obs) = 3%.

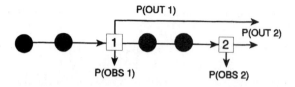

P(OUT 1)

P(OUT 2)

1

2

P(OBS 1)

P(OBS 2)

Figure 9-17. Quality flow for two inspection-test stations with different characteristics.

For station 1 p(out 1) is computed using Equation (9-1):

$$p(\text{out }1) = p(\text{obs}) \, [(1 - E)/E]$$

$$p(\text{out }1) = (.04) \, [(1 - .80)/.80]$$

$$p(\text{out }1) = (.04) \, [.2/.8] = (.04) \, (1/4)$$

$$p(\text{out }1) = .01 \, (1\%)$$

For station 2 p(out 2) is computed using Equation (9-1):

$$p(\text{out }2) = p(\text{obs}) \, [(1 - E)/E]$$

$$p(\text{out }2) = (.03) \, [(1 - .90)/.90]$$

$$p(\text{out }2) = (.03) \, [.10/.90] = (.03) \, (1/9)$$

$$p(\text{out }2) = .0033 \, (.33\%)$$

The final outgoing quality, p(out), is

$$p(\text{out}) = p(\text{out }1) + p(\text{out }2)$$

$$p(\text{out}) = .01 + .0033 = .0133 \, (1.3\%)$$

Case 2. This case involves two inspection-test stations checking same characteristics, which is the same as 200 percent inspection where the outgoing quality from the second station is a function of the observed quality at the first station and the effectiveness of the two stations as shown in the following illustration.

The formula for computing the outgoing quality of station 2, p(out 2) is.

$$p(\text{out }2) = p(\text{obs }1) \times \{ \, [(1 - E1) \times (1 - E2)] \, / \, E1 \}$$

where E1 is the effectiveness of station 1, E2 is the effectiveness of station 2, and p(obs 1) is the observed quality at station 1.

Quality Flow for Two Inspection-Test Stations with Same Characteristics. The preceding formula for p(out 2) is derived from applying Equation (9-1), principle 2, and since both stations are checking the same characteristics, the outgoing quality of station 1, p(out 1), equals the incoming quality of station 2, p(in 2). (See Figure 9-18.)

Applying Equation (9-1) to stations 1 and 2 gives

$$p(\text{out } 1) = p(\text{obs } 1) \times [(1 - E1)/E1]$$

$$p(\text{out } 2) = p(\text{obs } 2) \times [(1 - E2)/E2]$$

The observed quality of station 2, p(obs 2), is expressed as a function of the incoming quality at station 2 using principle 2, E2 = p(obs 2) / p(in 2). This gives

$$p(\text{obs } 2) = p(\text{in } 2) \times (E2)$$

Since p(in 2) = p(out 1) as stated previously, then substituting p(out 1) into the formula for p(obs 2) gives

$$p(\text{obs } 2) = p(\text{obs } 1) \times [\ (1 - E1) / E1\ ] \times E2$$

Substituting this result into the formula p(out 2) gives

$$p(\text{out } 2) = p(\text{obs } 2) \times [(1 - E2)/E2]$$

$$p(\text{out } 2) = \{\ p(\text{obs } 1) \times [(1 - E1)/E1] \times E2\ \} \times [(1 - E2)/E2]$$

$$p(\text{out } 2) = p(\text{obs } 1) \times \{\ [(1 - E1) \times (1 - E2)] / E1\}$$

Case 2 Example

Suppose for station 1 that E1 = .80 and p(obs 1) = 25% (.25). For station 2, E2 = .90.

The outgoing quality of station 2, p(out 2), is

$$p(\text{out } 2) = p(\text{obs } 1) \times \{\ [(1 - E1) \times (1 - E2)] / E1\}$$

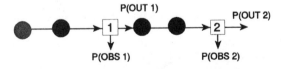

Figure 9-18. Quality flow for two inspection-test stations with same characteristics.

$$p(\text{out } 2) = (.25) \times \{[(1 - .80) \times (1 - .90)] \, / \, .80\}$$

$$p(\text{out } 2) = (.25) \times ([(.20) \times (.10)] \, / \, .80)$$

$$p(\text{out } 2) = (.25) \times \{[.02] \, / \, .80\} = (.25) \times \{.025\}$$

$$p(\text{out } 2) = 0.00625 \ (0.625\%)$$

By 200 percent inspection-test the incoming quality at station 1 is "sorted down" to .62 percent.

Case 3. This involves two inspection-test stations checking some characteristics that are different and some that are the same. The final outgoing quality from station 2 is the sum of the outgoing quality resulting from the different characteristics and the outgoing quality resulting from the characteristics checked at both stations.

The formula for the outgoing quality of station 2, p(out 2), is

$$p(\text{out } 2) = p(\text{out } 1A) + p(\text{out } 2A) + p(\text{out } 2B)$$

where 1A are the characteristics checked only at station 1, 2A are the characteristics checked only at station 2, and 2B are the characteristics checked at both stations.

Case 3 Example

Suppose station 1 has E1 = .80, p(obs 1A) = .02, and p(obs 1B) = .04 where 1A and 1B are two sets of characteristics where set 1B is checked again at station 2 by set 2B.

Suppose station 2 has E2 = .90, p(obs 2A) = .03.

Compute the outgoing quality for the set of characteristics checked only at station 1, 1A, using the following formula:

$$p(\text{out } 1A) = p(\text{obs } 1A) \, [(1 - E1)/E1]$$

$$p(\text{out } 1A) = (.02) \, [(1 - .80)/.80]$$

$$p(\text{out } 1A) = (.02)[.20/.80] = (.02)[1/4] = 0.005 \ (0.5\%)$$

Compute the outgoing quality for the set of characteristics checked only at station 2, 2A, using the formula:

$$p(\text{out } 2A) = p(\text{obs } 2A) \, [(1 - E2)/E1]$$

$$p(\text{out } 2A) = (.03) \, [(1 - .90/.90]$$

$$p(\text{out } 2A) = (.03) \, [.10/.90] = 0.0033 \ (.33\%)$$

Compute the outgoing quality at station 2, p(out 2B), for the characteristics checked at both stations using the formula:

$$p(\text{out 2B}) = p(\text{obs 1B}) \times ([(1 - E1) \times (1 - E2)]/E1)$$

$$p(\text{out 2B}) = (.04) \times \{[(1.0 - .80) \times (1.0 - .90)]/.80\}$$

$$p(\text{out 2B}) = (.04) \times \{[(.20) \times (.10)]/.80\}$$

$$p(\text{out 2B}) = (.04) \times \{[.02]/.80\} = (.04) \times \{.025\}$$

$$p(\text{out 2B}) = .001 \ (.1\%)$$

The outgoing quality for all characteristics at station 2 is computed as follows:

$$p(\text{out 2}) = p(\text{out 1A}) + p(\text{out 2A}) + p(\text{out 2B})$$

$$p(\text{out 2}) = (.005) + (.0033) + (.001)$$

$$p(\text{out 2}) = 0.0093 \ (.93\%)$$

Target Computations

The outgoing quality from an inspection-test station is the quality that is sent to the next station and is the incoming quality at the next station. For the last station in the final assembly of a product, the next station is the customer.

In order to meet the quality requirements of the customer, the maximum quality levels (maximum percent defective) allowed from each station must be established and met or exceeded. In particular, the observed quality level must be controlled at each station. The observed quality level is controlled by establishing a target for it. The target is the maximum value of the observed quality at each station that cannot be exceeded in order to meet or exceed the quality requirements of the next station.

From Equation (9-1) the outgoing quality level of a station can be determined. Solving this for p(obs) gives the formula to be used in establishing quality targets.

From Equation (9-1), $p(\text{out}) = p(\text{obs}) [(1 - E)/E]$.

Solving for p(obs) gives: $p(\text{obs}) = p(\text{out}) [E/(r - E)]$.

The formula for determining the quality target at each station is similar to the preceding p(obs) with the following notation changes.

Let the maximum outgoing quality allowed from a station be p(MOQA).

$$\text{Target} = p(\text{MOQA}) \times [E / (1 - E)] \tag{9-2}$$

The model described earlier is modified as shown in Figure 9-19.

Target Example.   Suppose the maximum outgoing quality allowed for a particular station is 2 percent. Suppose further that the effectiveness, E, is 0.8. The target for this station is as follows:

p(MOQA) = 0.02 (2%)

Target = p(MOQA) × [E / (1.0 − E)]

Target = (0.02) [.80/(1.0 − .80)] = (.02) [.80/.20]

Target = (0.02) (4.0) = .08 (8%)

The target value is larger than the outgoing quality allowed since a 100 percent inspection-test station is utilized. The quality is being "sorted" to ensure that the outgoing quality from the station does not exceed 2 percent by keeping the observed quality under 8 percent.

If 100 percent inspection-test is not utilized, then the incoming quality (outgoing from the previous station) would be required to be the same as the outgoing quality required. This will be described in more detail later.

Targets for a Sequence of Stations.   The computation of targets for a sequence of stations follows the cases presented in determining the outgoing quality for a sequence of stations.

Case 1.   For two inspection-test stations with different characteristics, the target must be computed for each station using the preceding formula. Since the final outgoing quality is the sum of the outgoing quality from both stations, the maximum outgoing quality allowed must be divided up

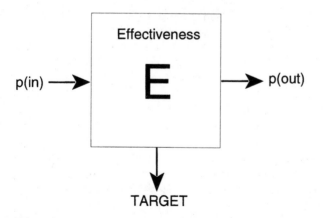

Figure 9-19. 100 percent inspection/test.

between the two stations. Two methods are presented for dividing or allocating the maximum outgoing quality allowed between the two stations, one based on no prior quality history and the other based on prior quality history.

Method for No Prior Quality History Available. The method recommended for no prior quality history available is based on the number of characteristics being checked at each station. Divide the maximum outgoing quality allowed proportional to the number of characteristics inspected at each inspection station.

Example

Suppose two inspection stations are inspecting different characteristics with station 1 checking 10 characteristics and station 2 checking 20.

Suppose the maximum outgoing quality allowed from these two stations is .03 (3 percent).

Determine the maximum outgoing quality allowed from each station when a total of 30 characteristics are inspected, 10 (or 33 percent) at station 1 and 20 (or 67 percent) at station 2. Divide the 3 percent allowed from these two stations as follows:

$$33\% \ (1/3) \text{ of } 3\% \text{ to station } 1 = 1\%$$

$$67\% \ (2/3) \text{ of } 3\% \text{ to station } 2 = 2\%$$

Method for Prior Quality History Available. When prior quality history is available, divide the maximum outgoing quality allowed proportional to the quality history (percent defective) for each inspection station.

Example

Suppose two inspection stations are inspecting different characteristics with a maximum outgoing quality allowed of 3 percent.

Suppose the prior quality history is as follows:

Station 1   0.04 or 4%
Station 2   0.02 or 2%

Determine the maximum outgoing quality allowed for each station as follows: The total percent defective from the two stations is 6, 4, or 67 percent of the total from station 1 and 2 or 33 percent of the total from station 2.

Divide the required quality of 3 percent as follows based on the prior quality history:

$$66\% \ (2/3) \text{ of } 3\% \text{ to station } 1 = 2\%$$

$$33\% \ (1/3) \text{ of } 3\% \text{ to station } 2 = 1\%$$

Case 1 Example

Suppose two inspection stations are inspecting different characteristics and the maximum outgoing quality allowed from the two stations is 0.03 or 3 percent.

Suppose further that the quality history is available for each station and is the same as in the previous example with 4 percent from station 1 and 2 percent from station 2.

Suppose the effectiveness is 0.80 for station 1 and 0.85 for station 2. Determine the targets for each station as follows: The maximum outgoing quality allowed from both stations is divided between the two stations according to the prior quality levels. 2 or 67 percent of the target is assigned to station 1 and 1 or 33 percent is assigned to station 2.

The maximum outgoing quality allowed from station 1 is

$$p(MOQA1) = 0.02$$

The maximum outgoing quality allowed from station 2 is

$$p(MOQA2) = 0.01$$

Using Equation (9-2), the target for station 1 is

$$\text{Target} = p(MOQA1) \, [E1/(1 - E1)]$$

$$\text{Target} = (0.02) \, [.80/(1.0 - .80)] = (.02)[.80/.20]$$

$$\text{Target} = (0.02) \, (4.0) = 0.08 \, (8\%)$$

Using Equation (9-2) the target for station 2 is

$$\text{Target} = p(MOQA2) \, [E2/(1 - E2)]$$

$$\text{Target} = [0.01] \, [.90/(1.0 - .90)] = (.01) \, [.90/.10]$$

$$\text{Target} = (0.01) \, (9.0) = .09 \, (9\%)$$

Case 2. For two inspection-test stations checking the same characteristics, the formula for the target at station 1 is

$$\text{Target station 1} = p(MOQA2) \times \{(E1)/[(1 - E1) \times (1 - E2)]\}$$

The formula for the target at station 2 is

$$\text{Target station 2} = p(MOQA2) \times [E2 \, / \, (1 - E2)]$$

The formula for determining the outgoing quality from two stations checking the same characteristics and solving this for p(obs) gives the basic formula for calculating the target at station 1 as follows:

$$p(\text{out 2}) = p(\text{obs 1}) \times \{[(1 - E1) \times (1 - E2)] / E1\}$$

$$p(\text{obs 1}) = p(\text{out 2}) / \{[(1 - E1) \times (1 - E2)] / E1\}$$

$$p(\text{obs 1}) = p(\text{out 2}) \times \{(E1)/[(1 - E1) \times (1 - E2)]\}$$

The formula for determining the target at station 2 uses the basic target formula stated in Equation (9-2).

Case 2 Example

Suppose two final inspection stations are inspecting the same characteristics.

Suppose the maximum outgoing quality allowed from these 2 stations is 0.02 (2 percent). Since both stations check the same characteristics, the maximum outgoing quality allowed from station 2 is 2 percent (MOQA2 = 0.02).

Suppose the effectiveness for each station has been determined as follows:

Station 1   E1 = 0.80 or 80%
Station 2   E2 = 0.85 or 85%

Target station 1 = $p(\text{MOQA2}) \times \{(E1)/[(1 - E1) \times (1 - E2)]\}$

Target station 1 = $(0.02) \times \{(.80)/[(1.0 - .80) \times (1.0 - .85)]\}$

Target station 1 = $(0.02) \times \{(.80)/[(.20) \times (.15)]\}$

Target station 1 = $(0.02) \times \{(.80)/(.03)\} = \{(0.02) \times (26.70)\}$

Target station 1 = 0.534 or 53.4%

Target station 2 = $p(\text{MOQA2}) \times [E2 / (1 - E2)]$

Target station 2 = $(0.02) \times [0.85 / (1.0 - .85)]$

Target station 2 = $(0.02) \times [(0.85) / (0.15)]$

Target station 2 = $(0.02) \times [5.67]$

Target station 2 = 0.113 or 11.3%

Note that the target values in this example are large values 53 percent at station 1 and 11.3 percent at station 2. This means that the incoming quality at station 1 could be as high as 68 percent and the maximum outgoing quality allowed from station 2 of 2 percent can still be met. However, this requirement would be met with 200 percent inspection and a large amount of rework and reinspection, a costly method to meet an maximum allowed outgoing quality of 2 percent.

A better way to meet the allowed quality is to reduce the incoming process defectives to the point where the incoming to station 1 is the same as the incoming to station 2 must be to meet 2 percent allowed outgoing from station 2. The incoming quality to station 1 would have to be 13.3 percent. When this happens, station 2 can be reduced to a sample or possibly eliminated.

Further, when the incoming process defectives to station 1 are reduced to 2 percent, then station 1 can be considered for reduction from 100 percent to sampling.

Case 3. This involves two inspection-test stations checking some characteristics that are different and some that are the same. As noted in computing the outgoing quality for case 3 in the section Sequences of Inspection-Test Stations, the outgoing quality for all characteristics at station 2 is computed by using the following formula:

$$p(\text{out } 2) = p(\text{out } 1A) + p(\text{out } 2A) + p(\text{out } 2B)$$

where $p(\text{out } 1A)$ is the outgoing quality for the characteristics checked only at station 1 and $p(\text{out } 2A)$ is the outgoing quality for the characteristics checked only at station 2.

$p(\text{out } 2B)$ is the outgoing quality at station 2 for the characteristics checked at both stations. The initial step in determining the targets is to divide the maximum outgoing quality allowed among the three areas described above (1A, 2A, 2B). The amount assigned to the characteristics checked twice (2B) should be the smallest.

The method for dividing the maximum outgoing quality allowed is the same as described in case 1. Once the outgoing quality allowed for each station is obtained, then Equation (9-2) can be applied to each station for characteristics that are different at each station (set 1A and 2A). The target formula for the characteristics checked at both stations can be applied to stations 1 and 2.

Determining Effectiveness

Effectiveness for computing quality targets is concerned only with the correct identification of defective pieces. False alarms do not influence the effectiveness used in quality target model.

Two ways provided to determine effectiveness for quality target models:

1. Inspection capability study—attribute data
2. In-process product audit

Inspection Capability Study—Attribute Data. If an inspection capability study is performed using the method described in the inspection capability study procedure, Effectiveness = 1-P (miss).

In-Process Product Audit. The following in-process product audit can be performed to determine the effectiveness for the quality target model. The pieces to be audited can be selected either before or after inspection.

Selecting Audit Pieces Prior to Inspection. The following steps outline the method for obtaining data for determining effectiveness when selecting parts to be audited prior to the inspection station.

1. Select sample size of 20 to 25 parts at random prior to inspection. One fourth to one half of the parts should be nonconforming parts.
2. Code the parts; measure and record the data.
3. Randomly provide the parts, conforming and nonconforming, into the production line.
4. Inspect the parts and record the results.
5. Compute the effectiveness as described subsequently.

Selecting Audit Pieces After Inspection. The following steps outline the method for obtaining data for determining effectiveness when selecting parts to be audited after the inspection station.

Obtain 20 to 25 parts that were inspected and dispositioned by the inspector. Select both accepted and rejected parts. The rejected parts should make up about one fourth to one half the sample. Inspect the parts and record the results. Compute the effectiveness (E) as follows:

$$E = (D - K) / (D - K + B)$$

where $D =$ number of rejects reported by the inspector
  $K =$ number of conforming units rejected by the inspector as determined by the auditor
  $B =$ number of nonconforming units accepted by the inspector as determined by the auditor

Effectiveness Calculation Example: Out of a sample of 25 parts, inspection rejected eight of the parts; two were conforming units and two parts were accepted that were nonconforming.

From the data provided, $D = 8$, $K = 2$, and $B = 2$. The effectiveness, E, is computed as follows:

$$E = (D - K) / (D - K + B)$$

$$E = (8 - 2) / (8 - 2 + 2) = 6 / 8 = 0.75 \text{ or } 75\%$$

Sampling/Control Charts/Audits

The material presented for calculation of targets at each station has assumed that 100 percent inspection is used and the targets are the maximum value that the observed quality can be at each station. It should be noted that the targets are larger than the maximum outgoing quality allowed. This is a result of 100 percent inspection sorting out defectives, a very expensive way to meet the outgoing requirements.

The process average fraction or percent defective is the quality level coming into a station and, if this is large, then a large amount of rework is required to meet the maximum outgoing quality allowed. A better strategy would be to get the process average at each station equal to or less than the maximum outgoing quality allowed for that station. However, until the incoming quality meets the maximum outgoing allowed, 100 percent inspection may be required. Once the process average is reduced, then sampling, control charts, or an audit may be used. The following criteria are offered for knowing when to switch between 100 percent inspection and some form of sampling:

If the incoming process average is greater than the maximum outgoing quality allowed, 100 percent inspection will be necessary.

If the incoming process average is equal to or less than the maximum outgoing quality allowed, then sampling, control charts, or audit can be used.

100 Percent Inspection. When 100 percent inspection is being used and the value of p(obs) is greater than the target, the maximum outgoing quality allowed is being exceeded. Additional inspection may be needed to meet the requirement until the value of p(obs) is equal to or less than target.

If p(obs) is less than target, 100 percent inspection will be continued until the value of p(obs) is equal to or less than the maximum outgoing quality allowed for that station. When this occurs, then the value of p(in) is monitored to determine when 100 percent inspection can be reduced. The value of p(in) can be monitored by using the formula:

$$p(in) = p(obs)/E$$

When p(in) is equal to or less than the maximum outgoing quality allowed, consider using sampling, control chart, or audit.

Sampling Plans. When a sampling plan is used, the selection of the plan is based on the AOQL parameter. Select a plan with an AOQL equal to the maximum outgoing quality allowed. The following will occur during the operation of the sampling plan.

If the incoming process average, p(in), is greater than the AOQL value, the sampling plan will reject almost every lot. This will result in having almost 100 percent inspection. When the incoming process average is less than the AOQL value, few lots will be rejected and maximum outgoing quality allowed will be met or exceeded. However, some short runs of product may exceed the maximum outgoing quality allowed, which is the usual risk associated with sampling.

Control Charts. Control charts are a very effective way to sample a process and meet the quality requirements. Several samples, called *subgroups,* are obtained periodically and the results are plotted on the appropriate control chart. The fraction defective control chart can be used to plot and track the quality level at each station. The value of the target at that station is the maximum outgoing quality allowed and is posted on the control charts.

If the control chart is in control and the process average fraction defective is less than or equal to the target, the target is being met. If the control chart is in control and the process average fraction defective exceeds the target, the target is not being met. If a control chart is to be used in place of 100 percent inspection, an action plan will be required to indicate how and when process average will be reduced to meet requirements.

Audits. If process average fraction defective is less than the target value or the maximum outgoing quality allowed, audits can be considered to monitor the station quality level or to verify what the true process average fraction defective is.

Target Computations. The target computations are modified slightly for a station not using 100 percent inspection. Once the maximum outgoing quality allowed is computed for the station, this becomes the target value to be placed on the control chart. When the process average fraction defective is equal to or less than target value at the station, the maximum outgoing quality allowed is met.

Target Economics

The costs or savings resulting from meeting the target are determined as described in this section. The total cost of meeting the targets includes the cost of inspection, scrap, rework, reinspection, and implementation. The formula can be defined as

$$\text{Total cost} = I + S + REW + REI + IMP$$

where
$$I = \text{cost of inspection}$$
$$S = \text{cost of scrap}$$
$$REW = \text{cost of rework}$$
$$REI = \text{cost of reinspection}$$
$$IMP = \text{cost of implementation}$$

The total cost is computed for the current process average percent defective and for the target process average that equals the maximum outgoing quality allowed for that station. When computing the cost for the current process average, the implementation cost is omitted. A worksheet follows to facilitate the computations of these costs. The savings or cost is the cost with the current process average minus the costs with the target process average.

- If the current process meets the target, then savings will be encountered when the process average decreases toward the maximum outgoing quality allowed. The savings are from reduced scrap, rework, and reinspection.

- If the amount of inspection is changed from 100 percent to sampling (control chart or other), then additional savings is encountered through reduced inspection.

- If the current process does not meet target with 100 percent inspection, then additional inspection or test may be needed. This will result in additional costs until the process average is improved to allow a reduction from 100 percent inspection.

Effects of Rework and Reinspection

When inspection is performed at any station, the defectives that are found must be repaired or reworked and then the repaired part is reinspected or retested. The effect of these actions on meeting the maximum outgoing

---

Worksheet for Computing Cost Savings

A. Inspection cost (annual) _____

B. Scrap
   (1) No. of pieces scrapped per month = _____
   (2) Cost per piece scrapped = _____
   (3) Annual scrap cost (1) × (2) × (12 months/yr.) = _____

C. Rework
   (1) No. of pieces per month = _____
   (2) Time (hours) per piece to rework = _____
   (3) Cost of repair person per hour = _____
   (4) Annual rework cost (1) × (2) × (3) × (12 months/yr.)
      = _____

D. Reinspection
   (1) No. of pieces per month reinspected = _____
   (2) Time to reinspect pieces/hour = _____
   (3) Cost per hour of inspector = _____
   (4) Annual reinspection cost (1) × (2) × (3) × (12) = _____

E. Cost to implement
   (1) Cost of new tools = _____
   (2) Cost of new equipment = _____
   (3) Cost of new supplies = _____

Total cost to implement = _____
Total cost _____

Figure 9-20.

quality allowed at each station will be discussed in this section. The rework and reinspection loop is added to the basic station model as shown in Figure 9-21.

The quality level going out of any station consists of "escapes" from the inspection process (100 percent inspection not 100 percent effective) or from the rework and reinspect process loop. The final outgoing quality of the inspection station can be expressed by the following formula:

$$p(out) = P(out\ station) + p(out\ reinspection)$$

The incoming quality to the rework process is 100 percent defective since all defectives are sent to rework. If the rework is 100 percent effective, then 0 percent goes to the reinspect or retest process and the outgoing

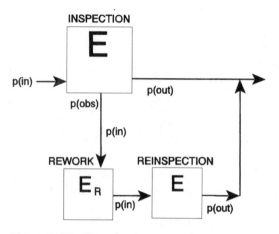

Figure 9-21. Rework reinspection loop.

quality from the retest station is 0 percent defective. This means that the outgoing quality is actually decreased by the rework and reinspection loop. The following example illustrates the effect of rework and reinspect on the outgoing quality from a 100 percent inspection-test station.

Suppose the effectiveness, E, of a station is 0.80 and the incoming quality, p(in), is 10 percent. The effectiveness of the reinspection is also 0.80. Suppose the rework has an effectiveness of 0.90. Consider the flow of 100 parts through the diagram shown below.

Eight parts will be sent to the rework and reinspect process, while 92 will go out of the station with two of those being defective: p(out) of 2/92 or 2.1 percent.

The eight parts reworked will have seven good and one defective go into reinspect, where the one defective has an 80 percent chance of being detected. If the one defective is detected, then seven good ones are added to the 90 good ones and two defectives, giving an outgoing quality of 2/99 = 0.20 or 2 percent. The outgoing quality has actually decreased.

Consider the example just given and change the rework effectiveness to 0.50. The eight parts to be reworked will result in four defective parts going to reinspection, three of which will be detected and one of which will not be detected.

Now the outgoing quality from reinspection is four good and one defective. This is mixed with the 90 good and two defectives missed to give 94 good and three defectives or 3/97 = 3.1% outgoing quality, an increase of 1 percent for a poor rework effectiveness (50 percent).

The preceding examples illustrate the following points.

If the rework is not perfect, then the outgoing quality from the rework station is not 0 and the rework and reinspection process increases the out-

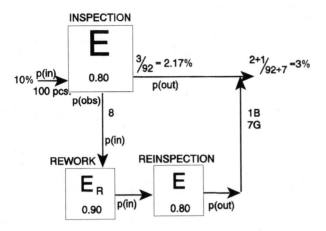

Figure 9-22.

going quality of the inspection test station. The amount that the outgoing quality is increased is related to the effectiveness of the rework and the quality level coming into the station, p(in).

If the rework effectiveness is 0.90 or higher, the overall outgoing quality is actually reduced. If the rework effectiveness is below 0.90, then the outgoing quality is increased. The lower the rework effectiveness, the more the outgoing quality is increased.

The incoming quality to the rework process is 100 percent defective. If the rework is 100 percent effective, then 0 percent goes to the reinspect or retest process and the outgoing quality from the retest station is 0 percent defective. When these parts that were repaired to good are added to the previously inspected parts, the overall outgoing quality is actually reduced.

Target Implementation Methods

The implementation of quality targets in a manufacturing operation involves providing a target monitoring system to ensure that the final outgoing quality of the product meets the customer requirement. If the requirement is not being met, then corrective and preventive actions should be implemented to adjust the processes to meet the customer requirement.

The following items are provided as a guideline to follow when implementing targets in a manufacturing area.

1. Identify the direct and indirect defects resulting from each inspection element.

2. Determine the effectiveness for the direct defects as a minimum.

3. Calculate the outgoing quality for each station using the effectiveness found previously, the actual observed quality, and the following formula:

$$p(out) = p(obs) \frac{1 - E}{E}$$

4. Combine the outgoing quality from each station to determine the overall outgoing quality to the customer.

5. Compare the customer requirement to the result of step 4 and identify where cost-effective improvements can be made to meet the requirement.

6. Establish a monitoring system for each station with the following items.

   a. Set up a percent or fraction defective chart with the station target as the centerline.

   b. Plot the weekly percent or fraction defective on the chart.

   c. Compute the standard deviation of the percent or fraction defective using the average (centerline) percent defective and determine the standard deviation zones based on the quantity of parts inspected each week. An illustration of how the chart would look is shown in Figure 9-23.

Use the following criteria to initiate required corrective or preventive actions at each station based on the percent or fraction defective chart. Refer to Figure 9-23 for the zone definitions.

   a. If the weekly percent defective is between +1 and −1 sigma from the centerline (zone A), take *no* action.

   b. If the weekly percent defective is between −1 and −2 sigma from the centerline (zone B2), implement audits to measure effectiveness.

   c. If the weekly percent defective is at −2 sigma or less (zone C2), initiate immediate audit to measure the effectiveness.

   d. If the weekly percent defective is between +1 and +2 sigma (zone B1), initiate a purge, repair, and reinspection, as well as developing an immediate preventive action plan.

   e. If the weekly percent defective is at +2 sigma or more (zone C1), initiate immediate preventive action plan development and stop production.

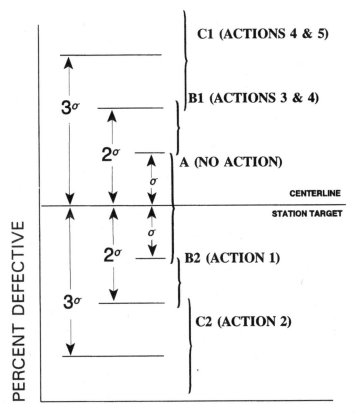

Figure 9-23. Target monitor.

Target Monitor

1. Implement audits to measure effectiveness.
2. Initiate immediate audit to measure effectiveness.
3. Purge, repair, and reinspect.
4. Develop an immediate preventive action plan.
5. Stop production.

Design of Experiments

One of the most powerful quality improvement tools available is the design of experiments (DOE). This technique allows you to study the effect of process input variables (factors) on the process output variables

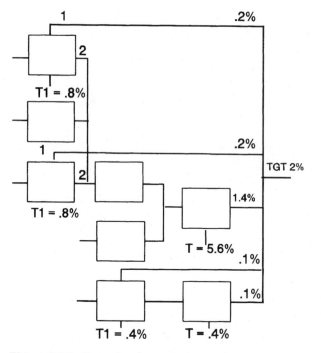

Figure 9-24. Example of an outgoing quality target using
the various cases.

(responses). By changing all of the factors in a methodical fashion, you can quantify the effect of each factor on the process output and identify any interactions between factors. Interactions occur when the effect of one factor on the process output depends on the level of another factor.

The power of DOE comes from the statistical properties of orthogonal matrices. It is not necessary to understand the statistical theory to use the tool. The mathematics involved make this technique more complex than SPC, but the concept is easily understood. By varying factors according to the randomized plan provided by the design of experiments, you can identify the statistically significant factors and interactions. Those factors are used to center the process with minimum variation. Randomization of the experimental runs removes bias by giving the effect of unknown process factors an equal chance of occurring throughout the experiment. Statistical significance of a factor occurs when the effect of that factor is large enough to provide at least a 90 percent confidence level of significance.

DOE is more efficient than the traditional approach of varying each factor while holding the other factors at constant settings. This one-factor-at-a-time type of experimentation is successful 10 to 25 percent of the time. DOE can increase this success rate to the 50 to 75 percent range.

DOE is an appropriate tool when you need to know the process in more detail and the process output can be measured. For example, a research and development team is creating a new process to manufacture a more advanced product. The effect of using different materials and changing operating temperature and pressure settings is unknown. DOE can enable the team to select the best combinations of material, temperature settings, and pressure settings quickly and reliably. In another example, a team is working on implementing SPC. The team wants to identify what changes in the process can cause the process output to change and become out-of-control. DOE can identify which factors are the most significant ones. When the SPC chart indicates an out-of-control condition exists, problem solving will focus on changes in those important process factors.

The advantages of DOE are:

- To differentiate the important process factors from the unimportant

- To optimize product designs

- To optimize manufacturing processes

- To reduce costs by reducing rework and scrap or enhancing yield

- To accelerate develop cycle and yield curves

- To troubleshoot problems with manufacturing processes

- To set process tolerances

There are no disadvantages when DOE is used appropriately. Unsuccessful experiments are usually due to selecting the wrong process factors for the experiment. When this occurs new factors should be selected for further experimentation.

The simplest form of experimental design is to study the effects of a single factor at two levels. This type of data can be analyzed by hypothesis testing for a qualitative factor (i.e., material A versus B or process A versus B). The null hypothesis is that no difference exists between the distributions resulting from each setting. The alternative hypothesis is that the one distribution is different from the other. The t-test determines differences in means, while the f-test determines differences in the variances of the distributions. For quantitative factors (i.e., temperature, pressure, or humidity), simple regression analysis provides a measure of statistical significance of the factor and fits a line or a curve to the data gathered in the experiment.

The next level of complexity is multiple-factor, two-level screening experiments. The goal of screening is to identify which of many factors (four or more) have a statistically significant effect on the process response being studied. Qualitative and quantitative factors can be used in the same experiment. For each quantitative factor, a low and high setting is selected. For qualitative factors, each case is identified (i.e., material A versus material B). Multiple regression analysis is used to analyze the data. As a result the significant factors and factor interactions are identified. This information can be used to direct SPC efforts and can be used to design optimization experiments.

Screening experimental designs are:

- *Two-level full factorials (one to four factors).* All possible combinations of the factors at low and high values. For example, a three-factor experiment would have eight runs.

- *Two-level fractional factorials (more than three factors).* A fraction of the full-factorial experiment is used to reduce the number of experimental runs. Fractions can be chosen to provide three types of information (the more information, the larger the experiment):

  1. Factor effects cannot be separated from the effect of two-factor interactions.

  2. Factor effects are defined, but the effect of two-factor interactions cannot be separated.

  3. Factor effects and two-factor interaction effects are defined.

- *Plackett-Burman.* Minimum number of runs to identify factor effects only. Used for a large number of factors.

The highest level of complexity is multiple factor, multilevel optimization experiments. The goal of optimization is to identify the best settings of the process to center the response, maximize or minimize the response, or minimize variation. As in screening experiments, qualitative and quantitative factors can be used in the same experiment. For quantitative factors, at least three setting levels are used; thus, enough information is provided to identify any curvature in the response. Multiple regression analysis examines the best fit of the data. Contour plots are prepared to map the process response. While this analysis can be performed manually, many computer programs are available to assist the experimenter with the analysis. As a result, optimum settings are identified. A conformation run verifies the results of the optimization experiment.

Optimization experimental designs are:

- *Central composite.* Multiple-level designs with center points and star points (points beyond the low and high values). Used to estimate quadratic models (curvature) of the response.

- *Box-Behnken.* Three-level designs with center points. Used to estimate quadratic models (curvature) of the response.

- *Three-level factorials.* All combinations of the process factors at three levels. Usually abandoned in favor of the above designs due to the larger number of runs involved.

- *D-optimal.* Computer-generated designs which are used when classical designs cannot be used. For example, when a qualitative factor is combined with quantitative factors at three levels.

If the experimenter uses screening and optimization sequentially, much information is provided to facilitate SPC efforts. This information can be used to develop Corrective and Preventative Action Matrices. It may also be used to identify locations where other SPC charts could be useful earlier in the process. Finally, DOE information allows you to characterize your process so that sources of variation are understood.

For a better understanding of DOE application, examples should be studied. Note that in these case studies, the entire problem is not solved through the use of DOE, but understanding of the process provided by DOE is essential to the solution.

Case Study 1

An American electronics manufacturer is interested in reducing costs to remain competitive with foreign competition. The company embarks on a quality improvement program that starts with training of management and continues on down to the line workers before the quality improvement tools are applied. When the required training is completed, teams are formed in each area of manufacturing. One of the teams is formed in the area responsible for photographing the circuit images on a photosensitive material, which is used as a mask to define the circuit layer at a later step.

The team begins meeting and decides to focus on reducing rework caused by not meeting specification limits for circuit line widths after the image has been developed. The team members gather data that indicate a rework rate of approximately 20 percent and that most of the rework is not due to defects, but rather for being beyond specification limits. They meet with the management SPC resource committee and present their findings along with a plan to reduce rework by raising the Cpk above 1.0. The resource committee agrees to the team's plan and work begins.

The team members meet twice a week for the first two weeks. They put together a flow chart of the process and a cause-and-effect diagram. Minutes of each meeting are issued to team members and the resource committee. The team decides to use a screening experiment to study the effect of eight process factors on the mean and standard deviation of the line width. The effects of any interactions cannot be separated from the effects of the factors in this fractional factorial design, so it is assumed that significant effects are due only to the factors. It takes two weeks to run the experiment and analyze the data.

Two factors are identified as significant for the mean of the response. Those factors are the incoming material surface index and the focus of the "camera." There are no significant factors found for the standard deviation.

Team members study the results of the experiment and find that the adjustment of focus is a problem for manufacturing. When the "camera" is shut down for adjustment, it takes several hours and is not always successful on the first attempt. It is also noted that the extreme values for the surface index were produced through a nonproduction process. This process has since been found not to be viable. The team is doubtful that normal process variation will cause an effect. A second experimental design is chosen to optimize line-width values. It takes much longer to run this experiment because it is difficult to produce the sample needs with the correct surface index values. The results of the experiment indicate that the process is completely controlled by the focus of the "camera."

This is good news to the team. Evaluation of the DOE results by team members causes them to reach a conclusion: The optimum focus is in a different location than previously thought; therefore, the mean of the line width varies excessively. By changing the focus specification target value and making the specification width larger, you will reduce mean variation of the line width and need to make adjustments less frequently. The team makes a confirmation run which verifies its assumptions.

The team members presents the results of their work to the resource committee and the committee agrees to change the focus specification. Process control charts are implemented for both the width of the line and the focus checks. Over the next few months, the results are impressive. Rework for not meeting specification is eliminated. The Cpk is 1.5. The equipment is up more often because focus adjustments are nearly eliminated. Product moves through the line faster and yields go up. The team effort is rewarded by recognition from management. The success of this project causes the formation of new teams and the improvement process continues.

Case Study 2

A medical device manufacturer uses injection molding to create their products. The regulatory agency has encouraged the manufacturer to implement SPC. The company gets some training and purchases software to create control charts. Over a period of time, the charts show that variation within a job order is acceptable for most specifications, but variation from job order to job order is excessive. The manufacturing group feels helpless to control their processes. The quality group and an external facilitator encourage the manufacturing group to perform experiments to identify the optimum process settings to achieve a centered process with minimum variation. Manufacturing concedes but is not comfortable with reducing the range of allowable settings of process input factors. Manufacturing has always felt that creating molding parts is an art and the intervention of statistical techniques causes apprehension.

Over a period of time, each product type undergoes an experimental design study. Most designs are full factorials with three factors (time, temperature, and pressure). Each product type is characterized and confirmation runs are performed. The process settings are fixed for the significant factors and specifications remain wide for factors that are found not to be significant. This provided a comfort zone for manufacturing. If one of the measurements is not centered, the mold is reworked to center that measurement.

The results of the experiments indicated improved Cpk's, reduced scrap, and more predictable, stable performance. Setup cycles were greatly reduced. Inspection costs were reduced because more job orders passed sampling inspection plans and 100 percent inspection was not required.

Case Study 3

A circuit board manufacturer is getting complaints from a customer because of excessive variability in the distance between two tooling holes. The company decided to use experimental design to determine the causes that affect the growth and shrinkage of the circuit board. Since there was no DOE expertise within their organization and no time to train, they sought the help of a consultant.

A team was formed to work with the consultant. At the first meeting the team selected 13 factors to study in 16 runs. The experiment was very difficult to perform because the material had to be identified and tracked throughout the entire process. The team members planned carefully and met daily to ensure that the material was processed correctly and not scrapped.

This screening design was intended to reduce the number of factors for more detailed study to follow. However, because the experiment was conducted throughout the entire process cycle, only three factors were eliminated. After the data were analyzed, it was noticed that seven of the ten remaining factors were qualitative. Lengthy discussions were conducted to decide which of the factors were most important and which level of the factor was most desirable. Priorities were established for three of the qualitative factors. Through the consideration of those priorities, two processes were selected for conformation runs. Since the initial design was highly fractionated, it was necessary to perform fairly large conformation runs for the chosen process settings. One of the two processes was very successful and produced a centered process with a Cpk above 1.1.

One more experiment was identified to examine the effect of using an additional board type and the possibility of reducing of a bake time. The process was optimized for both board types with minimum bake times. Variation was further reduced. Because the process was controlled, the output was more predictable. There was no longer a need to start more product than was required to fill the order. The circuit board manufacturer was pleased with the results of the study and proceeded with a plan to develop DOE skills within its organization.

Summary. Design of experiments provides a great deal of information about process variation in the most efficient manner. Success rates increase in a shorter period of time than when other methods are used. SPC implementation is much more successful when this tool is employed.

Analysis

Introduction and Review

Solving Problems with the Help of Statistics. Two types of problems exist. First, there is the problem of interpreting data which is already on hand. You must be sure the data provide a representative and random sampling. Using this method gives an estimated probability of success ≤0.25. The second type of problem involves deliberately doing something to get data. This involves such things as:

- Screening a large number of factors
- Probing into known causative factors
- Resolving specific questions
- Product improvement

In an effort of this nature, the estimated probability of success is ≥0.75.

Sample-to-Population Reasoning. Whenever we use "statistics" in our work, we're trying to reach some conclusion about a "population" by using a "sample" taken from that population. This will be the case for all that follows.

The possible variation in numerical data from one sample to another could possibly cause us to reach a wrong conclusion about the population. Therefore, when you reach a decision or draw a conclusion you should know the probability that you have made a wrong decision *because* of a sampling variation. This is such an important idea that it bears repeating:

- You're interested in the *population.*
- You get your information from a *sample.*
- Statistical theory can tell you the *risks* associated with your decision because of sampling variability.

One Variable

One Level. A variable such as "height" or "weight" can take on "values" such as 71 inches or 178 pounds. The "level" of the variable basically tells how many cases are being considered for that variable. For example, if our variable is height and we measure the height of six people in this room, then the variable "height" appears at one level with six measurements or values at that level. The case or level would be the *one* room.

Case 1*a:* The Standard Quality Control Problem

You see before you a "lot" containing, say, 500 parts which have been produced by your company and are ready to be shipped to a customer. You extract a "sample" from this "population" and perform a test on the items in the sample. If the sample results "look good" to you, you decide to ship the lot; if the sample results "look bad" to you, you decide to scrap the lot. You make a decision about the *population* based on what you find in the *sample*. The statistical phrasing of this problem is as follows:

Approach 1

Step 1: Formulate a null hypothesis (a "working assumption"), e.g., "The population in the box is OK."

Step 2: Take the sample *randomly* and perform the tests.

Step 3: Calculate the probability of getting sample results like ours, or "worse," under the assumption that the null hypothesis is true, i.e., that your "working assumption" is true. Call this probability $\propto$.

Step 4: Make your decision to ship the lot or not ship the lot. If the lot is, in truth, good and you decide not to ship it, $\propto$ denotes the probability that you just made a wrong decision! Statisticians call $\propto$ the "probability of a type 1 error."

An alternate approach (approach 2 which follows) is to fix a value for $\propto$ *before* you run the test. This lets you compute a "threshold" and you can set up a "decision rule," which is like a go/no-go gage: If the test results are "worse" than this threshold, then do not ship the lot; otherwise, ship the lot. Approach 2 is the usual philosophy behind "acceptance sampling" and has the disadvantage that it can be applied in an unthinking manner.

Case 1*b:* The Standard SPC Problem

This is an example of approach 2. Control charts have been installed on your process. Periodically, you take a sample, compute $\overline{X}$ and R, and plot them on the control chart. "Control limits" are on the chart. These "limits" are like "thresholds," and your decision rule is something like this: If the $\overline{X}$ falls within the control limits, then the process is still performing satisfactorily for you. If the $\overline{X}$ falls outside of the control limits, then something bad has happened to the process, and it is not performing satisfactorily.

The difficulty of this decision-making technique is the "sharpness of the decision line." If a point falls *barely* inside the line, the process is pronounced "healthy." If the point falls *barely* outside the line, the process is pronounced "sick." For this reason, approach 1 is favored, but approach 2 is more easily used "on the floor."

The statistical phrasing of this situation is as follows:

Approach 2

Step 1: Formulate a null hypothesis (a "working assumption"), e.g., "The process average is located satisfactorily close to the target value."

Step 2: Take a sample *randomly* and compute $\overline{X}$.

Step 3: Plot the computed $\overline{X}$ on the control chart and make your decision. With "standard control charts," $\propto = 0.0027$. That is, the control limits have been determined such that the probability that $\overline{X}$ falls outside of the control limits, given that the process average is on target, is $\propto = 0.0027$.

Step 4: If you reject the null hypothesis because the sample $\overline{X}$ fell outside the control limits, then you state that "the process average has shifted out of the 'satisfactory region'; therefore, 'the process is not in control.'" Your "false alarm probability" is $\propto = 0.0027$.

Two Levels. Case 1 consisted of comparing one sample against a "standard." In case 2 we want to compare two samples against each other. The most commonly used statistical method, called the *t-test*, proceeds as follows, assuming that the two samples (1 and 2) came from two populations (I and II), and that

1. Both I and II are normal (gaussian).

2. The standard deviation of population I is the same as the standard deviation of population II.

Approach 1

Step 1: Formulate a null hypothesis ("working assumption"): H_o: The mean of population I and the mean of population II are equal.

Step 2: Perform the tests (measurements) on the items in Sample 1 and Sample 2.

Step 3: Calculate the probability of getting sample means this far (or farther) apart, under the working assumption (that the null hypothesis is true); i.e., calculate $\propto$.

Step 4: Make your decision (to reject or to not reject H_o).

If you wanted to use Approach 2, you would proceed as follows, assuming that the two samples (1 and 2) came from two populations (I and II), and that

Approach 2

Step 1: The same.

Step 2: Select a value for $\propto$, which is indicative of your fear of making a Type I error (you say that the two populations are "different," when in truth they are not different.) The value of $\propto$ which you have selected allows you to obtain a "threshold."

Step 3: Perform the tests (measurements) on the items.

Step 4: If the sample data exceeds your threshold, you pronounce that the populations are different. Otherwise, you pronounce that they are not different.

Digression on Population Characteristics. There are three features which characterize a population: its *shape,* its *spread,* and its *location.* The assumptions preceding the four steps (usually tacit) must be made before you can "run a t-test." These two assumptions concern two of the three features: assumption 1 says that the two populations under consideration have the same *shape,* and assumption 2 says that they have the same *location.* If you accept (i.e., believe) the null hypothesis, then the two populations are identical; if you reject (i.e., don't believe) the null hypothesis, then the two populations have the same shape and the same spread, but are located at different places. End of digression.

Numerical Example: Day Shift/Night Shift. Are the day shift and the night shift producing the same thing? The numerical values are here in order to show you how to manipulate the data as you strive to reach a conclusion. It is most important that *you* be able to apply this technique to the problems that *you* face. This attitude will apply to every problem/example in this text.

Here is the numerical data you have:

Day shift people: 79, 84, 108, 114, 120, 103, 122, 120

Night shift people: 91, 103, 90, 113, 108, 87, 100, 80, 99, 54

$\bar{X}$Day = 106.25 ^{S}Day = 16.64
$\bar{X}$Night = 92.50 ^{S}Night = 16.82

The first thing to do in a study like this is to *plot the data* and *look at it.* (See Figure 9-25.)

First, that low value of 54 on the night shift looks out of line. A brief investigation of it is in order (i.e., make a few phone calls). Second, if that 54 were ignored, it appears that the night shift production is a bit lower than that of the day shift, but with a little less scatter. If the 54 is retained, the spread looks the same for the two samples, but the shift in location is more obvious. Let's "do the statistical analysis" twice, once with the 54 retained and once with it discarded, and see what the data tells us in the two cases.

Formula	Our example (with the 54 retained)	Our example (with the 54 discarded)
$t = \dfrac{\bar{x}_1 - \bar{x}_2}{{}^ap \sqrt{\dfrac{1}{n_1} + \dfrac{1}{n_2}}}$	$+ = \dfrac{106.25 - 92.50}{(16.74)\sqrt{\dfrac{1}{8} + \dfrac{1}{10}}}$	$+ = \dfrac{106.25 - 96.78}{(13.75)\sqrt{\dfrac{1}{8} + \dfrac{1}{9}}}$
$s_p^2 = \dfrac{(n_1 - 1)s_1^2 + (n^2 - 1)s_2^2}{n_1 + n_2 - 2}$	$= 1.73$	$= 1.42$
	$s_p^2 = \dfrac{(7)(16.64)^2 + (9)(16.82)^2}{16}$	$s_p^2 = \dfrac{(7)(16.64)^2 + (8)(10.60)^2}{15}$
	$= 280.27$	$= 189.14$

Using the t-table, with $\sqrt{} = 16$, we find that $\propto = 0.10$. Using $\sqrt{} = 15$, we find that $\propto = 0.18$.

If we wanted to use approach 2 in this problem, we could select (say) $\propto = 0.02$. Then our threshold would be 2.583 and our decision rule would be: If $1 + 1 > 2.583$, we'll say that the shifts are "different." Otherwise, we'll say that they're not different.

Whether you use approach 1 or approach 2, keep in mind *why* you do a statistical analysis. The basic use for statistics in the industrial situations like we encounter is to provide backup for our engineering decisions. In this example the engineering decision is "the shifts are different" or "the shifts are not different, and the perceived discrepancy between the two samples is just a random error." The values tell you the probability that you're seeing a "sampling variability," under our working assumption.

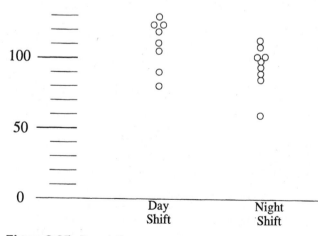

Figure 9-25. Day shift versus night shift.

Thus, the value provides you with another bit of information which you can include in your decision-making process.

Day Shift versus Night Shift Production

Test procedure

Variable: Y

Shift	N	Mean	Std.dev.	Std. error
1	8	106.25000000	16.63687814	5.88202468
2	10	92.5000000	16.82095254	5.31925224

Variances	T	DF	PROB >]T]	
Unequal	1.7338	15.2	0.1032	
Equal		1.7316	16.0	0.1026

For H_o: Variances are equal, $F' = 1.02$ with 9 and 7 DF
Prob $> F' = 1.0000$

Statgraphics
Two-sample analysis

Sample statistics:	Sample 1	Sample 2	Pooled	
Number of obs.	8	10	18	
Average	106.25	92.5		98.6111
Variance	276.786	282.944	280.25	
Std. deviation	16.6369	16.821		16.7407
Median	111	95	101.5	

Conf. interval for diff. in means: 95 percent
Sample 1 ÷ Sample 2

Hypothesis test for H_o: Diff. = 0 Computed T statistic = 1.73156
vs. alt: NE   Sig. level = 0.102583
at alpha = 0.05   so do not reject H_o.

Press the F2 key to plot or Enter to recalculate.
*1HELP 2BOXPLT3 4 5 6 7 8 9 10QU
PRINT   MON SEP 9 1985   02:23:00 PM   VERSION 1.0   REC:

Several Levels. In case 3 we want to compare several samples among themselves. The most commonly used statistical method, called the *F-test,* proceeds as follows, assuming that the several samples (1,2,3,4...) came from several populations (I,II,III,IV,...):

1. Each population is normal (gaussian).
2. The standard deviations of all populations are equal.

Step 1: Formulate a null hypothesis (a "working assumption").

Step 2: Perform the tests (measurements) on the items in the samples.

Step 3: Calculate the probability of getting sample means showing this much "dispersion," under the working assumption (that the null hypothesis is true), i.e., calculate ∝.

Step 4: Make your decision (to reject or to not reject Ho).

This is approach 1. If you prefer approach 2, then (as usual) you may select ∝ beforehand. Using ∝, √, and √₂, you may obtain what is called a *critical value of F* from the F-tables. Then you "reject your null hypothesis" if "your F" exceeds the "critical F."

Numerical Example: Daily Production Rates. Are the daily productions the same?

	Mon.	Tue.	Wed.	Thurs.	Fri.
	174.0	173.0	171.5	173.5	No data
	173.0	172.0	171.0	171.0	
	173.5	173.0	173.0	172.5	
$\overline{X}$	173.5	172.7	171.8	172.3	—
s	0.5	0.6	1.0	1.3	—

The first step, as usual, is to *plot the data* and *look at it.* (See Figure 9-26.) In Figure 9-26 it looks to me that Monday's production is a bit high, but the spread within each group is so large that we can't be sure.

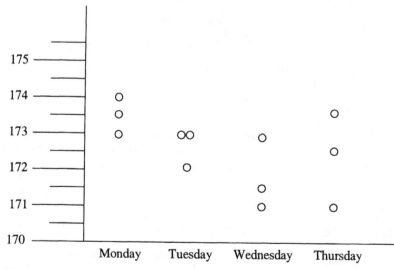

Figure 9-26. Daily production rates.

Ascertain $\propto$, which is the probability that a variation this bad (or worse) could be obtained by chance, under the assumption that there truly is no difference between the days as far as "production" is concerned.

In order to determine $\propto$, we need to compute a quantity called F, and then look up $\propto$ in a table.

Formula*	Our example
$F = \dfrac{\text{"between" variance}}{\text{"within" variance}}$	$F = \dfrac{1.548}{.825} = 1.876$
"Within" variance $= \dfrac{1}{k} \displaystyle\sum_{i=i}^{K} s_i^2$	"Within" variance $=$ $\dfrac{1}{4}(.25 + .36 + 1.0 + 1.69) = .825$
"Between" variance $= \dfrac{1}{k-1} \displaystyle\sum_{i=i}^{K} n(\overline{X}_i - \overline{x})^2$	"Between" variance $= 1.548$
k = number of samples	$k = 4$
n = number of items in each sample	$n = 3$
$\overline{x}$ = mean of the sample means	$\overline{x} = 172.575$

* *Warning:* These formulas apply *only* when there are the same number of items in each sample.

To find $\propto$, we use the F-tables. We need "the degrees of freedom for the numerator, V_1", and "the degrees of freedom for the denominator, V_2."

We use V_1 = number of samples -1

We use V_2 = sum of the degree of freedom in each sample

In our case, in the appropriate "cell" of the F-table, we see that $\propto = 0.30$.

Type I and Type II Errors

How Could You Go Wrong? You decide that the observed data shows that the population has had a change, such as "The process has gone out of control," or "The night shift is different from the day shift," or "The daily production rates are different," when in truth it has not changed. This is called a *Type I error,* or an *alpha error.*

How Else Could You Go Wrong? You decide that the observed data shows that the population has not had a change, such as "The process is still in control," or "The day and night shifts are essentially the same," or "The daily production rates are essentially the same," when in truth your population has changed. This is called *Type II error,* or *beta error.*

There is a relationship between the following "variables":

∝ $1 - \alpha$ is the confidence you have that you have *not* made a Type 1 error.

β $1 - \beta$ is the confidence you have that you have *not* made a Type 2 error, if the true mean is located δ units away from the hypothesized value XO. This is also called the *power of the test*.

σ   The standard deviation of the population.

n The number of items in your sample.

Rather than use a formula, I prefer to use the ANOVA tables to find the "correct" sample size or, if my sample size is fixed, to find the confidence levels at which I can work.

The use of the following ANOVA table to estimate the number of tests or experiments you need to perform requires you to decide, *before doing any test*, several things:

1. How large a Type 1 error ∝ will you tolerate?
2. If the population has shifted (slid) by an amount δ, then with what confidence 1-β do you want to detect that shift?
3. Do you know your population standard deviation σ?

If you know these things, then you may use the tables (taken from Davies[*] and a journal[†]) to get a recommended sample size n. In many cases, of course, your sample size is upper-bounded because of your limited resources (time, people, and money). In this case you may use these tables to "juggle" the values ∝, β, and δ/σ within the limits of your constraints.

One-Way ANOVA

General

Introduction. "Analysis of variance" is the name given to a fundamental analytical technique. This technique is almost ubiquitous, and is actually increasing in utilization because the computer is removing the drudgery of the computations involved. Our goal is to comprehend the concepts; in particular, we want to be able to understand the messages and information in the computer output, so that this statistical technique can contribute to our decision-making efforts.

---

[*] *The Design and Analysis of Industrial Experiments,* second edition, edited by Owen L. Davies, Longman Inc., New York, 1978 (published for the Imperial Chemical Industries Limited).

[†] Lloyd S. Nelson, "Sample Size Tables for Analysis of Variance," *Journal of Quality Technology,* vol. 17, no. 3, July 1985.

The basic concept of ANOVA is that *variation* is broken down in different ways, and the different breakdowns allow us to assess the statistical significance of various results.

Numerical Example: Three Furnaces.   Figure 9-27 is an illustration of "experimental variation."

Suppose the following experimental results were obtained by using three different furnaces in the tests.

Experimental Response Values Broken
Down According to Furnace

Furnace 1	Furnace 2	Furnace 3
6	8	8
4	6	8
5	6	6
5	8	8
4		10
6		

As usual, the first step is to *plot the data,* and *look at it.* (See Figure 9-28.)

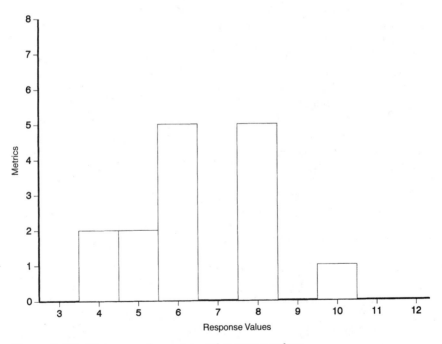

Figure 9-27. Histogram of experimental response values.

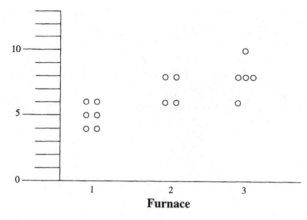

Figure 9-28. Furnace example.

Examination of the graphs in Figure 9-27 and Figure 9-28 (and contemplation of the data as broken down by furnace) seem to indicate that the total variability in response is partly due to a variability in the data from one furnace column to another (i.e., *between* furnace columns).

The question is: Is the perceived variability *between* furnaces statistically significant when compared against the inherent, God-given, "natural" variability of our system, as shown to us by the variability *within* each set of furnace data?

We need to know $\propto$, the probability that a between-furnace variability as large as (or larger than) the perceived between-furnace variability could occur by chance alone, working under the assumption that all the furnaces are equivalent.

In order to do this calculation (i.e., to find $\propto$), the working assumption that we make is that the data in each "furnace sample" came from the same population. The statistical details involved in the working assumption (i.e., the null hypothesis) are:

1. The various samples came from normal populations.

2. The populations all have the same variance (same "spread").

3. The populations all have the same mean (same "location").

Statistical texts usually present item 3 as the null hypothesis, while items 1 and 2 are regarded as conditions which must prevail before you can do an analysis of variance.

Assume, for the discussion, that items 1 and 2 are true. Now assume, as a working assumption, that item 3 is true. Under this working assumption, all of the samples were drawn from the *same* population (i.e., the populations giving rise to the samples all have the same location, the same spread, and the same shape). Then the "between" variability and the "within" variability should reflect that fact, i.e., the "between variance" should be the same as the "within variance."

The ratio

$$F = \frac{\text{"between variance"}}{\text{"within variance"}}$$

is named F in honor of R. A. Fisher, who conceived this analytical tool in the mid-1920s. Now, if item 3 is not true (i.e., if the populations are each normal) and if they each have the same spread (variance), but if they are *not* all located at the same place, then the "between variance" will be quite a bit larger than the "within variance," and that will be a signal to us that the working assumption 3 is questionable. In short, we can use the F ratio to furnish the value of $\propto$ that we seek.

We have previously discussed the way we calculate the numerator and the denominator of the F ratio. Each computer statistical package will present the computations in a slightly different format, so you must learn to translate between them and, most importantly, to interpret their message so that it may help you in your decision-making efforts.

The ANOVA Table. These computations are so frequent that they have been formalized into the ANOVA table. The ANOVA table for the "three furnaces" example looks like this:

Source of variation	Degrees of freedom	Variation (SS)	Variance (MS)	F	$\propto$
Between furnace	$2^{(7)}$	$25.73^{(6)}$	$12.867^{(8)}$	$9.65^{(9)}$	$.002^{(10)}$
Within furnace	$12^{(4)}$	$16.00^{(3)}$	$1.333^{(6)}$		
Total	$14^{(1)}$	$41.73^{(2)}$			

The table will be explained according to the parenthetical ID numbers shown in the columns:

(1) There are 15 data points. The grand mean is 6.53, which reduces the degrees of freedom to 14.

(2) The total variation is given by

$$(6 - 6.53)^2 + (4 - 6.53)^2 + ... + (8 - 6.53)^2 + (6 - 6.53)^2 + ... + (8 - 6.53)^2 + (8 - 6.53)^2 + ...$$

The "Furnace 1" Sample	The "Furnace 2" Sample	The "Furnace 3" Sample

The total of these values is 41.73, i.e., the "total sum of squares" is 41.73.

(3) The average of the furnace 1 values is 5. Within furnace 1, the sum of squares is

$$SS \text{ (furnace 1)} = (6-5)^2 + (4-5)^2 + (5-5)^2 + (5-5)^2 + (4-5)^2$$
$$+ (6-5)^2 = 4$$

The average of the furnace 2 values is 7. Hence,

$$SS \text{ (furnace 2)} = (8-7)^2 + (6-7)^2 + (6-7)^2 + (8-7)^2 = 4$$

The average of the furnace 3 values is 8. Hence,

$$SS \text{ (furnace 3)} = (8-8)^2 + (8-8)^2 + (6-8)^2 + (8-8)^2 + (10-8)^2 = 8$$

The total of these is the "within" variation, i.e.,

$$SS \text{ (within)} = 4 + 4 + 8 = 16$$

(4) Within furnace 1, there are six data points. We used the furnace 1 mean of 5 in the preceding computation, so there are only 5 degrees of freedom for SS (furnace 1). Within furnace 2, there are four data points. We used the furnace 2 mean of 7 in the preceding computation, so there are only 3 degrees of freedom (d.f.) here. Similarly, there are only 4 d.f. in the furnace 3 computations. Hence, there are $5 + 3 + 4 = 12$ d.f. associated with the "within-furnace" source of variation.

(5) The variance, or mean square (MS) is obtained by dividing the sum of squares by the degrees of freedom. Thus, $1.33 = 16.00 \div 12$.

(6) To get a feeling for the between-method variation, suppose your test results had been the following:

Furnace 1	Furnace 2	Furnace 3
5	7	8
5	7	8
5	7	8
5	7	8
5		8
5		

The grand mean is the same as before, but within variability is gone. Now, the total variability is

$$(5 - 6.53)^2 + \ldots + (5 - 6.53)^2 + (7 - 6.53)^2 + \ldots + (7 - 6.53)^2 + (8 - 6.53)^2$$
$$+ (8 - 6.53)^2 = 6(5 - 6.53)^2 + 4(7 - 6.53)^2 + 5(8 - 6.53)^2 = 25.73$$

This value represents the "between" variability, because all other variability has been artificially removed.

(7) There are only three different numbers involved (5, 7, and 8) and the grand mean 6.53 is used in the computation, so the d.f. associated with the "between" variation is 2.

(8) As before $12.867 = 25.73 \div 2$.

(9) The value for F is

$$F = \frac{\text{"between" variance}}{\text{"within" variance}} = \frac{12.867}{1.333} = 9.65$$

(10) Looking up $F = 9.65$, with 2 d.f. for the numerator and 12 d.f. for the denominator, we find

8.51   .005
9.65 $\propto$
13.0   .001

and hence (linear interpolation)

$$\frac{9.65 - 8.51}{13.0 - 8.51} = \frac{\propto - .005}{.001 - .005}$$

gives us

$$\propto = .004$$

Linear Contrasts.   The sketch and our analysis have indicated that furnace 1 is "different" from the other two. We already know that we could use a t-test on the two subgroups to get the $\propto$ that goes with this engineering conclusion. One subgroup consists of the six data points coming from furnace 1, and the other subgroup consists of the nine data points coming from furnace 2 and furnace 3. A formal way to compare the two subgroups using ANOVA techniques is to use the technique of linear contrasts. You can do this by forming the "linear contrast":

$$L_1 - \frac{6 + 4 + + 5 + 5 + 4 + 6}{6} - \frac{8 + 6 + 6 + 8 + 8 + 8 + 6 + 8 + 10}{9}$$

to see if this comparison is S-sig. Note that L_1 is the average of the furnace 1 data minus the average of the rest of the data. This technique for finding this will give you exactly the same results as a t-test. You might want to use this technique as part of your ANOVA, so it might be more convenient than a t-test.

In our case

$$L_1 = \frac{1}{6}(6) + \frac{1}{6}(4) + ... + \frac{1}{6}(6) - \frac{1}{9}(8) - \frac{1}{9}(6) - ... - \frac{1}{9}(10) = 9 - 0.256$$

$$\text{and } \frac{1}{6} + \frac{1}{6} + ... + \frac{1}{6} - \frac{1}{9} - \frac{1}{9} - ... - \frac{1}{9}(10) = -2.56$$

then

$$SSL = \frac{(-2.56)^2}{(1/6)^2 + ... + (1/6)^2 + (1/9)^2 + ... + (1/9)^2} = \frac{6.53}{.28} = 23.51$$

You determine your $\propto$ by the usual method:

$$F = \frac{\text{"between" variance}}{\text{"within" variance}} = \frac{23.51}{1.33} = 17.63$$

$$dfL = 1 \text{ and } df2 = 12$$

Hence $\propto = .001$.

This is one comparison we could make using the "method" data. Another comparison would be to check furnace 2 versus furnace 3 to make sure they weren't significantly different. We form another contrast:

$$L_2 = \frac{8 + 6 + 6 + 8}{4} - \frac{8 + 8 + 6 + 8 + 10}{5}$$

$$= \frac{1}{4}(8) + \frac{1}{4}(6) + \frac{1}{4}(6) + \frac{1}{4}(8) ... - \frac{1}{5}(10) = -1$$

Note that

$$\frac{1}{4} + \frac{1}{4} + \frac{1}{4} + \frac{1}{4} - \frac{1}{5} - \frac{1}{5} - \frac{1}{5} - \frac{1}{5} - \frac{1}{5} = 0$$

as is required for t "Linear contrast," and calculate its sum of squares.

$$SSL_2 = \frac{(-1)^2}{(1/4)^2 + ... (1/4)^2 + (1/5)^2 + (1/5)^2} = \frac{1}{0.25 + 0.20} = 2.22$$

These two contrasts may be put into our ANOVA table so as to document our two comparisons. The revised table looks like this:

Source of variation	d.f.	Variation (SS)	Variance (MS)	Variance ratio F	$\propto$
Furnace	2	25.73	12.87	9.65	.004
L_1 (1 vs. 2&3)	1	23.51	23.51	17.63	.001
L_2 (2 vs. 3)	1	2.22	2.22	1.67	.32
Error	12	16.00	1.33		
TOTAL	14	41.73			

This concept appears where it is used to determine the statistical signifi-cance of the "effects of the factors."

Confidence Limit Estimates. If you decide to "reject your null hypothesis," then you are really saying that the various populations involved all have the same *shape* and the same *spread*, but are *located* in dif-ferent places. In that case, we'd like to estimate the locations of the differ-ent populations. We could use the appropriate sample means as "point estimates" of the population means. We can form "interval estimates" of the population means using the variability shown to us by the data to determine the width of the intervals. The width of the intervals depends also on the confidence we want to have when we make the statement: "The population mean is somewhere in the interval."

Continuing with our example, from the revised ANOVA table, it appears that there are two populations involved: one which gave rise to the furnace 1 example, and the other which gave rise to the furnace 2 and furnace 3 data. Both of these populations are normal, and they both have the same spread (i.e., variance or, equivalently, standard deviation).

The estimated standard deviation of each population is

$$\hat{\sigma} = \sqrt{1.33} = 1.155,$$

$$= \sqrt{\text{error variance}}$$

The two populations differ in their locations. The location of a distribution is specified by its mean, so we want to be able to estimate the population means using our sample data, while taking into account the variability shown by our sample data. Our confidence limit estimates of the means of the two populations are as follows:

$$\hat{\mu}_1 = 5.00 \pm \left(\frac{^{+}\propto}{2}, 12\right)\left(\frac{(1.155)}{\sqrt{6}}\right)$$

$$\hat{\mu}_{2,3} = 7.556 \pm \left(\frac{^{+}\propto}{2}, 12\right) \cdot \left(\frac{(1.155)}{\sqrt{9}}\right)$$

where 5.00 is the mean of the furnace 1 data, 7.556 is the mean of the data in the furnace 2 and furnace 3 samples $^+\propto/2$, 12 is our usual $+$, with its degrees of freedom being the same as the d.f. of $\hat{\sigma}$.

n_e is the appropriate "sample size" and is given by:

$$n_e = 6 \text{ for the } \hat{\mu}_1 \text{ calculation}$$

$$n_e = 9 \text{ for the } \hat{\mu}_{2,3} \text{ calculation}$$

For example, if we use $\propto = .10$, then

$$\frac{^{+}0.1}{2}, 12 = 0.2179$$

$$\hat{\mu}_1 = 5 \pm \frac{(2.179)(1.155)}{\sqrt{6}}$$

$$3.97 < \mu_1 < 6.03$$

$$\hat{\mu}_{2,3} = 7.556 \pm \frac{(2.179)(1.155)}{\sqrt{9}}$$

$$6.72 < \mu_{2,3} < 8.59$$

Two Variables; Two-Way ANOVA

General. When we consider two variables, we find that there are two different kinds of experimental situations, or experimental designs, to consider. The first type we look at is an experimental plan called *fully crossed,* and the second type of plan is called *fully nested.* The analysis is different for the two types.

Fully Crossed, One Data Point Per Cell

Introduction. In a crossed design, each level of each factor appears in conjunction with each level of each other factor, as shown in Figure 9-29. This design actually has 12 experimental conditions. In each condition, one of the levels of B appears with one of levels of A. In order to analyze this type of design, we want to expand the ANOVA technique from one variable up to two variables, obtaining what we call *two-way ANOVA.* First, we consider the case where we have two variables, each at several levels, but only one data point per "cell." Later we will study the case where we have several points per "cell."

Numerical Example: Analysts and Thermometers, Crossed, One Data Point Per Cell. Suppose that we have three analysts (technicians) who

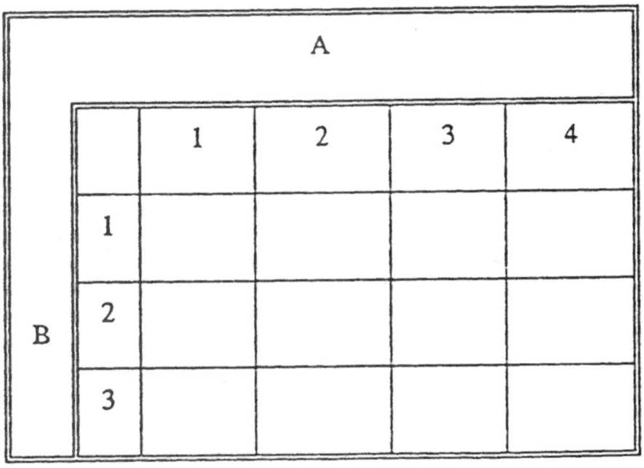

Figure 9-29. Crossed design.

take readings from each of four thermometers. Since each level of the variable "thermometers" appears in conjunction with each level of the variable "analyst" (that is, each technician actually uses each thermometer), this is a fully crossed design. We have 12 (= 3 × 4) readings.

| Thermometer | Data: Observed Melting Points (°C)* | | | | |
	A	B	C	D	Averages
Analyst I	174.0	173.0	171.5	173.5	173.0
Analyst II	173.0	172.0	171.0	171.0	171.75
Analyst III	173.5	173.0	173.0	172.5	173.00
Averages	173.5	172.67	171.83	172.33	172.58

* *Quality Control and Industrial Statistics*, p. 632, fourth edition, Acheson J. Duncan, Richard D. Irwin, Inc., 1974.

As usual, the first thing we do is to plot up the data and look at it. (See Figure 9-30.) From the graphs, it looks like analyst II is different from analysts I and III, and thermometer A is different from the others.

In order to check out this lead, we may do various "statistical things." This concept is, in my opinion, very important. I believe that the main role of statistics is to provide backup or support for engineering decisions. Thus, a typical problem would have the statistical analysis done *after* the engineering analysis, and would provide statistical credibility to the engineering analysis. A secondary role for statistics is to occasionally expose interesting results which might be "obscured by the noise."

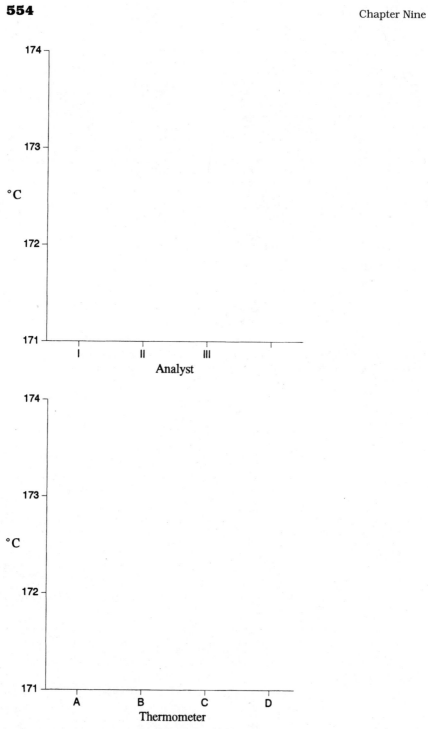

Figure 9-30. Analysts and thermometers example.

The calculations we now have are exactly the same in philosophy as before, but they are a bit more complicated. In particular, we have

$$SS_{AN} = 4(173.00 - 172.58)^2 + 4(171.75 - 172.58)^2 + (173.00 - 172.58)^2 = 4.167$$

$$SS_{TH} = 3(173.5 - 172.58)^2 + 3(172.67 - 172.58)^2 + 3(171.83 - 172.58)^2$$
$$+ 3(172.33172.58)^2 = 4.438$$

$$SS_T = (174.0 - 172.58)^2 + (173.0 - 172.58)^2 + \ldots + (172.5 - 172.58)^2 = 10.917$$

$$SS_E = SS_T - SS_{AN} - SS_{TH} = 2.312$$

As before, the results of our computations may be put in tabular form.

Source of variation	d.f.	Variation (SS)	Variance (MS)	Variance ratio F	$\propto$
Analysts (rows)	2	4.167	2.084	5.4130	0.05
Thermometers (columns)	3	4.438	1.479	3.8416	0.08
Residual or error	6	2.312	0.385		
Total	11	10.917			

ANOVA Table

Following the hints given us by the graphs, we want to find the $\propto$ which goes with the difference between the analyst II data and the remaining data. We get $\propto$ by forming a linear contrast, calculating its MS, finding its F, and looking $\propto$ up in the tables.

L_A = (mean of analyst II) − (mean of analysts I and III)

$$= \frac{173.0 + 172.0 + 171.0 + 171.0}{4} - \frac{174.0 + 173.0 + \ldots + 173.0 + 172.5}{8}$$

$$= 171.75 - 173.0$$

$$= -1.25$$

Next we calculate

$$SSL_A = \frac{L_A^2}{D}, \text{ where}$$

$$D = (1/4)^2 + \ldots + (1/4)^2 + (1/8)^2 + \ldots + (1/8)^2$$

$$= 4(1/4)^2 + \quad + 8(1/8)^{2 + (1/4) + (1/8)} = 3/8 = .375$$

We get $SSL_A = 4.17$, and $MSL_A = 4.17$, since d.f. = 1 (always).

We may also do the same analysis with respect to thermometer A and the other three thermometers. We form

L_T = (mean of thermometer A) − (mean of others)

$$= \frac{174.0 + 173.0 + 173.5}{3} - \frac{173.0 + 172.0 = \dots = 171.0 + 172.5}{9}$$

$$= 173.5 - 172.28$$

$$= 1.22$$

$$D = (1/3)^2 + (1/3)^2 + (1/3)^2 + (1/9)^2 + \dots + (1/9)^2$$

$$= 3(1/3)^2 + 9(1/9)^2 = (4/9) = .444$$

and calculate

$$SSL_T = \frac{(1.22)^2}{.444} = 3.35$$

Hence, $MSL_T = 3.35$, since d.f. = 1 (always).

We may want to compare the thermometer C results against the results from thermometers B and D. Then

$$^L T_2 = \text{(mean of thermometer C data)}$$
$$- \text{(mean of thermometer B and D data)}$$

$$= 171.83 - 172.5 = 0.67$$

$$^{SSL}T_2 = \frac{(0.67)^2}{0.5} = 0.89$$

This gives us a revised ANOVA table:

Source of variation	SS	d.f.	MS	F	$\propto$
Analyst (rows)	4.167	2	2.08	5.33	0.05
L_A	4.167	1	4.167	10.68	0.02
Thermometers (columns)	4.438	3	1.48	3.79	0.08
L_T	3.35	1	3.35	8.59	0.03
$^L T_2$	0.89	1		2.28	0.10
Residual	2.33	6	0.39		
Total	10.92	11			

These linear contrasts show us that the analyst variability is totally caused by how II differs from I and III, and the thermometer variability is mostly due to how A differs from B, C, and D; we also obtain the $\propto$ which goes with these decisions.

Since our MSE = 0.385, we have $\hat{\sigma} = \sqrt{.385} = .62$ with 6 degrees of freedom. Using $\propto = 0.05$, we get

$$\frac{{}^{+}0.05}{2}, 6 = 2.447$$

Hence, for the analysts,

$$\hat{\mu}_2 = 171.75 \pm (2.447)\,\frac{.62}{\sqrt{4}}$$

So, $170.22 < \mu_2 < 173.03$

$$\hat{\mu}_{1,3} = 173.00 \pm (2.447)\,\frac{.62}{\sqrt{8}}$$

and $172.46 < \mu_{1,3} < 173.54$

Similar confidence limit estimates can be made for the thermometers.

$$\hat{\mu}_A = 173.5 \pm \frac{(2.447)\,(.62)}{\sqrt{3}}$$

So, $172.62 < \mu_A < 174.38$

$$\hat{\mu}_{B,C,D} = 172.28 \pm \frac{(2.447)\,(.62)}{\sqrt{9}}$$

and $171.775 < \mu_{B,C,D} < 172.79$

Fully Crossed, More than One Data Point Per Cell (But an Equal Number)

Introduction.   Next we consider two-way ANOVA, with more than one but an equal number of cases in each class. We'll first look at the fully crossed design, with two readings per cell. Now we have 24 (= 3 × 4 × 2) readings.

Numerical Example: Analysts and Thermometer, Crossed, Two Data Points Per Cell

Thermometer	Data: Observed Melting Points (C°) A	B	C	D	Averages
Analyst I	174.0	173.0	171.5	173.5	
	173.5	173.5	172.5	173.5	173.12
Avg.	173.75	173.25	172.00	173.50	
Analyst II	173.0	172.0	171.0	171.0	
	173.0	173.0	172.0	172.0	172.12
Avg.	173.00	172.50	171.50	171.50	
Analyst III	173.5	173.0	173.0	172.5	
	173.0	173.5	173.0	173.0	173.06
Avg.	173.25	173.25	173.0	172.75	
Averages	173.33	173.00	172.17	172.58	172.77

NOTE: Duncan, p. 646.

Having more than one data value in each cell allows us to study the "within-cell" variability, which we could not do earlier. This lets us make an even more complicated breakdown of the data.

The idea is that any data value may be thought of as being "composed of" a part due to the grand mean, plus a bias depending on which thermometer was used, plus a bias depending on which analyst took the reading, plus a bias which depends on the simultaneous combination of analyst and thermometer, plus a random noise component. The bias which comes from the simultaneous combination of analyst and thermometer is called an *interaction*. Thus, the values in a particular cell will depend on which analyst takes the reading (a *row effect*) and on which thermometer was used (a *column effect*) and, in addition, may be dependent on some unique thing that happens when that particular analyst uses that particular thermometer. This interaction effect is like a nonlinearity, in a sense.

We can get a feel for this concept by looking at a breakdown of the data. Before we do that, let's look at Figure 9-31, which shows the data in graphical form.

Our ANOVA calculations are as follows. Note particularly how the variation (SS) due to the interaction $AN \times TH$ is "extracted" from the variation (SS) due to the cells. We will encounter this concept again and again.

$$SS_{TH} = 6((173.33 - 172.77)^2 + (173.00 - 172.77)^2 + (172.17 - 172.77)^2 + (172.58 - 172.77)^2) = 4.576$$

$$SS_{AN} = 8((173.12 - 172.77)^2 + (172.12 - 172.77)^2 + (173.06 - 172.77)^2) = 5.033$$

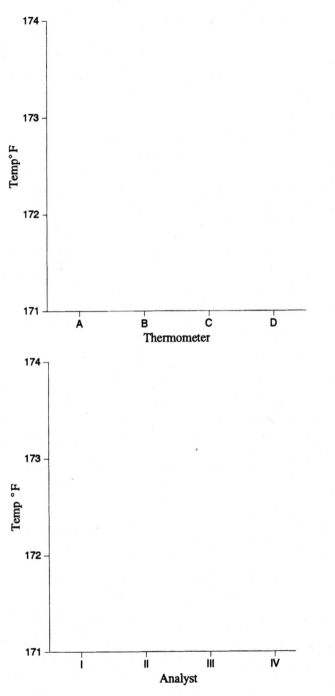

Figure 9-31. Thermometer and Analyst example.

$$SS_{CELLS} = 2((173.75 - 172.77)^2 + (173.25 - 172.77)^2 + ... + (172.75 - 172.77)^2)$$
$$= 12.365$$

N.B.

$$SS_{AN \times TH} = SS_{CELLS} - SS_{TH} - SS_{AN} = 2.756$$

$$SS_T = (174.0 - 172.77)^2 + (173.5 - 172.77)^2 + ... + (173.0 - 172.77)^2$$
$$= 14.990$$

$$SS_E = SS_T - SS_{AN} - SS_{TH} - SS_{AN \times TH} = SS_T - SS_{CELLS} = 2.625$$

This gives us our ANOVA table:

Source of variance	d.f.	Variation (SS)	Variance (MS)	Variance ratio F	$\propto$
Analysts	2	5.033	2.516	11.49	0.003
Thermometer	3	4.576	1.525	6.96	0.006
AN × TH	6	2.756	0.459	2.10	.10
Error	12	2.625	0.219		
Total	23	14.990			

The ANOVA $\propto$ values certainly support the observation that the data appears to come from different populations. The Figure 9-21 shown earlier prompt us to make linear contrasts looking into "A versus the other three" and "II versus the other two." We form

$$L_A = 173.33 - 172.58 = 0.75 \quad \text{and} \quad SS_{(LA)} = \frac{0.5625}{0.2222} = 2.5312$$

While we're at it, we can check thermometer C versus thermometers B and D. We form

$$L_C = 172.17 - 172.79 = 0.62 \quad \text{and} \quad SS_{(LC)} = \frac{0.3844}{0.2500} = 1.5376$$

For the analysts, we form

$$L_N = 172.12 - 173.09 = 0.97 \quad \text{and} \quad {}^{SS}(L_N) = \frac{0.9409}{0.1875} = 5.0181$$

The ANOVA table can be revised as follows:

Source of variation	d.f.	Variation SS	Variance MS	Variance ratio F	$\propto$
Analyst	2	5.033	2.516	11.49	0.003
L_H	1	5.0181	5.0181	22.91	0.001
Thermometer	3	4.576	1.525	6.96	0.006
L_A	1	2.5312	2.5312	11.59	0.005
L_C	1	1.5376	1.5376	7.02	0.025
AN × TH	6	2.756	0.459	2.10	0.10
Error	2	2.625	0.219		
Total	3	14.990			

From the revised ANOVA table, we see that essentially all of the analyst variability is due to II, and that the difference between A and B, C, D and the difference between D, B and D accounts for essentially all of the thermometer variability.

Confidence limit estimates of the means of the different populations may be calculated. We may use $\hat{\sigma} = \sqrt{0.219}$.

The estimated standard deviation $\hat{\sigma} = \sqrt{0.219}$ has 12 degrees of freedom associated with it (1 degree of freedom per cell) in this situation where we have two data points per cell. Using $\propto = 0.05$, we find that $t = 2.179$. Hence,

$$\hat{\mu}_A = 173.33 \pm \frac{(2.179)\sqrt{0.219}}{\sqrt{6}}$$

$$= 173.33 \pm 0.42$$

So
$$= 172.91 < \mu_A < 173.75$$

and

$$\hat{\mu}_{B,C,D} = 172.58 \pm \frac{(2.179)\sqrt{0.219}}{\sqrt{18}}$$

$$= 172.58 \pm 0.24$$

So
$$= 171.88 < \mu_{B,C,D} < 172.36$$

For the analysts, we get

$$\hat{\mu}_H = 172.12 \pm \frac{(2.179)\sqrt{0.219}}{\sqrt{8}}$$

So
$$= 171.76 < \mu_H < 172.48$$

and

$$\hat{\mu}_{I,III} = 173.09 \pm \frac{(2.179)\ \sqrt{0.219}}{\sqrt{16}}$$

$$= 173.09 \pm 0.255$$

So
$$= 172.835 < \mu_{I,II} < 173.345$$

DOE Concepts and Tools

Introduction. Lately, the DOE field has been invigorated by the appearance of the "Taguchi method." The increasingly widespread acceptance of this method by working engineers in the United States and its proven track record in Japan indicate to me that we practicing engineers ought to be acquainted with this method.

Some of us have, for a number of years, been practicing the DOE using what I now want to call "the classical method," which is best presented, in my opinion, in the book *Statistics for Experimenters* by George Box, W. G. Hunter, and J. S. Hunter, published by Wiley. The Taguchi method (TM) differs quite drastically from the classical method in several areas. Many of the more advanced concepts discussed in Box, Hunter, and Hunter will not be covered in this text. (The goal of this text is to present to the reader a tool which he or she can use immediately.)

Dr. Genichi Taguchi is a Japanese statistical engineer. He has worked in the Electrical Communication Lab from its beginning in the early 1950s. One of the first things he did there was to translate Cochran and Cox into Japanese. He rejected the classical approach as not very useful in industrial situations, and developed his own design of experiments methods. He won the Deming Award in 1960 for his statistical contributions. He visited the Bell Labs in 1980 to "repay debts," and thereby first brought his methods to the attention of American engineers.

The Taguchi method has been well accepted in Japan. Leonard Lamberson, head of the Industrial Engineering and Operations Research Department at Wayne State University furnished the following information:

- The method is presented in a 120-hour course in Japan.
- More than 31,000 engineers have taken this course (i.e., are trained in DOE).
- Course has been used by N.E.C., Hitachi, OKI, Fujitsu, Nippondenso, and others.
- Toyota requires *all* engineers to take the course.

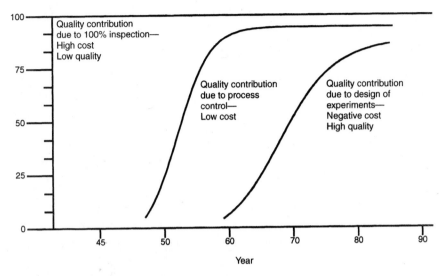

Figure 9-32. Quality in Japan.

■ In 1980, it was estimated that more than 1 million experiments using orthogonal arrays had been performed in Japan.

The Taguchi method has been well accepted in the United States at Bell Labs, Xerox, ITT, and Ford Motor, and its acceptance is rapidly increasing as engineers discover that "it works." ITT has a "manager of Taguchi methods" (Hassan Mutlu); I have heard informally that Ford Motor will require that all of their engineers be trained in the Taguchi method! (See Figure 9-32.)

Definitions. An experimental design matrix is a delineation of the experimental conditions which should prevail in each of the trials, or tests, or "runs" involved in the overall plan. In such a matrix, the columns represent your variables, or factors, and the rows represent the tests, or experiments, to be performed.

In the TM, a set of design matrices has already been prepared. The simplest of these is called L4, and looks like this:

	L4		
Run no.	Column 1	Column 2	Column 3
1	1	1	1
2	1	2	2
3	2	1	2
4	2	2	1

In this table, there are three columns, so three independent variables or factors can be studied. In each column appear the numbers 1 and 2, which denote the two levels that each variable can have. The four rows, labeled "run number," denote the four experimental conditions which are supposed to hold for the four different tests which this matrix calls for. Please note that the four tests are not necessarily performed in the order 1–2–3–4. Instead, the tests should be performed in a *random* order, unless engineering common sense requires that they be done in subgroups. If so, the runs within a subgroup should be done in a random order. Doing the runs in a random order is equivalent to taking a random sample, so your sample results may be extrapolated to the population from whence they came, using the laws of statistics.

For example, suppose that you want to study three factors:

	Factor	Level 1	Level 2
A	Rear bearing fit	Tight (<0.0010)	Loose (>0.0010)
B	Concentricity	Good (<0.0020)	Bad (>0.0020)
C	Type of bearing	NDH	Other

The L4 design matrix can be used, since it has three columns. The experimental design could look like the following:

	A (0.0001)	B (0.000)	C	
Run no.	1	2	3	Results
1	8	13	NDH	
2	8	35	Other	
3	11	10	Other	
4	14	50	NDH	

The four tests then consist of preparing four experimental units (in this case, they were compressors) and performing the trials in a random order. (In this case, the trial was "run to destruction," and the experimental output was "hours to failure").

Orthogonal Arrays and Linear Graphs. In addition to L4, which we discussed earlier in the chapter, there is an entire set of orthogonal arrays having more rows and columns. Associated with each orthogonal array is a "table of interactions," showing which columns represent the interaction of which other pairs of columns. The interactions may be represented by

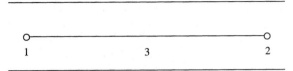

Figure 9-33. The linear graph for L4.

linear graphs to facilitate the assignment of variables to columns. Figure 9-33, for example, is the linear graph which goes with the L4 design. It says that column 3 represents the interaction of columns 1 and 2 by showing the number 3 on the line segment connecting the nodes 1 and 2.

In the design of the previous section, column 3 represents the interaction of A (rear bearing fit) and B (concentricity), *as well as* the variable C (type of bearing). This arrangement is good only if you believe that the AB interaction is negligible from an engineering point of view. Then, variable C is not confused with the two-factor interaction AB, and the experimental results may be interpreted without worry. If you cannot believe that the two-factor interaction AB is negligible, then you cannot obtain the effect of the variable C in a clear manner. This situation is resolved by using a larger design. The next larger design is L8, which has eight rows and seven columns. It has two linear graphs associated with it, both of which reflect the interactions between the columns. You may use whichever linear graph is best suited to your particular situation to help you assign the variables to the columns. (See Figure 9-34.)

Run no.	1	2	3	*L8* 4	5	6	7	Results
1	1	1	1	1	1	1	1	
2	1	1	1	2	2	2	2	
3	1	2	2	1	1	2	2	
4	1	2	2	2	2	1	1	
5	2	1	2	1	2	1	2	
6	2	1	2	2	1	2	1	
7	2	2	1	1	2	2	1	
8	2	2	1	2	1	1	2	

In the present example, it makes sense to assign A to 1, B to 2, and C to 4 (using the left-hand linear graph), whereupon column 3 represents the AB interaction, column 6 represents the BC interaction, and column 5 represents the AC interaction. Column 7 does not have a factor assigned to it, and therefore represents "random noise." (More properly, it represents the three-factor interaction ABC. Three-factor interactions are *extremely rare,*

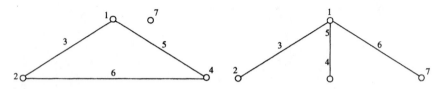

Figure 9-34. The linear graphs for L8.

and may be *ignored* from an engineering point of view at this stage in our statistical development. (If you ever think you have a three-factor interaction, "consult your local statistician.")

In the present example, the design now looks like this:

Run no.	A (0.0001) 1	B (0.0001) 2	AB 3	C (type) 4	AC 5	BC 6	E 7	Results
1	9 (1)	15 (1)	1	NDH (1)	1	1	1	457
2	8 (1)	13 (1)	1	Other (2)	2	2	2	112
3	6 (1)	35 (2)	2	NDH (1)	1	2	2	447
4	8 (1)	35 (2)	2	Other (2)	2	1	1	114
5	12 (2)	15 (1)	2	NDH (1)	2	1	2	485
6	11 (2)	10 (1)	2	Other (2)	1	2	1	163
7	14 (2)	50 (2)	1	NDH (1)	2	2	1	507
8	12 (2)	25 (2)	1	Other (2)	1	1	2	171

The experimental results (hours to failure) are shown. The "effect" of a variable (say, A) is given by

$$\text{Effect of A} = (\text{average of results when A} = 2) - (\text{average of results when A} = 1)$$

$$= 331.5 - 282.5 = 49.0$$

This definition also applies to the interactions. For example,

$$\text{Effect of AB} = (\text{average of results when AB} = 2) - (\text{average of results when AB} = 1)$$

$$= \frac{447 + 114 + 485 + 163}{4} - \frac{457 + 112 + 507 + 171}{4}$$

$$= 302.25 - 311.75 = -9.5$$

Later we will discuss what the −9.5 means.

Similarly, all of the effects of the variables and of the interactions, including the "effect" of the unassigned factor in column 7 may be calculated. The interpretation of the effects of A, B, and C are as follows. The effect of A is 49.0, which means that a change from "tight" to "loose" rear bearings will, on the average, result in an increase of 49.0 hours in "time to failure." The effect of B is –5.5, which means that a change from "good" concentricity to "bad" concentricity will, on the average, result in a decrease of 5.5 hours in "time to failure." The effect of C is –334.0, which means that changing from "NDH" to "Other" bearings will, on the average, cause a decrease of 334.0 hours in "time to failure."

Variable	Effect
A	49.0
B	–5.5
C	–334.0
AB	–9.5
AC	–5.0
BC	0.5
E	–6.5

Let us postpone until later the interpretation of the interaction and noise terms.

Statistical Significance of Effects.   In every experiment, there is an experimental error. This is not a blunder, but is the variability which results because of minute, uncontrollable changes in the conditions of the experiment. In our present example, this is illustrated by the fact that the noise variable in column 7 caused an effect of –6.5, which, incidentally, is larger than the effect of concentricity! This, of course, makes us suspect that concentricity really has no effect at all, and the observed effect pf – 5.5 is merely experimental error.

Using exactly the same logic as applied earlier, we need to ascertain which of our effects are statistically significant, and which are not. The easiest way to do this is by doing an ANOVA. For factors at two levels, as in the case here, the sum of squares is obtained using the linear contrast technique we studied earlier. Recall that the sum of squares for a linear contrast LA is given by

$$SS_{(LA)} = \frac{LA^2}{\text{sum of squares of the coefficients}}$$

and it is convenient to take

$$L_A = (\text{sum of results when } A = 2) - (\text{sum of results when } A = 1)$$

For example,

$$L_C = (112 + 114 + 163 + 171) - (457 + 447 + 485 + 507) = -1336$$

Using this method, the following ANOVA table may be obtained for the example problem we're presently looking at.

		ANOVA Table			
Source of variation	d.f.	Variation SS	Variance MS	Variance ratio F	$\propto$
A	1	4802.0		56.8	.09
B	1	60.5		0.7	
C	1	223112.0		2640.4	.02
AB	1	180.5		2.1	
AC	1	50.0		0.6	
BC	1	0.5		0.0	
E	1	84.5			
Total	7	228290.0			

Examining the F tables, we see that A is S-sig and C is highly S-significant, and all of the other effects (namely, B and the three interactions) are not S-sig. In a situation like this, many people will combine all of the statistically nonsignificant terms into what's called a *pooled error* term, by adding together the sums of squares (and the degrees of freedom). If we do that, the previous table becomes:

		ANOVA Table			
Source of variation	d.f.	Variation SS	Variance MS	(Pooled) Variance, ratio F	$\propto$
A	1	4802.0		63.9	.001
C	1	223112.0		2966.9	.001
(E)	5	376.0	75.2		
Total	7	228290.0			

Note the increased "sharpness" of the $\propto$ values.

Interpretation of Interaction. The two-factor-interaction terms represent the change in the effect of one variable, depending on the levels of another variable. The easiest way to perceive interactions is by graphical presentation. For example, consider the AB interaction in Figure 9-35.

A graphical presentation shows how the effect of rear bearing fit is slightly different, depending on the concentricity. The rear bearing fit has a slightly more noticeable effect when the concentricity is bad than it does when the concentricity is good. In this example, the two lines are almost parallel. When the lines are parallel, we say that there is no interaction, and the more nonparallel the lines are, the more interaction we have.

In the hypothetical example in Figure 9-36, a severe interaction is present. Clearly, the effect of temperature depends drastically on the level of the other variable, pressure.

When the graphed lines are parallel, no interactions are present. The effect of temperature is the same (an increase in tensile strength) whether pressure is low or high. (See Figure 9-37.)

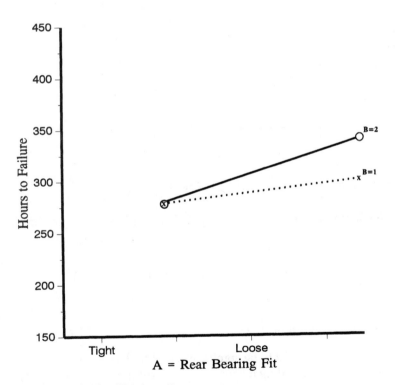

Figure 9-35. The AB interaction.

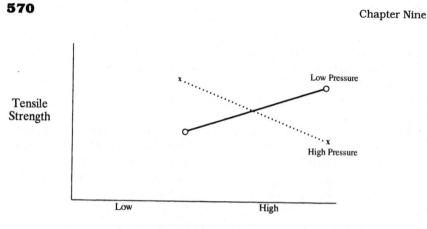

Figure 9-36. Hypothetical example (large interaction).

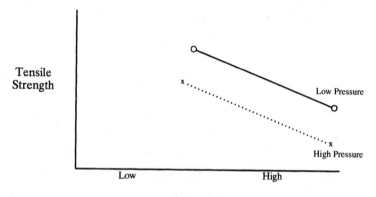

Figure 9-37. Hypothetical example (no interaction).

A Hypothetical Example. As a more specific example, suppose you have

      A = raw material (2 levels)

      B = annealing method (2 levels = 2 different furnaces)

      C = annealing temperature (2 levels)

      D = annealing time (2 levels)

You expect B × C and B × D to be significant, and the better raw material will be better for any annealing technology. Accordingly, you're interested only in the main effect of A (i.e., you're not interested in any of the interactions of A with any other variable).

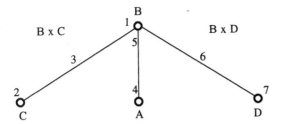

Figure 9-38. Assignment of variables to column of
L8 where 5 is an unassigned factor (UF).

You are interested in A, B, C, D, B × C, and B × D. (See Figure 9-38.)
It seems that L8 can handle this experiment, since only 6 columns are needed. Your experimental design then looks like this:

Run no.	B 1	C 2	BC 3	A 4	e 5	BD 6	D 7	Test results
1	1	1	1	1	1	1	1	14.53
2	1	1	1	2	2	2	2	18.66
3	1	2	2	1	1	2	2	11.84
4	1	2	2	2	2	1	1	16.42
5	2	1	2	1	2	1	2	11.83
6	2	1	2	2	1	2	1	16.42
7	2	2	1	1	2	2	1	13.23
8	2	2	1	2	1	1	2	18.79
Effect	−0.295	−0.290	−02.175	4.715	−0.360	−0.355	0.130	15.2150

Shown on the bottom line are the effects of the various variables and interactions corresponding to the test results given in the right-hand column.
The ANOVA Table for this example is as follows:

		ANOVA Table			
Source of variation	d.f.	Variation SS	Variance MS	Variance ratio F	∝
A	1	44.4625		171.5	0.04
B	1	0.1741			
C	1	0.1682			
D	1	0.0338			
BC	1	9.4613		36.5	0.10
BD	1	0.2521			
E	1	0.2592			
Total	7	54.8112			

After pooling, the ANOVA table becomes:

Source of variation	d.f.	*ANOVA Table* Variation SS	Variance MS	Variance ratio F	$\propto$
A	1	44.4625		250.49	0.001
BC	1	9.4613		53.30	0.025
Error	5	0.8874	0.1775		
Total	7	54.8112			

Having decided by ANOVA that A and BC are S-sig, their effects may be illustrated graphically, as shown in Figure 9-39.

If "high results are good," your conclusions from this experiment would be as follows:

- Use the raw material corresponding to "A is HI."
- Use either B is LO *and* C is LO or B is HI *and* C is HI as the combination of annealing method and annealing temperature.
- Annealing time has no effect, so use that value which is cheapest.

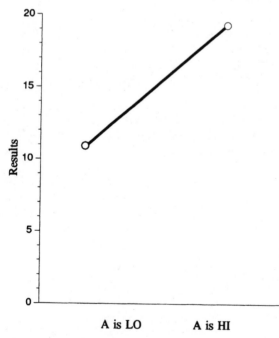

Figure 9-39. Effects of A.

Figure 9-40 showing the BC interaction is my preferred method for illustrating what an interaction is. The "effect" of the interaction is an indicator as to the amount of "nonparallelism" percent in your situation. Since the effect of BC was S-sig, we see we have a lot of nonparallelism. The effect of BD is small, i.e., not S-sig, so if we were to plot the BD interaction graph, we would expect to see the two lines be quite parallel.

An Example. For example, consider a situation where you have three types of units, each of which may be of small, medium, or large size, and to each of which you may apply a varying amount of oil. You are interested in the time needed to complete a certain task. Using L9 (and ignoring column 4), you might design the following experiment:

Column 1   Type: 1 = A, 2 = B, 3 = C
Column 2   Size: 1 = small, 2 = medium, 3 = large
Column 3   Number of drops: 1 = 0, 2 = 2, 3 = 4
Results   Number of seconds required to complete a task

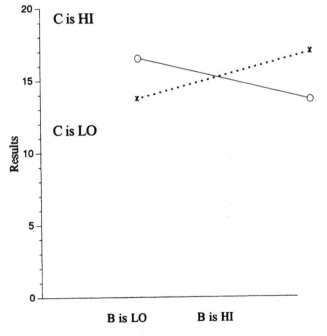
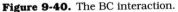

Figure 9-40. The BC interaction.

Run no.	1	2	3	4	Results
1	A	Small	0		13.6
2	A	Medium	2		14.6
3	A	Large	4		15.0
4	B	Small	2		14.0
5	B	Medium	4		18.8
6	B	Large	0		12.2
7	C	Small	4		18.6
8	C	Medium	0		11.2
9	C	Large	2		13.6

This experiment is identical to a $3 \times 3 \times 3$ Latin square, as is easily verified.

A Set of Orthogonal Arrays, Interaction Tables, and Linear Graphs. Presented here are the orthogonal arrays, interaction tables, and linear graphs from Appendix 4 of *Introduction to Off-Line Quality Control*, by Genichi Taguchi and Yu-in Wu, published by the Central Japan Quality Control Association. There are orthogonal arrays other than the ones given here, but they are of higher order, i.e., they call for experiments consisting of 32 or more trials. At this stage in our development, the arrays given here will probably suffice. If not, "consult your local statistician."

Before presenting the arrays, tables, and graphs, the following should be noted:

1. In order to easily lay out various kinds of experiments, some standard linear graphs are listed.

2. "No." and "Column" indicate experimental numbers and columns in a table, respectively.

3. Tables of interactions are listed for the purpose of obtaining the interactions between two columns.

4. In linear graphs, groups in a graph are indicated by the following symbols:

$L_{32}(2^{31})$		$L_{64}(2^{63})$		Other tables	
Symbols	Groups	Symbols	Groups	Symbols	Groups
○	Groups 1 and 2	○	Groups 1, 2, and 3	○	Group 1
◎	Group 3	◎	Group 4	◎	Group 2
⊙	Group 4	⊙	Group 5	⊙	Group 3
●	Group 5	●	Group 6	●	Group 4

	2^N Series $L_4 (2^3)$*		
	Column		
No.	1	2	3
1	1	1	1
2	1	2	2
3	2	1	2
4	2	2	1
	Group 1	Group 2	

* See Figure 9-41.

	$L_8 (2^7)$*						
	Column						
No.	1	2	3	4	5	6	7
1	1	1	1	1	1	1	1
2	1	1	1	2	2	2	2
3	1	2	2	1	1	2	2
4	1	2	2	2	2	1	1
5	2	1	2	1	2	1	2
6	2	1	2	2	1	2	1
7	2	2	1	1	2	2	1
8	2	2	1	2	1	1	2
	Group 1	Group 2		Group 3			

* See Figure 9-42.

Interactions Between Two Columns						
Column						
1	2	3	4	5	6	7
(1)	3	2	5	4	7	6
	(2)	1	6	7	4	5
		(3)	7	6	5	4
Columns			(4)	1	2	3
				(5)	3	2
					(6)	1
						(7)

Figure 9-41. Linear graph of L_4 table.

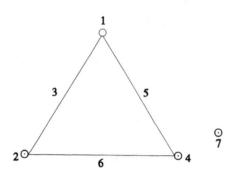

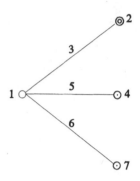

Figure 9-42. Linear graphs for L_8 table.

					$L_{12}(2^{11})$						
					Column						
No.	1	2	3	4	5	6	7	8	9	10	11
1	1	1	1	1	1	1	1	1	1	1	1
2	1	1	1	1	1	2	2	2	2	2	2
3	1	1	2	2	2	1	1	1	2	2	2
4	1	2	1	2	2	1	2	2	1	1	2
5	1	2	2	1	2	2	1	2	1	2	1
6	1	2	2	2	1	2	2	1	2	1	1
7	2	1	2	2	1	1	2	2	1	2	1
8	2	1	2	1	2	2	2	1	1	1	2
9	2	1	1	2	2	2	1	2	2	1	1
10	2	2	2	1	1	1	1	2	2	1	2
11	2	2	1	2	1	2	1	1	1	2	2
12	2	2	1	1	2	1	2	1	2	2	1
	Group 1					Group 2					

NOTE: The interaction components for two given columns are confounded with the remaining nine columns. Sequential analysis is necessary to find the interactions. This array should therefore not be used for experiments requiring interactions.

$L_{16}(2^{15})^*$

No.	Column														
	1	2	3	4	5	6	7	8	9	10	11	12	13	14	15
1	1	1	1	1	1	1	1	1	1	1	1	1	1	1	1
2	1	1	1	1	1	2	2	2	2	2	2	2	2	2	2
3	1	1	1	2	2	2	2	1	1	1	1	2	2	2	2
4	1	1	1	2	2	2	2	2	2	2	2	1	1	1	1
5	1	2	2	1	1	2	2	1	1	2	2	1	1	2	2
6	1	2	2	1	1	2	2	2	2	1	1	2	2	1	1
7	1	2	2	2	2	1	1	1	1	2	2	2	2	1	1
8	1	2	2	2	2	1	1	2	2	1	1	1	1	2	2
9	2	1	2	1	2	1	2	1	2	1	2	1	2	1	2
10	2	1	2	1	2	1	2	2	1	2	1	2	1	2	1
11	2	1	2	2	1	2	1	1	2	1	2	2	1	2	1
12	2	1	2	2	1	2	1	2	1	2	1	1	2	1	2
13	2	2	1	1	2	2	1	1	2	2	1	1	2	2	1
14	2	2	1	1	2	2	1	2	1	1	2	2	1	1	2
15	2	2	1	2	1	1	2	1	2	2	1	2	1	1	2
16	2	2	1	2	1	1	2	2	1	1	2	1	2	2	1
	Group 1	Group 2		Group 3				Group 4							

* See Figure 9-43.

Interactions Between Two Columns

							Column							
1	2	3	4	5	6	7	8	9	10	11	12	13	14	15
(1)	3	2	5	4	7	6	9	8	11	10	13	12	15	14
	(2)	1	6	7	4	5	10	11	8	9	14	15	12	13
		(3)	7	6	5	4	11	10	9	8	15	14	13	12
			(4)	1	2	3	12	13	14	15	8	9	10	11
				(5)	3	2	13	12	15	14	9	8	11	10
					(6)	1	14	15	12	13	10	11	8	9
						(7)	15	14	13	12	11	10	9	8
Column							(8)	1	2	3	4	5	6	7
								(9)	3	2	5	4	7	6
									(10)	1	6	7	4	5
										(11)	7	6	5	4
											(12)	1	2	3
												(13)	3	2
													(14)	1

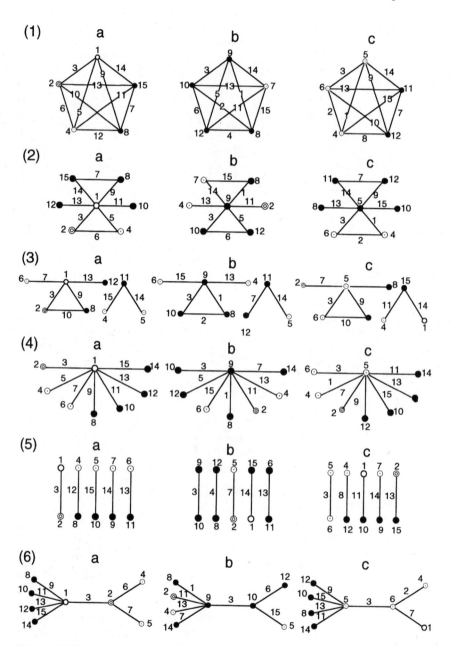

Figure 9-43. Linear graphs for L_{16} table.

	$L_9 (3^4)*$			
	Column			
No.	1	2	3	4
1	1	1	1	1
2	1	2	2	2
3	1	3	3	3
4	2	1	2	3
5	2	2	3	1
6	2	3	1	2
7	3	1	3	2
8	3	2	1	3
9	3	3	2	1
	Group 1	Group 2		

* See Figure 9-44.

	$L_{18} (2^1 \times 3^7)^{*\dagger}$							
	Column							
No.	1	2	3	4	5	6	7	8
1	1	1	1	1	1	1	1	1
2	1	1	2	2	2	2	2	2
3	1	1	3	3	3	3	3	3
4	1	2	1	1	2	2	3	3
5	1	2	2	2	3	3	1	1
6	1	2	3	3	1	1	2	2
7	1	3	1	2	1	3	2	3
8	1	3	2	3	2	1	3	1
9	1	3	3	1	3	2	1	2
10	2	1	1	3	3	2	2	1
11	2	1	2	1	1	3	3	2
12	2	1	3	2	2	1	1	3
13	2	2	1	2	3	1	3	2
14	2	2	2	3	1	2	1	3
15	2	2	3	1	2	3	2	1
16	2	3	1	3	2	3	1	2
17	2	3	2	1	3	1	2	3
18	2	3	3	2	1	2	3	1
	Group 1	Group 2			Group 3			

* See Figure 9-45.

NOTE: Interactions can be found without sacrificing columns, by using the two-way layout of columns 1 and 2. The interactions between three-level columns, however, are partially confounded with the remaining three-level columns.

	L_{27} (3^{13})*												
	Column												
No.	1	2	3	4	5	6	7	8	9	10	11	12	13
1	1	1	1	1	1	1	1	1	1	1	1	1	1
2	1	1	1	1	2	2	2	2	2	2	2	2	2
3	1	1	1	1	3	3	3	3	3	3	3	3	3
4	1	2	2	2	1	1	1	2	2	2	3	3	3
5	1	2	2	2	2	2	2	3	3	3	1	1	1
6	1	2	2	2	3	3	3	1	1	1	2	2	2
7	1	3	3	3	1	1	1	3	3	3	2	2	2
8	1	3	3	3	2	2	2	1	1	1	3	3	3
9	1	3	3	3	3	3	3	2	2	2	1	1	1
10	2	1	2	3	1	2	3	1	2	3	1	2	3
11	2	1	2	3	2	3	1	2	3	1	2	3	1
12	2	1	2	3	3	1	2	3	1	2	3	1	2
13	2	2	3	1	1	2	3	2	3	1	3	1	2
14	2	2	3	1	2	3	1	3	1	2	1	2	3
15	2	2	3	1	3	1	2	1	2	3	2	3	1
16	2	3	1	2	1	2	3	3	1	2	2	3	1
17	2	3	1	2	2	3	1	1	2	3	3	1	2
18	2	3	1	2	3	1	2	2	3	1	1	2	3
19	3	1	3	2	1	3	2	1	3	2	1	3	2
20	3	1	3	2	2	1	3	2	1	3	2	1	3
21	3	1	3	2	3	2	1	3	2	1	3	2	1
22	3	2	1	3	1	3	2	2	1	3	3	2	1
23	3	2	1	3	2	1	3	3	2	1	1	3	2
24	3	2	1	3	3	2	1	1	3	2	2	1	3
25	3	3	2	1	1	3	2	3	2	1	2	1	3
26	3	3	2	1	2	1	3	1	3	2	3	2	1
27	3	3	2	1	3	2	1	2	1	3	1	3	2
	Group 1	Group 2			Group 3								

* See Figure 9-46.

(1)

1   3,4   2

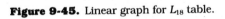

Figure 9-44. Linear graph for L_9 table.

(1)

1   2

Figure 9-45. Linear graph for L_{18} table.

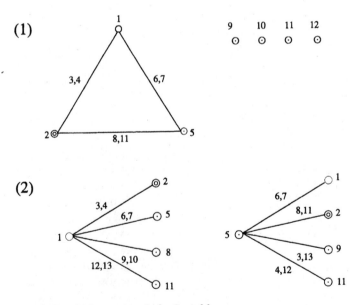

Figure 9-46. Linear graph for L_{27} table.

Interactions Between Two Columns

Column

1	2	3	4	5	6	7	8	9	10	11	12	13
(1)	3	2	2	6	5	5	9	8	8	12	11	11
	4	4	3	7	7	6	10	10	9	13	13	12
	(2)	1	1	8	9	10	5	6	7	5	6	7
		4	3	11	12	13	11	12	13	8	9	10
		(3)	1	9	10	8	7	5	6	6	7	5
			2	13	11	12	12	13	11	10	8	9
			(4)	10	8	9	6	7	5	7	5	6
				12	13	11	13	11	12	9	10	8
				(5)	1	1	2	3	4	2	4	3
					7	6	11	13	12	8	10	9
					(6)	1	4	2	3	3	2	4
						5	13	12	11	10	9	8
						(7)	3	4	2	4	3	2
							12	11	13	9	8	10
							(8)	1	1	2	3	4
								10	9	5	7	6
								(9)	1	4	2	3
									8	7	6	5
									(10)	3	4	2
										6	7	7
										(11)	1	1
											13	12
											(12)	1
												11

Column

Using the Tools for Experimental Designs

General. In this section, we will present the idea of selecting an orthogonal array, using linear graphs to help you assign your variables to the columns of the orthogonal array you have selected, taking into account the considerations facing the engineer. This idea was mentioned briefly earlier, but we will expand it in various ways in this section. When selecting an orthogonal array, the total degrees of freedom in your problem must be less than the total degrees of freedom represented in the array in order to even hope to set up your problem. The Taguchi philosophy (and good engineering sense) says to use the smallest orthogonal array which meets your needs.

Using the Table of Interactions. In case you have difficulty finding an "appropriate" linear graph, you fall back on the table of interactions to make sure that your assignments will not create any confusion. For example, suppose you have the following factors and are interested in the indicated interactions (the terms came from Fram air-laid media).

A = line speed
B = amount of wood pulp
C = pump speed
D = blender speed
E = latex viscosity
F = vacuum pull (CFM)
G = moisture
H = amount of fiber

BH
CD
CE
EG

Because of the interactions, we need to prepare a plan which "accommodates" the substructure shown in Figure 9-47. This "structure" represents the interactions of concern here. Every problem has its own structure. It is not necessary to use the linear graphs (they are there only to help you). We may directly use the table of interactions to make the assignments. There are many ways to do this, and only one of them is shown here. Another assignment incorporating other aspects of the variables is shown in the next section. (See Figure 9-48.)

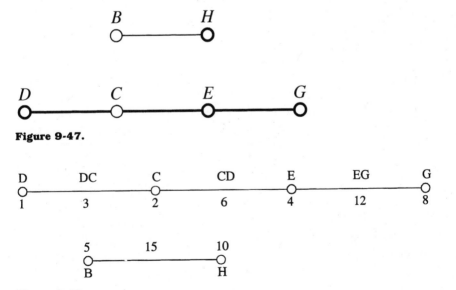

Figure 9-47.

Figure 9-48.

In the present case, we could proceed from left to right by assigning D to 1 and C to 2. Then the DC interaction will appear in 3. Next, we assign E to 4, whereupon the CE interaction will appear in 6. If we try to assign G to 7, we discover from the interaction table that the EG interaction will appear in 3, which already represents the DC interaction. To prevent this (and similar) confusions, we assign G to 8, whereupon the EG interaction appears in 12. The variable B may be assigned to 5, and H may be assigned to 10, giving us the BH interaction in 15.

Using the Groups of Columns. The columns of an orthogonal array can be put into different groups depending on how many changes of level there are in the column. For example, column 1 of L_{16} has only one change from 1 to 2 in the column, and the levels appear in blocks of eight. Columns 2 and 3 have only two or three changes, and the levels appear in blocks of four. Columns 4, 5, 6, and 7 have more changes still, and the levels appear in blocks of two. In the remaining columns, the levels alternate. If you thought of running the experiments in order, it would be convenient for you to assign to column 1 that particular variable which is hardest to change, and to assign to columns 2 and 3 that variable which is next-hardest to change, and so on. Obviously, if you run the experiments in a *random* order, then this feature of orthogonal arrays is of no use to you. I have heard some expositors of the TM say that if you assign your variables to columns within a group in a random way, that gives a satisfactory amount of randomness in your effort. For example, you may assign your next-hardest-to-change variable to either column 2 or column 3, so select the column at random. I support this idea, because it makes good engineering sense to me.

In the previous section, we studied an example from Fram concerning air-laid media. Let us reconsider that example, in light of the concepts of this section.

The variables we considered are listed again here, along with my opinion of how difficult they would be to reset during the experimental program. (If the opinions are wrong, I beg the indulgence of the Fram engineers.)

A = line speed	*Easy,* just adjust a control knob.
B = amount of wood pulp	*Difficult,* calls for making another formulation of ingredients.
C = pump speed	*Easy,* just adjust a control knob.
D = blender speed	*Easy,* just adjust a control knob.
E = latex viscosity	*Very difficult,* harder than B. It probably calls for a different batch of latex from a supplier.
F = vacuum pull (CFM)	*Easy,* just adjust a control knob.

| G = moisture | *Very* difficult, calls for re-setting the humidity in the room. |
| H = amount of fiber | *Difficult,* similar to B. |

If G is the hardest factor to change, it could be assigned to column 1; and if E is the next-hardest, it could be assigned to column 2 or column 3 (select at random, i.e., flip a coin—literally). Then the EG interaction will appear in column 3 or 2, so we must not confuse it with another main effect or another interaction of interest. Since C is easy, we can assign it to 7 and, similarly, D can be assigned to column 8. Since B is difficult, we would like to assign it to 5, and since H is difficult, we could assign it to 6, whereupon the BH interaction will appear in 11. This is not possible, however, because 5 is already used up as the EC interaction. We must assign B to one of the columns in group 4. Say we assign B to 12 and H to 6; then the BH interaction appears in column 10, and the final assignment is shown in Figure 9-49. Note that it is different from the assignment we did in the previous section, because we took into account the degree of difficulty of setting the variables.

Blocking Your Design. Following immediately from the "groups" of the variables, which we discussed in the previous section, we see that this structure makes it very easy to break an orthogonal array into blocks. If, for example, you needed to break L_{16} into two blocks, then column 1 does this for you automatically. Note that you should not use column 1 to represent a factor; it already represents a "blocking variable" and confusion (confounding) would result. If you needed four blocks, then column 2 provides the desired breakdown. Note that this renders column 1 useless, and column 3 useless, because they don't change within a block, so they are automatically confounded with your blocks. Within blocks, the concepts of the previous section apply. You should have as much randomness as possible in your program.

No A Priori Knowledge About Interactions. If you have no information about interactions before you set up your experiment, and if you

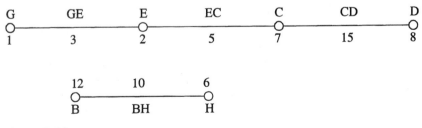

Figure 9-49.

want to design your study so that all interactions are available to you, then your best approach (in my opinion) is to restrict your factors to only two levels, and use L_4 if you have two factors, use L_8 if you have three factors, and use L_{16} if you have four or five factors. If you have six or more factors, you may either use L_{32} or else define a screening-type experiment involving at most 16 runs. I recommend the latter approach. For example, if you have four factors, each at two levels, the L_{16} is the best orthogonal array to use. Assign your factors to four nodes of the pentagon-shaped linear graph, and it will tell you the columns where the six interactions will appear. There will be five columns which represent neither main effects nor interactions; their use will be discussed later.

Screening Designs. You may use these designs to screen a set of variables. You *discard* those variables which are *not* statistically significant, and retain for future probing investigations those which are statistically significant.

Using Linear Graphs to Accommodate a Multilevel Factor. Occasionally, an experimental situation will involve some factors having two levels, and other factors having more than two levels. In cases like this, the linear graphs (and, of course, the orthogonal arrays) may be modified to handle the assignment problems.

Four-Level Factor in L_8. Suppose you have factors A (four levels), B (two levels), C (two levels), and D (two levels), and suppose you want to use the indicated linear graph. The revised linear graph is obtained by combining the first three columns into one new column which provides the four needed levels of A. You could just as well combine columns 1, 4, and 5 or columns 1, 6 and 7. Since a four-level factor has three degrees of freedom, you must use three "related" columns to handle it. If you use columns 1, 2, and 3, you replace those columns by a single column to represent the four levels of A. (See Figures 9-50 and 9-51.)

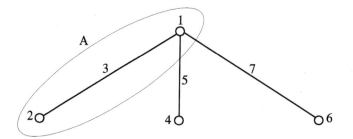

Figure 9-50.

Run	1	2	3	Level of A
1	1	1	1	1
2	1	1	1	1
3	1	2	2	2
4	1	2	2	2
5	2	1	2	3
6	2	1	2	3
7	2	2	1	4
8	2	2	1	4

Figure 9-51.

By this means, $L_8\,(2^7)$ becomes $L_8\,(4 \times 2^4)$ as shown in the following tables:

	Column						
Run	1	2	3	4	5	6	7
1	1	1	1	1	1	1	1
2	1	1	1	2	2	2	2
3	1	2	2	1	1	2	2
4	1	2	2	2	2	1	1
5	2	1	2	1	2	1	2
6	2	1	2	2	1	2	1
7	2	2	1	1	2	2	1
8	2	2	1	2	1	1	2

Run	A 123	B 4	UF 5	C 6	D 7
1	1	1	1	1	1
2	1	2	2	2	2
3	2	1	1	2	2
4	2	2	2	1	1
5	3	1	2	1	2
6	3	2	1	2	1
7	4	1	2	2	1
8	4	2	1	1	2

Suppose you have the following:

Run	A	B	e	C	D	Results y
1	1	1	1	1	1	15.53
2	1	2	2	2	2	17.66
3	2	1	1	2	2	22.84
4	2	2	2	1	1	21.60
5	3	1	2	1	2	21.83
6	3	2	1	2	1	22.42
7	4	1	2	2	1	17.23
8	4	2	1	1	2	18.79

With several factors, we might want to "run them through an ANOVA" to assess statistical significance before we look at the engineering significance. (See Figure 9-52.)

To show what A is doing, we draw a sketch, shown in Figure 9-53.

Source of variation	d.f.	Variation SS	Variance MS	Variance ratio F	(Pooled)	α
A	3	49.44	~~16.48~~	82	14.88	.02
B	1 ⎫	1.16 ⎫	1.16	5.80		
C	1 ⎬ 4	0.72 ⎬ 4.43	.72 1.1075	3.60		
D	1	2.35	2.35	11.75		
E	1 ⎭	0.20 ⎭	.20	(See Note 1)		
Total	7	53.87				

Figure 9-52. ANOVA table. Note: In the 1, 1 block of the F table, we see that the computed value of F has to be larger than 40 in order to be statistically significant. Thus, we see that neither B, C, nor D are statistically significant, and may be pooled with the noise.

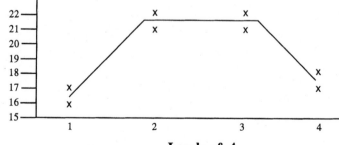

Results y

Figure 9-53.

Levels of A

Four Four-Level Factors in L_{16}. Suppose you have four different cars, four different drivers, and four types of tires. Since there are four different positions where the tire may be mounted on the car, we could have the following situation:

Car $C_1 \, C_2 \, C_3 \, C_4$

Position $P_1 \, P_2 \, P_3 \, P_4$

Driver $D_1 \, D_2 \, D_3 \, D_4$

Tire type $T_1 \, T_2 \, T_3 \, T_4$

This situation may be handled using a design which is derived from L_{16}. If we use the linear graph 5a, we can accommodate the four four-level factors quite easily. Suppose we use columns 1, 2, and 3 for T; columns 4, 8, and 12 for C; columns 5, 10, and 15 for P; and columns 7, 9, and 14 for D. (See Figure 9-54.)

Then L_{16} can be rewritten as follows:

Run No.	T	C	P	D	6	11	13	Wear
1	1	1	1	1	1	1	1	20
2	1	2	2	2	1	2	2	18
3	1	3	3	3	2	1	2	19
4	1	4	4	4	2	2	1	17
5	2	1	2	3	2	2	1	22
6	2	2	1	4	2	1	2	25
7	2	3	4	1	1	2	2	22
8	2	4	3	2	1	1	1	20
9	3	1	3	4	1	2	2	20
10	3	2	4	3	1	1	1	21
11	3	3	1	2	2	2	1	22
12	3	4	2	1	2	1	2	21
13	4	1	4	2	2	1	2	21
14	4	2	3	1	2	2	1	21
15	4	3	2	4	1	1	1	27
16	4	4	1	3	1	2	2	22

Suppose the "wear" figures represent test results. Then an ANOVA may be run to determine which (or if) any factors are S-sig.

Eight-Level Factor in L_{16}. A similar idea carries up to the eight-level factor case. Suppose A has eight levels. Then A has seven degrees of freedom, and seven "related" columns of L_{16} are needed to accommodate A. Thus, in one of the linear graphs for L_{16}, it is necessary to find a subgroup which shows seven related columns. This is done by finding the triangle which represents the related columns (note that you have to add another segment to account for column 7). (See Figure 9-55.)

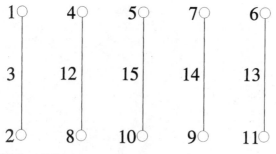

Figure 9-54.

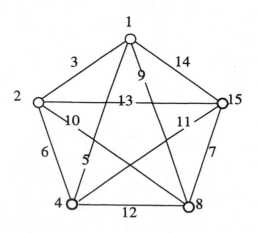

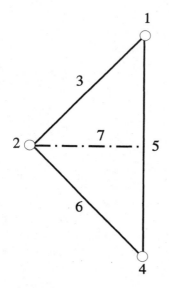

Figure 9-55.

Run no.	1	2	3	4	5	6	7	Levels of A
1	1	1	1	1	1	1	1	1
2	1	1	1	1	1	1	1	1
3	1	1	1	2	2	2	2	2
4	1	1	1	2	2	2	2	2
5	1	2	2	1	1	2	2	3
6	1	2	2	1	1	2	2	3
7	1	2	2	2	2	1	1	4
8	1	2	2	2	2	1	1	4
9	2	1	2	1	2	1	2	5
10	2	1	2	1	2	1	2	5
11	2	1	2	2	1	2	1	6
12	2	1	2	2	1	2	1	6
13	2	2	1	1	2	2	1	7
14	2	2	1	1	2	2	1	7
15	2	2	1	2	1	1	2	8
16	2	2	1	2	1	1	2	8

After this change, L_{16} becomes L_{16} (8×2):

Run no.	A	8	9	10	11	12	13	14	15
1	1	1	1	1	1	1	1	1	1
2	1	2	2	2	2	2	2	2	2
3	2	1	1	1	1	2	2	2	2
4	2	2	2	2	2	1	1	1	1
5	3	1	1	2	2	1	1	2	2
6	3	2	2	1	1	2	2	1	1
7	4	1	1	2	2	2	2	1	1
8	4	2	2	1	1	1	1	2	2
9	5	1	2	1	2	1	2	1	2
10	5	2	1	2	1	2	1	2	1
11	6	1	2	1	2	2	1	2	1
12	6	2	1	2	1	1	2	1	2
13	7	1	2	2	1	1	2	2	1
14	7	2	1	1	2	2	1	1	2
15	8	1	2	2	1	2	1	1	2
16	8	2	1	1	2	1	2	2	1

A Three-Level Factor in a Two-Level Design. For this case, the easiest way to proceed is to follow the method to obtain a four-level factor, and then replace the level 4 with level 1 (or replace it with level 2 or with level 3—the choice depends on the experimenter's fondness for some particular level). For instance, if level 1 denoted the "presently used level," then it should be used to replace "level 4," so that equal participation by the two new levels (2 and 3) will be attained.

If no such clear-cut choice is obvious, a rational way of assigning the three levels to the derived column is available. It is called *the idle-column method*, and goes as follows:

Step 1: Choose a line segment from a linear graph and choose one end point of the line segment to be the "idle" point.

Step 2: Remove the interaction column between the idle and the opposite point from the array.

Step 3: Insert a new column for the three-level factor A, and label its levels according to the following rules.

- When the idle column has a "1," allocate A_1 and A_2 corresponding to the 1s and 2s of the outer-point column.
- When the idle column has a "2," allocate A_2 and A_3 corresponding to the 1s and 2s of the outer-point column.

Step 4: Remove the idle column and the outer-point column from the array.

A Two-Level Factor in a Three-Level Design. If all but one of your factors are at three levels, and one of your factors is at two levels, one easy way to proceed is to select which level is the more important and let that level replicate. As in the preceding section, this will create an imbalance design, but this will cause you no trouble. From an engineering point of view, the graphical analyses will proceed as usual; from a statistical point of view, PROC GLM of SAS must be used (do not use PROC ANOVA), but your ∝ values will appear for you without any problem. The specific procedure is as follows:

Step 1: Select a column for the two-level factor from a three-level array.

Step 2: Select the level of the two-level factor which may be more important than the other.

Step 3: Rewrite the levels in the "two-level" column, doing the labeling such that the important level is replicated.

Systems Analysis

General. One of the levels of application of the Taguchi method is what used to be called "systems analysis." After a system design has been decided on, it is necessary to obtain numerical values for all of the system variables; this process is called "parameter design" in the Taguchi method, and it is basically concerned with the system's response to "noise," or disturbances. In this section we introduce the idea of repeated trials (with which we're all familiar) to bring out the idea that *both* the absolute magnitude *and* the variability of our system's responses are of interest. Then the ideas of parameter design and tolerance design are discussed.

Replicated Designs. An experimental program in which repeated trials are run provides more information about experimental error than the program in which each test is performed only once. A repeated test involves starting from scratch, as it were, while repeated measurements means just that, namely, repeated readings. The latter will show you the variability due to your measurement method; the former will, in addition, show you the variability due to slightly different experimental conditions.

As an example, suppose we repeated each run of an L_8 twice (i.e., suppose we did 16 tests in a random order) and obtained the following results:

Run no.	C 1	B 2	3	A 4	5	6	7	Results	y	s^2	d.f.
1	1	1	1	1	1	1	1	43.1, 32.6	37.85	55.125	1
2	1	1	1	2	2	2	2	31.9, 43.3	37.60	64.980	1
3	1	2	2	1	1	2	2	47.6, 38.9	43.25	37.845	1
4	1	2	2	2	2	1	1	36.9, 48.4	42.65	66.125	1
5	2	1	2	1	2	1	2	46.5, 39.7	43.10	23.120	1
6	2	1	2	2	1	2	1	38.3, 47.6	42.95	43.245	1
7	2	2	1	1	2	2	1	42.2, 32.2	37.20	50.000	1
8	2	2	1	2	1	1	2	30.3, 43.0	36.65	80.645	1
									40.156	52.6356	

If we regard the eight y values as the experimental output, the ANOVA table for this example looks like this:

Source of variation	d.f.	Variation SS	Variance MS	Variance ratio F	$\propto$
A	1	0.5625	0.5625	0.01	
B	1	0.8100	0.8100	0.02	
C	1	0.5625	0.5625	0.01	
AB	1	0.1225	0.1225	0.00	
AC	1	0.0100	0.0100	0.00	
BC	1	128.8225	128.8225	2.44	0.168
ABC	1	0.0000	0.0000	0.00	
Error	8	422.09	52.7613		
TOTAL	15	552.9800			

The error term reflects the variability we got when we replicated the experimental design. In particular, note that MS = 52.7613, which is essentially the same as the average S^2 shown above. There is more that we can observe, however, by studying the S^2 even though it has no effect on the $\bar{y}$ results. Thus, it is conceivable that B and C could be used to place the absolute level ($\bar{y}$) on target, and the factor A could be used to reduce the variability (S^2) without affecting the absolute level! The Taguchi method extends this concept in an organized way, which we will discuss next.

Signal-to-Noise (S/N) Ratio.   Classical assumes that variation "within" is the same. Signal-to-noise ratio uses a combination of both *average* and *variation.*

The target is the best formula to use:

$$S/N = 10 \log_{10} (\bar{y}^2/s^2)$$

For the case where "smaller is better" we use:

$$-10 \log_{10} \frac{1}{N} \Sigma\, y^2$$

and for the case where "larger is better" we use:

$$-10 \log_{10} \frac{1}{N} \Sigma \left(\frac{1}{y}\right)^2$$

Parameter Design.   The idea of replication has been extended in the TM. As part of the design process (i.e., determining the "best" numerical values for the parameters) the experimenter will deliberately subject his or her system to carefully planned "noise" inputs in order to determine the ability of the system parameter values to: (1) keep the system on target, and (2) do so *despite* the noise.

The TM calls this method "parameter design," and it yields a designed product which tends to have the same performance whatever the values of the noise which it encounters. Such a system is called *robust to noise.* Since noise includes those factors which are *not* under the control of the manufacturer, being robust to noise is an excellent characteristic.

In the following design, the L_8 represents factors which *are* under the control of the manufacturer, and is called the *inner array.* The L_4 represents factors which are *not* under the control of the manufacturer, and this is called the *outer array.* In this hypothetical example, the factors of the inner array (which are under our control) include:

A Amount of resin in the formulation

B Amount of organic material in the formulation

C Amount of nonorganic material in the formulation

D Mix layover time

E Cure press temperature

F Cure press pressure

And the factors in the outer array (which are under the control of our customer) include:

G Brake pedal pressure

H Amount of moisture in the environment

I Vehicle speed at start of brake application

The eight experimental conditions of L_8 form a study which investigates the effect on, say, stopping distance and peak deceleration due to the factors A through F. For each experimental condition (i.e., for each run of L_8), four tests are to be performed, according to the values of G, H, and I which are determined by the outer array L_4. The overall program plan may be depicted as shown in Figure 9-56.

There will be $4 \times 8 = 32$ test results obtained, which can be analyzed using the y values and the S/N values to study both the level of the response (small values of y = stopping distance are preferred) and the variability of the response (small values of S/N are preferred).

As a result of parameter design, the numerical values for the system parameters have been obtained by *optimizing* the S/N ratio. Thus, your system design has been realized in a way which provides you with opti-

								I	1	2	2	1
							H	1	2	1	2
							G	1	1	2	2
Run	E 1	F 2	e 3	A 4	B 5	C 6	D 7					
1	1	1	1	1	1	1	1					
2	1	1	1	2	2	2	2					
3	1	2	2	1	1	2	2					
4	1	2	2	2	2	1	1					
5	2	1	2	1	2	1	2					
6	2	1	2	2	1	2	1					
7	2	2	1	1	2	2	1					
8	2	2	1	2	1	1	2					

Figure 9-56. Parameter design example.

mum performance (i.e., optimum resistance to noise and staying on target). If at this time your actual performance is satisfactory (in a cost sense to be discussed in more detail later), then you have completed your job. If not, then you will have to improve your cost values by reducing variability, i.e., by tightening tolerances.

Tolerance Design.   The basic idea behind tolerance design comes from the desire to produce a product which is robust (insensitive) to deterioration and to outer noise, and which is made of material and methods which are as low-cost as possible. Since low-cost material and methods generally have fairly wide tolerances, you may not be able to achieve your objectives with these wide tolerances. You may use tolerance design to determine which of your low-cost tolerances must be replaces by higher-cost (narrower) tolerances in order to meet your objectives. A tolerance design study is performed only if your system cannot meet objectives using your low-cost tolerances.

Suppose you are studying four factors, A, B, C, and D, using L_8. Let factor A have tolerance limits which are ±20 percent (say) of the nominal value of A; let B have tolerance limits which are ±10 percent (say) of nominal; let C have tolerance limits which are +15 and −5 percent of nominal; and let D have tolerance limits which are +5 and −10 percent (say) of nominal. For each variable, there are three levels; suppose

1 = lower tolerance

2 = nominal, or "target"

3 = upper tolerance

This situation may be represented by L_9.

Run no.	A 1	B 2	C 3	D 4
1	1	1	1	1
2	1	2	2	2
3	1	3	3	3
4	2	1	2	3
5	2	2	3	1
6	2	3	1	2
7	3	1	3	2
8	3	2	1	3
9	3	3	2	1

This design may be used like the outer array to determine the robustness of your parameter design to the variability introduced by the low-cost tolerances as shown in Figure 9-57.

							D	1 2 3	3 2 1	2 3 1
							C	1 2 3	2 3 1	3 1 2
							B	1 2 3	1 2 3	1 2 3
							A	1 1 1	2 2 2	3 3 3

Run	A 1	B 2	e 3	C 4	e 5	e 6	D 7			
1	1	1	1	1	1	1	1		①	
2	1	1	1	2	2	2	2			
3	1	2	2	2	2	1	2			
4	1	2	2	2	2	1	1			
5	2	1	2	1	2	1	2			
6	2	1	2	2	1	2	1			
7	2	2	1	1	2	2	1			
8	2	2	1	2	1	1	2			

Figure 9-57. Tolerance design example.

This design calls for $8 \times 9 = 72$ experiments, the results of which are analyzed as before. The nominal levels of the factors in the study are determined by the L_8, and the tolerance variations are represented by the L_9. For example, the experiment indicated by the circled 1 is set up as follows:

A is at its low nominal level.

B is at its low nominal level minus the lower tolerance variation.

C is at its high nominal level.

D is at its high nominal level plus the upper tolerance variation.

At the end of this effort (which, incidentally, is only done as a last resort), your job is indeed complete. Your system's performance has been optimized via the parameter design effort, and your system's cost performance has been brought into line by a judicious tightening of tolerances in a systematic manner by use of the tolerance design study information.

Introduction to ISO*

International Organization for Standardization

ISO is usually recognized as an acronym standing for the International Organization for Standardization, with headquarters in Geneva, Switzerland. It is a useful connection, although not actually correct—ISO was adopted from the Greek root meaning *equal, homogeneous, uniform.*

* Contributed by Dan Pitkin.

The intention from the beginning of ISO efforts in developing quality standards has been to integrate and harmonize similar existing quality management standards into a single body of international quality standards which could universally apply to world trade and commerce. ISO is a world political body comprised of:

- 91 member nations belong to ISO
- 164 technical committees
- 2200 subcommittees
- ANSI, which is the United States member of ISO

ISO 9000-9004 Series of Quality Management and Quality Assurance Standards

Lineage of ISO 9000

1. In 1959, U.S. Department of Defense (DOD) established MIL-Q-9858, its designated quality management program.
2. In 1963, MIL-Q-9858 was revised and enhanced.
3. In 1968, NATO adopted provisions of MIL-Q-9858A in its Allied Quality Assurance Publication 1 (AQAP-1).
4. In 1970, U.K. Ministry of Defense adopted provisions of AQAP-1 in Management Program Defense Standard, DEF/STAN 05-8.
5. In 1979, the British Standards Institute (BSI) developed BS 5750, the first *commercial* Quality Management System standards.
6. In 1987, the International Organization for Standardization (ISO) adopted most of BS 5750 to create ISO-9000. Later, BS 5750 and ISO 9000 were harmonized to make them equivalent documents.
7. In 1987, the European Community (EC) adopted a harmonized standard and designated it European Norme EN 29000.
8. In 1987, the American Society for Quality Control (ASQC) and the American National Standards Institute (ANSI) established and published the Q-90 series, essentially identical to ISO 9000.
9. Members of the European Community (EC) and the European Free Trade Association (EFTA) have adopted ISO 9000 standards and will enforce strict compliance with these standards for suppliers of goods and services.
10. In 1992, the U.S. Department of Defense (DOD) adopted both ANSI/ASQC Q-90 series and the ISO 9000 Quality Assurance Standards to facilitate contracting with U.S. and foreign firms.

ISO 9000 Philosophy

1. Global trend toward higher customer expectations regarding product and service quality.

2. Technical specifications alone cannot guarantee conformance with customer requirements.

3. International quality system standards *complement technical specifications* in order to consistently meet customer requirements.

4. Quality system of an organization influenced by its vision, mission, and values, culture, management style, industry, product, and service. Therefore, *quality systems will differ* from one organization to another.

5. Purpose of Quality Standards Q90–Q94 is to *provide guidelines for developing effective quality systems* and not to standardize quality systems that are implemented.

Basic Elements and Terminology

- *Quality policy:* Overall quality intentions and direction of an organization as regards quality, as authorized and formally expressed by top management.

- *Quality management:* That aspect of the overall management function that determines and implements the quality policy.

- *Quality system:* The organizational structure, responsibilities, procedures, processes, and resources for implementing quality management.

- *Quality plan:* A document setting out the specific quality practices, resources, and sequence of activities relevant to a particular product, service, contract, or project.*

- *Quality control:* The organizational techniques and activities that are used to fulfill requirements for quality.

- *Quality assurance:* All those planned and systematic actions necessary to provide adequate confidence that a product or service will satisfy given requirements for quality.†

- Unless given requirements fully reflect the needs of the user, quality assurance will not be complete.

- For effectiveness, quality assurance usually requires a continuing evaluation of factors that affect the adequacy of the design or specification

---

* International Standard ISO 8402.

† ANSI/ASQC Q90-1987, *Quality Management and Quality Assurance Standards—Guidelines for Selection and Use,* American Society for Quality Control, 1987, p. 2.

for intended applications as well as verifications and audits of production, installation, and inspection operations. Providing confidence may involve producing evidence.

■ Within an organization, quality assurance serves as a management tool. In contractual situations, quality assurance also serves to provide confidence in the supplier.

Quality Objectives for Companies Installing ISO 9000

■ Achieve and sustain the quality of the product or service produced so as to continually meet the purchaser's stated or implied needs (*customer satisfaction*).

■ Organization provides confidence to its own management that the intended quality is being achieved and sustained (*internal quality assurance*).

■ Organization demonstrates confidence to the purchaser that the intended quality is/will be achieved in the delivered product or service (*external quality assurance*).

Five ISO 9000 Quality Standards Developed

1. *ISO 9000: Quality Management and Quality Assurance Standards—Guidelines for Selection and Use.* Set of guidelines for selecting and using appropriate quality system models, ISO 9001, ISO 9002, and ISO 9003.

2. *ISO 9001: Quality Systems—Model for Quality Assurance in Design/Development, Production, Installation and Servicing.* The most comprehensive model for quality assurance for companies which are involved in product design and development, production, installation, and servicing.

3. *ISO 9002: Quality Systems—Model for Quality Assurance in Production and Installation.* Model for quality assurance very similar to ISO 9001 except for companies not involved in the design and development or servicing of the product or service.

4. *ISO 9003: Quality Systems—Model for Quality Assurance in Final Inspection and Test.* Model for quality assurance for companies involved in final inspection and test; typically, distributors and value-added contractors.

5. *ISO 9004: Quality Management and Quality System Elements—Guideline.* Provides guidance on the technical, administrative, and human factors affecting the quality of products or services at all stages of the quality loop, from detection of need to customer satisfaction.

■ ISO 9000 and ISO 9004 provide guidance to all organizations for quality management purposes.

- ISO 9001, ISO 9002, and ISO 9003 are used for external quality assurance purposes in contractual situations.

- ANSI/ASQC Q90–Q94 are technically equivalent to ISO 9000–ISO 9004 standards, respectively, but use customary American language usage and spelling.

Worldwide Membership. By 1992, over 50 nations of the world adopted ISO 9000 as their official quality standards. These political, industrial, and economic powers include the countries of Germany, United Kingdom, France, the entire European Community and EFTA member countries, United States, Mexico, Canada, Israel, Australia, Japan, China, Taiwan, South Korea, Venezuela, Brazil, Argentina, and South Africa. It is estimated that these adopting nations now represent 90 percent of the industrial power and economic wealth of the world.

United States Involvement. The United States has its own National Committee for ISO deliberations which operates under the auspices of ANSI (American National Standards Institute). The U.S. Department of Defense has issued policy statements encouraging the adoption of ISO 9000, as have the NAFTA countries of Canada and Mexico. The U.S. medical products industry is considering the adoption of ISO 9001 as a replacement for its current "Good Manufacturing Practice" standards.

Major U.S. industries that recognize the importance of ISO 9000 as the dominant set of world quality standards include telecommunications, computers and software, electronics and semiconductors, aerospace, chemical, oil and gas production and hydrocarbon processing, toys, medical devices, industrial controls, safety equipment, and building products. It is noteworthy that many U.S. suppliers and subcontractors to these industries are being encouraged by their customers to install ISO 9000 quality systems even though their products and services are purchased within the United States.

Thousands of U.S. companies are now in the process of installing ISO 9000 quality management systems and advancing toward certification and registration. Even with a proliferation of consulting companies and register agencies in the United States to handle the growing demand for implementation and certification, there are extended waiting periods of many months because of limited resources. The January 1993 deadline imposed on suppliers by EC markets is driving much of this demand; however, there is substantial interest in installing ISO 9000 to become a more cost-efficient producer and a more competitive supplier of high-quality goods and services.

Significance of ISO 9000 Universal Quality Standards

ISO 9000 Influence on International Business and Trade. Throughout the world, ISO 9000 quality system standards are fast becoming the de facto international quality standards for commercial, industrial, and even military goods and services. Contractual requirements that supplier companies become ISO certified and registered are appearing more frequently in purchase orders for products manufactured in every country of the world, including the United States. This is especially true in regulated industries such as medical products, toys, safety equipment, and telecommunications.

- Standardization of quality standards throughout the world provides competitive opportunities to suppliers from all nations.

- ISO 9000 will facilitate access to global markets and open new markets, reducing the market influence of trade barriers and political alliances.

- Certification to ISO 9000 standards will reduce/avoid the extra costs and delays associated with supplier surveys, bid qualification procedures, supplier quality auditing, source inspection, and other quality assurance oversights.

- Issues such as product liability, safety, health, and environmental compatibility, international commerce terms and conditions, and packaging and shipping procedures will be more universally addressed by manufacturers and suppliers trading within ISO 9000 standards and guidelines.

Benefits to U.S. Companies. Aside from the global market incentive for ISO certification, U.S. industry leaders recognize that improved quality management systems and practices are essential for competing in today's dynamic, quality-demanding, price- and service-sensitive business environment. ISO 9000 serves as an excellent model for quality assurance throughout the organization, whether or not ISO certification is ever required by a company's customers and markets. In the event that certification becomes necessary later, the framework has already been established, and formal ISO 9000 certification becomes an easier task.

A sound and integrated quality management system, as represented by ISO 9000, is the foundation for initiatives such as Total Quality Management and the prestigious Malcolm Baldrige National Quality Award. The ISO Quality Management System is an ideal model for building a highly efficient organization committed to continuous improvement and world-class competitiveness.

Influence on Company Culture and Employees. Implementing an ISO 9000 quality management system will have a profound affect on the organization and the way people in all departments work. The discipline associated with developing and documenting procedures for each activity which affects quality makes everyone aware of the importance of each task and exactly how it must be performed to assure quality.

"Doing it right the first time" applies to all management processes, not just those used in production and operations. When company employees, who know the process better than anyone else, accept ownership of the process, there is pride in making sure the process works consistently and efficiently. Process control, measurement, and continuous improvement become a way of life.

Effect on Customers. Existing customers may already prefer suppliers who are implementing an ISO 9000 Quality Management System and who plan to obtain certification and registration. Certainly a supplier who is certified to ISO 9000 has an immediate competitive advantage over those suppliers who are not yet certified. Certification gives confidence to a customer that the supplier has already demonstrated to an objective third party that its quality management system conforms to the requirements of an internationally recognized set of standards, endorsed by nations, governments, and industries the world over.

Effect on Suppliers and Subcontractors. Suppliers and subcontractors are very much affected by their customers or purchasers who subscribe to ISO 9000 standards. The ISO requirements contain a comprehensive section for the purchaser on how its suppliers and subcontractors must assure quality of the process and product. It is often the case that suppliers and subcontractors are encouraged or even required to become ISO 9000 certified in order to remain a qualified source of supply and an approved bidder/supplier on new products.

Major Challenges. The major challenges of implementing an ISO 9000 Quality Management System and gaining certification and registration are not unlike those associated with instituting any significant organizational change. There is usually a built-in reluctance to abandon procedures and practices that are well entrenched and have served their purpose reasonably well for many years.

For those companies already operating under U.S. quality standards such as 10CFR50, Appendix B, Mil-Q-9858A, DOE 5700.6C, or ASME NQA-1, conversion to ISO 9000 will not be difficult. While overall system requirements are similar, the ISO 9000 standards tend to be more specific in prescribing exactly *how quality assurance shall be managed* (i.e., procedures, work instructions, documentation, forms, and records). The operative command in ISO standards is *shall*, indicating that the prescribed procedure is mandated, not optional.*

Companies operating under commercial quality standards may find that their quality management system will need complete overhaul or replacement in order to implement ISO 9000. The level of documentation required is likely to be much greater than that for an existing quality management system, and the structure of the appropriate ISO 9000 quality system model might be altogether different from that in place. In such cases, companies may be wise to develop a whole new quality management system following the ISO 9000 model and then make a phased or abrupt conversion. The findings of a quality management system *preassessment* conducted by a capable third party can help determine the means of converting the existing system to a compliant ISO quality management system.

Key Factors for Success. Likewise, any major organizational undertaking involving change in the way an organization conducts its routine operations and the way it organizes and administers procedures and work instructions will take time and extra effort to assimilate. There are many pitfalls and numerous examples of ISO 9000 implementation restarts, especially with U.S. companies not used to the extensive documentation, training, auditing, and record keeping. From actual experience with implementing ISO 9000 Quality Management Systems, companies report the following key factors for success.

Key Factors for Successful ISO 9000 Implementation

- Senior management leadership and commitment
- An active and supportive management steering committee
- Implementation teams that are properly trained
- Effective internal auditing, corrective action, and process improvement
- Organization, teamwork, and discipline to systematically follow proven methods for accomplishing work projects

ISO 9000 Planning, Implementation, and Training

Quality Management Plan. The starting point for ISO 9000 implementation is a quality management plan. The plan is initiated by top man-

---

* "A Critical Review of the ISO 9000 Series Quality Assurance Standards" by Jack Norris, ASQC Western Regional Conference, Oct. 1992.

agement of the company when it decides to install a comprehensive quality management system which conforms to ISO 9000 standards. The quality planning effort must be (1) initiated by top management, (2) visibly directed and supported by top management (by the executive staff or resource committee), and (3) ultimately integrated with the culture and managing and operating styles of the organization. In other words, ISO 9000 policies, procedures, and work practices must become an integral part of "the way we do things in our company."

ISO 9000 standards apply to the entire quality management system of an organization and therefore to all departments of the organization. For effective implementation, all departments and all people at different organizational levels must become aware of, and involved in, the implementation process. Certain functions and departments have specific responsibilities to accomplish, but no department can afford to ignore the change effects associated with ISO 9000 implementation. For example, the buyers, customer service people, field sales, and technical service people must all become knowledgeable representatives of the company in demonstrating their awareness of and full commitment to ISO 9000 quality management requirements.

The *quality management plan* contains the following key elements:

1. Company vision, mission, and values
2. Strategic direction and quality management policy
3. Quality management objectives
4. Quality system design
5. Organizing and chartering implementation teams
6. Training and deploying teams
7. Implementation schedule and milestones
8. Allocating resources
9. Progress review and reporting
10. Instituting the quality management system

Role of Chief Executive. The first two planning elements of the quality management plan form the launching pad for ISO 9000 implementation: (1) *company vision, mission, and values* and (2) *strategic direction and management policy.* The chief executive of the company or organization is ultimately responsible for business plans, policies, organization, and operating results. He or she must establish and communicate the vision, mission, and values of the organization, and must set the direction and strategies for the quality management system.

Most important, the ISO 9000 standards and quality management system should complement and support the business direction, competitive strategies, and policies established by the chief executive. The chief executive's role is one of leadership in bringing to life and role-modeling the vision, mission, and values. Successful implementation of ISO 9000 is difficult to achieve without personal involvement and active support from the chief executive and his or her executive staff.

Company strategies and policies should be expressed in terms of how they promote the success of the organization and welfare of its people. The purpose and benefits of ISO 9000 must be understood and communicated to everyone in the organization before training and implementation begin. It must be set in the context: *Why are we doing this? How will it be done? What is my part? How will it affect me and my job?* This ISO 9000 "implementation readiness" is usually accomplished through a series of presentations called "ISO 9000 Awareness." Open meetings, employee conferences, announcements, newsletters, bulletin board displays, and contests are all useful in getting the organization aware and ready for ISO 9000.

Role of Resource Committee or Steering Committee. The next four elements of the quality management plan are usually prepared by a resource committee appointed by the chief executive officer of the company:

3. *Quality management objectives*

4. *Quality system design*

5. *Organizing and chartering implementation teams*

6. *Training and deploying teams*

The resource committee will represent top management as well as the main functions or departments of the organization ... typically, operations, sales and marketing, engineering, design and development, human resources, and financial and administrative management. The committee operates as an advisory body to the chief executive, preparing specific recommendations on objectives, system design, organizing and chartering implementation teams, and providing special training for team members.

The resource committee will select and actively direct the implementation teams throughout the course of ISO 9000 implementation. This may involve several teams representing different functions, departments, and organizational levels, depending upon the size of the organization. Each team will be chartered by the resource committee to perform specific implementation tasks. Regular progress reviews will be scheduled by the resource committee.

Selection of members for the implementation teams is very important. Successful implementation will depend upon the performance of these teams. Members will be auditing work areas and processes throughout the company, and will meet resistance when advocating changes that are necessary to comply with ISO 9000. The resource committee should select team members with the following qualities:

- Team players committed to the entire company, not just their departments
- Analytical and objective, with problem-solving skills
- System and process experience or orientation
- Good communication and writing skills
- Development potential to supervision and management

Qualified candidates can come from any function or department, and it's usually a good idea to balance the teams with people from different disciplines and backgrounds. For example, administrative people from support departments like accounting, human resources, customer service, and marketing can make valuable contributions as team members. If the work force is multilingual, then members with appropriate language skills can facilitate implementation.

The resource committee's role is to plan, communicate, and direct the overall conversion from the existing quality management system to the selected ISO 9000 quality management system. It provides the implementation teams and the entire organization with leadership, top-management support, and resources needed for successful implementation of an ISO 9000 quality management system. It must meet on a regular basis, assess progress, and help clear roadblocks and resolve conflicts which the implementation teams encounter.

However, the resource committee is not a substitute for leadership from the chief executive. He or she must continue to demonstrate visible personal involvement and enthusiasm for a company transformation to an ISO 9000 quality management system. The chief executive must keep the organization focused on the grand purpose and benefits of implementing ISO 9000. This can be accomplished by continual emphasis on the company's vision, mission, and values—in meetings, informal conversations, company newsletters, and announcements. Citing examples of improvements made by using ISO 9000 procedures, no matter how small, will help encourage people to make the extra effort which successful implementation requires. Of course, open praise or recognition of work groups and individuals who are making progress in implementing ISO 9000 is an effective way to promote companywide cooperation and support.

Role of Implementation Teams. Implementation teams are formed and chartered to install the specific requirements of ISO 9000 throughout the organization. Members will be thoroughly trained in ISO 9000 system requirements, procedures, internal auditing techniques, and corrective action methods. Before team members are deployed throughout the organization, they must be thoroughly trained in all phases of implementation.

- ISO 9000 standards and system requirements
- Implementation planning

- Internal auditing
- Audit reporting and corrective-action planning
- Procedures and work instructions
- Forms, records, and documentation

Quality Management System, Policy, and Manual

ISO 9000 Family of Standards

- *ISO 9000* and *ISO 9004* provide guidance to all organizations for quality management purposes.
- *ISO 9001, ISO 9002,* and *ISO 9003* are used for external quality assurance purposes in contractual situations.
- *ANSI/ASQC Q90–Q94* are technically equivalent to ISO 9000–ISO 9004 standards, respectively, but use customary American language and spelling.

ISO 9000 Quality Loop.   The Quality Loop* helps illustrate the different functions or departments of a company which influence the quality of the product and service, and therefore are integral parts of the ISO 9000 Quality Management System. (See Figure 9-58.)

* ISO 9004: 1987, "Quality Management and Quality System Elements—Guidelines," Section 5.1.

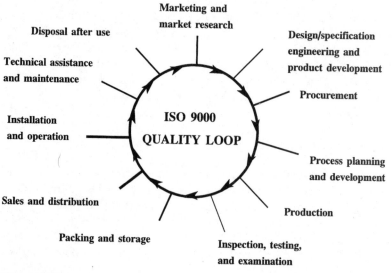

Figure 9-58.

ISO 9000 Quality System Models and Elements

ISO 9003	ISO 9002	ISO 9001
Management responsibility	Management responsibility	Management responsibility
Quality systems	Quality systems	Quality systems
Product identification and traceability	Product identification and traceability	Product identification and traceability
Inspection and testing	Inspection and testing	Inspection and testing
Inspection, measuring, and test equipment	Inspection, measuring, and test equipment	Inspection, measuring, and test equipment
Inspection and test status	Inspection and test status	Inspection and test status
Control of nonconforming product	Control of nonconforming product	Control of nonconforming product
Handling, storage, packaging, and delivery	Handling, storage, packaging, and delivery	Handling, storage, packaging, and delivery
Document control	Document control	Document control
Quality records	Quality records	Quality records
Training	Training	Training
Statistical methods	Statistical methods	Statistical methods
	Purchaser supplied product	*Purchaser supplied product*
	Corrective action	*Corrective action*
	Process control	*Process control*
	Purchasing	*Purchasing*
	Contract review	*Contract review*
	Internal quality audits	*Internal quality audits*
		Design Control
		Servicing

Selection of ISO 9000 Model. Selection of the appropriate quality system model of ISO 9000 (9001, 9002, or 9003) is an important decision. Because ISO 9001 is the most comprehensive model (broadest coverage of functions), it is often regarded as the most desirable quality system to install for competitive advantage. This is not necessarily the case.

ISO 9002 and ISO 9003 were designed to serve companies that may not perform all functions from initial product design to after-sale servicing. The guiding reference in selecting the ISO quality system model most appropriate to your company and its operational and business needs is *ISO 9004* (or ANSI/ASQC Standard Q90-1987): *Quality Management and Quality Assurance Standards—Guidelines for Selection and Use.*

Section 8.2.1 of ISO 9004 describes functional or organizational capability factors for selection of the quality assurance model:

- "ISO 9001 for use when conformance to specified requirements is to be assured by the supplier during several stages which may include design/development, production, installation, and servicing."

- "ISO 9002 for use when conformance to specified requirements is to be assured by the supplier during production and installation."

- "ISO 9003 for use when conformance to specified requirements is to be assured by the supplier solely at final inspection and test."

Section 8.2.3 of ISO 9004 describes six other factors to consider when selecting the appropriate model for a product or service:

- Design-process complexity (degree of difficulty in designing new product or service)

- Design maturity (extent to which total design is proven)

- Production-process complexity

- Product or service characteristics (number and criticality of interrelated characteristics)

- Product or service safety (risks and consequences of failure)

- Economics (system costs versus risks of nonconformities)

Oftentimes, the deciding factor for selecting the appropriate quality system model is which, if any, is specified by your major customers and which is used predominantly in the industry and markets you serve. It is very important to determine from your customers and markets (1) what model is preferred or specified by major customers, (2) how your certification will affect your standing with major customers and your overall competitive position in the marketplace, and (3) exactly what products/services are so affected by certification requirements and when.

Quality Policy. The cornerstone of the quality management system is the company's quality policy. This is a statement (usually published and widely communicated) describing the overall intentions and direction of the company as regards quality. Quality policy is usually formulated, authorized, and expressed by top management. Quality policy statements of clarity and brevity help people in the organization develop a common understanding about the aims of the organization in dealing with management processes and work activities that affect all aspects of quality performance.

Quality Manual. The company's quality manual is a general policy document which explains the manner in which the company intends to comply with the ISO 9000 standards. It is a broad document which will remain current and can be distributed to employees, customers, and suppliers for quality system guidance. In addition, the quality manual:

- Communicates company management's quality policy and quality objectives to employees, customers, suppliers.
- Provides general policy guidelines covering main sections of the ISO 9000 quality system.
- Describes overall quality management approach and is broad rather than specific in content (will not require frequent revisions).
- Does not contain proprietary or confidential company information.
- Is brief but understandable in reflecting how work is performed.
- After certification, cannot be changed without approval.

Levels of Documentation. Figure 9-59 depicts the various levels of ISO 9000 documentation.

Quality Policy
- Written by executive management and communicated
- Signed by current management
- Pertinent to current organization and activities
- Understood and implemented throughout the organization

Quality Manual Components
- Quality policy
- Business description
- Organizational description
- Quality assurance system description
- System for regular assessment of the quality assurance system

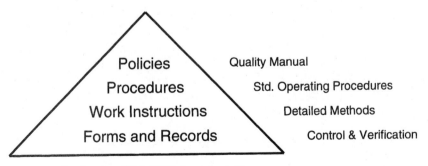

Figure 9-59. Levels of ISO 9000 documentation.

ISO 9000 System Implementation

Implementation Plan and Schedule.   A sample implementation plan is shown on the accompanying Gantt chart. The tasks, time durations, and sequence of events might differ slightly depending upon the organization and its readiness to implement an ISO 9000 quality management system. However, most elements will apply and deserve some explanation.

Select a Quality Management Consultant.   An experienced consultant can bring objectivity and independent thinking to your company's implementation planning and project management. Since your system and practices will be critically evaluated by an outside certification agency, it is helpful to gain an outsider's perspective early in the implementation process to avoid common pitfalls and false starts. A management consultant will ensure that the implementation effort is properly directed toward system compliance, and can provide management and employees with special training needed for effective ISO 9000 implementation and operation.

Conduct a Companywide Preassessment.   Determine how the present quality management system is designed to operate, how it actually operates, and the extent to which quality-related activities are properly documented, communicated, and followed in everyday practice. This preassessment by an outside consultant will set the baseline for system modification and improvement.

Assessment Report.   The consultant will prepare and present to company management a comprehensive report about findings in all categories of the relevant ISO 9000 quality system model. This report will highlight nonconforming areas and indicate the nature and extent of the implementation effort ahead. It can also help management determine the value of retaining the current body of operating procedures and work instructions, or else the need for replacing some entirely with ISO 9000 conforming documentation.

Executive Overview—ISO Awareness.   It is critical that senior company management understand the importance, benefits/risks, and resource requirements of implementing and operating according to ISO 9000 standards. ISO awareness training introduces quality management as a business strategy for organizational improvement and competitive advantage. Top-management awareness and active support from the very start is essential for successful implementation.

Draft the Company's Quality Policy.   This is the first task of senior management in embarking on a campaign for total quality management and ISO 9000.

Form and Charter the Resource Committee. This committee is comprised of company executives and managers who direct the implementation effort and actively manage key functions and departments which utilize the quality system. Functional heads and key department managers are usually included in its membership.

ISO Systems and Standards Training. A series of seminars and workshops which cover the detailed requirements of the quality management system and its standards. General seminars and workshops on ISO 9000 system implementation are worthwhile, but often the company will have special operations or unusual circumstances which require customized training and on-site facilitation by an experienced quality consultant.

Select and Charter Implementation Teams. Implementation teams are the ISO 9000 project teams whose members represent different functions and departments. These are the people who work closely with their associates to make sure that implementation is planned, executed per schedule, and effectively accomplished.

Training for Implementation Teams. Since team members are so critical to implementation success, they must be thoroughly trained in ISO 9000 and in their specific mission (analyzing processes, developing documentation, auditing, applying corrective actions, etc.). It is important that this training take place before implementation activities begin.

Company Awareness Presentations. Since implementation will inevitably involve most, if not all, company personnel and will necessitate change in the way work is performed throughout the company, employees need to be kept informed about implementation activities and the probable effects on their departments and their jobs. Open communication to all personnel at the start of implementation is essential.

Develop Implementation Plan and Schedule. Implementation teams need to develop a master plan and schedule which outlines the sequence of critical steps which lie ahead. Coordination of activities and their impact on the organization are important factors to consider. Also, a master plan helps anticipate problems, identify resource requirements, and provides a basis for directing and measuring progress during the course of implementation.

Submit Registrar or Agency Questionnaires (RFQs). ISO 9000 registrars must ultimately determine conformance with appropriate quality system criteria. These third-party quality assessment organizations, called *certification bodies* in Europe and *registrars* in the United States and Canada, are authorized to issue a "certificate of registration" when your system conforms to their interpretation of an ISO 9000 standard. Most registrars use a questionnaire as an information package to prepare a suitable proposal for audit review and certification. Typically, a company will ask several different registrars to make proposals.

Develop Quality Manual and Policies. The company's quality manual and policies can follow the same sequence of ISO 9000 system requirements, giving directional guidance to all employees who will operate the quality management system. The quality manual and policies can be distributed to employees, suppliers, customers, regulatory agencies, industry groups, etc., to describe the company's quality management system and explain how it assures quality of its products and services.

Review Proposals and Select Agency/Registrar. Select an ISO 9000 registrar as you would select an important service provider to your company . . . one who has the experience, capabilities, and formal qualifications to assess your quality management system and work with you toward full conformance and certification. Make sure that the scope and cost of their services are clearly stated and appropriate to your needs.

Develop/Review/Revise Procedures. Actual procedures need to be evaluated in terms of conformance to ISO 9000 standards, and revised/updated accordingly. Written procedures must reflect the actual process and methods in use, and must be well understood and readily available to all those who perform and/or supervise the particular work processes and methods.

Apply for Certification. This begins the certification and registration process. You will complete a formal application for assessment of the quality management system per ISO 9000 standards. The application form will be supplied by the registrar you select.

Develop/Review/Revise Work Instructions. Likewise, the detailed methods and practices used to perform a job or task must be properly documented to serve as helpful working instructions to employees and operators assigned to that particular job or task. These instructions can take the form of written sequential steps, flowcharts, graphs, sketches or diagrams, recipes, checklists, and so forth.

Install and Utilize Procedures and Work Instructions. Procedures and work instructions, especially those newly developed or revised, should be installed and applied in routine operations for an extended period, prefer- ably enough time to monitor, gather data, and evaluate their effectiveness. It is a rule of thumb that procedures and work instructions should be in effect for at least six months before the certification audits are conducted.

Submit Manual to Agency for Documentation Assessment. The regis- trar will first conduct an off-site audit of the company's quality manual and policies in order to determine the extent of readiness of the company and its quality management system.

Internal Auditing and Corrective-Action Training. Meanwhile, as an effective way of preparing the organization for the critical certification audits, the implementation project teams can conduct internal audits on a regular basis. In order to prepare team members for this important assign- ment, the teams are trained in the ISO 9000 audit process, auditing proto- col and techniques, detailed system requirements, interpretation of standards, reporting of nonconforming situations, and closed-looped cor- rective action. This internal auditing and corrective-action planning is key to continuous improvement so that the organization learns how to operate according to ISO 9000 standards and system requirements.

Conduct First Round of Internal Audits. Internal audits can begin as soon as the implementation teams are trained for this activity.

Audit Reporting and Corrective Action. Reporting of the findings of internal audits should be detailed enough to explain exactly what the non- conforming situation is and the governing ISO 9000 standards. It should provide specific recommendations for correcting nonconformities or ideas for modifying the procedures or documentation in order to reduce the likelihood of confusion and/or nonconformance. Definitive audit report- ing is one of the most useful skills in making the new quality system func- tion effectively in all areas of the company.

Conduct Second Round of Internal Audits. At monthly or quarterly intervals, as appropriate to the initial audit findings and corrective actions under way, another round of internal audits is conducted.

Audit Reporting and Corrective Action. Internal auditing, reporting, and taking corrective actions is an iterative process to help all employees get accustomed to new procedures, work instructions, and documentation requirements.

Schedule for Certification Audits by Agency/Registrar. This is the telling set of audits which will determine whether the company, its qual- ity system, and its people are qualified to consistently manage their system, perform their work, and operate their processes according to the high standards of ISO 9000. The discovery of nonconforming situations at this point can prove costly in time and money, so preparation and rehearsal are important measures to take before the certification audit.

System Assessments. System assessments are usually conducted as a series of functional audits, depending upon the structure of the organization. Each of the system elements is traceable to one or several functions or departments which are primarily responsible for the quality-assuring activities required. In planning a system assessment, it is helpful to identify where organizationally these activities occur. A matrix like the one shown in Figure 9-60 is used to plan the assessment process, allowing time to interview key people and observe the processes and procedures they use to perform their work.

Quality Audits by Function. In organizing an internal audit, it is helpful to understand the structure of the organization and which functions are involved with each particular element to the quality management system. Audits consist of questions and observations directed to the people and processes that perform a particular activity which affects quality. For example, audit questions about the purchasing function should be addressed to materials managers, purchasing agents, and buyers who normally work within the operations or materials management functions of the company.

Internal Quality Audits
*ISO 9001 Requirements for Internal Quality Audits**

- Planned and documented internal quality audits and follow-up actions scheduled and performed in accordance with procedures.
- Results of audits documented and reported to responsible management personnel.
- Timely corrective action taken on deficiencies.
- Audits verify quality system is effective.

Corrective Actions. ISO 9000 standards for corrective actions are excellent guidelines for developing an effective process that can be used throughout the company, and most effectively during the internal auditing process. In fact, this is one of the first procedures that the implementation teams should evaluate and revise according to ISO 9000 requirements.

---

* Section 4.17 of ISO 9001: 1987 (E), Quality Systems Model for Quality Assurance in Design/Development, Production, Installation and Servicing.

Section 4.0	Quality System Requirements	Company Management	Sales & Field Service	Marketing	Engineering and Dev.	Operations Materials, Quality	Finance & HR
4.1	Management Responsibility						
4.1.1	Quality Policy	✓	✓	✓	✓	✓	✓
4.1.2	Quality Organization	✓					
4.1.3	Management Review	✓					
4.2	Quality System	✓				✓	
4.3	Contract Review	✓		✓	✓		
4.4	Design Control				✓		
4.4.1	General			✓	✓		
4.4.2	Design and Development Planning			✓	✓		
4.4.3	Design Input				✓		
4.4.4	Design Output				✓		
4.4.5	Design Verification				✓		
4.4.6	Design Changes				✓		
4.5	Document Control						
4.5.1	Document Approval and Issue	✓			✓	✓	
4.5.2	Document Changes/Modifications					✓	
4.6	Purchasing					✓	
4.6.1	General					✓	
4.6.2	Assessment of Subcontractors				✓	✓	

Figure 9-60. Functional requirements of ISO 9001.

4.6.3 Purchasing Data				✓	
4.6.4 Verification of Purchased Product				✓	
4.7 Purchaser Supplied Product				✓ ✓	
4.8 Product Identification and Traceability				✓	
4.9 Process Control				✓ ✓ ✓	
4.9.1 General					
4.9.2 Special Processes	✓		✓	✓	✓
4.10 Inspection and Testing				✓	
4.10.1 Receiving Inspection and Testing				✓	
4.10.2 In-Process Inspection and Testing			✓	✓	
4.10.3 Final Inspection and Testing				✓	
4.10.4 Inspection and Test Records				✓	
4.11 Inspection, Measuring, and Test Equipment				✓ ✓	
4.12 Inspection and Test Status				✓	
4.13 Control of Nonconforming Product					
4.13.1 Nonconformity Review and Disposition				✓	

Figure 9-60. (*Continued*)

Section 4.0 Quality System Requirements	Company Management	Sales & Field Service	Marketing	Engineering and Dev.	Operations Materials, Quality	Finance & HR
4.14 Corrective Action	✓	✓	✓	✓	✓	✓
4.15 Handling, Storage, Packaging, and Delivery					✓	
4.15.1 General					✓	
4.15.2 Handling					✓	
4.15.3 Storage					✓	
4.15.4 Packaging					✓	
4.15.5 Delivery					✓	
4.16 Quality Records	✓	✓	✓	✓	✓	✓
4.17 Internal Quality Audits	✓	✓	✓	✓	✓	✓
4.18 Training	✓	✓	✓	✓	✓	✓
4.19 Servicing	✓	✓	✓	✓	✓	✓
4.20 Statistical Techniques	✓	✓	✓	✓	✓	✓

Figure 9-60. (*Continued*)

*ISO 9001 Requirements for Corrective Action**

- Procedures for investigating the cause of nonconforming product and corrective action needed to prevent recurrence

- Procedures for analyzing all processes, work operations, concessions, quality records, service reports, and customer complaints to detect and eliminate potential causes of nonconforming product

- Procedures for initiating preventative actions to deal with problems to a level corresponding to the risks encountered

- Procedures for applying controls to ensure corrective actions are taken and that they are effective

- Procedures for implementing and recording changes resulting from corrective action

Documentation and Record Keeping. ISO 9000 requirements for documentation and quality record keeping are reputed to be very demanding if not excessive. "A fundamental premise of ISO 9000 is that the company should be able to continue making its products at the same level of quality if all personnel were replaced. This requires that all procedures, work instructions, processes and related activities be exhaustively documented."[†] Complete and accurate documentation and records administration are needed to determine whether procedures are effective in assuring quality, and to provide evidence to a third-party assessment agency that procedures and work instructions are indeed being followed. Significant process and quality information should be collected, recorded, analyzed, and properly stored for reference.

ISO 9001 Requirements for Quality Records[‡]

- Procedures for identifying, collecting, indexing, filing, storing, maintaining, and disposing of quality records.

- Records demonstrate achievements of required quality and effective operation of quality system (including pertinent subcontractor quality records).

- Records safely stored and readily retrievable; retention times established and recorded.

---

[*] Section 4.14 of ISO 9001: 1987 (E), Quality Systems Model for Quality Assurance in Design/Development, Production, Installation and Servicing.

[†] "ISO 9000 and Europe's Attempts to Mandate Quality" by Michael J. Timbers, *The Journal of European Business,* published by Faulkner & Gray Inc.

[‡] Section 4.16 of ISO 9001: 1987 (E).

ISO 9000 Procedures and Work Instructions

Procedures and work instructions are the main body of directions about performing all company activities affecting quality. The ISO 9000 standards prescribe that certain procedures exist in writing, are firmly established in practice, are properly documented, controlled and made accessible, and are reviewed for effectiveness on a regular basis.

Procedures and detailed work instructions are restrictive in nature and can act to inhibit the efficient performance of work. Their purpose is to ensure that the operator of the process or person performing the task will be able to do it safely, efficiently, and effectively while meeting all quality requirements. If the absence of a procedure could adversely affect the safety, efficiency, and quality of work being performed, then a written procedure is appropriate. Again, ISO 9000 assumes that procedures are necessary to assure quality even if the entire workforce were to be replaced with less experienced people.

In order to make procedures and work instructions as effective and practical as possible, without undue restrictions on the operators, it is advised that the operator be allowed to develop his or her own procedures and work instructions, or at least be given the opportunity to contribute to their preparation and modification before they are issued and mandated. In this way, a certain amount of latitude and practicality can be preserved without compromising the basic purpose of the procedure or work instruction . . . to perform work safely and efficiently while assuring quality.

Purpose. The overall purpose of procedures and work instructions is to establish and maintain a documented quality system which will ensure that product conforms to specified requirements. Such documentation must be properly prepared according to ISO 9000 standards, and must be effectively implemented throughout the organization.

Content, Style, and Format. Content is often prescribed by the relevant ISO 9000 standard, whereas style and format are left to the discretion and needs of company management. The form in Figure 9-61 of a typical procedure illustrates a style and format which will serve the needs of most organizations as well as the requirements of the ISO standards.

Work Instructions. The next level below procedures in the hierarchy of ISO 9000 documentation is referred to as work instructions—those documents providing technical detail in how work is to be performed. According to the ISO 9000 standards, "work instructions are required where their absence would affect quality." While this is not a definitive

---

Procedure No: _____

Rev/Issue No: _____

Effectivity Date: _____

Title of Procedure

1. Purpose: *State the reason for establishing the procedure and its primary objective.*

2. Scope: *Describe the organizational units, activities, people, and processes affected by the procedure.*

3. Responsibility: *Name those people (by function or position title) responsible for installing and using the procedure and maintaining its practical application.*

4. Definitions: *Define terms and words that have special meanings or may not be easily understood to those using the procedure.*

5. Procedure: *List the elements of the procedure, normally in sequential steps, explaining: (a) what is to be done, (b) by what person or group, and (c) when, where, and why these steps are performed in this prescribed manner.*

6. References: *Cite references (manuals, forms, charts, and diagrams, etc.) helpful in understanding and following the procedure.*

7. Document Control: *State company policy and procedure on controlling the issuance, review, approval, change, replacement, distribution, confidentiality, and availability of this procedure. Refer to Section 4.5 of ISO 9001 for requirements concerning document review, approval and issue, and document changes/modifications.*

Originator Name & Dept: _____   Date: _____

Function/Department Head: _____   Date: _____

Authorizing Manager/Executive: _____   Date: _____

Figure 9-61. Example of style and format for a typical ISO 9000 procedure.

statement, it does imply that they are sometimes needed to provide additional detailed instruction and technical information beyond that provided by the relevant procedure. In this way, work instructions supplement and support procedures rather than take their place.

Work instructions include quality plans, raw and intermediate material specifications, engineering drawings, product specifications, special requirements, operation sheets, process flowcharts, servicing instructions, test methods, equipment operating instructions, inspection instructions, and similar detailed technical information. The style and format of a work instruction is dictated by its use. It should display technical information in a manner that is clear and understandable to the operator who must use it.

Forms, Records, and Documentation

The basic and most detailed level of quality system documentation consists of reference information and data which support procedures and work instructions. A well-structured system for handling process information and quality data is important for analyzing problems and performance trends, and for measuring the effect of changes and improvements. Good organization and records administration are key ingredients to keep this information and data current, accurate, complete, and useful.

While the use of a real-time computerized database might promote organization, analysis, and ready retrieval of information, it is not always the most practical and cost-effective records system. In practice, hard copy of forms and records kept at the operation or work area so that they are visible and accessible for checking and updating is often the most effective way to keep this documentation accurate and current. Bar-coding and portable data input/output devices are gaining popularity in making record systems user-friendly and more reliable.

Instituting System Improvement

Implementation of an ISO 9000 Quality Management System is a significant undertaking for most companies and represents a commitment to change the way things are done throughout the company. It is an ongoing process of learning new procedures and work methods and getting used to the extensive documentation requirements. Most companies find that it takes more than a year before the complete quality management system starts to function effectively as designed. There are constant revisions and adjustments that are made during implementation and these must be monitored for a time in order to determine their practicality. Continual system improvement is needed to deal with new operations, new products and services, more demanding customer requirements, and even changing ISO 9000 standards affecting certification renewals. As with Total Quality Management, the conversion to an ISO 9000 system takes a long-term commitment to maintain and improve the whole quality management system to better serve the changing requirements of the company and its customers.

Internal Auditing and Corrective Action. One of the important requirements of the ISO 9000 standards is management review. Section 4.1.3 of ISO 9001 states:

> *The quality system adopted to satisfy the requirements of this International Standard shall be reviewed at appropriate intervals by the supplier's management to ensure its continuing suitability and effectiveness. Records of such reviews shall be maintained.*

The means to accomplish this requirement are the processes of regular system assessments, internal audits, and closed-loop corrective actions. Management must stay actively involved in reviewing findings of assessments and audits and making sure that corrective actions and recommended improvements are installed.

Training and Implementation. Training in all aspects of the quality management system should precede implementation. Training will emphasize understanding of ISO 9000 standards and learning the planning and implementing skills needed to build an effective system. Much of the training should be devoted to preparing implementation teams to make things happen in all the operations and departments of the company. Workshops and practical exercises pertaining to actual operating situations is the best way to prepare the organization. The more people involved in the training, the easier implementation can proceed, because everyone understands the importance of the effort and the bearing it will have on competitive advantage, business growth, and job retention.

Process Measurement and Control. Process measurement and control are important components in the three quality system models, ISO 9001, 9002, and 9003. Although statistical techniques are not imposed as strict requirements, many of the elements of the ISO 9000 standards lend themselves to statistical analysis and statistical process control.

Section 4.10, Inspection and Testing, and Section 4.11, Inspection, Measuring and Test Equipment, contain many requirements on inspecting, measuring, and testing processes. Also, ISO 10012-1: 1992 (E), Quality Assurance Requirements for Measuring Equipment—Part 1: Metrological Confirmation System for Measuring Equipment, is applicable to *suppliers of products or services who operate a quality system in which measurement results are used to demonstrate compliance with specified requirements; this includes operating systems that meet the requirements of ISO 9001, ISO 9002 and ISO 9003.*

This is a comprehensive standard which requires the application of strict controls, disciplined verification procedures, and statistical techniques to ensure that measurements are made with the intended accuracy.

Problem Solving and Continuous Improvement. A simple diagram helps summarize the continuous improvement cycle (see Figure 9-62).

1. Identify the problem and objectively define it.
2. Describe the current situation from different points of view.

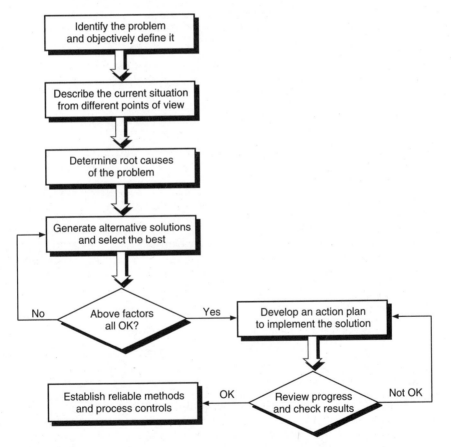

Figure 9-62. Problem solving and continuous improvement.

3. Determine root causes of the problem.

4. Generate solutions and select the best.

5. Develop an action plan to implement the solution.

6. Review progress and check results.

7. Establish reliable methods and process controls to prevent recurrence.

Tips on Implementation

1. Get involvement and firm commitment from the chief executive and the senior management staff from the onset. Don't attempt to implement ISO 9000 without it.

2. Enlist the help of an outside quality management consultant early in the process. Start with a comprehensive preassessment of the existing quality system, procedures, practices, and records.

3. Based upon the findings of the preassessment, develop a detailed plan of implementation emphasizing those areas needing improvement . . . typically, document control, process control, inspection and testing, quality records, and training.

4. Prepare a quality policy manual that is consistent with ISO 9000 standards and broad in nature (any policy manual change must be approved by registrar). Cross-reference supporting procedures, but keep them separate.

5. Inform the entire organization about ISO 9000 and implementation progress. Explain the significance of ISO 9000 to the company and how it will affect the quality system and how work is performed.

6. Charter project teams, with interdepartment representation, to develop procedures and work instructions. Don't overdo the documentation; use flowcharts, diagrams, and illustrations to make procedures and instructions easy to understand and user-friendly.

7. Make sure documentation reflects how people actually perform their work and keep it simple and practical. Employees must always know where to find and how to use the documentation that affects their work.

8. Procedures and methods should be reviewed for effectiveness and changed whenever appropriate. Keep track of changes and update documentation as improvements are made.

9. Every employee who affects the quality of the product or service must follow documented procedures and must have a written job description. Keep records in the human resources department on job descriptions, qualifications, and special training and operator certification.

10. Authorized signatures for document review/approval must appear on all ISO 9000 required documents.

11. Quality records can be kept on convenient forms. Make sure these forms are self-explanatory, contain all necessary instructions, and are properly completed and routed/filed. Procedures should indicate how records are used, routed, stored, disposed. All forms must be controlled documents.

12. Make sure nonconforming material or product is segregated, clearly marked, and not available for use. There is no ISO 9000 time limit on its disposition.

13. Internal audit reports must describe the nonconformity, cite references in standards or procedures, define corrective action, and explain how the corrective action can be verified.

14. ISO 9000 standards require close controls on suppliers, distributors, and even intercompany transfers. Suppliers must have complete documentation about your requirements and should be encouraged to obtain ISO 9000 certification themselves.

15. Make sure your reasons for certifying to ISO 9000 standards are valid for your company and your customers' needs, and select a registrar who will help you achieve your goals.

10
Customer Focus

Dan Pitkin

Purpose

The purpose of TQM marketing is to create a well-thought-out marketing plan and to execute the plan to increase sales and profits through existing products, new markets, and new products.

Objectives

The objectives are as follows:

1. Build a product and company reputation.
2. Make your products and company known to the marketplace.
3. Create a demand for your product.

Defining the Marketing Function

Marketing can be described as a business philosophy aimed at improving profits by identifying the needs of key customer groups and offering a product package that serves their needs more fully and effectively than does the competition. Marketing therefore aims at profit improvement, focuses on customer requirements, targets certain market segments or customer groups, and develops a product and service package that offers value to customers in its served markets.* Thus, there are four key dimensions of business marketing:

- Identifying customer requirements and expectations
- Targeting customer groups or segments to serve
- Developing the product or service package to satisfy customers
- Improving profit performance (earnings and profitability)

Defining Quality

Every aspect of the marketing function involves *quality*. It is certainly a key ingredient in improving profit performance through total quality management, continuous process improvement, and reducing the *cost of quality*. Customer requirements and expectations always have quality implications, and satisfying the customer takes product and service that meet or exceed the quality standards held by the customer. In designing, producing, and distributing the product package, the supplier must *build in* quality every step of the way. The final product package is the resultant combination of quality and value delivered to the customer.

Quality is a familiar term commonly used to describe a desirable attribute of a company's product or service offering. Here are some definitions that have been used by experts in the quality management field:

- "conformance to requirements" . . . Philip Crosby
- "fitness for use" . . . Dr. Joseph M. Juran
- "meeting or exceeding customer expectations at a cost that represents value to them" . . . H. James Harrington
- "totality of features and characteristics of a product or service that bear on its ability to satisfy stated or implied needs" . . . ISO 8402, ASQC, and ANSI

There is no single, official definition of quality which can universally apply to all types of companies in different industries, supplying all varieties of products and services. Each company must define its own meaning of quality as it serves its customers, employees, stockowners, markets, and community. The term will have a particular meaning to your company in its pursuit of total quality. Ideally, it will reflect your company's vision, mission, and values which your people can wholeheartedly embrace in a journey of *continuous improvement* toward *Total Quality Management*.

---

* B. Charles Ames and James D. Hlavacek, *The Ultimate Advantage*, Managerial Marketing Inc., Mountainside, N.J., 1984.

It is interesting to note how business leaders relate the importance of quality and quality management to marketing. Here are a few of their comments about quality and its connection with marketing.

E. S. Woolard, Chairman/CEO Dupont:

To compete and win, we must redouble our efforts . . . not only in the quality of our goods and services, but in the quality of our thinking, in the quality of our response to customers, in the quality of our decision making, in the quality of everything we do.

D. T. Kearns, Chairman/CEO Xerox:

Quality has changed the way we manage our business. We believe it has made us more competitive than anything else could have and that we would not exist today had we not changed.

R. M. Price, Control Data Corporation:

Quality is management's prime responsibility: (1) Quality can and must be managed; (2) Everyone has a customer; (3) Processes are the problem, not people. . . . At the strategic level, quality is synonymous with marketing. Strategy is made real through a detailed list of initiatives that address improvement in each process that is a part of achieving a goal.

TQM and Continuous Improvement

Total Quality Management (TQM) is both a philosophy and a set of guiding principles that form the foundation for a changing and continuously improving organization. TQM is the application of systematic processes and quantitative methods to improve: (1) the materials, information, and services supplied to an organization, (2) all the processes performed within an organization, and (3) the degree to which the needs of customers are understood and consistently met.

TQM integrates fundamental management techniques, company processes and operating procedures, and technical tools and information under a systematic and disciplined approach toward continuous improvement.

Thus, TQM is comprised of management systems and styles of operation embracing the principles of process improvement and organizational effectiveness. TQM is not simply a program, but rather an ongoing campaign that usually requires fundamental cultural changes in *the way we do things in our company* and the way we deal with our associates, suppliers, and customers.

Some of the attributes of Total Quality Management are:

- Relentless pursuit of improvement in all processes
- Never-ending process of understanding our customers' requirements and expectations, and then applying this knowledge in designing, manufacturing, and supplying products and services they need
- Satisfying our customer needs better than any other supplier
- Doing it right the first time and every time
- Striving to satisfy customer expectations 100 percent of the time
- Reengineering processes for improved service, quality, delivery and cost

Some other important points regarding total quality:

- It involves *everything* the company does, not just manufacturing.
- It involves *everyone* in the organization.
- It requires a *cultural change* to get people involved and committed.
- Real change and significant results can take *months* rather than years.
- It is a journey worth the effort.

Some key questions concerning total quality:

- What does it mean to your company?
- How will your company benefit?
- What does it mean to *you*?
- How will *you* benefit?

Traditional Role of Marketing

A traditional form of marketing (industrial product marketing) is one of product stewardship . . . that is, formulating marketing strategies, plans, and programs aimed at getting the right quantity and mix of products to the market in order to meet sales requirements. Product differentiation, opportunistic pricing, special promotions, and the like are techniques used to build market share and profitability, key measures of competitive success in the marketplace.

The marketing role is often defined by the people in the organization who carry *marketing* or *product* or *customer* in their position titles. Smaller companies often combine marketing with sales, with emphasis on providing the sales group with administrative support in activities such as product literature and price lists, advertising and trade shows, planning and forecasting, quoting and distribution, order processing and customer service. Larger companies typically organize the marketing function by product lines with several product managers reporting to a marketing manager. Often, product teams led by the product managers, with representatives from other functions, perform the traditional marketing duties as well as prepare strategic marketing plans for their product line.

The traditional marketing role is that of balancing served market needs with the capabilities of the company. Its universe consists of:

- An existing customer and application base plus potential new customers and applications in served markets

- A range of products and salable services, plus potential new products/services that the company is capable of developing

- An evolving distribution system which connects target markets with products offered

- An array of capabilities and potential capabilities for strategically linking markets, products, and distribution*

In the traditional marketing role, *market share and product-line profitability* are key measures of marketing effectiveness. Market share is usually tracked by product line and by region in specific markets served. For those with commanding market shares, market segmentation and product differentiation allow for opportunistic pricing so that high overall profitability can be maintained. Customer satisfaction is a secondary consideration and is only important as it affects significant shifts in market share. Competitive advantage is determined by manufacturing cost, technology, and relative market share. With a competitive advantage in a particular segment, a company can control the level of quality, value, and service offered.

Role of Marketing in Total Quality Management

With the advent of global markets and worldwide distribution channels, plus emerging international competitors with world-class quality and technology, once-protected market segments are now open to truly competitive enterprise. The wide range of available products and services with superior quality challenges the marketing power of established suppliers. Companies that don't pioneer paradigms or keep pace with quality, productivity, and technology advancements can lose market share dramatically. Profitability can evaporate not only through lost sales but with the extra burden of obsolete product and technology.

---

* Edward S. McKay, *The Marketing Mystique*, American Management Association, New York, 1972.

Total quality management is a key strategy for regaining and building market share and profitability by concentrating on:

- Developing market plans, strategies, and decisions on external information gained from customers in the marketplace nearest to the buying activity
- Identifying customer requirements and expectations for each market segment
- Developing product and service packages for each segment or customer group
- Selecting marketing, selling, and distribution methods most appropriate for each specific market segment or customer group
- Organizing and managing around the needs of the customers who ultimately make the buying decisions

Marketing strategies must respond to the realities of economic power in the marketplace. Peter Drucker of the Claremont Graduate School in California describes the recent power shift to retailers and distributors who are closest to the buying action:

> Since I first said it in my 1954 book, *The Practice of Management,* it has become commonplace to assert that results are only in the marketplace; where things are being made or moved, there are only costs. Everybody these days talks of the 'market-driven' or the 'customer-driven' company. But as long as we did not have market information, decisions (especially day-to-day operating decisions) had to be made as manufacturing decisions. They had to be controlled by what goes on in the plant, and they had to be based on the only information we had (or believed we had) on manufacturing costs.
>
> Now that we have real-time information on what goes on in the marketplace, decisions will increasingly be based on what goes on where the ultimate customers, whether housewives or hospitals, take buying action. These decisions will be controlled by the people who have the information—retailers and distributors. Increasingly, decision making will shift to them.

Ten Key Principles for Marketing Success*

Principle 1. Create customer "want" satisfaction.

Principle 2. Know your buyer characteristics.

Principle 3. Divide the market into segments.

Principle 4. Strive for high market share.

---

* Fred C. Allvine, *Marketing: Principles and Practices,* Harcourt Brace Jovanovich, New York, 1987.

Principle 5. Develop deep and wide product lines.

Principle 6. Price position products and upgrade markets.

Principle 7. Treat channels as intermediate buyers.

Principle 8. Coordinate elements of physical distribution.

Principle 9. Promote performance features.

Principle 10. Use information to improve decisions.

Marketing with TQM Characteristics

Marketing in a Total Quality Management environment is characterized by a number of telling qualities:

- It knows exactly where it stands against competition in served markets, and where it and its top competitors are headed.
- It constantly plans, implements, measures, adjusts for competitive improvement and competitive advantage.
- It gathers information in the marketplace where buying action takes place. It analyzes what customers need and expect, and measures to what extent its products serve customer needs and meet their expectations.
- It knows the capabilities of the company and how to build upon them for competitive advantage; it also recognizes weaknesses as areas for necessary improvement.
- It focuses and coordinates company resources on market growth opportunities and on customers who lead their respective industries.
- It emphasizes high quality and product innovation.
- It embraces Total Quality Management as a company philosophy and key marketing strategy, and leads in commitment and action.

Marketing Imperatives

Old	New
Market manipulation	Market creation
Promotion	Education
Marketing as one department's function	Marketing as the way a company does business
Customers at the end of the process	Customers integrated into the process
Goal is control	Goal is leadership

Differences Between Sales and Marketing

Many organizations combine the sales and marketing functions simply because they appear to have similar objectives and activities, mostly involving customers. Yet in most companies, the roles are quite different, requiring a different set of expertise, special skills, and business perspectives. Combining these two important functions into one department or group often results in an unhealthy preoccupation with immediate business priorities, slighting longer-range strategic planning, and product development needs. Market analysis, strategic planning, and product development are essential elements for significant improvement in competitive position and cannot be neglected.

Selling is:	Marketing is:
1. Highly selective	1. Broad in perspective
2. Thinking in terms of customers and their immediate needs	2. Thinking of industries, applications, and trends
3. Relating to people with buying influence, who and where	3. Relating to product and service, what type and how much
4. Taking information and opportunities to and from the customer	4. Analyzing information and the environment for opportunities creating
5. Developing personal relationships	5. Impersonal and analytical
6. Discovering who buys, when, and why (inflow of business is salesperson's lifeblood)	6. Unconcerned about who buys, so long as enough buying is done
7. Emphasizing customers, units, and dollar volume purchased	7. Concerned with *market segments,* units and dollar volume purchased
8. Synonymous with tactics and action planning	8. Synonymous with strategy and planning
9. Tactics designed to go after every piece of business and never lose a customer	9. Strategy which might include losing/bypassing some customers, markets, applications
10. Managing a sales territory with finite boundaries	10. Understanding and covering the entire market area
11. Decentralized and regionally located to serve customers in the marketplace	11. Centralized with other business area functions at headquarters
12. Predominantly people-oriented, skilled in interpreting customers' wants and needs	12. Predominantly product-oriented, skilled in analyzing, planning, and coordinating

TQM Marketing Strategy for Competitive Advantage

Introduction

Every company competing in an industry or market does so with a marketing strategy. It may consist of well-defined policies and plans that are understood and followed throughout the organization, but it is more likely a strategy that has evolved over several years as a composite approach in the way managers conduct business in a particular industry. With the advent of Total Quality Management, there is an opportunity to use TQM principles to formulate a marketing strategy aimed at achieving competitive advantage.

There are practical boundaries of marketing strategy with many options available within those boundaries. In formulating a marketing strategy for competitive advantage, there are four main limiting influences:

1. *The broad geopolitical and economic environment.* Significant developments and trends taking place in different industries around the world.
2. *The competitive environment.* Industry opportunities and threats (economic and technical risks and rewards).
3. *Company strengths and weaknesses.* Profile of capabilities and resources relative to competitors.
4. *Organizational values and effectiveness.* Capabilities of management to develop goals and policies which can be implemented.

Outline for a TQM Marketing Strategy*

1. What is the business doing now?
   a. Current strategy (explicit or implicit)
   b. Implied assumptions by management: How is the current marketing strategy based upon assumptions about industry and company's relative position? How does the current strategy make sense?
2. What is happening in the environment?
   a. Industry analysis: What are the key factors for competitive success and the significant industry opportunities and threats?
   b. Competitor analysis: What are the capabilities and limitations of existing and potential competitors and their probable future moves?
   c. Societal analysis: What important governmental, social, and political factors will present opportunities and threats?
   d. Strengths and weaknesses: Given an analysis of industry and competitors, what are the company's strengths and weaknesses relative to present and future competitors?

* Michael E. Porter, *Competitive Strategy*, The Free Press, New York, 1980, pp. xviii–xx.

3. What should the business be doing?
   a. Tests of assumptions and strategy: How do the assumptions embodied in the current strategy compare with the analysis in item 2? Is the current strategy (1) internally consistent, (2) in keeping with the industry and broad environment, (3) a match for the company's resources and its ability to change, and (4) supported by organization's capability to understand and implement?
   b. Strategic alternatives: Based upon the above analysis, what are the feasible strategic alternatives?
   c. Strategic choice: Which alternative best relates the company's situation to external opportunities and threats?

Formulating Strategy for Competitive Advantage. Michael E. Porter contends that the success or failure of any firm depends on its competitive advantage—"delivering the product at lower cost or offering unique benefits to the buyer that justify a premium price."* The cornerstone to gaining and sustaining a competitive advantage is developing a *competitive strategy* that (1) targets opportunities in growing and profitable industries and (2) leverages the company's ability to supply superior value to its buyers. "Value is what buyers are willing to pay, and superior value stems from offering lower prices than competitors for equivalent benefits, or providing unique benefits that more than offset a higher price."*

While a company can have many strengths and weaknesses vis-à-vis competition, there are two basic competitive advantages that matter: *low cost* and *differentiation*. And there are two basic pathways to leverage these advantages: *overall* cost leadership, or differentiation in a broad range of industry segments, and *focused* cost leadership, or differentiation in a narrow range of industry segments. (See Figure 10-1.)

A company that cannot choose a clear and singular competitive strategy appropriate to its markets and organizational capabilities will become stuck in the middle. Achieving competitive advantage through any of these pathways takes the courage to select a strategy consistent with the company's vision, mission, and values and staying the course. When clear choices cannot be made about how to compete, management signals become mixed and marketing actions become fragmented and ineffective.

Industry Analysis

The first fundamental determinant of a firm's profitability is industry attractiveness. Competitive strategy must grow out of a sophisticated

* Michael E. Porter, *Competitive Advantage,* The Free Press, New York (Macmillan, Inc.), 1985, p. 7.

Overall Differentiator

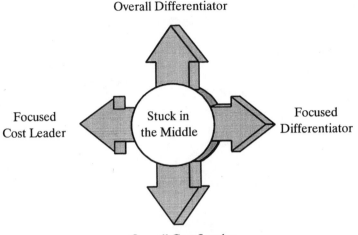

Focused
Cost Leader

Stuck in
the Middle

Focused
Differentiator

Overall Cost Leader

Figure 10-1. Strategies for competitive advantage.

understanding of the rules of competition that determine an industry's attractiveness.* (See Figure 10-2.)

Five Competitive Forces That Determine Industry Profitability

There are five forces that determine industry profitability because they influence the prices, costs, and required investments of firms in an industry—the elements of return on investment. The strength of each competitive force is a function of industry structure, or the underlying economic and technical characteristics of an industry.

Rivalry of Competitors[†]

- Industry growth
- Fixed costs and value added
- Intermittent overcapacity
- Product differences
- Brand identity

---

* Michael E. Porter, *Competitive Advantage—Creating and Sustaining Superior Performance,* The Free Press, New York (Macmillan, Inc.) 1985, pp. 4–10.

† Michael E. Porter, *Competitive Advantage—Creating and Sustaining Superior Performance,* The Free Press, New York (Macmillan, Inc.) 1985, p. 6.

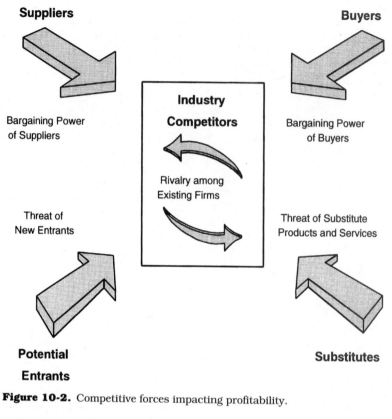

Suppliers

Buyers

Bargaining Power
of Suppliers

Industry
Competitors

Bargaining Power
of Buyers

Rivalry among
Existing Firms

Threat of
New Entrants

Threat of Substitute
Products and Services

Potential

Substitutes

Entrants

Figure 10-2. Competitive forces impacting profitability.

- Switching costs
- Concentration and balance
- Informational complexity
- Diversity of competitors
- Corporate stakes
- Exit barriers

Bargaining Power of Buyers
Bargaining Leverage
- Buyer concentration versus firm concentration
- Buyer volume
- Buyer-switching costs versus firm-switching costs
- Buyer information
- Ability to backward-integrate

- Substitute products
- Pull-through

Price Sensitivity
- Price/total purchases
- Product differences
- Brand identity
- Impact on quality/performance
- Buyer profits
- Decision-makers' incentives

Substitutes
- Relative price performance of substitutes
- Switching costs
- Buyer propensity to substitute

Potential Entrants
- Economies of scale
- Proprietary product differences
- Brand identity
- Switching costs
- Capital requirements
- Access to distribution
- Absolute cost advantages
  Proprietary learning curve
  Access to necessary inputs
  Proprietary low-cost product design
- Government policy
- Expected retaliation

Suppliers
- Differentiation of inputs
- Switching costs of suppliers and firms in the industry
- Presence of substitute inputs

- Supplier concentration
- Importance of volume to supplier
- Cost relative to total purchases in the industry
- Impact of inputs on cost or differentiation
- Threat of forward integration relative to threat of backward integration by firms in the industry

Significance of Five Competitive Factors

The five-forces framework allows a firm to see through the complexity and pinpoint those factors that are critical to competition in its industry, as well as to identify those strategic innovations that would most improve the industry's—and its own—profitability. The five-forces framework does not eliminate the need for creativity in finding new ways of competing in an industry. Instead, it directs managers' creative energies toward those aspects of industry structure that are most important to long-run profitability. The framework aims, in the process, to raise the odds of discovering a desirable strategic innovation.*

Developing a clear and straightforward marketing strategy for gaining competitive advantage is an imperative for TQM marketing. Since the strategy a company selects will drive the priorities of all functions and departments, greatly influence its operating style, and ultimately determine its competitive position in the marketplace, the strategy must be carefully developed from a comprehensive analysis of market/industry structure and a clear understanding of the capabilities and aspirations of the company (vision, mission, and values).

There is probably no other company strategy so vital to its survival and health than a well-developed and clearly articulated marketing strategy for gaining/sustaining competitive advantage.

Competitive Intelligence and Benchmarking

Market Information and Competitive Intelligence

The starting point for the formulation of marketing strategy is understanding the structure and dynamics of the marketplace, specifically in those market areas and segments in which you have the capability and interest to

---

* Michael E. Porter, *Competitive Advantage—Creating and Sustaining Superior Performance,* The Free Press, New York (Macmillan, Inc.), 1985, p. 7.

compete. The analysis of markets, their structures and dynamics, is an ongoing responsibility of the marketing function. To be useful in formulating and adjusting strategy for gaining/sustaining competitive advantage, the information collected must be market/customer/competitor-specific, timely, and accurate. It helps to have a consistent format and measure of market indicators so that trends can be discerned over a time interval.

Information is quite different from intelligence. Market information can be gathered from a multitude of sources and yield little competitive insight until it is selectively processed, analyzed, interrelated, and interpreted. An effective competitive intelligence system can provide vital information about customer requirements and attitudes, competitive activity and strategies, market conditions and business environment.

The competitive intelligence system consists of five basic steps:

1. Determine categories of intelligence desired.
2. Identify appropriate information sources and collection methods.
3. Gather, analyze, and interpret the information.
4. Develop competitive profiles based upon intelligence.
5. Formulate competitive strategies and plans.

Some categories of intelligence might include:

Customers

- Organizations, strategies, policies, procedures, and practices
- Purchasing methods and procedures
- Ranking of key buying motives: quality, delivery, service, price, product performance, reliability, after-sale support, etc.

Markets

- Size, location, demographics
- Requirements to become qualified, certified supplier
- Expectations to become preferred supplier
- Opinions and attitudes about competing suppliers, their composition by industry, application, product
- Sales representation (direct sales, distributors, manufacturing reps, agents, wholesalers, mass merchandisers)
- Distribution channels
- Growth and technology trends by segment
- Market size and share trends
- Profitability trends by segment, product type, application

Competitors

- Form of organization, ownership, and financial strength
- Capabilities: quality, productivity, innovation, service
- Strategies and emphasis by market
- Product and service offerings
- Sales and distribution methods
- Bidding tactics, pricing, and commercial terms
- Customer service policies and procedures

Industry

- Industry leaders by market segment
- Industry groups and trade associations
- Industry standards and practices
- Growth and profitability trends within industry
- Acquisitions, alliances, partnerships

Environment

- Economic trends by industry by region
- Social and political influences
- Environmental, health, and safety requirements
- Legislative and regulatory requirements
- Introduction of new technologies, materials, and processes

Dimensions of Competitive Position

A company's competitive position in a market segment can be described by five primary dimensions or parameters. There may be many secondary factors which contribute to your competitive position in a particular market, but their effect is usually reflected in these five dimensions:

1. *Relative market share.* The ratio of your competitive share of a particular market segment to that of the leading competitor (if you have 20 percent share and the leading contender has 30 percent share, your relative market share is 0.67).

2. *Applied technology.* The extent that your product technology is distinctive, proprietary, and/or protected.

3. *Manufacturing cost.* The comparative efficiency and productivity of the operation which produces your product and service package.

4. *Sales and distribution.* The comparative capacity and cost effectiveness of delivering the product and service to market.

5. *Quality and reliability of product/service.* As perceived by the market, industry leaders, and major customers.

Strategic Market Analysis

Strategic market analysis is a systematic determination of where your company stands in the marketplace vis-à-vis competition in supplying the needs of the market and specific customer groups. Market intelligence provides the picture of where you stand in a particular market compared to your top competitors. And market research provides an understanding about the changing needs, preferences, and expectations of customers who populate that market. You can now get a clear and realistic profile of your company's strategic market position, commonly called *competitive position.* (See Figure 10-3.)

Benchmarking Process

Benchmarking is a powerful process for improving your competitive position. It looks beyond the prevailing conditions and practices of a particular market segment. Benchmarking helps you identify and study processes, techniques, and practices of innovative market leaders often operating in different industries and markets. You can then apply those proven processes, techniques, and practices to the problems and opportunities that you face in your markets, thereby improving your competitive position. It can be a valuable source for ideas and approaches that have worked well elsewhere but never quite adapted to your business.

Benchmarking

- Searching for best industry practices that indicate superior performance of a company in its class
- Emulating the best of the best
- Constantly seeking and applying new ideas, methods, practices, and processes used by industry leaders

Basic Elements

- Know your organization and operations.
- Know the industry leaders or top competitors.
- Seek out and incorporate the best; gain superiority.

DIMENSIONS SUPPLIERS	RELATIVE MKT. SHARE	APPLIED TECHNOLOGY	LOW COST OF MANUFACTURE	SALES & DISTRIBUTION	PRODUCT QUALITY AND RELIABILITY	WEIGHT	TOTAL VALUE
OUR COMPANY							
COMPETITOR A							
COMPETITOR B							
COMPETITOR C							

Figure 10-3. Rankings of dimensions for competitive position. Comparative ratings are made from 1 to 10, with 10 being the clear leader in that category and 1 being the very weakest. In this way, categories can be weighted in terms of relative importance, and competitors can be ranked overall using a composite, weighted rating.

Approaches for Market Research, Competitive Analysis, and Benchmarking

	Market research	Competitive Analysis	Benchmarking
Approach	Analyze industry segments or customer groups	Analyze competitive strategies and capabilities	Analyze industry leaders, their strategies and tactics
Focus	Market requirements and customer needs	Competitors' strategies and market position	Business practices that seem to work effectively
Application	Products and services	Competitive offerings and market share	Business practices as well as products and services
Limitations	Past and current market trends	Market activities and current market share	Not limited to past or current conditions; study of innovative trends
Information sources	Customers	Customers, competitors, trade shows, literature	Firsthand experience with industry leaders with proven success

Benchmarking Process Steps

1. Identify what business areas or functions are to be benchmarked.
2. Identify appropriate industry leaders in similar or different industries.
3. Determine examination methods and collect data.
4. Project future performance levels.
5. Report benchmark findings and gain organization acceptance.
6. Establish definitive functional goals.
7. Develop action plans for implementation.
8. Activate action plans and monitor progress.
9. Recalibrate benchmarks.
10. Attain leadership position.
11. Integrate processes into company system and continuously improve.

Reasons for Benchmarking	
Without benchmarking	With benchmarking
Becoming competitive	
Internal focus	External focus on competition
Slow, evolutionary change	Acceptance of new ideas, methods
Low commitment	Higher enthusiasm and commitment
Adopting best-of-class practices	
Not-invented-here syndrome	Active search for new ideas
Myopic vision	Expanded horizons into other industries
Self-imposed limits	Breakthrough possibilities
Understanding customer requirements	
Based on history or intuition	Based upon market realities
No interest in exploring new trends	Constant search for significant trends
Establishing clear goals & objectives	
Reactive to historical standards	Reaching for dramatic improvement
Based upon catching up with industry	Aimed at industry leadership
Improving productivity	
Acceptance of status quo	Passion for continuous improvement
Low receptivity to new initiatives	No tolerance for coasting on past success

Benchmarking Myths

- Benchmark targets should be established first and practices investigated later.
- There is only one way to benchmark: against direct product competitors.
- Benchmarks are only quantitative, financially based statistics.
- Benchmarking investigations are focused solely on operations showing a performance gap.
- Benchmarking is something that needs to be done occasionally and can be accomplished quickly.
- There is a single company, somewhere, most like our firm, only much better, that is "the benchmark."
- Staff organizations cannot be benchmarked.
- Benchmarking is a target-setting stretch exercise.
- Benchmarking can most effectively be accomplished through third-party consultants.
- It is difficult to decide what to benchmark for each business unit.
- Processes don't need to be benchmarked.

- Internal benchmarking between departments and divisions has only minimal benefits.
- There is no benefit in qualitative benchmarking.
- Benchmarking is comparing our firm to the dominant industry firm and emulating them six months later.
- Benchmark practices are the same as enablers.

Guiding Principles for the Benchmarking Process

A high degree of ownership exists for the benchmarking process and findings.

- Management directed
- Function supported

The benchmarking effort is continuous.

- A competency center exists to provide consulting expertise for the benchmarking process.
- All potentially relevant information is gathered and cataloged, on an ongoing basis, for possible future reference.

Benchmarking is eventually performed by those who will be responsible for implementing the findings.

- Benchmarking is the responsibility of line managers.
- Benchmarking expertise is available to assist in effective use of the process.

The organization is encouraged to have pride in benchmarking because it makes the work easier.

- The operation wants to be the best at each task.
- It sees benchmarking as the means to achieve the best of the best.

Operations holds regular and mandatory reviews on benchmarking progress in all areas.

Figure 10-4 illustrates the relationship of benchmarking to business planning.

Types of Benchmarking

Internal benchmarking. A comparison of internal operations.

Competitive benchmarking. Specific competitor-to-competitor comparisons for the product or function of interest.

Functional benchmarking. Comparisons to similar functions within the same broad industry or to industry leaders.

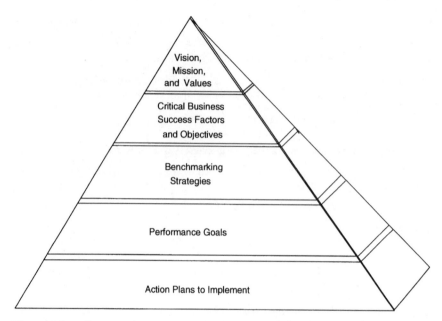

Figure 10-4. Strategic business and tactical planning relationship to benchmarking.

Generic benchmarking.   Comparison of business functions or processes that are the same regardless of industry.

Candidates for Benchmarking

Customer requirements	*Critical success factors*
Products	Customer satisfaction level
Services	Delivery performance
	Unit costs
	Asset utilization

Products manufactured	*Products purchased*
Finished assemblies	Components
Repair parts	Material-handling equipment

Services provided	*Processes used*
Repair services	Order entry
Financing	Customer inquiry/problem
	Warehouse fulfillment
	Billing
	Collection

Quality and Benchmarking

Quality process.　Meeting internal and external customer requirements.

Benchmarking.　Establishing objectives based on industry best practices.

Performance teams.　Involving employees in solutions to work practices.

Performance management.　Communicating objectives, setting goals, and recognizing employees for achievement of results.

Managing Behaviors Important to Benchmarking

- Provide supportive leadership in planning and organizing the benchmarking efforts.
- Gain agreement on the benefits to be achieved, the partnership companies, the approach to be used in the investigations, the roles of each member of the benchmarking team, and determining the barriers to effective benchmarking.
- Foster the viewpoint that benchmarking is the way more effective work is done, not extra work.
- Ensure that the benchmark findings are adequately understood and accepted.
- Ensure that performance levels needed and strategies pursued are based on benchmark practices.
- Ensure that performance is projected and periodically recalibrated based on benchmark findings.
- Ensure that a communications process is agreed on that will inform the organization of its progress toward benchmark targets and goals.
- Integrate benchmarking findings with the organization's objective setting, performance appraisals, and operating-plan processes.
- Seek out successful case history examples which demonstrate how the process is used and how the steps of benchmarking are applied.

Information Sources
Getting Started

- Focus on an area/element that needs to be pursued. Examples:
  Order entry
  Service dispatching
  Warehouse storage
- Contact a business library.
  Request a search of information produced in the last three to five years for your topic of interest.

Library will identify articles/sources from external reports, public magazines, industry journals, annual reports.

- Contact internal experts.

  Market research

  Competitive analysis

  Functional experts

- Survey internal reports/studies.

  Special studies

  Surveys

  Market research

Specific Functional

- Subscribe to and monitor trade periodicals.

- Use professional associations.

  Newsletters

  Seminars (especially speakers/tours)

  Bibliographies

  Special libraries

- Consult service agencies/bureaus in the business function.

  Asking if they can share anonymous experiences from their client companies

  Industry practices or methods

  State-of-the-art methods/practices

- Use consulting firms.

  Functional experts

  Asking if they are aware of particular breakthrough/practices in specific area

- Tap industry experts.

  Department heads of noncompetitors

  Teachers/professors at schools and universities

- Consult systems software development and hardware vendors.

  Asking about their experiences working in your functional area

Determining the Best Competitor

- Consider "competitor" in the broadest terms.

  What firm, function, or operation has the best industry practices?

  Look for comparable operations where best practices, methods, or processes are used.

- Ensure comparability.

  High-customer-satisfaction firms should be measured against high-customer-satisfaction firms.

  Product characteristics should be generic for the process. That is, packaged goods should be measured against packaged goods.

- Stay within the same industry.

  Define the industry broadly. (The electronics industry is an example.)

- Where are business practice breakthroughs found, or likely to occur?

  Uncover innovative practices wherever they exist, even in dissimilar industries.

Benchmarking Exercise

- What is most critical to business success?

  Customer satisfaction?

  Inventory turns?

  Expense to revenue ratio?

- What areas are causing the most trouble?

- What are the major deliverables of this area? Its reason for existence?

- What products are provided to customers?

- What factors are responsible for customer satisfaction?

- What problems have been identified in the operation?

- Where are competitive pressures being felt?

- What performance measurements are being tackled?

- What are the major cost components?

Planning

- Decide what to benchmark.

- Determine who to benchmark.

- Establish how to collect data.

Analyzing

- Are others better?

- Why are they better?

- How much better?

- What can be adopted?

Doing

- Commit to findings.

- Communicate new direction.

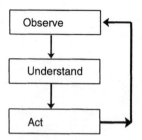

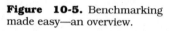

Figure 10-5. Benchmarking made easy—an overview.

- Take action to change.
- Recalibrate.

See Figure 10-5 for an overview.

Benchmarking Process Checklist

Step 1. Identify Benchmark Output

- Was the benchmarking study topic an outgrowth of the function's mission and deliverables?
- Was the subject selected critical to the success of the operation?
- Were practices benchmarked as well as key performance measurements?
- Was the subject and purpose of the benchmarking study reviewed with functional management and customers for their concurrence?

Step 2. Identify Comparative Companies

- Were the comparative companies selected the best competitors or functional industry leaders?
- Were all types of benchmarking considered identifying functional, industry best leaders?

Step 3. Determine Data-Collection Methods

- Was a questionnaire prepared prior to gathering the data?
- Were the questions pretested by answering them for the internal operation?
- Were internal sources researched for data and information?
- Were existing public data and information sources researched?
- Were original sources and investigations, including direct-site visits, considered?
- Were all research methods reviewed before the benchmarking investigations were conducted?
- Was the basis for information sharing reviewed before the research was conducted?

Step 4. Determine the Correct Competitive Gap
- Did the benchmark findings identify the differences in practices?
- Did the practices show for what reasons the differences resulted?
- Was a gap identified? Negative? Parity? Positive?

Step 5. Project Future Performance Levels
- Did the projection of the gap consider the best industry knowledge of trends?
- Was the gap understood in terms of tactical and strategic actions required?

Step 6. Establish Functional Goals
- Were the findings communicated to the affected organizations?
- Were all methods for gaining acceptance considered?
- Was there concurrence and commitment to the findings from the affected organization or customers?

Step 7. Develop Functional Action Plans
- Were functional goals reviewed to incorporate benchmark findings?
- Were the benchmark practices clearly delineated to show how the industry best accomplished their results?

Step 8. Implement Specific Action Steps
- Did the action plans clearly show how the gap would be closed?
- Was the action plan implemented?

Step 9. Monitor and Report Progress
- Were benchmarks incorporated with the management and financial process?
- Was an inspection process implemented?

Step 10. Consider Recalibration and Maturity
- Is there a plan for recalibration?
- Has benchmarking become institutionalized?
- Has a leadership position been attained?

See Figure 10-6 for helping and hindering forces.

Goal Setting and Action Planning

Set S M A R T Goals
- Significant—accomplishment, improvement, result
- Measurable—quantitative/qualitative

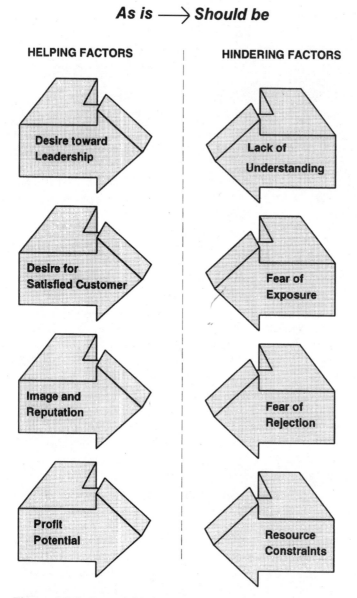

Figure 10-6. Force-field diagram of implementation—helping and hindering forces.

- Attainable—within reach with concerted effort
- Realistic—consistent with capabilities
- Timely—building confidence and sense of urgency

Action Planning
- Express goal as definitive statement of accomplishment.
- Visualize desired outcome and attendant rewards.
- Identify obstacles likely to be encountered.
- Develop alternative solutions for each obstacle.
- Identify sequential steps and measurable milestones.
- Monitor progress as you proceed; adjust plans accordingly.
- Record results and review process for improvement.
- Affirm your commitment and celebrate achievements.

Managing Customer Satisfaction

The focus of marketing in a TQM organization is satisfying the customer and creating new markets and products. The total quality of the product and service delivered to the customer must clearly meet the customer's requirements and exceed the customer's expectations in order to achieve customer satisfaction. While it may be difficult to fully satisfy each and every customer all the time, it is certainly a worthy objective to embrace as a company standard.

The reference to our customers is universal . . . it applies to internal as well as external customers. Other departments depend upon the services, information, reports, instructions, specifications, and products that you supply them in the normal course of transacting company business. If the internal supply line breaks down in delivering important requirements, then the satisfaction of the ultimate external customer is put at risk. For a company to become customer-focused and dedicated to customer satisfaction, all people in the company must recognize their important link in the "chain of customers," including their internal customers in the next department or work area.

Why is it so important to satisfy our external customers and how can we do it in an organized and systematic way? To appreciate the importance of satisfying customers, consider the typical profile of a satisfied customer:

- Prefers doing business with your company and regularly orders products and services from you

- Regards you as a valuable supplier and business partner . . . keeps you informed about industry/market developments, and his or her changing needs and requirements

- Relies upon your products and services . . . gives candid feedback on supplier performance, how you stand competitively, and identifies opportunities for improvement

- When supply problems arise, contacts you immediately to work together to resolve the problem

It is estimated that it takes five times as much cost and effort to replace the business of a dissatisfied customer than to convert him or her to a satisfied customer. Conversion is not easy, but is well worth the effort.

The outcome of total quality marketing is a satisfied customer who prefers doing business with your company for many reasons. Within the organization, internal customers are being satisfied by their internal suppliers, thus enabling satisfaction of the external customer. Working associates can rely upon the information and services which are provided by another department acting as internal supplier. Effective marketing depends upon the interdepartmental communication and cooperation used to satisfy internal customer needs and, ultimately, external customer requirements. In this case, marketing is really a value-adding process which makes goods and services valuable to the ultimate customer or "end user" at the end of the "chain of customers."

There are some basic definitions which describe this marketing process:

Customer. *Any* receiver of a product or service; may be *internal* or *external*.

Supplier. *Any* group which provides or delivers a product or service; may be an individual, department, business unit, or company.

Processes. Value-added transactions by means of which suppliers serve customers. These are typically *internal, numerous, and outside of manufacturing.*

Total quality. The continuing quest to select (only) the most productive undertakings and carry them out via optimum value-added transactions (processes) among all suppliers and all customers in the chain, thereby conforming to every client's expectations.

Listening to the Customer*

To become customer-focused and market-driven, people throughout the company must develop an orientation and sensitivity to people who are

---

* Listening notes adapted from Human Synergistics, "Project Planning Situation."

their customers, whether in the next department or across the country. Where once marketing alone acted as the "eyes and ears" of the company—interpreting customer needs and defining expectations which the company must meet—now all functions must relate their activities to their customers and ultimately to the customers and markets served by the company. Listening to the market is now part of everyone's job.

Listening to the market requires excellent listening, communicating, and interpreting skills, rarely taught in school or emphasized in company training and development activities. Yet, listening for and understanding customer expectations is a key and essential part of Total Quality Management. There are some time-tested principles for effective market and customer listening:

- Let the customers know the subject and why you're interested in talking with them; if the moment is not convenient for them, arrange another time.

- Outline the information and opinions sought. Prepare questions but do not follow a rigid script. Ask open-ended questions; take notes and ask for clarification when you need it.

- Anticipate where the conversation is headed; keep it cordial but focused.

- Record and objectively weigh the evidence being presented.

- Periodically review and summarize what has been said.

- Pay attention to nonverbal behavior as well as the verbal.

A key factor in *poor communication* is the tendency to critically judge and evaluate the speaker and his or her expressions during the conversation. Not only is the listener indulging in "selective listening," but it often leads the speaker to attempting to justify, rationalize, defend, or protect his or her position.

The remedy is *empathetic listening,* whereby the listener tries to understand the speaker from the speaker's point of view. If this is practiced, the listener will find that he or she is able to understand points quicker and remember them longer and more vividly.

Another good listening technique is *supporting.* Supporting means the ability to acknowledge the specific merits of the speaker's ideas and to build upon them. In supporting, we need to do the following:

- Assume the customer has useful ideas, information, opinions.

- Listen carefully and attentively.

- Mention the specific point which you find useful.

- Build rapport on these useful points.

Determining Requirements and Expectations

A vital part of Total Quality Management is converting sometimes vague and generalized customer needs into fully defined marketing requirements and then into product/service specifications. This is a process of collecting customer and market information through a multitude of sources, and then systematically categorizing and evaluating this information to derive a composite view of requirements and expectations for each customer group or market segment.

It is appropriate to treat requirements and expectations differently. Requirements, if properly defined, can be useful "go/no-go" filters. If a certain attribute of a product or service can be defined as a "specification" or "requirement," then the determination of whether you can meet or exceed these stipulations is usually straightforward. Therefore, be careful in defining a customer requirement or specification. It may screen out alternative solutions that can offer ultimate value well beyond the candidate ideas that conform to conventional requirements. When in doubt, express a requirement or specification as a "want" (expectation) rather than a "must" (requirement). In that way, the range of possibilities for satisfying customer needs is broader.

Analyzing and Measuring

Analysis and measurement of customer requirements and expectations is essential in achieving customer satisfaction. Although not every attribute can or must be reduced to a quantifiable and measurable item, the discipline of measuring is very useful in indicating direction of effort and progress. After all, continuous improvement connotes a movement from a current state to a more desirable state, and definitive measures are helpful in gauging progress and celebrating significant gains.

There is nothing wrong with qualitative measurement. The idea is not to reduce all activity and accomplishment to numbers. Quite the contrary, the fulfillment of Total Quality Management and continuous improvement is the realization that progress has been made not only in measurable results but in the attitudes and behaviors of those involved. So measurement is an ally and an important tool in Total Quality Management, an indicator of moving in the right direction and instituting changes to which the organization is committed. If a certain attribute or objective defies quantitative measurement, and yet is important to the entire TQM effort, by all means describe and measure it in qualitative terms.

Serving Customers, Internal and External

The essence of Total Quality Management is delivering product and service to our customers which satisfies their needs and expectations. At every step of the way, we are trying to add value to our customers, helping them achieve their objectives more easily, more quickly, more consistently, and more effectively.

Suppliers serve customers via value-added transactions called *processes*. When we commit to serve our customers with full conformance to requirements and expectations, we are pledging to add value to their business or work product. *Value* can be defined as that ingredient that helps the internal or external customer improve the output of his or her efforts. (See Figure 10-7.)

Managing Customer Relationships.   No matter what the industry or product, the daily inquiries and conversations that are a part of transacting business are personal contacts between people. Whether face-to-face discussions or voice-mail messages across time zones, the common denominator is people communicating with people. Customer relations are built upon familiarity and trust. Slow or impersonal responses to customers can be deadly. Anyone interacting with a customer is an important extension of your marketing department. And anyone dealing with customers on a regular basis needs to understand and appreciate the customer.

Some Important Points About Customers

- They are not outsiders, but part of our business.
- They are not an interruption of our work, but the purpose for it.
- We are not doing them a favor by serving them; they give us an opportunity to do so.
- They do not depend upon us; rather, we depend upon them.

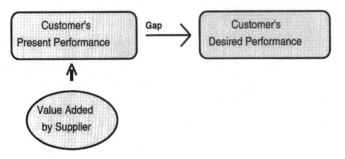

Figure 10-7. Satisfying customer expectations.

- Customers are not statistics or market data; they are real people with feelings, emotions, ambitions, worries, biases like our own.

- They are our most important visitors, whether they come in person, by phone, fax, or mail.

Determining Customer Requirements with Quality Function Deployment

Quality Function Deployment (QFD) is a system for translating customer requirements into appropriate technical requirements at each stage in the product development process and the engineering tools it specifies. Effective use of QFD will yield reduced development time cycle and associated costs, and much greater productivity with assigned people and development resources.

QFD deploys the voice of the customer, i.e., customer requirements are defined by detailed consultation, brainstorming, feedback mechanisms, and market research. This is accomplished through the total product development process involving translating customer requirements at each stage of product development and production.

Developing a top-notch project team is one of the most challenging and potentially rewarding aspects of the process. All areas involved in product development should be represented on the team, i.e. marketing, product planning, product design, engineering, prototyping and testing, process development, manufacturing, assembly, sales, and service. Each team should be working toward a shared goal: a customer-defined product to be completed by a specific date and at a specific cost.

The process begins with the customer requirements, which are usually loosely stated qualitative characteristics, such as "looks good," "easy to use," "works well," "feels good," "safe," "comfortable," "long lasting," "luxurious." These characteristics are important to the customer but defy quantification and are difficult to act on.

During product development, customer requirements are converted into internal company requirements called *design requirements*. These requirements are generally global product characteristics (usually measurable) that will satisfy the customer requirements if properly executed.

Before QFD can become operational, we must identify who the customer is. In most cases, more than one customer exists. For example, the end user, the company for whom the product is being produced, and the assembly operator who will be putting the product together are all customers. In almost all cases, there will be both *external* and *internal* customers. Both need to be taken into account. Should conflicts arise, the *internal* customers usually defer to the *external* customers so that end-user customers get what they want.

The next step, "determining customer requirements," is accomplished through a series of charts and matrices that may seem very complex at first glance. However, when broken into elements, the process becomes easy. In fact, the premise for determining customer requirements is similar to that of Management by Objectives (MBO). Emphasis is placed on what needs to be done and how to go about doing it.

Building a House of Quality. The following summary of the components of a "house of quality" (see Figure 10-8) will help clarify the content and function of each room. Touring the rooms will help us understand the house itself. But first, let's examine a recurring QFD theme: from "what" to "how" to "how much."

The theme is based on an input-output strategy. The process begins with a list of loosely stated objectives ("what we want to accomplish" list). These "what" items are the basic customer requirements. They will prob-

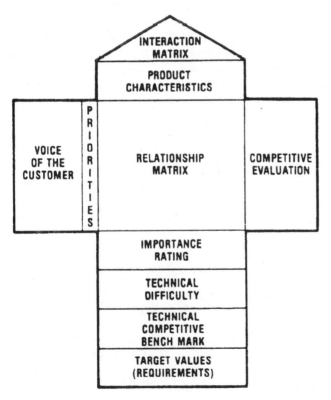

Figure 10-8.

ably be vague and require further definition. One such "what" item might be an excellent cup of coffee. Every coffee drinker wants this, but providing it requires further definition.

To provide further definition, each "what" item is broken into one or more "how" items. This process is similar to the process of refining marketing specifications into system-level engineering specifications. Customer requirements are actually being translated into design requirements. The excellent cup of coffee requirements, for example, would be translated into "hot," "eye opener," "rich flavor," "good aroma," "low price," "generous amount," and "stays hot." If the cup of coffee were being served in a restaurant, "service with a smile" and "free refills" might also be customer requirements or "how" items. Customer requirements and conditions of use are correlated.

The "how" items will also usually require further definition, and are thus treated as new "what" items that are broken into additional "how" items. This is similar to the process of translating system-level specifications into parts-level specifications. Using the cup of coffee example, new "how" items might be "serving temperature," "amount of caffeine," "flavor component," "flavor intensity," "aroma component," "aroma intensity," "sale price," "volume," and "temperature after serving."

This refinement process is continued until every item on the list is "actionable." Such detail is necessary to ensure successful realization of a customer requirement. This process is complicated by the fact that some of the "how" items affect more than one "what" item and can even affect one another.

Product Attributes (S U R P A S S)

- Styling
- Usability
- Reliability
- Performance/price
- Availability
- Serviceability
- Supportability

Customer Expectations

Since customers provide the revenues that represent the lifeblood of the business, achieving customer satisfaction should be the primary focus of business. A customer's buying decision is based on the expectations that are created before the time of sale; these expectations come from many sources: previous satisfaction with your product, experience with competitors' products, and advertising and sales claims about your product. After ordering a product, the customer will be satisfied only to the extent that no disappointments occur. Therefore, the definition of customer satisfaction could be stated as the *absence of disappointments or the absence of the elements of dissatisfaction.*

In a dynamic, competitive environment, it is doubtful that a perfect state of customer satisfaction can exist. Customer expectations continuously rise as technology and competitive offerings improve. A business strives to improve as rapidly as its competitive environment improves, but always seems to lag because of the response of the system. Further satisfaction depends on whether the business can improve faster than its competitors, thereby commanding a greater share of the market.

It is virtually impossible for any given company to always be first with new technology. Customers will continue with a name brand if they have confidence in their supplier; this is referred to as *customer loyalty.* Loyalty is built up over time if customers are consistently satisfied with a supplier's product. Competitors may occasionally offer higher performance/price or lower cost through new technologies, but most customers maintain name-brand loyalty and know that their supplier will respond shortly. Loyalty does not last forever, though, and businesses must always work toward product improvement to ensure an acceptable level of customer satisfaction in order to retain customer loyalty.

The *acceptable* level of customer satisfaction, however, may be far below the *possible* level of customer satisfaction. For example, a closed economic environment with only a few competitors existed in the U.S. auto industry 20 to 30 years ago. Then, customers were satisfied at an acceptable level with the products that American car manufacturers offered. The styling, the quality, the performance/price, and the low fuel efficiency fit the expectations of American culture.

Then the environment changed dramatically—as in 1973 with the fuel shortage, the story changed. The Japanese auto industry entered U.S. markets with more fuel-efficient cars, which the American public desperately needed.

This new need gave the Japanese a foothold in the American market. Once there, the Japanese demonstrated higher levels of quality, performance/price, and fuel efficiency. They set new standards for customer expectations. The Japanese demonstrated that a higher level of satisfaction was possible. To regain the lost market share, American manufacturers had to do more than improve fuel economy, and only now—20 years later—are some American auto manufacturers showing signs of recovery.

The Chronology of Customers' Expectations. When customers buy a product, they have a set of expectations for that product. Customers are neither satisfied nor dissatisfied at any one particular time. Full or partial satisfaction or dissatisfaction is determined over the life of the product, and begins when customers decide to buy the product. For example, customers won't decide to buy the product if they are not already certain or reasonably assured that the product is going to satisfy their needs. This decision is influenced by the level of communication that takes place between the prospective buyer and the seller.

The degree of customer satisfaction will be the result of the product being received and used, and then failing to measure up (dissatisfaction) or succeeding (satisfaction) over the useful life of the product. This occurs over a long period of time, referred to as the *customer life cycle*. Let's consider some customer expectations over the customer life cycle.

Before the Sale

- The specifications must not be ambiguous; they must relate to the product's intended application.
- That the product will function correctly in the customer's intended environment must be clear.
- The regulatory requirements that the product meets must be clearly defined, i.e., safety, health, environmental standards.
- All product capability must be made clear.
- The delivery information must be reliable.

After Delivery

- The product was received by the customer when promised.
- The shipment contains everything expected and needed to use the product.
- Operating and setup instructions are clear and complete.
- The product is received defect-free.
- Product use is easy to learn.
- The product functions as claimed and as expected.

After Setting Up, Learning to Operate, and Ready to Use

- The product meets specifications and stays that way.
- There is easy verification of continued conformance to specifications.
- The product is reliable over time.

As the Product Ages

- Preventative maintenance is clear, easy, and economical.
- Product repair is economical.
- Factory or service-center repairs are handled promptly.

- Spare parts are easily available and reasonably priced.
- Spare parts are available for the life expectancy of the product.

These are just a few of the expectations a customer may have of a supplier's product. Each business should modify this list to reflect its own customer base and to identify any product shortcomings. Once customer expectations are known, a business must satisfy them by implementing the necessary organizational processes. Where some customer dissatisfaction still exists, an improvement process should be implemented to increase the level and consistency of customer satisfaction.

Customer Satisfaction Growth Model.   Product or customer satisfaction growth modeling is an effective tool for understanding and managing the relationships between customer expectations and product performance. A growth model is also valuable for assessing the appropriate organizational needs and determining measures of customer satisfaction. Figure 10-9 illustrates a model that can be used to satisfy this need. It should serve for most manufacturing businesses and, with a little imagination, can be adapted to other types of businesses as well.

Customer Requirements.   The customer's requirements are based on needs. If the need exists, the customer will search for a product until it is found. The customer's search is also determined by past experience in the marketplace with similar products. Customer requirements are enhanced by known technological advances and by competitors' advertising claims. This set of requirements is relative to the customer's need and will not change unless the need changes or technological progress creates a new customer awareness of product capability.

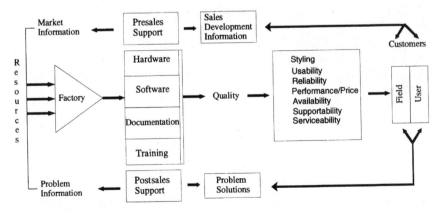

Figure 10-9. Customer satisfaction growth model.

Customer Expectations.   The customer's expectations are based upon wants as well as needs. Meeting a customer's set of requirements does not necessarily result in customer satisfaction. Since customer satisfaction is the ultimate aim, it is important to define those product/service attributes which may not be essential for product performance but which may be essential to full customer satisfaction. Customer requirements include the basic essentials for acceptable product performance . . . the must-have deliverables. On the other hand, customer expectations include those features and attributes that distinguish the product/service as excellent or extraordinary . . . the nice-to-have deliverables.

The competitive reality in today's global marketplace with world-class product offerings, is that meeting a set of rigorous product performance requirements is simply an ante for qualifying as a prospective supplier—it no longer differentiates you as a top contender. The differentiating criteria have become expectations above essential requirements, and business is often awarded on the basis of the perceived value of these "extra satisfiers."

Presales Support

The presales support function is part of the marketing interface responsibility. It encompasses customer contacts, product advertising, and other communication forms that influence customer requirements and expectations. The market interface function (marketing) is also responsible for understanding customer needs, customer requirements, special expectations, and competitive products. The information is collected, analyzed, and forwarded to the design group. This information is then used to define the external requirement specifications (ERS) of new products, thus contributing to continued product improvement and customer satisfaction.

The Factory.   The factory is made up of design, manufacturing, and support processes. All new product information is created in design and is handed over to manufacturing when the design work is complete according to the ERS. Manufacturing will receive all of the documentation, specifications, and tooling required to consistently build and ship the product according to the ERS. This will include specifications for SURPASS, an acronym for *S*tyling, *U*sability, *R*eliability, *P*erformance/Price, *A*vailability, *S*erviceability, *S*upportability.

Manufacturing develops the processes to build the product as designed and to alter it as required by changes in the environment. The changes may come from a number of sources: vendor part performance, part obsolescence, process modifications, or product enhancements. The shipped item may include various elements . . . for example, a computer product may include hardware, special software, and user information.

Postsales Support

The postsales support group is normally part of the marketing/customer interface group. This support organization represents the service departments and other customer service groups. It is responsible for providing solutions to the set of customer dissatisfiers; sometimes it does so profitably by selling its services. This group also collects information about customer problems (dissatisfaction) and passes it back to the factory. The information is then used to identify needed changes to improve the product or processes, thereby improving customer satisfaction.

The Product.   The elements of a product, as previously defined, represent all or part of the solution to a customer need. When they are provided together as the solution, they collectively represent the product. For example, consider a complex computer product where the solution may involve hardware, software, documentation, and training. A simple product could be defined by only one of the elements.

Quality.   The quality of the product is determined by how effectively the entire product meets the customer's requirements. When measuring the quality of a complex product, the SURPASS measure must be compared to the total product; however, SURPASS can be applied to an individual element of the product, as will be explained later.

SURPASS represents a generic set of product attributes that a customer uses to make buying decisions. They are introduced here to facilitate the definition of the mode. We will limit our present discussion to the definitions of SURPASS and how they apply to customer satisfaction. Specific ways to measure and evaluate the relative values of SURPASS are discussed later.

Customer Satisfaction.   Once the customer's requirements are set and the item is received, the product attributes should provide sufficient quality to satisfy the customer. If they don't, a set of dissatisfiers will appear for which compensation must be made. A business generally staffs a support organization to provide prompt response to these problems.

Closing the Loop

When both the presales and postsales groups actively pass information along to the factory—and when the factory systematically improves the product—a state of continuous improvement exists for the product. If this product improvement process occurs faster than the customer requirements change, then a state of growth exists for customer satisfaction. It is important to point out that product improvement occurring at the same rate as the change of customer requirements does not result in the growth of customer satisfaction. A business that continuously improves its product can still lose its market share if the improvement process is too slow.

SURPASS Definitions.   As previously defined, a complex product has a set of measures that relates to customer satisfaction; these are the SUR-PASS measures. When considering the product (hardware, software, documentation, or training), it is important to view each part as making a contribution to the whole. A product manual, for example, may stand on its own in some applications (like a how-to booklet sold for profit) and have its own quality measures. But, when taken as part of the product, it must be viewed by the contribution it makes to the overall effectiveness of the product.

This is true for each element of the product—its styling, usability, reliability, etc. When considered part of a bigger product, the only important factor is the contribution that it makes to the product's SURPASS attributes. As such, a product manual becomes an extension or an enhancement to the product.

Product manuals containing operating instructions and/or installation procedures have a somewhat universal application, and therefore make an excellent example for defining the elements of SURPASS. In the following definitions, the SURPASS attributes relate to more complex product information, but they can apply to any product that requires documentation.

Styling.   The function of the product manual is to communicate information about a product to the purchaser, installer, operator, and maintenance person. The style used to convey this important product information makes an initial impression on the user. The style also may determine how clearly this information is communicated and how effectively it is understood and applied. Use of photographs, color graphics, diagrams, training tapes, tutorials, and the like, are examples of styling qualities of modern product manuals. In any particular industry, there is usually a leading supplier who sets the standards for product manual and product information styling. Since this information remains on record with the user over the useful life of the product, its styling can contribute to customer satisfaction. A product manual will usually contain:

- Specifications
- Product features, capability, and intended use
- Installation/assembly procedures
- Operating-performance verification tests
- Troubleshooting information

- Operating procedures
- Replacement parts lists
- Repair procedures

Usability. The product manual should be user friendly and easy to read and follow. Typical considerations for usability include:

- Table of contents
- Index
- Glossary
- Prioritization/organization
- Optimum quantity of information
- Correct literacy level (vocabulary relative to intended users)
- Telephone and fax number for additional information

Reliability. The accuracy of the product manual can affect the perceived reliability of the product if it misleads users on the suitable application, operation, specifications, and diagnostic information of that product. The reliability of the documentation is determined by the following:

- Typographically correct information
- Accurate and clear diagrams, pictures, charts, etc.
- Diagnostically correct information (accurate identification of failures)
- Correct and validated operating procedures, installation procedures, etc.
- Correct and clear information throughout

Performance/Price. A product manual should explain to the installer and user how best performance of the product on a particular application can be achieved. The manual should make it easy to accomplish the tasks that help ensure top product performance:

- Assembly and installation
- Product learning cycle
- Task accomplishment (productivity of use)
- Diagnostics/troubleshooting efforts

Availability. Product manual availability relates to both new and old products. If a producer is in the habit of shipping new products without adequate documentation (preliminary documentation, in some cases), the customer will not get the full benefit of the product. If customers lose old manuals, they need to have access to manuals that reflect their product's vintage, not the new manuals currently available. The opposite is also true; when a product is updated, the manual also needs to be updated.

Serviceability. User costs to keep product operating and manuals current include those related to repairs, calibration, and preventative maintenance. The elements of the manual that can influence these costs are:

- Type of manual update service provided; e.g., change sheets, replacement pages, new inserts
- Completeness of service instructions
- Accuracy of parts lists and repair information

Supportability. Depending upon the industry and product, the product life cycle may be 5 to 50 years. Long-term supportability for products that are expected to remain in service for extended periods of time is an important consideration for customer satisfaction. In this example, the product manual for a particular product model must be continuously updated to new regulations, applications, and end-use requirements.

Conclusion. Customer requirements are continuously on the rise. A business will be successful to the extent that its product growth rises faster than its customer requirements. This is known as customer satisfaction growth. There are many aspects to customer satisfaction that take years to materialize. Businesses must be organized, with resources allocated to provide for long-term customer satisfaction. The measures of customer satisfaction referred to as SURPASS represent a generic set of attributes that can be used to evaluate a product's merits relative to its competitive product offerings.

Measuring Customer Satisfaction

Organizational Metrics

This entails the measures used to plan for and evaluate the business effectiveness. A business is successful when:

- It is better than its competition at delivering products with appropriate requirements.
- It is more productive than the competition.

Four Categories of Metrics

1. *Product requirements performance.* Measures how effectively a product meets the needs of the market/customer.

2. *Market interface performance.* Measures how effectively a business interfaces with the market to understand present and future needs, and how effective it is at generating awareness, sales, and distribution of the business product.

3. *Capacity performance.* Measures the production capabilities of the business.

4. *Process productivity performance.* Measures how efficiently a business uses its resources in providing for the other performance measures.

Product Requirements (SURPASS)

- Styling
- Usability
- Reliability
- Performance/price
- Availability
- Serviceability
- Supportability

Styling. Measurement of customer satisfaction regarding styling may be difficult but is nonetheless important for most products, even industrial products that are used in manufacturing processes. For example, electric power cords used for industrial and construction applications must have similar designs for functionality reasons, but the product can be differentiated by color and packaging. A major U.S. manufacturer of electric cord sets uses colors of local sports teams to style their product to appeal to end users in different regions of the country. This styling differentiation has been enough to sustain sales growth in relatively flat markets.

Usability. Metrics are applied to usability by considering the product cost components that are associated with proper product use during a particular period of time (often annualized). In the following example of process equipment used in a manufacturing operation, the percentage of the initial product acquisition cost which is represented by first-year usability costs is evaluated. Usability costs can prove substantial with certain capital equipment, and should be measured for the product on a regular basis.

Metrics for Usability Example	
Product selling price	$5000
Installation or preparation cost	180
Installation and operating maintenance cost	30
Cost of user training	200
Average cost of time to exercise a function	5
Total usability cost	335

$$\text{Usability index} = \frac{335}{5000} = 0.067 \ or \ 6.7\% \ of \ selling \ price$$

Reliability.   Reliability is a measure of the product's ability to perform to its specifications over a period of time. Some typical metrics used for evaluating a product's reliability include the following:

Mean Time Between Failures (MTBF)

Mean Time Between Repair (MTBR)

Mean Time to Diagnose (MTTD)

Performance/Price.   The measures for product performance relate to the product's specifications as they apply to the functionality of the product. Performance measures are usually clearly stated and subject to exact testing procedures and quantitative measurement. The challenge is often involved in devising test methods and measuring performance under both typical and extreme operating conditions and application environments. Performance testing and measurement have become extremely powerful indicators of a product's market acceptability and degree of customer satisfaction.

Performance measurements are often dictated by trade associations, industry qualification agencies such as UL, CSA, NEMA, FAA, etc., and leading competitive suppliers of products as they promote their most favorable performance features. When devising performance test methods and deciding on critical product performance parameters, it is essential that performance measurement be selected in light of customer applications, simulating actual operating uses and conditions.

Performance measurements specific to product price are often used to evaluate different product offerings. Such measurements are sometimes called *specific-performance.* Examples include computer memory speed and bytes per dollar of cost, commercial lease property measured in usable work area per dollar of monthly lease rate, process filtration capacity to a specified clarification level in gpm per operating dollar cost. Again, such specific performance/price measurements must be tuned to what the end-use customers require and expect from operating your product in their environment for their particular application purposes.

Availability. Availability applies to both new products under development and current products under production.

1. *New products.* The date of release and shipment to customers. Release dates for new products is a very market-sensitive subject, especially in the computer and electronics industry. Once committed, release dates must be met in order to satisfy those potential customers for the new product. They represent a window of opportunity when introducing new/improved products, and hitting release dates as promised is an early builder of customer satisfaction. Availability means concurrent development and release of operator manuals, qualification test data, and other documentation necessary for proper product application.

2. *Current products.* The average delivery time from the order date. When dealing with a broad mix of products supplying distributors or mass merchandisers, order fill rate becomes an important determinant of product availability and customer satisfaction.

Serviceability. Serviceability is a measure of the owner's cost to maintain the continuous services of the product (cost of ownership). Some of the components of serviceability are illustrated in the following example of determining a serviceability index. The formula is as follows:

Cost of service documentation + average annual cost of repair
    + average annual cost of calibration or preventive maintenance
    + annual cost of downtime [annual failure rate
    × (days to repair/250 days) × S.P.] ÷ selling price = *Serviceability Index*

Supportability. The metrics for supportability deal with the potential expansion, upgrade, or replacement costs for products incorporating new technology. In computer hardware, supportability is enhanced when equipment is modularized and can be updated or expanded by replacing certain component boards or chips. In computer software, supportability is enhanced when upgraded programs are offered to registered users at moderate costs. IBM and Digital Equipment have earned much of their customer loyalty through continued emphasis on product supportability.

New Product Development

Product and service innovation is the lifeblood of most companies who aspire to be industry leaders. In many cases, basic survival is dependent upon continually improving the net value of the product/service package supplied to your customers. Global competition and foreign government and industry emphasis on R&D investment and new technology funding has made product/service innovation an imperative for U.S. companies in order to remain competitive. Other business factors that have made product/service innovation so important are:

- The weakening of international patent protection and its enforcement
- The dramatic time compression of product development cycles, led by Japanese and Asian electronics manufacturers
- The redistribution of capital investment to countries with more plentiful, lower-cost labor and resources
- The steady reduction of international trade restrictions
- The rapid growth of international distribution networks around the world, linking suppliers and consumers in all corners of the globe

What is needed is a streamlined product development cycle that is tuned to the customers and markets it must ultimately serve. Ideas must be imagined from customer needs, behaviors, and attitudes that are recognized and understood. The benchmarking process is an excellent way to get fresh ideas from an industry leader in another field which has successfully innovated in its particular markets. Benchmarking against competitors helps determine the special capabilities and distinctive competencies that set your company apart and therefore offer an avenue of innovation.

While ideas from the marketplace provide raw opportunities for innovation, there needs to be a system to nurture ideas through a disciplined development process. The product development cycle is presented as a process diagram with seven phases from idea generation to commercialization. (See Figure 10-10.) As the idea proceeds through the process phases, more cumulative time and money is committed, so it is important to make go/no-go decisions at every phase of the process, preferably before the project phase when funding is decided.

The development funnel narrows considerably as the ideas and project candidates progress, such that 100 initial ideas might result in only five development projects which in turn might result in the launch of only one commercial venture at the end of the development funnel. The critical parameter of this development process is how quickly each phase can be accomplished so that the overall development cycle time is shortened for greater competitive advantage in fast-changing world markets. Market timing is often a critical success factor for any new product/service introduction, and market expectations are much like hitting a moving target from a moving platform; it helps when you're moving in the same general direction and at comparable speed.

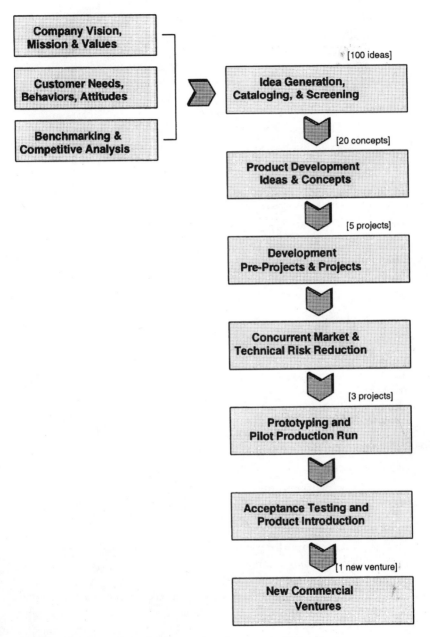

Figure 10-10. Product development cycle process diagram.

Rather than delay introduction until the ideal product can be perfected, it is often wise to introduce a first-generation model followed by successive upgrades and improvements. A prime example of this strategy is software development and release. There are many case studies of successes and failures due to premature versus late product introductions and the compromises these tough management decisions entail. Customer preferences, market dynamics, and competitive factors must be considered during every phase of product development because, ultimately, they determine the success of any new product/service offering.

Phase 1. Idea Generation, Cataloging, and Screening

The cornerstone of new product development and product and service innovation is the generation, collection, and processing of raw ideas. Such ideas can come from all quarters, but it is marketing's responsibility to make sure ideas for new products and services are:

- Actively solicited from customers, prospects, sales representatives, suppliers, and industry leaders
- Actively solicited from company executives, managers, supervisors, and operating employees
- Collected from trade shows, industry journals, competitive literature, and advertising
- Systematically categorized, grouped for synergistic effect, and recorded in a log for tracking and response
- Screened for initial compatibility with the company's vision, mission, and values, and potential contribution to Critical Business Success Factors
- Sorted and dispositioned in a timely manner, with response to the idea initiator, acknowledging receipt and screening of the idea and whether it is under consideration for further development

Typically, a small committee of three to five people is assigned the responsibility of soliciting ideas from employees, customers, distributors, and sales and customer service representatives, and then cataloging and screening them. Membership of this product innovation committee might be comprised of a representative from sales, marketing, design, engineering, and operations. The main responsibilities of the committee are as follows:

- Encourage a companywide and marketwide inflow of ideas regarding new product or service opportunities for the company.
- Catalog and screen ideas with a set of predetermined evaluation criteria.

- Develop a common format for idea submittals.
- Acknowledge receipt of each idea and thank the contributor.
- Establish an appropriate recognition program for people submitting ideas and contributing to the assessment process.
- Advance the most promising ideas meeting criteria to the next phase.

Phase 2. Product Development (Ideas and Concepts)

- Develop promising ideas into product and service concepts.
- Consider variations and extensions of central idea in light of design and technical requirements, styling, functionality, marketability, manufacturability, and so forth. Integrate ideas and variations into a whole product/service concept.
- Conduct preliminary (cursory) assessments on market acceptance and technical feasibility; rank risk as high, medium, or low.
- Select most promising product/service concepts and advance to next phase of preprojects and projects.

Phase 3. Product Definition and Development (Preprojects and Projects)

- Refine product/service concepts and conduct further assessment of marketability and technical feasibility. Consult with a wider circle of advisers.
- Review product/service concept with focus group of customers and/or distributors, and review comments and recommendations gained from them. Determine ideal performance characteristics that would most appeal to the market and target customer groups.
- Conduct market and competitive research to identify similar or potential competing technologies and product/service offerings.
- Derive a set of ideal product/service performance characteristics and chart all similar competitive offerings against the ideal (performance gap analysis).
- Plot candidate product/service preprojects on a grid comparing technical risk and market risk factors.
- Review preproject candidates with company management and recommend advancing one or more to project status (budgeted development funding).
- Develop a comprehensive project development plan for the approved project.

Marketing Links for Project Definition. Development projects are based upon an ideal model of a product that will satisfy customer needs, called a *product definition report.* The technical specifications and performance characteristics of the ideal new product are derived from a marketing specification which in turn is based upon perceived needs and requirements of the market as represented by certain customer groups. This is not a static model but, rather, it can change quickly and often unpredictably as market conditions shift. The project manager's responsibility is to maintain this linkage during the course of the development project so that the product definition remains aligned with the actual customer and market needs. While the focus on development is to achieve (as quickly and directly as possible) the specifications and performance characteristics that have been set as a target, the overall project mission is to keep development efforts aligned with the current and future needs of the market. (See Figure 10-11.)

Phase 4. Market and Technical Risk Reduction

Each preproject or candidate project must be evaluated on its inherent market and technical risks. *Market risk* is defined as the risk associated

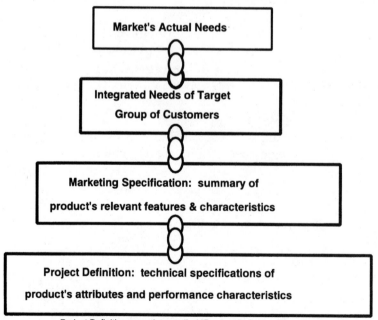

Project Definition sometimes called "Product Definition Report"

Figure 10-11. Market links for project definition.

with inadequate acceptance by the market, assuming that technical challenges can be met. Conversely, *technical risk* is defined as the risk associated with not being able to meet technical challenges while assuming market acceptance.

Plotting various preprojects and project candidates on a profile grid as illustrated in Figure 10-12 (with project size denoting relative profit potential) can help in the management decision process as to which candidates are worth funding as new product development projects.

Management must understand the risk/reward profile for each candidate project before committing resources. It must also understand the magnitude of associated market and technical risks and the strategies that will be undertaken to reduce those risks to acceptable levels. While risk reductions can be done concurrently, market risk reduction is usually less costly, and therefore is often started ahead of technical risk reduction.

Market risk usually has an important timing component or "window of opportunity" that is dictated by the market and which must be gauged when planning a product development project. Because the pace of technical risk reduction is usually less predictable than that of market risk reduction, both risks must be reduced concurrently in order to keep the project on schedule and commercially viable.

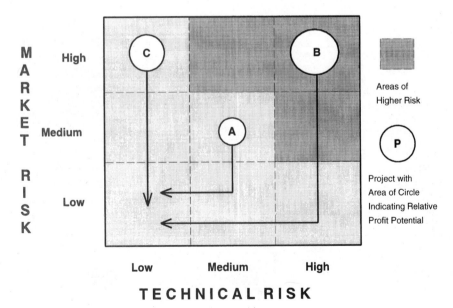

Figure 10-12. Market and technical risk profile.

Phase 5. Prototyping and Pilot Production Run

Prototyping is an important process for fast-track product development and technical risk reduction. The technical challenges faced by the development team are better understood when concept models and prototypes are made and tested and analyzed during the course of development. Prototypes can provide early insight on manufacturability and selection of appropriate processes, work methods, and test procedures to ensure quality and productivity.

Prototyping can also help in market risk reduction by confirming that customer requirements and expectations are fully understood and effectively translated into new product designs. Focus groups can examine concept models and prototypes to give initial market reactions and test the market specification and whether it still reflects the key criteria for customer/market acceptance.

Pilot production runs serve to evaluate technical issues of material and process selection, producibility, quality assurance, and manufacturing cost. Pilot runs can also be used to produce test specimens representative of production runs so that acceptance testing can be done.

Phase 6. Acceptance Testing and Product Introduction

The main purpose of acceptance testing is to ensure reliable and safe product performance in a full range of applications, under different and adverse operating conditions. Product acceptance testing is vital for quality assurance and customer satisfaction. It usually extends beyond qualification testing to an industry standard or to customer specifications. Acceptance testing helps anticipate potential risks of product liability or unsuitable application so that these situations can be avoided by safeguards before market introduction.

Both acceptance testing and new product introduction must be carefully planned and executed. While acceptance testing is more concerned with the minimization of technical risk, product introduction is more concerned with market timing and the minimization of market risk. Product introduction is planned to provide immediate customer notification and response to the new product/service offering so that marketing strategies can be adjusted if necessary. Early market reaction is very important to observe and measure so that the effectiveness of different introduction strategies can be assessed.

Examples of product introduction strategies are as follows:

- Special release to specific customer or customer group for whom it was developed and who may have contributed to its design and development

- Restricted release to target applications that represent greatest opportunities

- Controlled and limited release to targeted customers who can provide candid and confidential feedback prior to general release

- Introduced as a complementary product or improved-performance version of a current product in use

- Introduced as a higher-value replacement for an existing competitive product

- Offered with extended warranties and exceptional service and support

- Announced and introduced to the industry via special promotion or trade show

Phase 7. New Commercial Venture

Any new product launch is a major undertaking involving company management and several key team members devoting full time to its planning and execution. It is especially important that the first new product launch from a systematic product development process be done methodically and thoroughly. For this reason, a smaller, lower-risk project may be selected as the first one managed through the development funnel as a project. It may take longer than planned because the process itself is new and people are still learning and adjusting to the methodology. But a successful launch of an initial new product will give the organization confidence in its future.

After several successful new product developments using this system, the company can work on improving and streamlining the process for faster development cycle time or "time to market." Process phases can be compressed or combined to expedite development efforts. The ultimate goal is to continually improve the company's ability to develop products and services which closely match the requirements and expectations of its customers.

11
Train the Trainer

Purpose

The purpose of training the trainer in delivery techniques for TQM courses is to make the company self-sufficient and replicate the training to sustain the results of the techniques and continuance of the culture change.

Objectives

- To understand the adult learner and how to stimulate thinking and comprehension
- To learn practical steps to a successful presentation
- To become effective at handling audience questions
- To develop skills to competently manage difficult participants
- To utilize planning tools to create a strong presentation
- To learn the correct use of an overhead projector and flipchart

Introduction

Congratulations! Your organization has gone through the first difficult steps of the TQM implementation process. Now it is time to plan for the future of the program. Continuous improvement, utilizing the Production/Quality Improvement Process methodology must become enshrined within the corporate culture.

There are several ways in which this culture change will occur. First, management must role model, i.e., "walk the talk." Employees must see the management putting TQM principles to work. Second, the corporate culture must recognize and reward employees for value's behaviors and results. Recognition and rewards reinforce value's behaviors and achievement of results; they are, in effect, a way of walking the talk. Third, the organization must develop a cadre of trainers who can "proselytize" the philosophy and techniques of TQM.

Many TQM programs are launched with the guidance and encouragement of professional consulting organizations. These organizations can be useful to get a program off the ground because they bring a fresh perspective and (if the organization has done their homework) a successful implementation methodology. A good consulting organization will typically employ very polished instructors who are able to impart essential TQM techniques while engaging in TQM proselytizing. In other words, the trainers must be TQM salespersons.

Identifying, training, and rewarding internal trainers is crucial to the ongoing success of a TQM program. The internal trainers must master the technical techniques of TQM along with spreading the TQM philosophy. A potential pitfall for organizations is to appoint trainers who are the most proficient practitioners of specific TQM techniques such as SPC. There is nothing wrong with being a proficient user of TQM techniques; however, the "proselytizing" aspects of the position cannot be ignored. TQM is a truly revolutionary operating method. Do not let people get "turned off" by boring technocrats droning on about the arcane aspects of a specific technique.

This chapter contains very practical tips on how to be an effective trainer. Study the techniques and practice them. They will contribute to improving the effectiveness of TQM training within an organization; however, do not assume that these principles apply only to TQM presentations. These techniques apply to TQM presentations, presentations to the shareholders, or presentations to the Rotary Club.

Getting Started

So, you're going to give a presentation. It's important to note that different types of presentations require different actions on your part.

Most principles regarding effective presentations are the same; however, your role changes depending on the type of presentation in which you are involved.

Presentation. The "talk." This could be done in front of coworkers, subordinates, social organizations, or the like. Normally you provide information or your thoughts about a particular topic. There is little interaction with the audience except perhaps a question and answer period.

Seminar. Your role is instructor. You are teaching others about a particular topic. There is more interaction between you and the participants. You may even ask for feedback from the group.

Workshop. Your role is instructor and facilitator. The key word here is *experiential.* The participants are not only listening to information and asking questions, they are doing hands-on learning—actually *doing* what they are learning. The learners are highly involved in the process. You provide information, but what is more important, you facilitate their learning process. This is generally within the context of a training class.

Qualities of a Good Classroom Instructor

- Technical competence
- Classroom communication
- Management techniques
- Interactive skills
- Resourcefulness
- Evaluation and feedback skills
- Planning
- Professionalism

Primer on Adult Learning

Dr. Malcolm Knowles equated what was known about adult learning to a trip up the Amazon:

> It is a strange world that we are going to explore together, with lush growth of flora and fauna with exotic names (including fossils of extinct species) and teeming with savage tribes in raging battle. I have just made a casing-the-joint trip up the river myself, and I can tell you that my head is reeling.

Why do we need to know?

Sensory Stimulation Theory. How adults learn is based on stimulating their senses:

75%	Sight
13%	Hearing
12%	Touch, smell, taste

Andragogy*—The Study of How Adults Learn

- Encourage active participation.
- Introduce past experiences.

---

* Malcolm S. Knowles, *The Modern Practice of Adult Education,* 2d ed., Cambridge Book Co., New York, 1980.

- Collaborative.
- Mutual planning.
- Mutual evaluation.
- Experiential.

Ways to Involve the Adult Learner
- Allow maximum participation early on.
- Use more questions.
- Allow learners to analyze learning goals.
- Develop two-way communication.
- *WIIFM ("What's in It for Me?").* Tell your audience what they will get out of this presentation. Some examples are:

  Improve production on the job.

  Make your job easier.

  Increase your skills and assist in your promotion possibilities.

  Save your time on the job.

- *Dyads (pairs).* Pairing up participants in a presentation is a great way to involve everybody. For example, all you have to do is say, "Turn to the person on your left (or person behind you, or on your right) and discuss _____ , or practice _____ . This encourages thought and discussion. Then you can have the members share information with the entire group.
- *Triads (trios).* The most common use of the triad is to have two members role-playing or practicing a skill with the third member observing and taking notes. This can also be used by having to have one member present an idea while the other two critique. With the triad, you need to rotate roles so everyone has a turn.

Components of a Successful Presentation

Defining Your Purpose

Any presenter, regardless of whether you are starting from scratch or taking over a prepared presentation, needs to have a clearly stated objective prior to facing a classroom of expectant trainees. Effective trainers need to fill in the blank in the following sentence:

At the end of my presentation, the attendees will be able to _____

_____ .

Objectives must be measurable.   Examples of properly written, measurable objectives are:

"At the end of my presentation, the attendees will be able to compute an average and a range for a series of values."

"At the end of my presentation, the attendees will authorize the expenditure of $1,000,000 for the purchase of a new computer system."

"At the end of my presentation, the attendees will voluntarily join TQM improvement teams in their respective departments to begin solving long-standing problems."

Objectives Need to Be Written and Given to Participants.   It is probably obvious to most readers why the objective needs to be written down prior to the presentation. Documented, concrete objectives provide a goal to strive for and a measuring stick with which to evaluate oneself after the presentation.

A basic tenet of goal setting is that one must write down the goal. Writing down the goal helps to internalize it, thereby focusing our efforts more accurately.

Besides aiding us in achieving the goal, writing down the objective provides a measuring stick for self-evaluation. It is amazing how we can cause the objective to fit the outcome when we have not bothered to document the original goal of the presentation.

Preparation

Research what the audience wants. What are their concerns and fears? What arguments are they likely to have against you or what you're presenting?

Research the topic you are presenting. Be sure you know what you are talking about. Especially when you are giving a presentation to people who are also knowledgeable about your subject, be sure that you (and they) realize that your only "authority" for standing in front of them is your expertise in the particular subject area you are presenting. Your audacity to stand in front of a group is not based on your superior personality, mentality, or morality, but simply on the fact that you have done some research in order to have a specialized area of knowledge from which they may benefit.

Gather your data from *all* available sources. Don't overlook interviews, "pretests" to determine group's background knowledge, or simply asking questions when you see participants face-to-face.

Screen your data according to its importance to the audience! Do not be tempted to put something in just because you like it, or you want to "prove" how much work you have done.

Arrange your data in order of importance to the audience. What do they want or need to know first? At the end of the presentation have each participant create an action plan—how they will apply what they learned once they return to their jobs.

Practice. Rehearse your presentation out loud! Do it in front of an audience, with video- or audiotape, or in front of a mirror to get feedback. Include use of visuals in your practice.

Analyzing/Controlling Your Environment

Effective presenters do a lot of intelligence gathering and legwork prior to a presentation. The old adage, "an ounce of prevention is worth a pound of cure," really applies to giving presentations.

Few things can be more frustrating than being well prepared, getting to the designated room, and finding that the bulb in the overhead transparency projector is burned out. Running around minutes before a presentation trying to find a suitable bulb hurts your credibility and makes you more nervous than you already were. (Yes, even experienced presenters feel nervous before a presentation.)

The first step in the process of analyzing and controlling your environment is to learn about your presentation environment. Make sure that you do the following tasks.

Analyze by Learning About Your Physical Environment

1. Examine the room. Are there potential blind spots for the audience?

2. Check the equipment. Learn how to use and adjust the overhead projector. Know where spare bulbs are kept. Learn how to change bulbs. Remember Murphy's law: if it can go wrong, it will. (The author once gave a presentation in which a bulb for an overhead projector and a bulb for a 16mm movie projector burned out within 15 minutes of each other.)

3. Know the time of day for the presentation. Presentations after lunch require an extra dose of enthusiasm on the presenter's part.

4. Know how many people will be attending the presentation. Have sufficient copies of printed materials.

5. Find out the locations of bathrooms, vending machines, and water fountains.

6. Find out if you can control the temperature inside the room.

Analyze by Learning About Your Audience

1. What is the education/vocabulary level of the participants? Most presentations should be at the level of the first year of a community college. Learn any special lingo with which the participants may be familiar; then use these terms during your presentation.

2. What is on an audience's mind? Have there been recent audits or layoffs which may be weighing on participant's minds?

3. What are the knowledge levels and opinions of the participants regarding the subject? Do you need to explain basic terms and vocabulary, or is the audience versed on common terms within the field?

4. What are the opinions of the audience regarding you and your organization? Are you part of the dreaded executive staff (also known as the "scourge of line departments")?

5. What is the relationship of the participants to you, the presenter? Are they your subordinates? Are they your peers? Are they ranked higher than you in the organization?

6. Who are the key participants? How do these key participants like their information—as bottom-line facts? Are they data processors who require systematic evidence or team-oriented individuals who feel comfortable with consensus decision making?

Control Your Environment (If Possible)

1. Control the location. For example, avoid presenting in long, narrow rooms; people in the back of the room often feel left out.

2. Control the time of the presentation. People are most alert early in the morning. Presentations immediately after lunch can be murder.

3. Control time constraints. Are there participants who punch a time clock? Will their shift end in the middle of your presentation? Are there participants in carpools who have to leave at a designated time?

4. Control the audience. If you predict a hostile reception for your ideas, "infiltrate" some allies.

5. Control seating arrangements. If you know that certain participants are very friendly and prone to talk among themselves, separate them.

Organizing with A-I-D-A

The most effective way to present any form of human communication is by using a method called A-I-D-A. By following this format, you can write better letters, reports, memos, brochures, articles, and more. You can also gain more positive response to oral presentations given to large groups, and even in one-on-one personalized situations. A-I-D-A is an acronym for the following:

A   *Attention.*   The best way to get anyone's attention is to ask the group a question. We have all been conditioned to answer questions. An example for a class on *"Customer Service Effectiveness"* could be: "What problems have you had in providing excellent customer service?" Use chart paper to record class members' responses. By doing this you (1) get their attention, (2) get them involved, and (3) find out their views on the topic you are presenting.

I   *Interest.*   To develop the audience's interest in the message you have to present, tell your audience what benefits they will receive from listening to your message. Everybody is tuned to radio station WIIFM: What's in it for me?

D   *Details.*   Facts and figures should be presented in the simplest possible form. Use tables, charts, graphs, pictures, and itemized lists whenever possible. Build in "memory hooks" to ensure that your audience will remember and use the information you present.

A   *Action.*   What do you want them to *do* with the information you are presenting? Don't make them guess—tell your audience directly what action you would like them to take (but make it in the form of an invitation, rather than a command)—"What did you learn and how are you going to put it into practice?"

In your opening sentence strive to use the word "You," relating to the audience from their point of view, telling them how the information you are about to present will benefit them.

Open with energy and enthusiasm. Who wants to listen to a dry, boring speaker? Right from the beginning, project your interest in the subject, your energy and vitality. Stimulate your audience by challenging them.

Give an overview and a summary. Have you ever heard of the "tell 'em" principle? "Tell 'em what you're going to tell 'em, tell 'em, and tell 'em what you told 'em." Set the groundwork by this overview and define key terms or phrases. At the end of the training, recap what participants learned for reinforcement.

Use a logical sequence. The information you're presenting needs to be given in an order that makes sense and that enables participants to apply what they learned. This can be topical, chronological, or by putting simple skills and principles first and then building on them with more complexity.

Be specific. Communicate as effectively as you can. The clearer and more specific you can be, the easier it is for your participants to understand the information. Making the information come alive through analogies, metaphors, and examples or demonstrations is key.

Use a common language. Make sure you are speaking at a level your audience can understand. Avoid jargon, technical terms, and highly complex words. It may show you are intelligent, but it may also lose your audience and their attention.

Time it. No one will be angry with you for making your presentation shorter than originally planned (unless you are paid for a specific length of talk or the idea is not covered thoroughly enough), but if your timing is off so that you keep talking on and on you will spoil any good image with which you may have started. Remember that talking from the audience's point of view includes not wasting their time (the biggest flaw of off-the-cuff speaking). Know how long your presentation will take—then adjust the amount of material or question time you can allow.

Relax! Relaxing and feeling confident before speaking in front of a group is a necessary element of a successful presentation. Speaking in front of a group used to be the number one fear among adults. Anxiety and nervousness release adrenaline into the body. Some adrenaline is necessary to give you energy, and yet too much can make it difficult for you to get your point across. Follow these suggestions to help you relax and "wow" your audience:

Rehearse. Practice, practice, practice: in front of a mirror, friends, coworkers. Don't memorize *everything* you want to say. Instead, remember the flow of ideas and specific points you want to make and the examples that make the concept or principle come alive.

ENVISION yourself in front of the class. Visualize how you will look, how the room is laid out, where your materials and audiovisual equipment will be. See yourself as a success. Or, as a great philosopher once said, "Fake it till you make it!"

LOOK confident. "Never let 'em see you sweat." Remember, most likely only you will realize when you skip over an example or topic. The class won't know, so remain calm and look confident. However do not make it obvious you are skipping over transparencies.

Breathe deeply. Before beginning, take long, deep, slow breaths. Pauses are OK and are preferable to fillers "uh," "OK," "hmmm."

Xenomania? "A madness and enthusiasm or a mania for foreign customs." OK. So it was difficult to find a word that started with X. Be open to the members of your class and show enthusiasm for what they know. You don't have to be the know-it-all. Use your imagination!

Effectively Handling Questions

Step 1. The first step to answering questions is to paraphrase the question to check your understanding of what the participant is asking. Then, depending on the type of question, you can continue with step 2.

Step 2.   In general, whether you are an expert in the topic you are presenting or not, technical questions might best be answered by you. Course logistics should be included here too.

"Opinion" or nontechnical questions can be effectively handled by:

1. Referring the question back to the questioner: "What do you think?" Let students discover answers of which they are capable but may have forgotten at the moment.

2. Referring the question to the group: "What does the group think?" This keeps their attention on class *content* instead of on the almighty instructor. Remember the principles of adult learning—involve the learners.

Step 3.   Last, *always* get back to the questioner (the participant who asked the question) to make sure the question was answered to their satisfaction.

Questioning the Group

■ Ask questions of the group in an irregular pattern. Don't go clockwise or counterclockwise or in alpha order. Asking questions in an irregular pattern prompts students to answer questions in their own minds since they might be called upon at any time.

■ In *Teaching Tips,* W. J. McKeachie warns against asking questions that obviously have only one right answer. To get a discussion going, questions "need to get at relationships, applications, or analysis of facts and materials."

■ Directive questions that have one right answer are good for content review—not for discussions.

■ Ask "What questions do you have?" *not* "Any questions?" The former assumes that there *are* questions and that you expect the participants to ask questions of you.

Managing Difficult Participants

While most participants are cooperative and manageable, there may be times when you encounter someone whose behavior is particularly difficult. While there are no easy answers, the following tips should provide some help.

Shy Sam.   Since training is an opportunity for growth, we must find some method to "draw him out."

- Ask easy questions from time to time and give Sam positive feedback.
- Use dyads and triads to keep him involved.
- Socialize with him during breaks. This provides encouragement if he can warm up to the instructor.

Monopolizing Mona. These types can gobble up all available time if given the opportunity.

- Tactfully interrupt and ask others to give their comments.
- In a firm and polite tone say, "Let's get another opinion about this" or "Others of you haven't had a chance to respond yet—what do you think?" or "Let's talk about that on the break."

Negative Ned. Ned stays on the gloomy side of things. He's an expert on dredging up past complaints.

- The best approach is to ask if he can find anything positive in the situation at hand.
- Say, "You may have a point; however, we are looking for constructive ways to deal with these problems."

Clyde the Clown. Your task is to keep the group moving in a productive manner and his antics may hinder group process.

- Call on Clyde for a "serious" dialogue—a question that relates to the topic at hand.
- Give positive reinforcement when he joins in constructively.
- Avoid embarrassing Clyde; however, another way to handle side conversation is to stop talking and just look at him. Often, the group will look over too, and he and his partner in crime will realize that there isn't any other discussion going on.

Arguing Amanda. Avoid getting baited into a debate with her.

- One approach is to let the group deal with her: "Anyone want to respond to that?"
- Use empathy to diffuse the anger.
- Focus on areas of agreement.
- Last, paraphrase her belief: "I understand your position. Let's agree to disagree and move on to another topic."

Know-It-All Nora. She has a tremendous need to be heard.

- Treat Nora politely but say, "That's interesting, and yet we need to move on."
- Give some respect by stating something she has done well and then add, "We do have to get to our main issue, which is . . ."

Hostile Henry. Remain calm and don't get drawn into the conflict.

- "I see you have some strong feelings on this subject. Let's hear from the group."
- Focus on facts.
- Close the discussion quickly and, if necessary, agree to discuss the topic during a break.

Worksheets

Prepare to Present

1. Who is your audience? (List specific names and/or job titles, departments.) _____

2. List everything you know about your audience below:
   Age_____ Occupation(s) _____
   Education_____
   Other pertinent data_____
   Hobbies_____

3. In one sentence, what do you want your audience to do with the information you will present? (What is the purpose of your presentation or what do you want them to be able to do after the training?) _____
   _____

4. How could your audience possibly benefit from the message you have to present? (How could they be healthier, wealthier, wiser?) How will it make their jobs easier, the company more profitable? _____
   _____

5. What are your three most important ideas to present? _____
   _____
   _____
   _____

6. Which key idea will be of most interest or benefit to your audience?
   _____

7. Put your most important idea into a strong opening sentence or question, tying in the benefit to your audience. Give them both the main idea and the "WIIFM" statement all at once. Write your complete opening statement below. _____

   _____

   _____

8. Go back to item 7 and underline those words you will emphasize when you say them.

9. Put your three key ideas from item 5 into the order you will present them. _____

   _____

   _____

10. Write out your complete closing statement, being sure to include your call for action (what you want them to do with the information they have just heard). _____

    _____

    _____

Planning Resources.   Figure 11-1 (the resistance/complexity chart) can be a useful tools when planning your presentation. Use it before each presentation, seminar, workshop, or training session that you present.

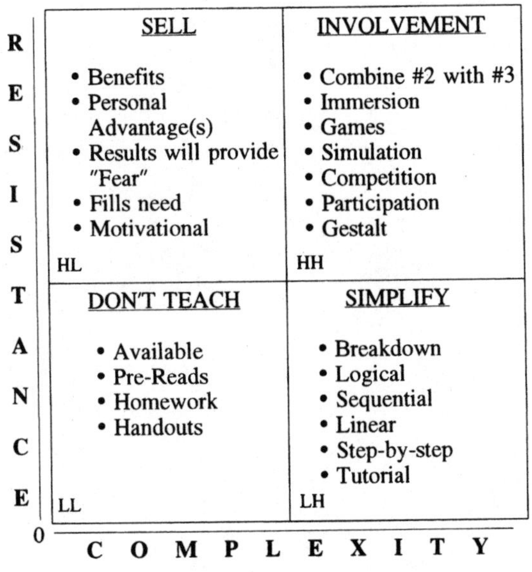

	SELL	INVOLVEMENT
R		
E	• Benefits	• Combine #2 with #3
S	• Personal Advantage(s)	• Immersion • Games
	• Results will provide "Fear"	• Simulation • Competition
I	• Fills need	• Participation
S	• Motivational	• Gestalt
	HL	HH
T	DON'T TEACH	SIMPLIFY
A	• Available	• Breakdown
	• Pre-Reads	• Logical
N	• Homework	• Sequential
	• Handouts	• Linear
C		• Step-by-step
		• Tutorial
E	LL	LH
0	C O M P L E X I T Y	

Figure 11-1. Resistance/complexity chart.

Self-Evaluation Worksheet.   After you've practiced and after the first one or two times you have taught a course, it's a good idea to evaluate your performance. Use the following worksheet to check on what you are doing well and what you can improve upon.

Area	Needs much improvement	Excellent
Relaxed	1	10
Showed interest in participants	1	10
Good communication skills	1	10
Enthusiastic	1	10
Well prepared	1	10
Involved participants	1	10
Effectively handled questions	1	10
Asked discussion questions	1	10
Preparation and organization	1	10
Gave overview and summary	1	10
Logical sequence	1	10
Quality visuals	1	10
Useful visuals	1	10
Quality handouts	1	10
Useful handouts	1	10
Variety of teaching methods	1	10

Appendix A

The Overhead Projector

The overhead projector can be an excellent aid to your presentations. However, its usefulness can be outweighed by improper usage.

Some advantages are:

1. It allows the presenter to face the audience at all times. You can maintain eye contact, check for reactions, and respond accordingly.

2. You can project in a fully lighted room. Often, when the lights are dimmed, participants can tune out and get drowsy (especially after lunch!). It's not necessary to turn the room lights on and off.

3. The horizontal surface provides flexibility. You can write on the surface or even overlay two to four transparencies on top of one another to add information to a concept.

4. You control what is on the screen at all times. It doesn't require another person to operate.

5. Color can be used at very little cost. Thermal and PPC/Xerographic films make it possible to prepare multicolor transparencies.

Correct Use of the Overhead Projector

1. Use the on/off switch. Put the transparency on the overhead before turning the projector on and turn off projector before removing. Another trick is to tape a 3 × 5 card to the top of the lens and just flip it up and down to control the light on the screen rather than turning the projector off and on. *Never* leave the projector on when there isn't a

transparency on. The bright white screen detracts from your presentation and it is hard on the eyes. Also, if you've moved on to another topic, turn off the projector or the audience's attention will still be drawn to the image on the screen.

2. Remember, the overhead is useful because you can present ideas and still keep eye contact with and face the audience. When pointing to topics on the transparency, point directly at the transparency, *not* the screen. If you point at the screen, you won't be facing the audience. Also, another tip is to use a pencil or pen and lay it down on the transparency as a pointer. This leaves your hands free for expression.

3. Rehearse. Sometimes it gets tricky flipping between one transparency and another. Make sure you feel comfortable with the use of the projector to give your participants the most professional presentation.

4. Focus the projector before class. Position the projector where it will give you optimal size on the screen and place a few of your transparencies on it to focus the image. Then all you have to do while presenting is to follow steps 1 and 2!

5. Stand on the left-hand side of the screen when facing the screen. This enables you to point to the beginning of each line.

6. Make eye contact with the participants. A guide would be "one thought, one person"—then change to a second person.

7. Your feet should be planted so that you are erect and balanced, gesturing with your arms and hands.

8. When using visuals, touch, turn, and talk.

9. Clear the visual by speaking each line before discussing each subject in depth.

Appendix **B**

Create Great Transparencies

Keep it simple.   Maximum five words per line, five lines per transparency.

Use key words and phrases only.   Transparencies are used to augment a presentation and help participants remember major points. Don't put your entire presentation on the transparency. They'll be too busy trying to read it to focus on what you have to say.

Letters need to be large to be readable.   Regular type size or font you would use for correspondence is unacceptable. A good rule of thumb is that the transparency itself (without projection) should be readable from 6 feet. If you are approximately 6 feet tall, drop the transparency on the floor and see if you can read it without straining. A good size for type is 18-point (¼ inch high).

Present a "revelation."   Use the "revelation technique" by placing a sheet of paper over transparency and revealing information line by line. This keeps the audience's attention.

Provide visual stimulation.   Use pictures and visuals to supplement words—symbols, maps, diagrams, and sketches—anything that will enhance the participant's understanding. Another idea is to copy cartoons and fill them in with color.

Create great borders.   Mount your transparencies or use an Instaframe. The projector stage is about 10 by 10 inches and standard transparencies are 8½ by 10½ or 11 inches. Without mounting, a large blank space is left on each side of the transparency, which doesn't look professional. Mounts make it easy to handle the transparencies and the borders make great places to put notes and key phrases. Only you can see these notes!

Use tinted backgrounds.   Tints are available in a variety of colors. They are easier on the eyes than the bright white background and can prevent eyestrain. Pastels are preferable because any additional information can be written on the transparency and is easily visible.

Use color.   Colors will make the transparency alive.

Appendix C
Flipchart Tips

The flipchart (large chart paper on an easel) is another invaluable tool. Utilize some of the following tips to enhance your presentation.

Chart paper with a grid pattern is easiest to use because you can keep lines of text straight and even.

Use a variety of colors just as you would with overhead transparencies.

Ahead of time, mark lightly in pencil the key words or phrases that you would like to use to enhance your presentation. Then write over that text with markers. Or you can have text in marker and your "notes" (key words or phrases) in pencil to help you keep on track. The audience doesn't see these notes! Tricky, huh?

Keep it simple—six to eight words per line, six to eight lines per page.

Use Post-It notes on the outside edge of every page that you are using to identify what is on that page. Stagger the Post-Its much like the dividers in a binder. This helps you keep track of your presentation.

Keep a blank page between each page of text. Chart paper is usually quite thin and letters can bleed through to the top page, distracting participants.

Put pieces of chart paper and markers at tables for participant use during small group discussions. They can write down their ideas to discussion questions and then present them to the group.

Appendix **D**

Speaking Styles

1. Alliteration is the stringing together of similar consonant sounds such as "tip-top techniques for top trainers." Used sparingly, this can be an effective tool for aiding memory—theirs and yours.

2. Reference to experience should cause your audience to identify with your topic and with you. This is especially true if reference is made to a common experience which both the audience and the presenter share.

3. Cumulation is the piling up of facts to prove a point. Generally three to four facts are sufficient.

4. Restatement is the simple repetition of a phrase more than once. This technique drives home the point and it aids memory.

5. Volume can be used as a presentation tool. It allows you to be heard; it helps to emphasize points; and, it can help make the transition from one point to another.

6. Pauses can be used to emphasize a point, prepare audiences for the next point, and it allows the audience to grasp what has just been said.

7. Pitch is a technique that is most frequently noted for its absence rather than its inclusion in a presentation. There are few things more boring than a presentation given in a flat, lifeless monotone. Novice presenters who rely on semiscripted material need to be aware of this failing. A good way to avoid a monotone presentation is to tape oneself and then listen to the tape.

8. Speed is another pitfall for many novice presenters. Frequently, they are so well prepared (memorized) that they present their material in a machine-gun delivery. Go to semiautomatic or single shot.

Index

About the Author

John L. Hradesky, P.E., president and founder of National Summit Group, Inc., has more than 30 years of experience in quality and manufacturing, including employment with American Hospital Supply, Eastman Kodak, General Motors, Johnson & Johnson, and Xerox. He has held positions from production operator through Director of Quality and Director of Manufacturing. Mr. Hradesky holds a BS in Industrial Engineering from the University of Pittsburgh, and an MS in Applied Statistics from the Rochester Institute of Technology, and is a member of the ASQC. A writer of a wide range of articles on productivity and quality, he is a frequent speaker at professional societies. He has been an invited speaker at Dr. Deming's seminar and is frequently a panel member with Dr. Deming's service. Mr. Hradesky is the author of *Productivity and Quality Improvement* (McGraw-Hill), a practical guide to the implementation of a quality/productivity improvement process for continuous improvement.